T0310708

Philosophy of Artificial Intelligence and Its Place in Society

Luiz Moutinho
University of Suffolk, UK

Luís Cavique
Universidade Aberta, Portugal

Enrique Bigné
Universitat de València, Spain

A volume in the Advances in Human and Social Aspects of Technology (AHSAT) Book Series

Published in the United States of America by
IGI Global
Engineering Science Reference (an imprint of IGI Global)
701 E. Chocolate Avenue
Hershey PA, USA 17033
Tel: 717-533-8845
Fax: 717-533-8661
E-mail: cust@igi-global.com
Web site: http://www.igi-global.com

Library of Congress Cataloging-in-Publication Data

Names: Moutinho, Luiz, editor. | Cavique, Luís, DATE- editor. | Bigne, Enrique, editor.
Title: Philosophy of artificial intelligence and its place in society / edited by Luiz Moutinho, Luís Cavique, Enrique Bigne.
Description: Hershey, PA : Engineering Science Reference, [2023] | Includes bibliographical references and index. | Summary: "Philosophy of Artificial Intelligence and Its Place in Society evaluates various aspects of artificial intelligence including the range of technologies, their advantages and disadvantages, and how AI systems operate. Spanning from machine learning to deep learning, philosophical insights, societal concerns, and the newest approaches to AI, it helps to develop an appreciation for and breadth of knowledge across the full range of AI sub-disciplines including neural networks, evolutionary computation, computer vision, robotics, expert systems, speech processing, and natural language processing"-- Provided by publisher.
Identifiers: LCCN 2023024049 (print) | LCCN 2023024050 (ebook) | ISBN 9781668495919 (hardcover) | ISBN 9781668495926 (paperback) | ISBN 9781668495933 (ebook)
Subjects: LCSH: Artificial intelligence--Moral and ethical aspects. | Artificial intelligence--Social aspects
Classification: LCC Q334.7 .P45 2023 (print) | LCC Q334.7 (ebook) | DDC 006.301--dc23/eng20230919
LC record available at https://lccn.loc.gov/2023024049
LC ebook record available at https://lccn.loc.gov/2023024050

This book is published in the IGI Global book series Advances in Human and Social Aspects of Technology (AHSAT) (ISSN: 2328-1316; eISSN: 2328-1324)

British Cataloguing in Publication Data
A Cataloguing in Publication record for this book is available from the British Library.

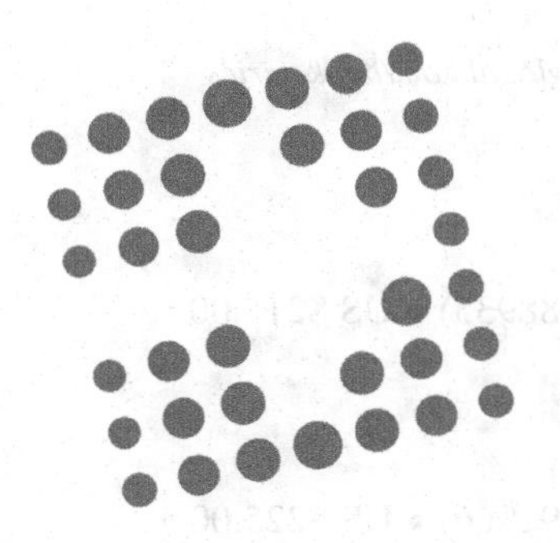

Advances in Human and Social Aspects of Technology (AHSAT) Book Series

Mehdi Khosrow-Pour, D.B.A.
Information Resources Management Association, USA

ISSN:2328-1316
EISSN:2328-1324

Mission

In recent years, the societal impact of technology has been noted as we become increasingly more connected and are presented with more digital tools and devices. With the popularity of digital devices such as cell phones and tablets, it is crucial to consider the implications of our digital dependence and the presence of technology in our everyday lives.

The **Advances in Human and Social Aspects of Technology (AHSAT) Book Series** seeks to explore the ways in which society and human beings have been affected by technology and how the technological revolution has changed the way we conduct our lives as well as our behavior. The AHSAT book series aims to publish the most cutting-edge research on human behavior and interaction with technology and the ways in which the digital age is changing society.

Coverage

- ICTs and human empowerment
- Technology Adoption
- Human Development and Technology
- ICTs and social change
- Computer-Mediated Communication
- Gender and Technology
- Cyber Behavior
- Human Rights and Digitization
- Philosophy of technology
- Digital Identity

IGI Global is currently accepting manuscripts for publication within this series. To submit a proposal for a volume in this series, please contact our Acquisition Editors at Acquisitions@igi-global.com or visit: http://www.igi-global.com/publish/.

The Advances in Human and Social Aspects of Technology (AHSAT) Book Series (ISSN 2328-1316) is published by IGI Global, 701 E. Chocolate Avenue, Hershey, PA 17033-1240, USA, www.igi-global.com. This series is composed of titles available for purchase individually; each title is edited to be contextually exclusive from any other title within the series. For pricing and ordering information please visit http://www.igi-global.com/book-series/advances-human-social-aspects-technology/37145. Postmaster: Send all address changes to above address.

Titles in this Series

For a list of additional titles in this series, please visit: www.igi-global.com/book-series

Cyberfeminism and Gender Violence in Social Media
Deepanjali Mishra (KIIT University, ndia)
Information Science Reference • © 2023 • 442pp • H/C (ISBN: 9781668488935) • US $215.00

Investigating the Impact of AI on Ethics and Spirituality
Swati Chakraborty (GLA University, India & Concordia University, Cnada)
Information Science Reference • © 2023 • 230pp • H/C (ISBN: 9781668491966) • US $225.00

Applied Research Approaches to Technology, Healthcare, and Business
Darrell Norman Burrell (Marymount University, USA)
Information Science Reference • © 2023 • 455pp • H/C (ISBN: 9798369316306) • US $285.00

Analyzing New Forms of Social Disorders in Modern Virtual Environments
Milica Boskovic (Faculty of Diplomacy and Security, University Union Nikola Tesla, Serbia) Gordana Misev (Ministry of Mining and Energy Republic of Serbia, Serbia) and Nenad Putnik (Faculty of Security Studies, University of Belgrade, Serbia)
Information Science Reference • © 2023 • 284pp • H/C (ISBN: 9781668457603) • US $225.00

***Adoption and Use of Technology Tools and Services by Economically Disadvantaged Communities Implications** for Growth and Sustainability*
Alice S. Etim (Winston-Salem State University, USA)
Information Science Reference • © 2023 • 300pp • H/C (ISBN: 9781668453476) • US $225.00

Advances in Cyberology and the Advent of the Next-Gen Information Revolution
Mohd Shahid Husain (College of Applied Sciences, University of Technology and Applied Sciences, Oman) Mohammad Faisal (Integral University, Lucknow, India) Halima Sadia (Integral University, Lucknow, India) Tasneem Ahmad (Advanced Computing Research Lab, Integral University, Lucknow, India) and Saurabh Shukla (Data Science Institute, National University of Ireland, Galway, Ireland)
Information Science Reference • © 2023 • 271pp • H/C (ISBN: 9781668481332) • US $215.00

Handbook of Research on Perspectives on Society and Technology Addiction
Rengim Sine Nazlı (Bolu Abant İzzet Baysal University, Turkey) and Gülşah Sari (Aksaray University, Turkey)
Information Science Reference • © 2023 • 603pp • H/C (ISBN: 9781668483978) • US $270.00

701 East Chocolate Avenue, Hershey, PA 17033, USA
Tel: 717-533-8845 x100 • Fax: 717-533-8661
E-Mail: cust@igi-global.com • www.igi-global.com

Table of Contents

Detailed Table of Contents

Judea Pearl's ladder of causation framework has dramatically influenced the understanding of causality in computer science. Despite artificial intelligence (AI) advancements, grasping causal relationships remains challenging, emphasizing the causal revolution's significance in improving AI's understanding of cause and effect. The work presents a novel taxonomy of causal inference methods, clarifying diverse approaches for inferring causality from data. It highlights the implications of causality in responsible AI and explainable AI (xAI), addressing bias in AI systems. The chapter points out causality as the next step in AI for creating new questions, developing causal tools, and clarifying opaque models with xAI approaches. The work clarifies causal models' significance and implications in various AI subareas.

This chapter explores the use of artificial intelligence (AI) in market research and its potential impact on the field. Discuss how AI can be used for data collection, filtering, analysis, and prediction, and how it can help companies develop more accurate predictive models and personalized marketing strategies. Highlight the drawbacks of AI, such as the need to ensure diverse and unbiased data and the importance of monitoring and interpreting results and covers various AI techniques used in market research, including machine learning, natural language processing, computer vision, deep learning, and rule-based systems. The applications of AI in marketing research are also discussed, including sentiment analysis, market segmentation, predictive analytics, and adaptive recommendation engines and personalization systems. The chapter concludes that while AI presents many benefits, it also presents several challenges related to data quality and accuracy, algorithmic biases and fairness issues, as well as ethical considerations that need to be carefully considered.

Artificial intelligence in tourism activities opens a bundle of emerging applications for tourists and companies. This chapter aims to delineate the stages of the tourist journey and the usage of four types of intelligence suggested in the literature: mechanical, analytical, intuitive, and empathetic. Based on these two ideas, the authors propose a useful framework for disentangling the different types of current and future applications of AI in tourism. Each stage involves multiple suppliers with different types of AI applications, and its adoption will ultimately rely on tourist trust and, therefore, willingness to share data and the use of robotics and other AI forms. The chapter ends with some trends and reflections on the expansion of AI in tourism that pivot around these ideas: job replacement and flexible operations; mobile-centric approach; data integration and analytics; revenue management and customer interactions tension; (v) neuroscientific tools for AI in tourism.

The proliferation of content generated by tourists, in parallel with the exponential growth of social media is causing a paradigm shift in research. Traditional surveys cannot be necessary to obtain users' opinions when scholars can access this valuable information freely through social media. In the domain of tourism, online tourists' reviews (OTRs) shared on online travel communities stand out. The aim of this study is to demonstrate the usefulness of OTRs in analysing the image of a green hotel. The authors also examine the possible differences in the content of green hotel online reviews across Anglos and European tourists. The data source are 28,189 reviews by tourists shared on TripAdvisor regarding the 82 green hotels of the city of Amsterdam. The findings showed that tourist's culture significantly determine the content of the OTRs. The results show preferences and opinions from the tourist's perspective, which can be useful for hotel managers to promoting sustainability practices.

A machine-based research reading methodology specific to the academic discipline of marketing science is introduced, focused on the text mining of scientific texts, analysis and predictive writing, by adopting artificial intelligence developments from other research fields in particular materials and chemical science. It is described how marketing research can be extracted from documents, classified and tokenised in individual words. This is conducted by applying text-mining with named entity recognition together with entity normalisation for large-scale information extraction of published scientific literature. Both a generic methodology for overall marketing science analysis as well as a narrowed-down contextualised method for delimited marketing topics are detailed. Automated literature review is discussed as well as potential automated formulation of hypotheses and how AI can assist in the transfer of marketing research knowledge to practice, in particular to startups, as they can benefit from AI powered science-based decision making. Recommendations for next steps are made.

Artificial Intelligence technology has advanced tremendously in recent years, and it is now widely used in a variety of fields, including energy, agriculture, geology, information processing, medicine, defence systems, space research and exploration, marketing, and many more. The introduction of artificial intelligence technology has ushered in a new era of renewable energy systems and smart power grid modernization. It assists in attaining the intended system availability, reliability, power quality, efficiency, and security goals through optimal resource utilization and cost-effective electricity. Automated power generation systems, energy storage control, wind turbine aerodynamic performance optimization, power generator efficiency enhancement, health monitoring of renewable energy generation systems, and fault detection and diagnose in a smart grid subsystem are just a few of the applications. The main aim of this proposed chapter is to demonstrate how artificial intelligence techniques play a significant role in renewable energy systems with their diverse applications.

As the volume and complexity of data streams continue to increase, exploratory cluster analysis is becoming increasingly important. In this chapter, the authors explore the use of artificial neural networks (ANNs), particularly self-organizing maps (SOMs), for this purpose. They propose additional methodologies, including concept drift detection, as well as distributed and collaborative learning strategies and introduce a new open-source Java ANN library, designed to support practical applications of SOMs across various domains. By following our tutorial, users will gain practical insights into visualizing and analyzing these challenging datasets, enabling them to harness the full potential of our approach in their own projects. Overall, this chapter aims to provide readers with a comprehensive understanding of SOMs and their place within the broader context of artificial neural networks. Furthermore, we offer practical guidance on the effective development and utilization of these models in real-world applications.

This chapter explores personalization and its connection to the philosophical concept of the person, arguing that a deeper understanding of the human person and a good society is essential for ethical personalization. Insights from artificial intelligence (AI), philosophy, law, and more are employed to examine personalization technology. The authors present a unified view of personalization as automated

met in various contexts to create an optimized IA system in service settings. In this respect, AI agents are not just tools but rather co-creators of value that can influence human agents' learning cycles. Hence, humans' effective interaction with AI agents produces a learning effect that can empower humans' interpretative capability. This chapter focuses on IA and shows that IA is not only a theoretical paradigm but also serves as a platform to facilitate the transition from smart services to wise service innovation to the benefit of both the multi-agent system benefitting service organizations and the consumers too. Potential challenges are also discussed from a societal viewpoint.

In the sports industry, artificial intelligence has become a powerful tool for sports managers interested in getting private sponsorships and for DMOs interested in branding a place. In this scenario, two main objectives guide this chapter (1) to generate a ranking of the leading Spanish marathons based on their presence on the four most important social networks in Spain (Facebook, Twitter, Instagram, and YouTube) and (2) to measure the engagement on social networks generated by the first of the marathons identified in the ranking. The official profiles of the accounts of the 10 marathons with the highest number of finishers in 2022 in Spain have been monitored on the social networks listed (Facebook, Twitter, Instagram, and YouTube). As the results show, a marathon can generate high network engagements. The destination's image can be highly favoured thanks to small local events (such as marathons) capable of generating a lot of movement on social networks. However, not all social networks work equally well in promoting sporting events capable of generating engagement.

Despite the increasing implementation of artificial intelligence (AI), it is puzzling why consumers are still resistant towards it. Part of the problem is how to create systems that appropriately meet consumer demand for good quality and functional AI. The chapter addresses this issue by providing the much-needed understanding of how AI technologies can shape a satisfactory customer experience. Results are clear in showing that easy-to-use and high-quality AI systems form positive attitudes, and consumers are willing to use such technology again. Functional and enjoyable interaction enhanced the experience and thus attitude formation. These results have been substantiated statistically only for the high satisfaction group. By contrast, for low satisfaction group, consumers have not enjoyed the experience they had with the AI system. They found the interaction to be unpleasant, and the system to be useless. The outcomes are summarised in a framework for designing appropriate AI systems shaping consumer journey beyond the traditional marketing context.

Preface

Welcome to *Philosophy of Artificial Intelligence and Its Place in Society*, edited by Luiz Moutinho, Luís Cavique, and Enrique Bigné. In this rapidly evolving era of the 2020s, Artificial Intelligence (AI) has once again taken center stage, rekindling the fervor and intrigue that surrounded it in previous decades. Despite enduring the trials of two AI "winters" in 1970 and 1990, the field of AI has emerged triumphant, delivering spectacular results, particularly over the past ten years. The driving forces behind this resurgence are the convergence of two critical elements: the unprecedented availability of vast datasets and the accessibility of high-level computational power capable of analyzing this data.

However, as we delve deeper into the realms of AI, we confront a significant limitation that looms large—the data that fuels AI systems can carry inherent biases, reflecting societal inequities and the implicit biases of those who design and input the data. These biases often manifest in the outcomes generated by AI, raising important questions about fairness, ethics, and societal implications.

The rise of artificial intelligence has sparked divergent opinions among futurists. Some predict that AI will usher in an era where most people's lives will markedly improve over the next decade, while others express profound concerns about how AI advances may redefine the essence of human productivity and the exercise of free will.

The Importance of Research and Problem-Solving

It is within this dynamic and complex landscape that research takes center stage. Research, in all its forms and facets, is the beacon that guides us through the intricate maze of challenges and opportunities presented by AI. We firmly believe that research is not only essential but also indispensable in addressing the multifaceted problems that AI brings to the fore.

1. **Technological Facets of AI:** Research in this domain is crucial as it enables us to understand the nuances of AI technologies better. It empowers us to navigate the advantages and disadvantages, helping us harness AI's potential while mitigating its pitfalls. Through rigorous investigation and experimentation, researchers can develop AI solutions that are not only efficient but also ethically sound.
2. **Philosophical Insights and Social Concerns:** Philosophy and ethics are at the heart of AI's relationship with society. Research in this area allows us to place AI in an ethical context, critically examining its impact on our lives, values, and principles. By delving into the philosophical underpinnings of AI, researchers can help forge a path toward ethical AI that contributes positively to society.

3. **Applications of AI:** Research here is transformative, as it unlocks the doors to innovation and practical solutions. Through rigorous study and experimentation, researchers can develop AI applications that address real-world challenges, from improving healthcare outcomes to enhancing business efficiency and bolstering law enforcement efforts. Research-driven insights pave the way for responsible AI deployment.

This research book is tailored primarily for the academic audience, including scholars, researchers, and doctoral students. However, in recognition of the growing popularity of AI-related courses at the Master's level, we are also considering a student edition that distills the core concepts for a wider readership. Moreover, professionals in the corporate world, such as systems analysts and corporate researchers, will find valuable insights within these pages.

As you embark on this intellectual journey with us, we invite you to recognize the profound importance of research in solving the outlined problems and shaping the future of AI. It is through research that we gain the knowledge, insights, and tools to harness AI's potential while mitigating its risks. Research is the linchpin that empowers us to create an AI-powered future that is not only technologically advanced but also ethically grounded, equitable, and beneficial to society at large.

The chapters within this book are a testament to the dedication and commitment of researchers worldwide who are tirelessly working to unravel the complexities of AI and illuminate its path forward. We hope this reference book will inspire you to engage with these critical issues, contribute to the ongoing dialogue, and ultimately play a role in the transformative power of research in the realm of Artificial Intelligence.

Chapter Overview

Causality: The Next Step in Artificial Intelligence

Author: Luis Cavique, Universidade Aberta, Portugal

This chapter delves into the pivotal role of causality in the realm of Artificial Intelligence (AI). Despite AI advancements, understanding causal relationships remains a formidable challenge. The chapter draws upon Judea Pearl's influential ladder of causation framework to illuminate the significance of causality in improving AI's understanding of cause and effect. It presents a novel taxonomy of causal inference methods, offering clarity on diverse approaches for inferring causality from data. Additionally, the chapter underscores the implications of causality in Responsible AI and explainable AI (xAI), particularly in addressing biases within AI systems. Ultimately, it highlights causality as the next frontier in AI, capable of generating new questions, fostering the development of causal tools, and enhancing the transparency of opaque models through xAI approaches. The chapter also elucidates the relevance of causal models across various subareas of AI.

Impact of Artificial Intelligence on Marketing Research: Challenges and Ethical Considerations

Authors: Laura Sáez-Ortuño, Javier Sanchez-Garcia, Santiago Forgas-Coll, Rubén Huertas-García, Eloi Puertas-Prat (University of Barcelona, Spain)

This chapter explores the transformative impact of Artificial Intelligence (AI) on marketing research. It delves into the potential applications of AI in data collection, filtering, analysis, and prediction within the marketing domain. The chapter emphasizes how AI can enhance the development of more accurate predictive models and personalized marketing strategies. However, it also discusses the challenges AI presents, including the need for unbiased data and the importance of result monitoring and interpretation. The chapter covers various AI techniques employed in marketing research, such as machine learning, natural language processing, computer vision, deep learning, and rule-based systems. It explores AI's applications in sentiment analysis, market segmentation, predictive analytics, and adaptive recommendation engines. Additionally, the chapter underscores the ethical considerations associated with AI in marketing research, including data quality, algorithmic biases, and broader ethical implications.

Artificial Intelligence in Healthcare

Authors: Kun-Huang Huarng, Tiffany Hui-Kuang Yu, Duen-Huang Huang (Various Taiwanese Universities)

This chapter provides an insightful overview of the applications of Artificial Intelligence (AI) in healthcare. It commences with a brief history of AI development and introduces machine learning as a prominent AI advancement. The chapter showcases AI's successful applications in diverse healthcare areas, including medical diagnosis and long-term care. It also discusses the noteworthy ChatGPT series of systems and their exceptional performance. The chapter concludes by examining debates and future expectations surrounding AI in healthcare, highlighting the significant role AI plays in enhancing healthcare quality and delivery.

Causal Machine Learning in Social Impact: Using AI in Impact Assessment

Authors: Nuno Lopes, Luís Cavique (Universidade Aberta, Portugal)

This chapter explores the integration of causal machine learning in assessing social impact, with a focus on utilizing AI to evaluate the consequences of social interventions. It delves into the significance of causal models and their applications in understanding how interventions affect outcomes. By leveraging AI-driven causal analysis, the chapter sheds light on how to better assess and optimize social impact initiatives.

The Incorporation of Large Language Models (LLMs) in the Field of Education: Ethical Possibilities, Threats, and Opportunities

Author: Paul Aldrin Dungca (Salesians of Don Bosco, Philippines)

This chapter delves into the ethical considerations surrounding the integration of Large Language Models (LLMs), exemplified by GPT-3.5, within the field of education. It explores the potential advantages of LLMs in personalized learning experiences and improved accessibility while addressing concerns related to data privacy, bias, and the implications of replacing human instructors. The chapter aims to provide a comprehensive understanding of the ethical implications of using LLMs in educational settings.

Artificial Intelligence in Tourism

Author: Enrique Bigné (Universitat de Valencia, Spain)

This chapter explores the applications of artificial intelligence in the tourism industry, unveiling a framework for understanding how AI can enhance various stages of the tourist journey. It introduces the concept of different types of intelligence, including mechanical, analytical, intuitive, and empathetic, and examines their application in AI-powered tourism. The chapter also discusses the role of trust and data sharing in AI adoption within the tourism sector and highlights trends in AI's expansion within tourism.

The Influence of Culture on Sentiments Expressed in Online Reviews of Eco-Friendly Hotels: The Case Study of Amsterdam

Authors: Estefania Ballester Chirica, Carla Ruiz-Mafé, Natalia Rubio (Various Spanish Universities)

This chapter investigates the impact of culture on sentiments expressed in online reviews of eco-friendly hotels in Amsterdam. It draws upon a dataset of 28,189 reviews from tourists on TripAdvisor to uncover how cultural factors shape the content of online reviews. The findings provide valuable insights for hotel managers seeking to promote sustainability practices based on tourist preferences and opinions influenced by their cultural backgrounds.

Artificial Intelligence Method for the Analysis of Marketing Scientific Literature

Authors: Antonio Hyder, Carlos Perez-Vidal, Ronjon Nag (Various Universities)

This chapter introduces a machine-based research methodology tailored for the analysis of marketing scientific literature. It employs text mining techniques, including named entity recognition and entity normalization, to extract and classify marketing research from published scientific documents. The chapter discusses the potential applications of automated literature review and hypothesis formulation, along with how AI can facilitate the transfer of marketing research knowledge to practice.

Artificial Intelligence for Renewable Energy Systems and Applications: A Comprehensive Review

Authors: Manikandakumar Muthusamy, Karthikeyan Periyasamy (Various Indian Universities)

This chapter offers a comprehensive review of the role of Artificial Intelligence (AI) in renewable energy systems and their applications. It explores how AI technology can optimize renewable energy systems, enhance power generation, improve efficiency, and ensure grid reliability. The chapter examines various AI applications in renewable energy, including power generation, energy storage control, and fault detection, emphasizing the transformative impact of AI in the renewable energy sector.

Exploratory Cluster Analysis Using Self-Organizing Maps: Algorithms, Methodologies, and Framework

Authors: Nuno Marques, Bruno Silva (Portugal)

This chapter explores the use of artificial neural networks, specifically self-organizing maps (SOMs), for exploratory cluster analysis. It introduces methodologies, including concept drift detection, distributed learning, and collaborative learning strategies, along with the presentation of an open-source Java ANN library designed for practical SOM applications. The chapter provides practical insights into visualizing and analyzing complex datasets, enabling readers to harness the potential of SOMs in real-world applications.

Persons and Personalization on Digital Platforms: A Philosophical Perspective

Authors: Travis Greene, Galit Shmueli (National Tsing Hua University, Taiwan)

This chapter delves into the philosophical underpinnings of personalization on digital platforms. It explores the relationship between personalization and the concept of the person, emphasizing the importance of ethical considerations in shaping personalized experiences. The chapter also discusses the ethical implications of personalization within the context of European AI and data protection law and proposes principles for aligning personalization technology with democratic values.

Mind Uploading in Artificial Intelligence

Authors: Jason Wissinger and Elizabeth Baoying Wang (Waynesburg University, USA)

Mind uploading is the futurist idea of emulating all brain processes of an individual on a computer. Progress towards achieving this technology is currently limited by society's capability to study the human brain and the development of complex artificial neural networks capable of emulating the brain's architecture. The goal of this chapter is to provide a brief history of both categories, discuss the progress made, and note the roadblocks hindering future research. Then, by examining the roadblocks of neuroscience and artificial intelligence together, this paper will outline a way to overcome their respective limitations by using the other field's strengths.

Ethical Issues of Artificial Intelligence (AI): Strategic Solutions

Authors: Sara Shawky, Park Thaichon, and Lars-Erik Casper Ferm (Various Australian Universities)

Ethical issues of AI have become a huge concern dominating government, media, and academic discourse. This book chapter sheds light on some of the most pressing ethical issues that result from the adoption of AI-powered tools. Increasing inequality, widening social and economic gaps, compromising privacy and data protection, outsmarting humans and impacting human rights, lack of accountability, liability and reliability, and lack of empathy and sympathy are considered the most pressing challenges that need to be addressed concerning AI and big data. This chapter also provides insight into strategies that are currently in place to overcome adverse implications of AI in the public and private sectors. Providing insight into these ethical challenges along with the governing solutions makes a significant contribution to the ongoing discourse and urges for bringing forth sustainable solutions that are necessary for the ethical application of these technologies in different fields.

Intelligence Augmentation via Human-AI Symbiosis: Formulating Wise Systems for a Metasociety

Author: Nikolaos Stylos (University of Bristol, UK)

This chapter delves into the concept of Intelligence Augmentation (IA) and its role in creating symbiotic interactions between humans and AI agents. It explores how IA can optimize collaborative integration across multi-agent systems and the potential benefits for service organizations and consumers. The chapter addresses the challenges and opportunities presented by IA, fostering a deeper understanding of its implications for societal progress.

Artificial Intelligence in Sports: Monitoring Marathons in Social Media

Authors: Natalia Vila-Lopez, Ines Kuster-Boluda, Francisco Sarabia-Sanchez (Various Spanish Universities)

This chapter explores the use of Artificial Intelligence (AI) in monitoring marathons through social media. It ranks Spanish marathons based on their social media presence and assesses the engagement generated by these events. The chapter highlights the potential of small local events, such as marathons, to enhance destination branding through social network engagement. It provides insights into the role of AI in leveraging social media for sports events.

AI-Driven Customer Experience: Factors to Consider

Author: Svetlana Bialkova (Liverpool Business School, UK)

This chapter addresses the challenges and considerations surrounding AI-driven customer experiences. It explores how consumers perceive and interact with AI systems, emphasizing the importance of usability and quality. The chapter examines the factors that contribute to a satisfactory customer experience, highlighting the role of functional and enjoyable interactions. It presents insights into consumer attitudes and behaviors related to AI systems.

Impact of Artificial Intelligence in Industry 4.0 and 5.0

Authors: Luis Cavique, Luiz Moutinho (Universidade Aberta, Portugal; University of Suffolk, UK)

This chapter delves into the impact of Artificial Intelligence (AI) in the context of Industry 4.0 and the evolving concept of Industry 5.0. It explores how AI technologies, such as the Internet of Things (IoT), Big Data, and Cloud Computing, are reshaping manufacturing and production processes. The chapter demonstrates how AI tools enhance predictive maintenance, production optimization, and customer personalization in Industry 4.0. It also emphasizes the transition to Industry 5.0, which focuses on human-centric production and sustainable manufacturing development. The chapter highlights the pivotal role of AI in optimizing industrial processes, reducing risks, and increasing profitability.

These chapters collectively offer a diverse and comprehensive exploration of Artificial Intelligence's multifaceted impact across various domains, shedding light on its potential, challenges, and ethical considerations. Each chapter contributes unique insights and perspectives to the broader discourse on AI's transformative influence on society, technology, and human experiences.

In Summary

As we bring this edited reference book to a close, we reflect on the multifaceted landscape of Artificial Intelligence (AI) and its profound impact on our society and various domains. The chapters within this volume have collectively illuminated the remarkable advancements, challenges, and ethical considerations that shape the world of AI.

Throughout this book, we have witnessed the evolution of AI from a niche field to a ubiquitous force, influencing industries as diverse as healthcare, marketing, education, tourism, and beyond. AI has not only transformed the way we live and work but has also opened up new horizons for research and innovation.

The exploration of causality in AI, as exemplified by Judea Pearl's influential framework, has shown us that understanding cause and effect remains a foundational challenge. This newfound clarity on causal inference methods and the implications of causality in Responsible AI and explainable AI (xAI) has the potential to pave the way for more transparent and accountable AI systems.

In the realm of marketing research, AI's capabilities for data analysis and prediction have reshaped the industry. However, we have also delved into the complexities of bias, data quality, and ethical considerations that must be navigated as AI continues to redefine marketing practices.

AI's entry into healthcare has opened up possibilities for more accurate diagnoses and personalized care, promising to improve the quality of life for individuals worldwide. Meanwhile, the incorporation of Large Language Models (LLMs) into education has unveiled new dimensions of personalized learning, though not without ethical dilemmas that require thoughtful consideration.

Tourism, too, has been transformed by AI, offering new experiences and opportunities for both tourists and companies. The ability to analyze sentiments expressed in online reviews and tailor recommendations based on individual preferences has the potential to reshape the industry's future.

From the ethical implications of AI in various fields to the role of AI in renewable energy systems and the practical applications of self-organizing maps in data analysis, this book has covered a wide spectrum of AI-related topics.

As we close this volume, we are reminded that AI's journey is far from over. With each passing day, new breakthroughs and challenges emerge, demanding our continued attention and collaboration. AI's impact on industry, society, and human lives is undeniable, and the responsible and ethical deployment of AI technologies remains a paramount concern.

We hope that this edited reference book has provided researchers, scholars, students, and practitioners with a valuable resource for understanding the complexities of AI and its profound place in our ever-evolving world. It is our sincere wish that the insights shared within these chapters serve as a foundation for further exploration, critical thinking, and the responsible advancement of AI for the betterment of our society and the world at large.

As editors, we extend our gratitude to the contributing authors who have shared their expertise, and to the readers who embark on this intellectual journey with us. Together, we continue to shape the future of Artificial Intelligence and its place in our ever-changing society.

Luiz Moutinho
University of Suffolk, UK

Luís Cavique
Universidade Aberta, Portugal

Enrique Bigné
Universitat de València, Spain

Chapter 1
Causality:
The Next Step in Artificial Intelligence

Luis Cavique
https://orcid.org/0000-0002-5590-1493
Universidade Aberta, Portugal

ABSTRACT

Judea Pearl's ladder of causation framework has dramatically influenced the understanding of causality in computer science. Despite artificial intelligence (AI) advancements, grasping causal relationships remains challenging, emphasizing the causal revolution's significance in improving AI's understanding of cause and effect. The work presents a novel taxonomy of causal inference methods, clarifying diverse approaches for inferring causality from data. It highlights the implications of causality in responsible AI and explainable AI (xAI), addressing bias in AI systems. The chapter points out causality as the next step in AI for creating new questions, developing causal tools, and clarifying opaque models with xAI approaches. The work clarifies causal models' significance and implications in various AI subareas.

1. INTRODUCTION

The causal revolution heralded by Judea Pearl [Pearl 2000], [Pearl, Mackenzie 2018], [Pearl 2019] caught the attention of disciplines such as causal discovery and causal inference. Causal discovery aims to infer a causal structure based on observable data. In other words, given a dataset, find the causal model usually represented by a direct acyclic graph. Causal inference comprises a set of tools that allow data analysts to measure cause-and-effect relationships. In a complex world, causal inference helps establish the causes and effects of the actions studied, for example, the impact of minimum wage increases on employment or the influence of legislation on the number of enrolled students.

In the current era of Big Data, the evaluation of data-driven models favors more explanatory models than just predictive ones. The difference between correlational and causation is at the heart of the controversy over prediction and explanation. These two tasks must be distinguished for Artificial Intelligence (AI), giving rise to new disciplines such as explainable AI (xAI) [Belle, Papantonis 2020].

DOI: 10.4018/978-1-6684-9591-9.ch001

Objectives

This work has two objectives. Firstly, clarify the concept of causality, exploring the cause-and-effect relationship between variables. Then, the research aims to highlight how grasping causality impacts the current advancements in AI.

Contributions

Beyond the contribution of an integrated view of causal models with xAI, two incremental contributions are proposed. The first is related to the new taxonomy of causal inference approaches. The second regards identifying the current and emergent groups of techniques in xAI.

Organization

The structure of the rest of the paper can be summarized as follows. Section 2 covers some background information. Section 3 introduces the subject of causality and relevant definitions. Causal discovery is presented in Section 4. Causal inference is developed in Section 5. In Section 6, responsible AI and explainable AI are presented as a consequence of causality. Finally, in Section 7, we draw some conclusions.

2. BACKGROUND INFORMATION

This work aims to present a comprehensible data science maturity model that includes the well-known business intelligence and analytics areas, the new practices in business experimentation [Thomke 2020], and Pearl's latest ideas on causality [Pearl 2019]. The proposed pipeline can be scratched as BI → BA → BE. The proposed maturity model named IABE is the Intelligence, Analytics, and Business Experimentation acronym [Cavique et al. 2023].

In this section, three crucial components of Data Science have been detailed: Business Intelligence (BI), Business Analytics (BA), and Business Experimentation (BE), which includes causality concepts.

Data Science is the current term for the science that analyzes data, combining statistics with machine learning/data mining and database technologies to respond to Big Data's challenges.

The modern Data Science developed in the 2010s corresponds to the merges of several areas during that time [Davenport 2014]:

- in the 1960s, Machine Learning, ML,
- in the 1970s, Decision Support Systems, DSS,
- in the 1980s, Executive Information Systems, EIS,
- in the 1990s, Online Analytical Processing, OLAP,
- in the 2000s, Business Intelligence and Analytics, BI&A,
- in the 2010s, Big Data and IoT,
- and finally, in the 2020s, the rise of BE and causality.

2.1. Business Intelligence and Analytics

The work of Sharda, Delen, and Turban [2017] follows part of the Gartner Analytic Ascendancy Model (GAAM), dividing Business Analytics into three types: descriptive, predictive, and prescriptive. Unlike GAAM, the diagnostic type is not mentioned in this taxonomy. Moreover, the prescriptive type includes the areas of optimization, simulation, decision modeling, and expert systems. This type aims to find the best possible decisions and answer questions like "what should I do?" or "what is the best option?".

A classic taxonomy in Management Science [Eppen et al. 1998] split the equivalent BA types into two dimensions: reasoning and stochastic views. In reasoning, dimension models can be considered more deductive or inferential. Moreover, the stochastic dimension models are determinist or probabilistic.

Deductive modeling is a symbolic model in which variables, parameters, and algorithmic relationships are assumed from prior knowledge. This approach tends to be top-down and involves few data items. Deductive modeling can also be called decision-making and is equivalent to the prescriptive type of Sharda et al. [2017]. In this work, we use the term decision making instead of prescriptive. On the other hand, inferential modeling is a symbolic model in which variables, parameters, and algorithmic relationships are estimated by data analysis. This approach tends to be bottom-up and involves hundreds of data items, and inferential modeling is strongly related to data analysis. Table 1 shows a table with two entries with four types of models, given the possible pair of dimensions (reasoning, stochastic).

Table 1. Types of analytic models

	Deterministic modeling	Probabilistic modeling
Deductive modeling (decision making)	optimization	decision analysis, simulation
Inferential modeling (data analysis)	database query	statistical analysis, forecasting

The data-driven areas of Artificial Intelligence have changed names over time, from machine learning to knowledge discovery, data mining, and machine learning for the second time. The same happens in Operations Research (OR); after using the name Management Science, the Institute for Operations Research and the Management Sciences (INFORMS) promotes the new title of OR and Analytics research.

2.2. Business Experimentation

Fisher [1966] introduced the experimental design, which uses randomized trials with an experiment and a control sample, also known as A/B testing. Since then, this method has become a standard in science.

Although the origins of the earliest experimentalists are associated with health care, the social and economic laboratories became prominent in the 1930s. A relevant development in the structure of social networks came from an experiment by the American psychologist Stanley Milgram [1967]. Milgram's experiment consisted of sending letters from people in Nebraska in the Midwest to be received in Boston, on the East Coast, where people were instructed to pass the letters by hand to someone they knew. An average of six people passed on the letters that reached the destination. Milgram concluded that the experiment showed that, on average, Americans are no more than six steps away from each other. This experiment led to the six-degree concept of separation and the notion of the small world in the analysis of social networks.

The rise of the internet opened new opportunities for digital companies in the global market. The design of interfaces led to the knowledge of rapid prototyping and a new experimentation culture. B2C companies accustomed to rapid prototyping experiences provide new insights instead of circular conversations based only on opinion [Bland, Osterwalder 2020].

The scientific method has been used in large technological companies such as Google, Facebook, and Netflix in randomized trials with millions of participants. Companies look at experiments to understand better user behavior instead of wasting millions of dollars yearly on advertising campaigns.

In search of excellence in customer experience, developing new products, or trying new models, a new discipline arises – business experimentation [Luca, Bazerman 2020], [Thomke 2020]. The experimentation follows the classic deductive process, which includes three phases: firstly, generate testable hypotheses, then run disciplined experiments, and finally, learn meaningful insights.

Several myths about business experimentation tend to disappear in the most recent approaches. We choose three examples from Thomke [2020]:

i) The need for many transactions to run experiments in brick-and-mortar companies can be simplified using small samples.
ii) The Big Data from GAFAM (Google, Amazon, Facebook, Apple, Microsoft) is unnecessary to find cause and effect relationships, as small samples can infer causality.
iii) Companies that carried out customer experiments without prior consent are currently more restricted following the publication of the General Data Protection Regulation (GDPR). On the other hand, fully informing customers can lead to problems of emotional contagion. It is necessary to find a balance between giving up the innovation provided by experiments and permission to carry out tests.

3. CAUSALITY

In this section, we first introduce the concept of causality notation and the significant contributions made by Judea Pearl [2019]. Subsequently, we explore the dichotomy between causal discovery and causal inference.

3.1. Definitions

Analytical models discover or describe exciting patterns and make predictions based on good fits of historical data. In evaluating data-driven models' recent movements are presented in favor of explanatory models. The difference between correlational analysis and causality is at the heart of the controversy over prediction and explanation. In data science, two tasks must be distinguished: prediction and explanation.

In prediction, two variables are used: the independent variable X and the dependent variable Y. The original data is divided into the training and testing data sets to find the Y=f(X) function, where X is a covariate, and Y is the outcome.

A new variable type should be included: the intervention/treatment T. In this task, outcome Y of treatment T is the subject of the study. For this purpose, test and control datasets are used for treatment accomplished T=1 and not accomplished T=0. In analogy with Y=f(X), the explanatory function uses three variables, Y=f (T, X).

Judea Pearl contributed to a new view of causality from the point of view of computer scientists, summarized by the ladder of causation [Pearl, Mackenzie 2018]. He argues that artificial intelligence still does not master causal-and-effect relationships.

Pearl [2019] proposes three levels of causality: association, intervention, and counterfactual. The association has no causal consequences, contrary to the intervention and counterfactuals. Association corresponds to the predictive approach Y=f(X), usual in Machine Learning. The intervention is exemplified by the A/B testing, where treatment T appears in the equation Y=f(T, X). Finally, counterfactual (or unavailable data) involves imaginary worlds and specific approaches that compare treatments.

Since the counterfactual level includes questions with 'what if?' and 'why?', we add a new rung on the ladder, which makes the reason for the working title "Book of Why" [Pearl, Mackenzie 2018]. The different levels are associated with specific questions, as follows:

- Association: What is the relation between X and Y?
- Intervention: What if they do treatment T?
- Counterfactuals: What if they had acted differently? (T=1 instead of T=0)

Similarly, the work of Hernán et al. [2019] advocates three tasks in data science: description, prediction, and causal inference, which are applied to a training program in data science. The students learn to differentiate the three tasks and then generate and analyze data for each task. The students also learn to ask scientific questions for each task.

The Direct Acyclic Graphs, DAG, are a handy tool in causal representation; they describe the causal assumptions of each study [Pearl, Glymour 2016]. The nodes correspond to the variables (treatment T, covariates X, and outcome Y), and the arrows are the eventual association between the nodes.

Moving up the second rung of Pearl's causality ladder, the intervention corresponds to an experiment trial with randomly chosen test and control groups. Intervention answers questions like 'what is the effect of T on Y?' or 'what if I do treatment T?'.

In the study of causality, the data description and the DAG must be presented before the modeling phase. The assumptions represented in the causal diagrams are drawn before the conclusions [Hernán 2017].

One of the biggest causality problems in observational studies is spurious or confounding relationships. If we associate the relation T→Y, a third variable X, can be a mediator between T and Y (T→X→Y) or a covariate that influences the two variables (T←X→Y), depending on the direction of the relation. The Pearl's back-door path is any path from T to Y that starts with an arrow pointing to T, *i.e.*, X→T→Y. If we lock the back door X→T, the variables T and Y will not be confounded.

In many situations, a linear regression model can be used to estimate the treatment effect. This is often done in the context of an observational study or a randomized controlled trial (RCT). The treatment effect is the outcome difference between the treatment and control groups. In the context of a simple linear regression model, the model might look something like this:

$$Y = a + b.T + c.X + e$$

where e represents the error and slope b measures the causal effect [Angrist, Pischke 2015]. For the special case of the bivariate regression:

$$Y = a + b.T + e$$

The conditional expectation Y, given the treatment T, takes two values, as follows:

$$E(Y|T=0) = a$$

$$E(Y|T=1) = a + b$$

and then $b = E(Y|T=1) - E(Y|T=0)$ is the difference in expected Y with treatment T.

Using this notation $E(Y|T) = E(Y|T=0) + [E(Y|T=1) - E(Y|T=0)]. T = a + b. T$. So, $E(Y|T)$ is a linear function of T, with slope b and intercept a. The regression slope measures the difference in expected Y with treatment T switched on and off.

The fundamental problem of causal inference states that it is impossible to observe both potential outcomes of an individual treated and not treated [Holland 1986]. Since we only observe the outcome, the non-observable or unrealized outcome is called the counterfactual.

Counterfactual analyses have become popular since the philosophical developments in the 1970s. The best-known counterfactual analysis of causation is the theory of David Lewis [1973]. Counterfactual truths are fictional since they occur in a different world. A counterfactual world is Spatio-temporal disconnected from our world, and there is no interaction between worlds, so we cannot check its existence empirically.

The core idea behind the counterfactual theory of causation is causal dependence. A hypothetical scenario can define causal dependence: given that T and Y are different events, Y causally depends on T if and only if the counterfactual "if T were not to occur, Y would not occur" is a true sentence. The same reasoning is addressed in law, considering that jurists have searched for a direct test of the defendant's guilt called 'but-for causation' or objective cause for centuries. This case is also known as a *sine qua non* or necessary condition.

3.2. Causal Discovery and Causal Inference

As statistical learning and probabilistic reasoning, causal discovery (or causal learning) and causal reasoning (or causal reasoning) are two related concepts in the field of causality. Observations and outcomes support statistics and probabilities. On the other hand, causal discovery and inference include the treatment variable.

Causal discovery refers to identifying causal relationships, or dependencies between variables, from observational data without prior knowledge or assumptions about the underlying causal structure. It aims to uncover the underlying causal connections among variables based on statistical patterns and dependencies observed in the data.

Causal inference, on the other hand, involves using existing knowledge or assumptions about causal relationships to conclude the causal effects of interventions or treatments. It focuses on estimating the causal effect of the treatment variable T on outcome Y by considering the causal structure that is already known or assumed.

Causal discovery and causal inference work sequentially. Causal discovery techniques can help identify potential causal relationships between variables, which can be used as inputs to causal inference models. Causal inference, on the other hand, often relies on additional assumptions or experimental data to estimate causal effects more accurately. In the following sections, this topic is discussed.

4. CAUSAL DISCOVERY

Despite Pearl's levels of causality [Pearl 2019], there are two fundamental disciplines in the area of causality: causal discovery and causal inference.

In causal discovery (or causal structure learning), given a dataset, the objective is to draw a causal diagram or a DAG (Directed Acyclic Graph), where "X→ Y" can be read as "X causes Y", as shown in Figure 1. The causal inference methods are used to evaluate the causal effects after establishing the relation of the covariates, the interventions, and the outcomes.

Figure 1. Dataset and directed acyclic graph (DAG)

A	B	C	D
0.8	1.2	2.1	2
2.0	2.0	1.1	3.2
3.9	1.2	2	2.1
3.14	1.25	2.14	2
...	...	...	...

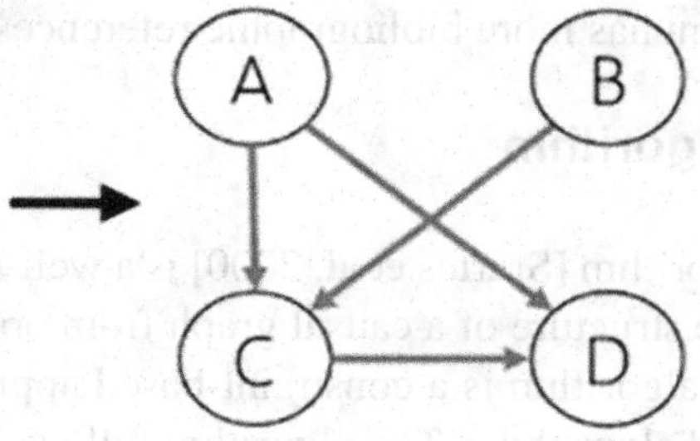

For DAG degeneration, there are three approaches: (i) functional causal models, (ii) score-based methods, and (iii) constraint-based methods [Zanga et al. 2022], [Molak 2023].

(i) Functional Causal Models

Functional Causal Models (FCMs) assume that the relationships between variables can be represented by structural equations involving the direct causes of each variable. LiNGAM (Linear Non-Gaussian Acyclic Models) focuses explicitly on the data's linear relationships between variables and non-Gaussianity. LiNGAM utilizes non-Gaussianity properties, such as the Independent Component Analysis (ICA), to identify the causal ordering of variables. By iteratively estimating the structural coefficients, LiNGAM aims to discover the underlying causal structure of the linear FCM.

(ii) Score-based methods

Score-based methods have also been used to discover causal structures. Score-based methods include the Greedy Equivalence Search (GES), Fast GES, and K2 algorithms. The Greedy Equivalence Search (GES) algorithm starts with an empty graph and iteratively adds and removes directed edges to maximize the improvement of the scoring function. The Bayesian Information Criterion (BIC) or the Akaike Information Criterion (AIC) are examples of scoring functions. Fast GES is an improved and parallelized version of GES. K2 performs a constructive heuristic search for the parents of each node. Max-Min Hill Climbing finds the skeleton of the Bayesian network followed by a heuristic to orient the direction of the edges.

(iii) Constraint-based methods

The Inductive Cause and PC algorithms are among the most common constraint-based algorithms. IC (Inductive Causation) returns the equivalent class of the DAG based on the estimated probability distribution of random variables and an underlying DAG structure. PC (Peter-Clark, named in honor of the authors Peter Spirtes and Clark Glymour) is produced by iteratively checking the conditional independence conditions of adjacent nodes given a partially directed acyclic graph. In causal discovery, the PC algorithm has more bibliographic references.

4.1 PC Algorithm

The PC algorithm [Spirtes et al. 2000] is a well-known algorithm used in causal discovery. It is used for learning the structure of a causal graph from observational data.

The PC algorithm is a constraint-based approach that combines statistical independence tests with graph-based algorithms. The algorithm follows a two-step process:

Step 1: *Determine the Graph Skeleton*

- Given a complete undirected graph, the algorithm identifies the skeleton of the causal graph by testing for statistical independence between variables. The algorithm checks the independence of pairs (X, Y) of variables conditioned by a third variable, Z, $X \perp\!\!\!\perp Y \mid Z$. If variable Z explains the correlation, the relation is called d-separated, and the edge is removed.
- The result of this step is an undirected graph that does not specify the directionality of the causal links.

Step 2: *Determine the Orientation of the Edges*

- The algorithm applies a set of orientation rules based on the concept of "immorality" in the graph. There are different possibilities to draw three variables: the chain $X \rightarrow Y \rightarrow Z$ or $X \leftarrow Y \leftarrow Z$; the fork $X \leftarrow Y \rightarrow Z$, and the collider $X \rightarrow Y \leftarrow Z$.
- Using the conditional independence tests again, if $X \perp\!\!\!\perp Y \mid Z$, we are in a fork chain. Otherwise, if $\sim X \perp\!\!\!\perp Y \mid Z$, it is the case of a collider. The algorithm adjusts the orientations of colliders to satisfy the global Markov property, a fundamental assumption in causal graphical models.

The PC algorithm's first step removes correlated edges given a third variable, and the second step studies what came first: the egg or the chicken. The PC algorithm is widely used in causal discovery due to its efficiency and ability to handle large datasets. It provides a valuable tool for understanding causal relationships and inferring causal structures from observational data.

5. CAUSAL INFERENCE

Causal inference is determining cause-and-effect relationships between variables in a system. It involves understanding how changes in one variable (treatment T) cause changes in another (outcome Y). Researchers often use experimental and statistical techniques to establish causality in causal inference.

The effect of the treatment (T) on the outcome (Y) can be expressed as the difference between the potential outcomes when the treatment is applied (Y(1)) and when it is not applied (Y(0)). The referred difference is known as the Average Treatment Effect (ATE):

$$ATE = E[Y(1) - Y(0)]$$

A widely used technique in experimental studies is randomized controlled trials, RCTs, randomly assigning treatments to study participants (like A/B testing), allowing the causal effect estimation.

A/B testing is the most common intervention, but it is impossible to apply A/B testing consistently. In some areas of social science and healthcare, the A/B test is not considered adequate because it is considered unfeasible or unethical. For instance, when studying the effect of smoking, it is unethical to ask users to consume tobacco to perform a controlled trial. Moreover, historical data does not present test and control samples in many economic activities. Instead of sampling and experimental design, the focus is on quasi-experimental design, mainly observational studies.

Given many causal inference techniques for observational studies, we present a taxonomy in the following subsection.

5.1. Taxonomy of Causal Inference Techniques

Unlike A/B testing, observational studies involve observing subjects in their natural environment without any manipulation or intervention from the researcher. In data science, the chosen technique is also very dependent on the available data. The presented taxonomy considers the accessible data. In Table 2, a synthesis of the causal inference techniques is presented. The following books are good references: Cunningham [2021] and Huntington-Klein[2022].

Table 2. Taxonomy of the causal inference techniques

category	variables	techniques
treatment T and outcome Y	X, T, Y	Propensity Score Matching and Stratification
previous variables with a constant	X, T, Y, cutoff	Regression Discontinuity and Difference-in-differences
previous variables with a third variable	X, T, Y, Z/M	Instrumental Variables and Mediation

Given the variables X (covariates), T (treatment), and Y(outcome), Propensity Score Matching can balance the covariates between treatment and control groups, mimicking a randomized control trial. Stratification, another method based on propensity scores, involves dividing the data into distinct strata based on the propensity scores and estimating the treatment effect within each stratum.

The second group of techniques adds a constant to the previous variables. Regression Discontinuity can be used when the assignment variable has a specific cutoff. Difference-in-differences is used when we have repeated cross-sectional data over time and are interested in the effect of a treatment applied at a specific time. It compares the outcome change over time between a group that received the treatment and a group that did not.

The last group of techniques adds a third variable to variables X, T, and Y. When a third variable, Z, is added. Instrumental Variables can be used when there is concern about unmeasured confounding variables affecting the treatment-outcome relationship. Adding a variable M, mediation analysis can be used to understand the indirect effects of the treatment on the outcome via the mediator.

Propensity Score Matching and Stratification

Given the fundamental problem of causal inference that it is impossible to observe both potential outcomes of an individual treated and not treated and to avoid confounding in the presence of many covariates, matching seems advisable. Propensity Score Matching involves pairing a treated individual with one non-treated individual with similar characteristics, given by the propensity score. Propensity Score Matching may not limit the matching to a 1:1 ratio. The 1:many matchings, where one treated unit is matched with several control units with similar propensity scores.

Stratification (or subclassification) is another method used in causal inference to deal with confounding variables when estimating the effect of a treatment. In propensity score stratification, individuals are divided into mutually exclusive groups (or "strata") based on their propensity scores. For example, one might create ten strata such that all individuals with propensity scores between 0 and 0.1 fall into the first stratum for ten deciles. Stratification is a relatively straightforward method of using propensity scores to estimate treatment effects. Stratification can be implemented with parametric (like Logistic Regression or Linear Discriminant Analysis) and non-parametric models (Machine Learning approaches like Decision Trees).

Regression Discontinuity and Difference-In-Differences

A significant advance in the operability of counterfactual reasoning took place in econometrics [Angrist, Pischke 2015]. Some counterfactual questions can be answered using appropriate statistical techniques, named counterfactual impact evaluation methods. Regression Discontinuity Design and Difference-in-differences allow an intuitive graphical representation [Crato, Paruolo 2019].

Given a set of observations and intervention, the Regression Discontinuity Design measures the impact of the intervention given by the discontinuity between two regressions. Since the observations contributing to identifying the counterfactual effect are mainly those around the intervention, the method may require large sample sizes.

The difference-in-differences method is a statistical technique that measures the impact evaluation of the relative impacts of two sets of observations, mimicking an experimental research design by studying the differential effect of an intervention on a treatment and control group. The difference in the slopes of the linear regressions gives the counterfactual change.

Instrumental Variables and Mediation

Since observational data is not always the outcome of a random trial, the instrumental variables [Angrist, Pischke 2015] approach is the most common technique to estimate causal effects using non-experimental data. Instrumental variables estimate causal relationships when controlled experiments are not feasible or when a treatment is not successfully delivered to every unit in a randomized experiment.

The most relevant cause-effect relationship mechanism is known as the 'why?' which can be found in controlled experiments and observational studies.

One of the first controlled experiments was carried out by the physician James Lind in 1747, who administered six different treatments to sailors with scurvy, concluding that citrus fruits cured patients. The correspondent DAG is Citrus → ~Scurvy, that is, citrus are the cause of non-scurvy, or citrus avoid scurvy. However, the reason for the cure is not evident, and the answer would only come in the 1930s with the discovery of vitamin C. On the path (Citrus → Vitamin C → ~ Scurvy), the variable Vitamin C mediates the relation. Therefore, mediation analysis is the tool that allows a better understanding of the association of two variables.

Baron and Kenny [1986] define the principles for detecting measurement between three variables, making it one of the 33 most cited papers. Following the proposal in their seminal work, to test for mediation, one should estimate three regression equations: a) mediator M is explained by the independent variable T; b) dependent variable Y is explained by the independent variable T; c) dependent variable Y is explained by the independent variable T and the mediator M.

After the regressions are calculated, to establish mediation, the following conditions must be fulfilled: a) the independent variable T must affect the mediator M; b) the independent variable T must show to affect the dependent variable Y; c) the mediator M must affect the dependent variable Y.

6. RESPONSIBLE AND EXPLAINABLE AI

Causality and explainable AI are related because they provide insights into complex systems' inner workings. The why questions of Pearl and the right-to-explanation are on the same path.

Causality is concerned with identifying the causal relationships between variables in a system, such as the relationship between a particular input and its effect on an output. In the context of AI, causal inference can help us understand how the treatment feature influences the outcome.

Explainable AI, on the other hand, is focused on developing AI systems that can provide clear and understandable explanations for their outputs or predictions. Causal tools are more and more critical in situations where the decisions made by AI systems have significant consequences, such as in healthcare, finance, or legal contexts.

Causal inference is an essential tool of responsible AI since we can better understand how AI systems make their predictions and provide more trustworthy and transparent explanations. Causality helps build trust in AI systems, improve performance, and ensure they are used ethically and responsibly.

6.1 Responsible AI

As AI becomes more prevalent in various domains, upholding fairness, transparency, explanation, and responsibility becomes increasingly vital to ensuring that AI benefits all and does not lead to unintended negative consequences. Responsible AI refers to the ethical and accountable development, deployment, and use of artificial intelligence technologies.

Moniz Pereira argues the need to design machine ethics in the theoretical field, paving the way for a new scientific area beyond computational ethics [Pereira, Lopes 2020]. Machine ethics involves creating AI systems that can recognize morally relevant situations, understand the consequences of their actions, and make choices consistent with ethical principles. The goal is to build AI to navigate complex moral dilemmas and contribute positively to society.

Dignum [2019], during her talk, generalized with humor how AI is used in different countries like the United States of America, China, and European countries. In the United States, AI is widely used for commercial purposes and generating revenue. The Chinese government invested heavily in AI for social control and surveillance. European Union (EU) has taken a more cautious approach to AI, focusing on regulatory frameworks to address ethical, legal, and privacy concerns.

In Europe, AI is employed in various sectors, much like in the USA and China, but with a greater emphasis on adhering to regulatory guidelines. The aim is to balance encouraging innovation and transparency in AI. From 2018 onwards, the EU extended this requirement by imposing the so-called right-to-explanation in algorithm decision-making [European Commission 2020].

As a result of the EU's regulatory effort, the EU is further ahead than most countries. Its advances are reflected in the so-called Brussels Effect [Bradford 2012]. The Brussels Effect occurs when other countries replicate the regulatory decisions of the EU. Examples of this are the global application of the General Data Protection Regulation (GDPR) by Facebook or the impact of the European Union's Emissions Trading System on aviation services and industries.

Currently, AI is facing significant controversy over whether to prioritize regulation or certification as a means of governing AI technologies. For example, the Responsible Artificial Intelligence Institute (RAI Institute) provides the first independent, accredited certification program for responsible AI systems. Both regulation and certification have their advocates and detractors, each offering different perspectives on addressing the challenges posed by AI.

Those in favor of AI regulation argue that the rapid advancement of AI technologies requires clear and enforceable rules to mitigate potential risks. They believe comprehensive regulations can set ethical standards, ensure transparency, and protect against bias and discrimination. Proponents of regulation stress the need for government intervention to prevent AI misuse and promote Responsible AI development.

On the other hand, advocates of AI certification argue that a flexible, industry-led approach is better suited to promoting innovation and adaptation. They believe heavy regulation could stifle AI progress and hurt competition. Instead, they propose voluntary certification programs that establish best practices and guidelines, allowing companies to demonstrate their adherence to AI ethical principles without imposing strict legal requirements.

Explainable AI can be seen as a subset of Responsible AI. In the framework of Responsible AI, explainability is a critical factor in ensuring transparency and accountability. Without explainability, holding AI systems accountable for their decisions or ensuring their fairness would be challenging, as we would not understand the reasoning behind their actions, decisions, or results. Similarly, explainability is critical for users and stakeholders to trust the system, providing them with the necessary insight into its decisions.

6.2. Explainable AI (xAI)

The interest in the interpretability of machine learning algorithms increased significantly after the publication of the 'Book of Why' [Pearl, Mackenzie 2018]. Concepts of interpretability are more oriented to white-box models such as decision trees, decision rules, and linear regression. On the other hand, explainable approaches focus on black-box methods, like neural networks, clarifying individual predictions with SHAP and LIME procedures [Molnar 2020].

A clear xAI taxonomy is presented by Belle and Papantonis [2020], differentiating algorithms (transparent and opaque), models (agnostic and specific), and a group of techniques: feature relevance explanation, local explanations, explanation by model simplification, and visual explanations.

Model-agnostic techniques are general approaches that can be applied to any machine learning model, regardless of its architecture. These methods provide insights into a model's functions without requiring specific knowledge of its internal workings. On the other hand, model-specific techniques are designed to explain the decisions of particular types of models. They exploit specific characteristics of the model to provide explanations.

The approaches to explain opaque algorithms use the reduction of the input data (feature or instances) or the simplifications of the model. Some of the most well-known groups of techniques for explanation include:

- Feature relevance: This technique involves analyzing the contribution of each input feature to the model's output. It can help identify the most critical features the model relies on to make predictions. Example: SHAP (SHapley Additive exPlanations).
- Local or counterfactual explanations: This technique involves generating alternative input scenarios that would result in a different output from the model. This approach is also known as sensitivity analysis or what-if analysis. It can help identify the factors most influential in the model's decision-making process and can also be used to test the model's robustness.
- Model simplification: This technique involves training a simpler, more interpretable model on a subset of the training data, aiming to approximate the original model's behavior in a specific region of the input space. It can help to provide insight into the decision-making process of the original model for a particular instance or set of instances. Example: LIME (Local Interpretable Model-agnostic Explanations).

Causality is not in the listed techniques since it is an exploring approach in xAI. Causality is already mentioned in recent works like Belle and Papantonis [2020] and Masís [2021].

Causality can be crucial in mitigating bias in xAI systems. The xAI goal is to provide explanations for model predictions and ensure that those explanations are fair, transparent, and unbiased. By incorporating causality into the xAI process, we can address potential sources of bias and achieve more reliable and interpretable results as follows:

- Identifying confounding variables: Causality allows us to distinguish between correlation and causation. Often, AI model biases result from correlations that might not be directly causal. We can use causal models to identify and control confounding variables, which are the hidden factors influencing the input features and the model's output.
- Counterfactual observations: Causality enables one to deal with counterfactual observations. By exploring counterfactuals, we can better understand the effect of a particular treatment variable. Counterfactual analysis can help identify and mitigate bias by understanding the causal relationships between features and outcomes and assessing fairness through counterfactual scenarios.

From the business point of view, in July 2023, billionaire Elon Musk announced the debut of a new AI company, XAI, intending to 'understand the true nature of the universe'. XAI stands for Explainable AI and sets itself apart from models like ChatGPT through several vital factors. Its primary focus is on explainability, ensuring it can provide clear justifications for its decisions and responses. Context awareness is another crucial aspect, allowing XAI to accurately understand and respond to nuanced conversations. Moreover, XAI emphasizes ethical considerations, aiming to align with ethical standards, address privacy and bias issues, and promote Responsible AI deployment.

Still in the business area, Gartner, a leading technology analysis company, included causal AI alongside the popular generative AI in its Hype Cycle for New Technologies 2023.

7. CONCLUSION

Judea Pearl has significantly impacted the understanding of causality in computer science through his ladder of causation framework [Pearl, Mackenzie 2018], Pearl [2019]. He emphasizes that despite Artificial Intelligence (AI) advancements, the field still struggles to understand causal relationships and their implications. Pearl's work highlights the significance of the causal revolution, emphasizing the need to improve AI's capacity for understanding cause-and-effect relationships. By doing so, we can unlock the potential for developing more robust and reliable AI systems that have profound and meaningful applications in the real world.

Business Experimentation (BE), as proposed by Thomke [2020], involves vital players GAFAM (Google, Amazon, Facebook, Apple, Microsoft) who play a significant role in this domain. The theoretical foundation of BE is intricately linked to the concept of intervention, as introduced by Pearl [2019]. The insights drawn from BE and its ties to Pearl's intervention concept contribute to the advancement of data-driven strategies and improvements in the overall business landscape.

This work describes the complementarity between causal discovery and causal inference. Moreover, introduces a novel taxonomy of causal inference methods based on the available data in the dataset. The taxonomy reveals three distinct groups: i) involving treatment T and outcome Y, ii) incorporating variables T and Y along with a time variable, and iii) including variables T and Y with an additional third variable, such as an instrumental variable or a mediator. This taxonomy clarifies the diverse approaches to infer causality from data and offers valuable insights for advancing causal analysis methodologies.

AI has rapidly advanced in recent years, transforming various industries and aspects of daily life. In Europe, it has developed several regulatory guidelines to balance innovation and transparency in AI, placing it ahead of many countries, reflected in the Brussels effect. In this work, after detailing the concept of causality, we described its implications in AI, namely in Responsible AI and explainable AI (xAI).

We identified the current groups of techniques in xAI and pointed out that the emergence of causality is crucial to mitigate bias in AI systems.

In summary, we believe that causality is the next step in AI in at least three ways:

- by creating new questions (and finding answers) that go beyond the classic machine learning prediction but measure the effect of the treatment variables in BE environments;
- by creating causal tools to respond to the right-to-explanation and the EU regulation;
- and by clarifying the opaque models with xAI approaches.

Finally, this paper clarifies causal models (discovery and inference) and shows the future implications of causality in different AI subareas.

REFERENCES

Angrist, J. D., & Pischke, J.-S. (2015). *Mastering Metrics: The Path from Cause to Effect*. Princeton University Press.

Baron, R. M., & Kenny, D. A. (1986). The moderator–mediator variable distinction in social psychological research: Conceptual, strategic, and statistical considerations. *Journal of Personality and Social Psychology*, *51*(6), 1173–1182. doi:10.1037/0022-3514.51.6.1173 PMID:3806354

Belle, V., & Papantonis, I. (2020). *Principles and practice of explainable machine learning*. arXiv:2009.11698.

Bland, D. J., & Osterwalder, A. (2020). *Testing Business Ideas: A Field Guide for Rapid Experimentation*. John Wiley and Sons.

Bradford, A. (2012). The Brussels Effect. *Northwestern University Law Review, 107*. https://ssrn.com/abstract=2770634

Cavique, L., Pinheiro, P., & Mendes, A. B. (2023). (accepted)). Data science maturity model: From raw data to Pearl's causality hierarchy. *WorldCIST*, 2023.

Crato, N., & Paruolo, P. (2019), Data-Driven Policy Impact Evaluation: How Access to Microdata is Transforming Policy Design. Springer. doi:10.1007/978-3-319-78461-8

Cunningham, S. (2021). *Causal inference: the mixtape*. Yale University Press.

Davenport, T. H. (2014). *Big Data at Work: Dispelling the Myths, Uncovering the Opportunities*. Harvard Business School Publishing Corporation. doi:10.15358/9783800648153

Dignum, V. (2019). The responsibility is ours. In *Artificial intelligence: applications, implications and speculations*. Fidelidade-Culturgest Conferences and Debates.

Eppen, G. D., Gould, F. J., Schmidt, C. P., Moore, J. H., & Weatherford, L. R. (1998), Introductory Management Science: decision modeling with spreadsheets. Prentice-Hall International.

European Commission. (2020), *White Paper on Artificial Intelligence: a European approach to excellence and trust* (White Paper No. COM(2020) 65 final), European Commission. https://ec.europa.eu/info/publications/white-paper-artificial-intelligence-european-approach-excellence-and-trust_en

Fisher, R. A. (1966). *The design of experiments* (8th ed.). Hafner Publishing Company.

Hernán, M. A. (2017), Causal Diagrams: Draw Your Assumptions Before Your Conclusions. EDX. https://www.edx.org/course/causal-diagrams-draw-your-assumptions-before-your.

Holland, P. W. (1986). Statistics and Causal Inference. *Journal of the American Statistical Association, 81*(396), 945–960. doi:10.1080/01621459.1986.10478354

Huntington-Klein N. (2022). *The effect: an introduction to research design and causality.* Chapman and Hall/CRC.

Lewis, D. (1973). Causation. *The Journal of Philosophy, 70*(17), 556–567. doi:10.2307/2025310

Luca M., & Bazerman, M. H. (2020). *The Power of Experiments: Decision Making in a Data-Driven World.* MIT Press.

Masís S. (2021). *Interpretable Machine Learning with Python: Learn to build interpretable high-performance models with hands-on real-world examples*. Packt Publishing.

Milgram, S. (1967). The Small World Problem. *Psychology Today, 1*(1), 60–67.

Molnar C. (2020). *Interpretable Machine Learning: a guide for making black box interpretable*. lulu.com.

Pearl, J. (2000). *Causality: models, reasoning, and inference*. Cambridge University Press.

Pearl J. (2019), The seven tools of causal inference, with reflections on machine learning. *Communications of the ACM*. ACM.

Pearl J., & Glymour, M. (2016). *Causal Inference in Statistics: A Primer*. Wiley.

Pearl, J., & Mackenzie, D. (2018). *The Book of Why: The New Science of Cause and Effect*. Basic Books.

Pereira, L. M., & Lopes, A. (2020). Machine Ethics: From Machine Morals to the Machinery of Morality, book series Studies in Applied Philosophy, Epistemology and Rational Ethics, SAPERE. Springer.

Sharda, R., Delen, D., & Turban, E. (2017). *Business Intelligence, Analytics, and Data Science: A Managerial Perspective* (4th ed.). Pearson.

Spirtes, P., Glymour, C. N., & Scheines, R. (2000). *Causation, prediction, and search*. MIT Press.

Thomke S.H. (2020). Experimentation Works: The Surprising Power of Business Experiments. *Harvard Business Review Press*.

Zanga, A., Ozkirimli, E., & Stella, F. (2022). A Survey on Causal Discovery: Theory and Practice. *International Journal of Approximate Reasoning, 151*. Doi:10.1016/j.ijar.2022.09.004

KEY TERMS AND DEFINITIONS

Artificial Intelligence (AI): A branch of computer systems capable of performing tasks that typically require human intelligence, such as learning, reasoning, problem-solving, and decision-making.

Business Experimentation (BE): A strategic approach that systematically tests and analyzes different business strategies, processes, or innovations.

Causal Discovery: Aims to identify cause-and-effect relationships between variables in datasets, helping to uncover underlying patterns and dependencies.

Causal Inference: The process of determining cause-and-effect relationships between variables by analyzing data and employing statistical methods to understand how changes in one variable influence another.

Causality: Refers to the relationship between cause and effect, where one event (the cause) leads to another event (the effect).

Explainable AI (xAI): An approach in artificial intelligence that focuses on developing machine learning models and systems that provide transparent and interpretable explanations for their decisions and predictions, enhancing trust and understanding of AI systems by humans.

GAFAM: An acronym representing five of the largest and most influential technology companies in the world: Google, Amazon, Facebook, Apple, and Microsoft.

IABE: An acronym representing the Intelligence, Analytics, and Business Experimentation maturity model.

Responsible AI: This refers to the ethical and accountable development and deployment of artificial intelligence systems, considering factors like fairness, transparency, privacy, and social impact.

Chapter 2
Impact of Artificial Intelligence on Marketing Research:
Challenges and Ethical Considerations

Laura Sáez-Ortuño
https://orcid.org/0000-0001-6660-9458
University of Barcelona, Spain

Javier Sanchez-Garcia
https://orcid.org/0000-0002-7865-0076
Jaume I University, Spain

Santiago Forgas-Coll
University of Barcelona, Spain

Rubén Huertas-García
University of Barcelona, Spain

Eloi Puertas-Prat
University of Barcelona, Spain

ABSTRACT

This chapter explores the use of artificial intelligence (AI) in market research and its potential impact on the field. Discuss how AI can be used for data collection, filtering, analysis, and prediction, and how it can help companies develop more accurate predictive models and personalized marketing strategies. Highlight the drawbacks of AI, such as the need to ensure diverse and unbiased data and the importance of monitoring and interpreting results and covers various AI techniques used in market research, including machine learning, natural language processing, computer vision, deep learning, and rule-based systems. The applications of AI in marketing research are also discussed, including sentiment analysis, market segmentation, predictive analytics, and adaptive recommendation engines and personalization systems. The chapter concludes that while AI presents many benefits, it also presents several challenges related to data quality and accuracy, algorithmic biases and fairness issues, as well as ethical considerations that need to be carefully considered.

DOI: 10.4018/978-1-6684-9591-9.ch002

1. INTRODUCTION

The advent of new technologies has gradually transformed the customer journey into an experiential process (Batat and Hammedi, 2022; Hadi and Valenzuela, 2019). This journey involves consumers engaging with a continuous flow of information about a product or service across multiple touchpoints and channels, starting before purchase, continuing during purchase, and culminating after purchase (Lemon & Verhoef, 2016). For instance, thanks to digital devices, consumers can instantly access details in the prepurchase phase about any product, its substitutes and complements, read customer reviews, and even find out opinions of peers through social networks (El-Shamandi Ahmed et al., 2022; Hill et al., 2015). The assumption is that with all this readily available information, consumers will make fully informed decisions. However, the digital environment has limitations, and not all the information consulted is reliable. It also provides such a large volume of information that it can lead to consumer fatigue (Moon et al., 2021).

As a result of this dynamic transformation, market researchers must reflect on whether there is a paradigm shift and, consequently, whether traditional methodologies are still viable (Batat & Hammedi, 2022) or, on the contrary, if they need to find new tools better suited to the new context. This is a scenario in which consumers can combine information from the online environment to finalize a purchase decision in a physical store or vice versa (Batat, 2022).

This chapter discusses the need to incorporate new market research tools based on Artificial Intelligence (AI) to scrutinize the consumer journey and experience in online and hybrid environments. Unlike the physical environment, where capturing information requires the use of standardized protocols such as questionnaires, the online environment generates information spontaneously through comments posted on blogs, social networks, platforms, etc. Much of this information is of an unstructured typology. Thus, an initial task for market researchers is to update information acquisition protocols and adapt them to informal environments. For this purpose, AI-based protocols have been adopted (Sáez-Ortuño et al., 2023a). In other words, the new protocols must collect information from different sources and formats, which must then be sifted and structured (e.g. by coding) so that analytical tools can be applied (Sáez-Ortuño et al., 2023b). Although online and hybrid market research allows academics and practitioners to obtain valuable information, only a few studies in the marketing field have utilized AI methodologies. This chapter details some of them.

This chapter outlines procedures for adapting information collection protocols in online environments using artificial intelligence, as well as proposed tools for analyzing the data. Taking a holistic approach to online market research, it provides an overview of methodologies that market researchers can implement, the most pertinent applications, and challenges that may arise in the future. Specifically, this chapter addresses three important areas: a description of artificial intelligence tools for data collection and analysis, the areas of the customer experience journey where these tools have been applied, and some future challenges. The aim is to serve as a starting point for future research on utilizing artificial intelligence for market research in digital or hybrid environments.

2. THE EVOLUTION OF MARKETING RESEARCH WITH AI

Modern market research is an integral part of marketing management and contributes to a greater understanding of the process that runs through the customer journey to experience (Lemon & Verhoef,

2016). Studying the customer journey to experience involves monitoring their multiple touch points about a product or brand, ranging from initial awareness before purchase through the purchase process to post-purchase. The information gathered and analysed can help companies understand how consumers interact with their brand at each customer journey stage and identify possible weak areas where they can improve their experience (Lemon & Verhoef, 2016).

Although market research and marketing management have a strong relationship, it is important to distinguish their different roles. While market research is descriptive, which involves collecting, sifting, and analysing information on behaviour, seller and consumer preferences, and market trends, marketing management is prescriptive. It involves making decisions about new product design and development, determining target audiences, market positioning and developing marketing strategies (Iacobucci, 2016). In other words, the role of market research is to provide information to managers to facilitate their tasks.

Market research as a subject has its roots in the transition from the nineteenth to the twentieth century and responds both to advances that took place in various disciplines (psychology, sociology, anthropology, and economics) and to the context of economic growth that followed the second industrial revolution (Stewart, 2010). One of the earliest known market research projects was carried out by the advertising agency NW. Ayer & Son. In 1897, they telegraphed state offices requesting information on expected grain production in their respective states, and all this information was used to inform a client who manufactured agricultural equipment and could make its demand forecast more effective (Bogart, 1957). However, already in the 1860s, some researchers in psychology, such as Gustav Fechner, had made essential contributions in techniques that would later be applied in market research, such as experimental design, to measure attention and memory and in the application of statistical methods to analyse the results (Heidelberger, 2004). The foundations of psychometrics were laid, as a branch of psychology linked to experimentation, to measure mental constructs such as intelligence, personality, or emotions and to establish correlations between external stimuli and human reactions.

However, it was not until the second decade of the 20th century that market research acquired the status of an academic discipline, thanks to the publication of two books, Frederick (1918) *Business Research and Statistics* and Duncan (1919) *Commercial Research: An Outline of Working Principles*, which sought to document best practices in market research (Steward, 2010). Another significant milestone was the establishment in 1923 of the A.C. Nielsen Company, one of the world's most powerful market research companies and famous for its retail panel, which initially engaged in test marketing of new products and in measuring in-store sales to estimate the market share of competitors in various product categories. Subsequently, Nielsen moved into audience studies by proposing a national radio index in 1942 and a television rating service in 1950 (Stewart, 2010).

Although there were significant advances in statistical methods, psychometric measures, experimental design, analysis of variance, and measurement theory during the 1920s, their extension to research occurred much later. Therefore, the methods used during this period were mainly surveys, individual and group interviews, and observation through ethnographic studies (Kartajaya, Setiawan, & Kotler, 2021). The development of the methods and techniques used today (the basic toolbox of marketing research: focus groups, survey research and sampling, experimental design, and multivariate data analysis) took a long time, estimated between 1925 and about 1980 (Stewart, 2010).

Perhaps the most significant development in recent decades has been the growth in the use of the Internet and all the interaction spaces (websites, blogs, social media, etc.) that have developed within it. The network not only provides a new vehicle for market research but also requires new metrics and tools to capture and analyse the enormous volume of information consumers generate. It is an environment where all movements and comments of consumers are recorded, so it generates enormous volumes of information. Although this might seem a huge advantage at first glance, there is the disadvantage that the perception of anonymity that consumers feel generates the proliferation of false information that is difficult to manage (Sáez-Ortuño et al., 2023b). However, the Internet is not only a new channel of communication, distribution, or a new source of gathering data, but it is also creating new ways for consumers to interact with each other that have never been seen before. The network allows the exchange of information between people from different countries, even with the most distant and culturally different ones, creating a social fabric without historical precedent. Some authors, such as McLuhan & Fione (1968), already predicted this by proposing the idea of the "global village", as new technologies would allow the exchange of information between people from different countries and cultures, apparently making the world more open, but at the same time leading to a consequent process of cultural dissolution. On the other hand, the Internet, through blogs or social networks (Facebook, Twitter, etc.), enables a business-customer relationship never seen before (Hoffman & Novak, 2009). In other words, the Internet has not only given rise to a new space in which to conduct empirical research and consequently poses technological and methodological challenges for the collection and analysis of information, but it also presents a new social environment in which consumers relate to each other, which represents a new paradigm and theoretical challenges for market researchers.

Market researchers are using AI-derived tools to try to meet the challenges posed by the new digital social environment. It is not clear the origins of artificial intelligence (AI). Some authors pointed to a period between the 1940s and 1950s when developments in machine learning algorithms made automatisms capable of obtaining rational solutions (Kaplan, 2022). Although the transfer of AI algorithms to market research has occurred during the first decades of the 21st century, it remains open to question whether this movement will have the same impact on market research in the following generations as the incorporation of psychometric techniques developed in the 1920s therefore, whether we are facing a new methodological revolution.

According to published information, the growth prospects for using AI techniques in market research are up-and-coming. According to a report, the application of AI tools in market research is expected to grow by 615.8% between 2018 and 2023. That is, from USD 214 million in 2018 to USD 1318 million in 2023 at a compound annual growth rate of 43.5% (MarketsandMarkets, 2018).

Huang, Rust, & Maksimovic (2019) define AI as the use of computing machines to replicate human skills, ranging from physical tasks (movement and displacement) and cognitive skills (thinking and reasoning) to emotional skills (feelings and affect). The goal of AI applications, whether in robotics or market research, is to emulate human abilities to act, reason and feel. Although work is underway to design machines equipped with instruments and AI capable of performing multiple tasks simultaneously, this challenge is quite difficult to achieve in the medium term (Andriella et al., 2022). While it has been possible to design machines replicating human forms, reproducing their behaviour, thinking, and feelings is much more challenging (Forgas-Coll et al., 2023). In other words, in line with Huang and Rust (2018), we can classify types of AI into mechanical, thinking and feeling AI according to the intrinsic difficulty of achieving it in a machine.

AI algorithms offer excellent opportunities for market researchers to work on the three forms delineated by Huang and Rust (2018). They can be applied to data collection, filtering, analysis, and prediction. Thus, AI Mechanical algorithms can be used for classifying, aggrupation in clusters, and for reducing dimensions from multiple factors; AI Thinking algorithms to process extensive data and recognise patterns and regularities in their behaviour; and finally, Feeling AI algorithms could be used to analyse two-way interaction between consumers, qualitative research techniques can be applied, such as natural language processing (NLP) techniques, the use of recurrent neural networks (RNN), or the analysis of data from nominal scales to detect affective signals (McDuff & Czerwinski 2018). All these tools make them possible to market researchers to analyse large amounts of data on the spot coming from various sources, such as social networks and websites (Jacobsen, 2023). Moreover, these tools not only allow to speed up tasks of collecting information but can also be used to clean data, do quick analyses, and identify patterns and trends that enable diagnoses to make decisions when needed. In other words, it makes it possible to develop more accurate predictive models that may be invisible to human researchers (Li & Zhang, 2022). AI tools can be very efficient, as they can perform tasks in a few minutes that would have taken researchers several hours, and they are also effective, as they do not miss relevant information that researchers might have missed (Sáez-Ortuño et al., 2023a). AI opens the possibility of developing effective marketing strategies, such as micromarketing, allowing companies and/or brands to personalise their marketing efforts by providing information about the preferences and behaviours of individual consumers (Belhadi et al., 2023). An example is Netflix's personalised movie or Amazon's cross-selling recommendations (Chung et al. 2016).

However, applying AI algorithms in market research is not the panacea, and they have some drawbacks that should be improved. On the one hand, some algorithms collect data from unstructured sources, i.e., blog comments, not from a structured questionnaire with closed questions. Therefore, the accuracy of the data analysed, particularly those used to train AI algorithms, must be ensured (Jacobsen, 2023). If the data is biased or incomplete, the information provided by the AI may be inaccurate or misleading. Therefore, it is essential to ensure that the data used to train AI algorithms are diverse, representative, and unbiased (Fernández-Cardenete, Molina-García, & García-García, 2020; Jacobsen, 2023). On the other hand, incorporating these algorithms reduces the more routine part of the researcher's task, leaving them to monitor and interpret the results. Above all, monitoring the data to train the algorithms is very important, as this helps avoid biased results (Bingley et al., 2023).

In conclusion, the emergence of the digital world has transformed how modern society relates to each other, and studying its behaviour requires new market research tools. Adopting AI algorithms may represent a market research revolution like the introduction of psychometrics in the 1920s. These algorithms can collect, filter, analyse and generate results from unstructured data sources, and given the volume of data analysed, they can provide valid, fast, accurate and value-added information to decision-makers to develop effective marketing strategies (Belhadi et al. al., 2023). However, while AI has made market research more efficient, reducing the time and cost of conducting it, it is not a panacea, and weaknesses still need to be addressed and corrected.

2. ADVANTAGES OF AI IN MARKETING RESEARCH

The application of AI algorithms to digital market research is revolutionising market research by providing tools to measure vast volumes of information quickly and efficiently about market buzz, the company itself, competitors, and customers (Huang & Rust, 2021). From these analyses, predictions can be made about which market segment is the most attractive for our product, which product design consumers are most likely to choose, which version of the advertisement generates the most clicks, and so on. As noted above, each type of AI is best suited to perform certain functions, whereby mechanical AI is applied for data collection, cognitive AI for market analysis, and emotional AI for customer understanding (Huang

& Rust, 2018). This section discusses the advantages of applying AI algorithms in market research, including handling large and complex datasets quickly and efficiently, increased personalisation and targeting capabilities, and the potential for cost savings and improved ROI. In addition, the results generated by these systems help decision-makers better understand customers and can propose innovative and competitive designs.

Undoubtedly, one of the main advantages of applying AI to market research is the possibility of working with large databases from different sources, which allows for a much more complete view of customer behaviour and preferences than possible with traditional methods (Zohdi et al., 2022). When working with large databases, the law of large numbers applies, which describes the fact that by experimenting many times, the results obtained get closer and closer to the expected value (Dekking et al., 2005). In addition, since the same client uses several networked devices, algorithms allow for collecting information about their activities in different locations and then cross-checking to corroborate them (Cooke & Zubcsek 2017). Moreover, not only can they process large databases, but they do so quickly, with a degree of accuracy and reliability that was unthinkable only a few decades ago (Manivannan, Prabha & Balasubramanian, 2022). That is, while data collection methods may seem repetitive and less sophisticated than more traditional methods (Wedel & Kamakura, 2000), the sheer volume of data they collect and analyse makes them efficient tools at scale. It helps to generate results with less likelihood of errors and biases compared to results obtained through traditional research (Manivannan, Prabha & Balasubramanian, 2022) and, therefore, provides relevant information on patterns and trends that can be used to guide marketing proposals (Pereira et al., 2022).

Another advantage is the ability to approach analytics personally, entering one-to-one marketing. AI algorithms can analyse a single customer's data and design personalised experiences or targeted marketing campaigns. For example, predictive analytics can create personalised product recommendations (e.g., Netflix or Amazon) based on the customer's previously expressed behaviour or preferences (Chung et al. 2016). Another example of personalisation is using chatbots or virtual assistants, which can provide a personalised product recommendation service that can increase customer satisfaction (Kaushal & Yadav, 2022).

In addition, all these operational advantages and efficiency gains that AI applications achieve also translate into financial results. Getting results quickly, economically, and reliably means savings in research costs and, consequently, an improved return on investment (ROI) (Haleem et al., 2022).

Finally, having reliable information available quickly also has consequences for strategic marketing management (Huang & Rust, 2021). On the one hand, decision-makers can make better-qualified decisions about product and service innovations, communication campaigns, etc. On the other hand, faster decision-making and shorter time-to-market of products or services improve the company's competitive advantage (Mo et al., 2023).

Despite these benefits, it is crucial to recognise the potential ethical considerations arising from using AI in market research. For example, given that researchers use historical data to train the optimising process of AI algorithms, there is a risk that the algorithm may perpetuate biased traits and/ or discriminatory performances. Another underlying concern arises through how data is collected and analysed on the Internet and social media. In some cases, it is not evident that the purpose of a website or an online game is to capture leads (Sáez-Ortuño et al., 2023b), and, in other cases, parts of social media conversations are taken without asking permission from the interlocutors, all of which may raise privacy concerns and/or breaches of data protection laws (Duhaim et al., 2023). To mitigate these risks, companies must prioritise ethical considerations and ensure transparency and accountability in using AI.

It includes regular audits of AI models for potential bias or discrimination behaviour, being transparent with customers about how data is collected and used, and allowing customers to freely opt-in or opt out of information provision (Fernandez-Quilez, 2022).

Recent studies emphasize the need for companies to take on Corporate Digital Responsibility (CDR) when utilizing AI in market research (Lobschat et al., 2021; Wirtz et al., 2023). CDR entails addressing ethical and fairness considerations throughout the entire data lifecycle, including its collection, storage, processing, and utilization. Companies must carefully assess the impact of their AI practices on consumer privacy and information rights. They should also monitor that algorithms do not replicate harmful biases or generate discriminatory outcomes for specific groups. An approach rooted in CDR can assist companies in striking a balance between the benefits of AI and its ethical and socially responsible use.

3. TYPES OF AI IN MARKETING RESEARCH

AI algorithms provide researchers with new ways of collecting, cleaning, and analysing large volumes of redundant data and conducting experiments or assessing experiences through surveys. Software such as SurveyMonkey or SurveyCake enables automation of the survey design and data collection (Huang & Rust, 2021). A wide range of AI techniques can extract and analyse information of different natures (quantitative or qualitative) from the thick web of the Internet and social networks, either by capturing leads or snippets of comments or ratings posted on blogs or social networks. This section describes some of these relevant AI techniques widely applied to market research, including machine learning, natural language processing, computer vision, deep learning, and rule-based systems. Moreover, it will also focus on some algorithms with the most significant impact, such as neural networks and decision trees.

Machine learning (ML) algorithms are used in exploratory studies when researchers do not have a clear idea of the relevant factors and processes. ML works in two stages. In the first stage, sample data is used to train the algorithm to generate a predictive model. In the second stage, new data is fed into the model and the algorithm learns from experience and provides dynamic results with predictions to aid practitioners in making decisions (Ngai & Wu, 2022). In general, exploratory research is used to study problems that are not clearly defined at a preliminary stage, or there is little understanding of the nature of consumers' or users' relationships with some product, service, process, or event, and therefore theories or models need to be built rather than theories tested (Wilson & Vlosky, 1997). Qualitative studies are often carried out at this research stage and often generate inconclusive results (Alam, 2002). However, ML methodology works with such a large volume of data that even if qualitative, the results can be used to select variables, considering many variables a priori and discovering which ones should be retained and which ones discarded, and to establish models of relationships between them (Dzyabura & Yoganarasimhan, 2018). In summary, ML is a tool that analyses large volumes of information, determines the most relevant factors, builds analytical models automatically, and offers marketers opportunities to gain new insights and improve the performance of marketing operations (Cui et al., 2006). Examples of the application of ML include Sarker (2021), who used ML to develop a model that can predict consumer preference for different types of mobile phone brands, and Mohammad et al. (2019), who applied ML to identify customer churn behaviour in a telecommunications industry.

Natural language processing (NLP) is another AI technique that studies human-computer interactions but has also been applied to online market research. According to Goldberg (2016), until recently, the algorithms used came from optimisation-based machine learning (linear models, vector machines or logistic regression), but recently they have been moving towards using neural networks. NLP encompasses many possibilities, such as language and acoustic modelling, using neural networks for machine translation, multimodal applications combining language and other signals such as images and videos (e.g., subtitle generation), etc. Although these programmes aim to develop skills capable of understanding, interpreting, and generating human language, applications have been developed that can analyse the content of documents, comments, or posts, including the contextual nuances of the language, and accurately extract the information contained, categorising and organising it for analysis (Goldberg, 2016). Examples of NLP's application in market research include its use in sentiment analysis, which helps marketers understand how customers feel about brands, products, or services (Rahman et al., 2023), or the work of Roy (2023), who used NLP to analyse online hotel reviews and identified the most critical factors leading to customer satisfaction. However, significant challenges facing this knowledge area include natural language processing, which often involves speech recognition, understanding and the possibility of generating it from artificial systems (Andriella et al., 2022).

Computer vision (CV) is an AI technique that allows computers to capture, analyse and interpret images, such as photographs or videos, extracting meaningful pieces of information to help practitioners propose recommendations. CV has also been applied to conduct market research quite successfully, e.g., Rahman et al. (2023), although it is still early. Huang, Rust, & Maksimovic (2019) consider that if the abilities of cognitive AI is to replicate human thinking and reasoning, CV applications for machine vision will allow them to go further and, thanks to the possibilities of seeing, observing, and understanding, come to interpret and reproduce emotive abilities. Many technologies use machine vision, including object recognition, event detection, scene reconstruction (mapping) and image restoration (Klette, 2014). However, it is essential to note that computer vision models do not see the content of an image but rather detect a series of points, and a mathematical algorithm deduces the image's content (Szeliski, 2011). As with all AI techniques, for the CV application to recognise an object, it must be trained with a large dataset of examples (e.g., an image of a dog) and the associated label (e.g., all these images are of a "dog") (Nanne et al., 2020). During the training process, the CV algorithm captures information that it transforms into parameters and configures a map of multilayer points related to the assigned label. Thus, once trained, the algorithm may be able to recognise a particular image (a dog) from a large volume of images at an astonishing speed and can also determine how many times that image appears, for example, in a large sample of Instagram content (LeCun, Bengio, & Hinton, 2015). This technique has been used in market research, as, based on the images used in training the algorithm, it can analyse visual content on social networks and identify trends and patterns in customers' visual preferences. For example, Rahman et al. (2023) used a CV to analyse images shared on social media platforms to identify popular trends.

Deep learning (DL) is an AI that applies artificial neural network algorithms for shape representation using computer vision, speech recognition, natural language processing, or automatic translation. Depending on the control over DL, learning can be supervised, semi-supervised or unsupervised (LeCun, Bengio, & Hinton, 2015). The term "deep" indicates that the algorithm uses multiple layers in the network (e.g., combining lines and curves), as during early work on DL software based on the linear perceptron, it was tough to make universal object classifications. However, multiple non-polynomial activation functions with a hidden layer of unlimited width can be used. Moreover, the layers are also allowed to be heterogeneous and deviate widely from biologically informed connectionist models for efficiency, trainability, and comprehensibility (Cireşan et al., 2012). For example, if a DL algorithm must detect the shape of an elephant in a sample of photographs and tries to find it using a model of combinations of linear shapes, it will have a much harder time finding it than if the model combines linear shapes, curves, irregular shapes, etc., in different layers. The latter combination allows the DL algorithm to reproduce the standard shape of an elephant much better and, therefore, detect its location. Some studies have compared the object recognition ability of DL algorithms with a sample of people, and in some cases, performance is superior to that of human experts (LeCun, Bengio, & Hinton, 2015). Market research has used this technique to develop predictive models to forecast customer behaviour and preferences. For example, Lee, Yoon, and Kim (2021) use DL to analyse raw data from online video streams and compare the results of these predictions with actual inventory to forecast future advertising inventory needs more accurately.

Rule-based systems (R-BS) are AI algorithms that use rules created by humans to store, sort, and manipulate data. The aim is to replicate humans' information management processes, which are not always rational. For example, the dual decision-making process theory proposes that individuals can decide based on a fast heuristic or a slow, elaborate analytical process (Evans, 2010). The first process, considered primitive, autonomous, intuitive, and not requiring memory work, is called system 1, and the second, deliberative, analytical, reflective, requiring memory work, cognitive decoupling, and mental stimulation, is called system 2 (Pennycook & Rand, 2019). Therefore, if System 1 is to be replicated in a machine, R-BS must detect the patterns and associations of information that lead humans to make such decisions.

In contrast, if System 2 is to be replicated, RG-S can use an analytical and systematic mechanism, more typical of AI algorithms. R-BS algorithms are formed by 'if-then' coding statements (i.e., if X performs Y, Z is the result). These models involve two crucial elements: "a set of rules" and "a set of facts", both of which are applied to achieve some goal. These RG-S models are deterministic, meaning they operate with the simple but effective "cause and effect" methodology (Davis & King, 1984). For example, Khodadadi, Ghandiparsi, & Chuah (2022) used R-BS to automate the classification of customer queries and improve customer service efficiency.

Regarding the algorithms, artificial neural networks (ANNs) are considered. ANNs are algorithms that attempt to reproduce the neural circuitry of the human brain through a system of nodes and connections through which information flows. However, ANNs have substantial differences concerning biological brains because while ANNs are static and symbolic constructs, biological networks are dynamic and analogical (Marblestone, Wayne, & Kording, 2016). ANN is based on a network of connected units called artificial neurons, where each connection (called edges) can transmit a signal to other neurons. However, when an artificial neuron receives input from other neurons (e.g., a natural number), the edge processes them through a non-linear function (e.g., summing the inputs) and sends them as signals to the neurons connected to it. Each neuron and edge are generally assigned a weight; as the algorithm works, this weight is adjusted with learning (Krogh, 2008). This algorithm has been used in market research to develop predictive customer behaviour and preferences models. Examples include Goode et al. (2005), who used an ANN model to predict customer satisfaction derived from mobile phone usage in the UK, and Asgar et al. (2020), who applied ANN to develop a model that can predict customer satisfaction with online shopping.

Another relevant algorithm is the decision tree (DT). It is an algorithm used in classification and regression models (Friedman, 2001). DTs use a hierarchical distribution to classify subjects according to some self-reported or self-recorded criteria (Sáez-Ortuño et al., 2023b). An example of DT is XGBoost (Chen & Guestrin, 2016). This algorithm creates a set of weak (provisional) decision trees and then intelligently combines them to create a more robust model. Given its ability to handle a large volume of data with many features (e.g., sparse data), it can be used to organise clusters of databases. Market research has used this technique to develop decision models that can help marketers understand customer behaviour and preferences. For example, Higueras-Castillo, Liébana-Cabanillas & Villarejo-Ramos (2023) used decision trees to identify the factors that drive customer satisfaction with online shopping.

The techniques and algorithms presented in this chapter are just a sample of the many options that exist and are being used to conduct market research. Other techniques include expert systems and AI algorithms that can provide expert advice or recommendations based on rules and knowledge. For example, Almomani et al. (2022) discussed expert systems that recommend personalised travel packages to customers based on their travel preferences. While among algorithms, fuzzy logic systems and genetic algorithms are considered. Fuzzy logic algorithms use mathematics to handle data in decision-making processes in situations of uncertainty and imprecision (Martin-Rodilla, Pereira-Fariña, & Gonzalez-Perez, 2019). An example of its application in market research is Nilashi et al. (2002), who used a fuzzy logic system to segment customers based on their purchasing behaviour and preferences. The other algorithms considered are genetic algorithms, inspired by the process of natural selection and genetics (Liu et al., 2010), which have been applied in market research to solve complex optimisation problems, such as product design, pricing strategies and supply chain management (Hajipour et al., 2023).

After describing five IA systems and two algorithms, this section concludes that the field of AI applied to market research is constantly evolving, and therefore new algorithms, each with their strengths and limitations, can be expected to emerge and their suitability for the study of online data will have to be considered. As stated in the market research literature, no multifunctional algorithms exist (Wedel & Kamakura, 2000). Therefore, depending on the specific characteristics of the data, their virtues must be tested and validated to be applied to specific functions (Andriella et al., 2022). In studying information published online, correctly applying the correct algorithm can improve data collection and analysis efficiency and generate more accurate results.

4. APPLICATIONS OF AI IN MARKETING RESEARCH

4.1 Exploring the Market Through Sentiment Analysis

Although the development of AI is considered to have begun with the advent of the first computer in the first half of the 20th century, its incorporation as a tool for conducting market research in a meaningful way is relatively recent (Mariani, Perez-Vega & Wirtz, 2022). Since then, there have been various applications in all phases of marketing management, be it to explore the market or to analyse the different components of the STP or the 4Ps (Huang & Rust, 2021). For example, qualitative research techniques such as sentiment analysis and social listening have been used to explore the market; for STP, techniques for market segmentation and sales forecasting; and, for the 4Ps, applications for one-to-one marketing with personalised recommendations, among others. This section will consider some of these recent applications.

The automated sentiment analysis in social media is grounded in social network theory (Granovetter, 1973) and information search theory (Bettman, 1979). Online social networks provide a rich source of qualitative data that consumers organically generate as part of their interactions. Through AI, brands can harness this data, which would otherwise be inaccessible, substantially reducing their costs of gathering information about consumer attitudes and preferences. This consumer generated data on social networks allows brands to explore the market.

To prospect the market, applications have been used to collect and analyse the sentiments expressed by consumers on social networks. Social networking sites have become a new forum where participants comment and exchange opinions on multiple topics, including praising or criticising products or brands they know or have purchased (Zachlod et al., 2022). Hewett et al. (2016) call this forum a reverberating "ecoverse", formed by complex feedback loops through which pieces of information ('echoes') flow, both from the 'universe' of communications generated by companies and brands, from news published by the media, and from user-generated comments (Hewett et al., 2016). Moreover, this "ecoverse" acts as a dynamic living organism, changing as it is fed with information, whether from online word-of-mouth or other sources. However, one of the most worrying elements of the "ecoverse" is its inability to regulate the flow of information, which is increasingly giving rise to a spiral of false and negative comments to such an extent that it is endangering the very communication of companies and brands on social networks (Hewett et al. 2016; Sáez-Ortuño et al., 2023a; Zhang et al., 2022).

AI can help collect and analyse information in the 'ecoverse' to identify customers' opinions and sentiments about brands, products, or services (Zachlod et al., 2022). An example is provided by Humphreys & Wang (2018), who use an NLP algorithm to analyse psychological and sociological constructs emanating from consumer comments on social networks. Using theories and linguistic methods standard in qualitative consumer research, they propose an algorithm for automated text analysis. Although they caution that it cannot be used to study all phenomena, they believe it is a valuable and efficient tool for extracting patterns in text, which researchers would be unable to extract without the help of the algorithm (Humphreys & Wang, 2018). However, it is essential to pay attention to the criticisms made by Ordenes et al. (2017). These authors question whether the sentiment analysis of comments posted on social media is measured by simply counting positive and negative words and propose a stepwise evaluation where different degrees of sentiment intensity are considered. For Ordenes et al. (2017), a graded measure would allow for a deeper analysis of the links between the expressions of sentiment made in comments and subsequent online consumer behaviour. For the empirical study, they used consumer reviews on three sites (Amazon.com, Bn.com, tripadvisor.com) and analysed them through Monzenda, a web scraping software service (Ordenes et al., 2017).

4.2 Market Segmentation and Targeting

Market segmentation and targeting are the first two stages of STP (Segmenting, Targeting and Positioning) and remain one of the fundamental phases of strategic marketing (Iacobucci, 2016). Segmentation consists of dividing the potential market of consumers of a product into groups that are as homogeneous as possible and, at the same time, sufficiently different from the rest of the groups, which makes it possible to carry out marketing actions that are more line with their needs and desires (Stead et al., 2007). Although it is one of the most studied marketing activities in the academic literature and most applied by marketers (DeSarbo & Grisaffe 1998; Wedel & Kamakura, 2000), AI can contribute to improving it by offering the possibility to analyse large amounts of data that were difficult to analyse with traditional tools and means (Chen & Guestrin, 2016). Examples include the study by Chen, Iyengar, and Iyengar (2017), who use AI algorithms to segment consumers more efficiently based on stated preferences by applying conjoint analysis. They argue that a multimodal continuous heterogeneous distribution typically represents consumers' preferences, but when assigned to different segments, it is transformed into a heterogeneous, unimodal distribution. Using optimisation and sparse learning algorithms, they manage, in the first stage, to divide consumers into several segments. In the second stage, they use each candidate segmentation to develop a set of heterogeneity representations at the individual level, and the optimal heterogeneity representation is selected by cross-validation. Finally, they compare the results obtained by their AI algorithms with the results of others and thus show the goodness of their proposition. Another example, with a completely different approach, is that of Valls et al. (2018), who study British tourists' motives for visiting a tourist area in Catalonia and group them into segments using ontologies that allow them to analyse these factors from a semantic perspective.

4.3 Predictive Analytics for Market Demand Forecasting

Predictive analytics or market demand forecasting is a fundamental element of upstream analysis in marketing management (Iacobucci, 2016). Predictive analytics consists of collecting quantitative or qualitative information to which statistical techniques, mathematical models or expert systems are applied to make forecasts about the future (Acciarini et al., 2023). A forecast makes it possible to accurately anticipate market trends (e.g., estimated sales or consumption for the following year), and decision-makers can use this information to adjust budget allocations and design marketing policies (Liu et al., 2016). For example, if R-BS is used for forecasting, the system can combine judgments or assessments generated from different sources, either the forecaster's own, the manager's or the management teams, along with historical reviews. R-BS assigns different weights to each of the forecasts made by different methods so that the final forecast combines qualitative rules to combine the results of the quantitative forecasts (Adyaa, Armstrong, Collopy & Kennedy, 2000). AI forecasting techniques applied to collecting information on online social platforms pose two significant challenges: First, the information collected presents multiple formats (text, images, audio, and video), whereas traditional systems use structured formats. Secondly, the volume of information is so large that standard information collection and analysis procedures do not work in this environment.

Nevertheless, some proposals have been made that point to a paradigm shift in perspective techniques. One example is the work of Liu et al. (2016). In this study, they forecast sales and consumption work by combining cloud computing, machine learning and text mining methods. To illustrate the procedure, they analyse two billion Tweets and 400 billion Wikipedia pages.

4.4 Adaptive Recommendation Engines and Personalisation Systems

AI-powered recommendation engines can analyse customer data to make personalised product recommendations, improving customer experience and increasing sales (Chung et al. 2009; Chung et al. 2016; Dzyabura & Hauser 2019; Liebman et al. 2019). Adaptive personalisation systems developed using AI aim to follow the logic of the decision process that consumers follow when choosing a product or service. In studies on consumer preferences, consumers are considered to use a learning system in preference formation that determines the most relevant attributes in the decision process and the weights they assign to those attributes (Netzer et al., 2008). For example, a consumer searching the Internet for information about cars might find the sight of cars with a sunroof very appealing, but when trying it out at the dealership, they might discover that they underestimated the boot space. In other words, what they initially valued and attached great weight to before the purchase changed entirely during the purchase, which responded to the dynamics of the consumer journey towards experience (Lemon & Verhoef, 2016). Therefore, personalisation and recommendation systems using AI must cope with a dynamic optimisation logic, in the sense that each time a product is searched, the weights assigned to preference-shaping attributes are updated and affect the expected utility of all products and the value of the subsequent optimal search (Dzyabura & Hauser 2019). As noted above, the ability of recommendation and personalisation engines to provide recommendations to customers based on their past consumption and the consumption trends of all customers contributes to improving the customer experience and increasing sales (Rathore, 2023). The most common example is Amazon, which uses purchase history and browsing behaviour to make recommendations (Gupta, Singhvi & Granata, 2023). Other examples of using recommendations and adaptive personalisation systems include Chung et al. (2016). These authors propose to test the effectiveness of mobile adaptive personalisation systems and how these systems affect consumer decision processes in choosing a service/product. To this end, the authors develop and implement an adaptive personalisation system to personalise mobile news based on the analysis of consumer behaviour and information published on their social network. The results indicate that using search engines that combine different sources (consumer behaviour plus posts in their social network) improves personalisation performance and may be a promising approach to make personalisation more effective (Chung et al. 2016).

Another example is Liebman, Saar-Tsechansky & Stone (2019), who studied automated personalised music recommendation services. They criticise that previous systems have used song and artist preferences without considering the temporal and sequential context. Therefore, they propose a system that is adaptive to consumers' circumstances and, to test it, they propose to analyse data from actual playlists combined with a field experiment with human listeners (Liebman, Saar-Tsechansky & Stone, 2019).

In addition, real-life examples of AI applications in companies within the field of marketing research include:

- Airbnb's system learns from guests' previous stays and feedback as well as external factors to offer customized accommodation recommendations.
- Amazon extensively utilises AI, including customer segmentation, personalised recommendations, predictive marketing, and campaign optimisation. They analyse every step users take on their platform.
- Banco Santander implemented a natural language processing-based virtual assistant to interact with customers and provide personalised assistance 24/7. It also generates information that the bank uses in its market research.
- Disney+ tailors homepage and "continue watching" sections according to profile configurations and watch histories.
- IKEA developed an app that uses augmented reality and computer vision to let customers virtually see how furniture looks in their homes. It provides valuable data on interior design preferences.
- Mercedes-Benz implements AI to study driver behaviors data and offer adaptive vehicle recommendations for new purchasers.
- Netflix uses machine learning algorithms to analyse user viewing data and recommend personalised content that matches their preferences. They also predict which types of movies or series will be successful among their subscribers.
- Spotify employs AI to study users' musical preferences and generate personalised recommendations and playlists. They also analyse musical trends to identify emerging artists and which songs will be successful.
- Tinder applies algorithms to users' swipes left and right to refine potential matches presented in an ongoing manner.

- TripAdvisor uses machine learning on traveler reviews/preferences to recommend sights, attractions and itineraries tailored for individuals.
- Walmart applies machine learning to its real-time sales and customer data to make decisions such as personalised offers, price tagging, in-store inventory, etc.
- Zalando's AI recommendation engine learns from shoppers' cart additions, sizing preferences, and category interests to present fitting, relevant apparel.

In conclusion, the use of AI for digital market research is still in its infancy, as numerous applications are being generated to provide better answers to traditional marketing problems, such as providing information to configure the STP and providing information that helps marketing decision-makers to manage better the 4Ps (Huang & Rust, 2021). In this section, we have provided a few of the many applications being implemented, such as qualitative research in the online era, searching for criteria for customer segmentation and selection, automatic forecasting processes and marketing personalisation systems. Numerous applications remain outside the scope of this chapter. However, a list of the most recent ones can be found in Huang & Rust (2021).

5. CHALLENGES OF AI IN MARKETING RESEARCH

The use of artificial intelligence (AI) in market research also presents several challenges related both to data quality and accuracy, algorithmic biases and fairness issues, as well as those concerning reliability issues, such as privacy and security over the fate of information, and ethical issues that need to be carefully considered.

5.1 Data Quality and Accuracy

One of the most important challenges relates to data quality and accuracy. As new technologies are driving the proliferation of unstructured data, this presents a challenge for using AI market research. However, despite all the efforts to combine different sources to generate accurate results (Chung et al., 2016; Liu et al., 2016), most companies still consider their potential untapped (Balducci & Marinova, 2018). On the other hand, as noted by Hewett et al. (2016), the proliferation of false data is jeopardising the quality and accuracy of the information collected for analysis, and this poor quality can endanger the reliability of the techniques applied to them, resulting in biased or inaccurate estimates (Sáez-Ortuño et al., 2023b). It can negatively affect the information decision-makers provide and consequently affect marketing strategies. Therefore, it is essential to develop algorithms that use cross-data from different sources and reliable and accurate information to train AI algorithms and generate valuable information (Rana *et al.*, 2022).

On the other hand, as explained in personalisation and recommendation systems using AI, they must cope with dynamic optimisation logic (Dzyabura and Hauser 2019), as the initial data may carry a preference and fairness bias. Preference biases can be progressively corrected by adopting consumer learning mechanisms; however, fairness biases may stem from mimetic data replication that may point to stereotypical outcomes not accepted in our society. For example, Adams & Loideáin (2019) analyse the gender stereotypes reproduced by virtual personal assistants, which use female names, voices and characters that seem to reproduce stereotypes about the role of women in society and the type of work they do. According to these authors, personal assistants such as Apple's Siri or Amazon's Alexa use these practices to convey that women are subordinate to men and exist to be "used" by men. In addition to criticising the reproduction of negative gender stereotypes in virtual personal assistants, these authors explore the possible legal consequences that could arise from these practices, which go against the work of the Committee against Discrimination against Women and, in terms of human rights, against the UN Guiding Principles on Business and Human Rights (Adams & Loideáin, 2019). Therefore, it is essential to monitor and mitigate bias in AI algorithms (Satornino *et al.*, 2023).

5.2 Algorithmic Biases and Fairness

Nevertheless, it is not only the stereotypical biases arising from the application of AI algorithms that are of concern, but also the ethical considerations arising from how the data provided on online sites are collected, used, and shared (Sáez-Ortuño et al., 2023b). It is an issue not only for consumers and businesses but also for society in the form of public opinion. The ubiquity and growing presence of the "ecoverse" is changing and impacting the experience of consumers as individuals and as a social group. For example, simple social bots transform public opinion and the political landscape. A study examining bots' impact in the 2016 US presidential election found that disseminating erroneous or misleading information via Twitter and Facebook has effectively manipulated and misled users and disrupted democratic political discussion (Bessi & Ferrara, 2016).

5.3 Privacy, Security, and Ethical Considerations

Consequently, companies should consider the ethical implications of using AI in market research and ensure that their practices align with ethical standards, including greater transparency about data collection methods and data use, obtaining customer consent, and ensuring that data is used ethically and fairly (Liu, Gupta, & Patel, 2021; Sáez-Ortuño *et al.*, 2023).

The growing ethical concerns surrounding the use of customer data are closely tied to the theory of privacy (Laufer & Wolfe, 1977). This theory posits that privacy is a multifaceted concept, extending beyond mere data confidentiality and encompassing individuals' ability to exercise control over the collection and use of their personal information.

For example, a recent survey on using Conversational Agents in education revealed that people showed a positive attitude towards this technology if privacy issues were addressed. However, the authors found that the meaning of privacy differed between children and adults (Latham & Goltz, 2019). Therefore, in ethical management, providing this knowledge should condition the system's design as it should focus on protecting the values and interests of the most vulnerable consumers. In addressing the ethical issues of AI-derived applications, researchers and decision-makers should be open to context-based, flexible, and pluralistic use to identify solutions appropriate to specific contexts rather than proposing some standard principles. Therefore, to develop AI programmes capable of capturing consumer information or establishing a conversation, applying ethical and responsible conventions should be about contextual and pluralistic approaches over abstract principles (Ruane, Birhane, & Ventresque, 2019).

In conclusion, the use of AI in market research offers many benefits but also presents several challenges linked to data quality and accuracy, bias and fairness of algorithms, and ethical considerations, involving not only regulatory compliance but keeping up with rapid changes in both technological advances and changes in societal values.

6. CONCLUSION

Market research has been developing for over a century and has continually adapted to technological changes like radio, TV, mail and email (Stewart, 2010). It can also adapt to challenges from the growth of the internet, websites, blogs and social networks by relying on new AI-driven data collection tools (Jacobsen, 2023). As shown, AI enables analyzing different marketing planning stages like macro-environment, market prediction, segmentation, targeting and positioning, and the 4Ps (Huang & Rust, 2021). This is possible due to algorithms handling vast, complex, unstructured data (Liu et al. 2016; Sáez-Ortuño et al. 2023b) in replicable analyses (Manivannan, Prabha & Balasubramanian, 2022). Some applied AI techniques like sentiment analysis and deep learning are early-stage and need improvements (Ordenes et al. 2017; LeCun, Bengio, & Hinton, 2015). External validity of results also requires more study (Dekking et al., 2005). While social media provides wealth of consumer data, anonymity yields falsified information challenging AI (Sáez-Ortuño et al. 2023b; Rana et al. 2022). As McLuhan & Fione (1968) predicted, technology shapes society in ways like cultural dissolution, presenting new challenges. Personalization through majority trends may neglect minority wants (Adams & Loideáin, 2019). As AI adoption grows, ethical issues like privacy, bias and transparency are increasingly important to ensure responsible, beneficial use (Zhang et al. 2019).

In the realm of future research directions, there exists a compelling need to advance AI techniques that can capture more intricate nuances in consumer sentiment analysis within social media, including the incorporation of semantic and emotional context. Further exploration into the validity and generalizability of insights derived through AI across diverse industries and markets is essential, accompanied by rigorous assessments of replicability (Sáez-Ortuño et al., 2023a; Wirtz et al., 2023). Additionally, the design of algorithms capable of identifying and mitigating biases in training data is crucial to eliminate harmful stereotypes within recommendation systems (Sáez-Ortuño et al., 2023c). Research should also focus on understanding the impact of AI in market research on minority groups and address the disproportionate emphasis on majorities, thereby fostering the development of more inclusive AI. Ethical applications of AI that empower consumers by enhancing transparency and control over their data warrant investigation, as well as the exploration of ways to integrate ethics and values into the design of AI systems applied in market research (Lobschat et al., 2021; Sáez-Ortuño et al., 2023b; Wirtz et al., 2023).

REFERENCES

Acciarini, C., Cappa, F., Boccardelli, P., & Oriani, R. (2023). How can organisations leverage big data to innovate their business models? A systematic literature review. *Technovation*, *123*, 102713. doi:10.1016/j.technovation.2023.102713

Adams, R., & Loideáin, N. N. (2019). Addressing indirect discrimination and gender stereotypes in AI virtual personal assistants: The role of international human rights law. *Cambridge International Law Journal*, *8*(2), 241–257. doi:10.4337/cilj.2019.02.04

Adya, M., Armstrong, J. S., Collopy, F., & Kennedy, M. (2000). An application of rule-based forecasting to a situation lacking domain knowledge. *International Journal of Forecasting*, *16*(4), 477–484. doi:10.1016/S0169-2070(00)00074-1

Alam, I. (2002). An exploratory investigation of user involvement in new service development. *Journal of the Academy of Marketing Science*, *30*(3), 250–261. doi:10.1177/0092070302303006

Almomani, A., Saavedra, P., Barreiro, P., Durán, R., Crujeiras, R., Loureiro, M., & Sánchez, E. (2022). Application of choice models in Tourism Recommender Systems. *Expert Systems: International Journal of Knowledge Engineering and Neural Networks*, *40*(3), e13177. doi:10.1111/exsy.13177

Andriella, A., Huertas-Garcia, R., Forgas-Coll, S., Torras, C., & Alenyà, G. (2022). "I know how you feel" The importance of interaction style on users' acceptance in an entertainment scenario. *Interaction Studies: Social Behaviour and Communication in Biological and Artificial Systems*, *23*(1), 21–57. doi:10.1075/is.21019.and

Asghar, M. Z., Subhan, F., Ahmad, H., Khan, W. Z., Hakak, S., Gadekallu, T. R., & Alazab, M. (2020). senti-esystem: A sentiment-basedesystem-using hybridised fuzzy and deep neural network for measuring customer satisfaction. *Software, Practice & Experience*, *51*(3), 571–594. doi:10.1002pe.2853

Balducci, B., & Marinova, D. (2018). Unstructured data in marketing. *Journal of the Academy of Marketing Science*, *46*(4), 557–590. doi:10.100711747-018-0581-x

Batat, W. (2022). What does phygital really mean? A conceptual introduction to the phygital customer experience (PH-CX) framework. *Journal of Strategic Marketing*, 1–24. Advance online publication. doi:10.1080/0965254X.2022.2059775

Batat, W., & Hammedi, W. (2023). The extended reality technology (ERT) framework for designing customer and service experiences in phygital settings: A service research agenda. *Journal of Service Management*, *34*(1), 10–33. doi:10.1108/JOSM-08-2022-0289

Belhadi, A., Kamble, S., Benkhati, I., Gupta, S., & Mangla, S. K. (2023). Does strategic management of digital technologies influence electronic word-of-mouth (ewom) and customer loyalty? empirical insights from B2B platform economy. *Journal of Business Research*, *156*, 113548. doi:10.1016/j.jbusres.2022.113548

Bessi, A., & Ferrara, E. (2016). Social bots distort the 2016 US Presidential election online discussion. *First Monday*, *21*, 11–17.

Bettman, J. R. (1979). *An Information Processing Theory of Consumer Choice*. Addison-Wesley.

Bingley, W. J., Curtis, C., Lockey, S., Bialkowski, A., Gillespie, N., Haslam, S. A., Ko, R. K. L., Steffens, N., Wiles, J., & Worthy, P. (2023). Where is the human in human-centered AI? insights from developer priorities and user experiences. *Computers in Human Behavior*, *141*, 107617. doi:10.1016/j.chb.2022.107617

Bogart, L. (1957). Opinion research and marketing. *Public Opinion Quarterly*, *21*(1), 129–140. doi:10.1086/266692

Chen, T., & Guestrin, C. (2016, August). Xgboost: A scalable tree boosting system. In *Proceedings of the 22nd ACM SIGKDD international conference on knowledge discovery and data mining* (pp. 785-794). ACM 10.1145/2939672.2939785

Chen, Y., Iyengar, R., & Iyengar, G. (2017). Modeling multimodal continuous heterogeneity in conjoint analysis—A sparse learning approach. *Marketing Science*, *36*(1), 140–156. doi:10.1287/mksc.2016.0992

Chung, T. S., Wedel, M., & Rust, R. T. (2016). Adaptive personalisation using social networks. *Journal of the Academy of Marketing Science*, *44*(1), 66–87. doi:10.100711747-015-0441-x

Cireşan, D., Meier, U., Masci, J., & Schmidhuber, J. (2012). Multi-column deep neural network for traffic sign classification. *Neural Networks*, *32*, 333–338. doi:10.1016/j.neunet.2012.02.023 PMID:22386783

Cooke, A. D. J., & Zubcsek, P. P. (2017). The connected consumer: Connected devices and the evolution of customer intelligence. *Journal of the Association for Consumer Research*, *2*(2), 164–178. doi:10.1086/690941

Davis, R., & King, J. J. (1984). The origin of rule-based systems in AI. *Rule-based expert systems: The MYCIN experiments of the Stanford Heuristic Programming Project*.

Dekking, F. M., Kraaikamp, C., Lopuhaä, H. P., & Meester, L. E. (2005). *A Modern Introduction to Probability and Statistics: Understanding why and how* (Vol. 488). Springer. doi:10.1007/1-84628-168-7

DeSarbo, W. S., & Grisaffe, D. (1998). Combinatorial optimisation approaches to constrained market segmentation: An application to industrial market segmentation. *Marketing Letters*, *9*(2), 115–134. doi:10.1023/A:1007997714444

Duhaim, A. M., Fadhel, M. A., Alnoor, A., Baqer, N. S., Alzubaidi, L., Albahri, O. S., Alamoodi, A. H., Bai, J., Salhi, A., Santamaría, J., Ouyang, C., Gupta, A., Gu, Y., & Deveci, M. (2023). A systematic review of trustworthy and explainable artificial intelligence in Healthcare: Assessment of quality, bias risk, and data fusion. *Information Fusion*.

Duncan, C. S. (1919). *Commercial Research: An Outline of Working Principles*. McMillan Co.

Dzyabura, D., & Yoganarasimhan, H. (2018). Machine learning and marketing. In *Handbook of marketing analytics* (pp. 255–279). Edward Elgar Publishing. doi:10.4337/9781784716752.00023

El-Shamandi Ahmed, K., Ambika, A., & Belk, R. (2023). Augmented reality magic mirror in the service sector: Experiential consumption and the self. *Journal of Service Management*, *34*(1), 56–77. doi:10.1108/JOSM-12-2021-0484

Evans, J. S. B. (2010). Intuition and reasoning: A dual-process perspective. *Psychological Inquiry*, *21*(4), 313–326. doi:10.1080/1047840X.2010.521057

Fernandez-Quilez, A. (2022). Deep Learning in Radiology: Ethics of data and on the value of algorithm transparency, interpretability and explainability. *AI and Ethics*, *3*(1), 257–265. doi:10.100743681-022-00161-9

Forgas-Coll, S., Huertas-Garcia, R., Andriella, A., & Alenyà, G. (2023). Social robot-delivered customer-facing services: An assessment of the experience. *Service Industries Journal*, *43*(3-4), 154–184. doi:10.1080/02642069.2022.2163995

Frederick, J. G. (1918). *Business Research and Statistics*. D. Appleton and Co.

Friedman, J. (2001). Greedy function approximation: A gradient boosting machine. *Annals of Statistics*, *29*(5), 1189–1232. doi:10.1214/aos/1013203451

Goldberg, Y. (2016). A primer on neural network models for natural language processing. *Journal of Artificial Intelligence Research*, *57*, 345–420. doi:10.1613/jair.4992

Goode, M. M., Davies, F., Moutinho, L., & Jamal, A. (2005). Determining customer satisfaction from mobile phones: A neural network approach. *Journal of Marketing Management*, *21*(7-8), 755–778. doi:10.1362/026725705774538381

Granovetter, M. S. (1973). The Strengh of Weak Ties. *American Journal of Sociology*, *78*(6), 1360–1380. doi:10.1086/225469

Grover, V., Chiang, R. H., Liang, T. P., & Zhang, D. (2018). Creating strategic business value from big data analytics: A research framework. *Journal of Management Information Systems*, *35*(2), 388–423. doi:10.1080/07421222.2018.1451951

Gupta, S., Singhvi, S., & Granata, G. (2023). Assessing the Impact of Artificial Intelligence in e-Commerce Portal: A Comparative Study of Amazon and Flipkart. In *Industry 4.0 and the Digital Transformation of International Business* (pp. 173–187). Springer Nature Singapore. doi:10.1007/978-981-19-7880-7_10

Hadi, R., & Valenzuela, A. (2019). Good vibrations: Consumer responses to technology-mediated haptic feedback. *The Journal of Consumer Research*, *47*(2), 256–271. doi:10.1093/jcr/ucz039

Hajipour, V., Niaki, S. T., Tavana, M., Santos-Arteaga, F. J., & Hosseinzadeh, S. (2023). A comparative performance analysis of intelligence-based algorithms for optimising competitive facility location problems. *Machine Learning with Applications*, *11*, 100443. doi:10.1016/j.mlwa.2022.100443

Haleem, A., Javaid, M., Asim Qadri, M., Pratap Singh, R., & Suman, R. (2022). Artificial Intelligence (AI) applications for marketing: A literature-based study. *International Journal of Intelligent Networks*, *3*, 119–132. doi:10.1016/j.ijin.2022.08.005

Heidelberger, M. (2004). *Nature from within: Gustav Theodor Fechner and his psychophysical worldview*. University of Pittsburgh Press.

Hewett, K., Rand, W., Rust, R. T., & Van Heerde, H. J. (2016). Brand buzz in the echoverse. *Journal of Marketing*, *80*(3), 1–24. doi:10.1509/jm.15.0033

Higueras-Castillo, E., Liébana-Cabanillas, F. J., & Villarejo-Ramos, Á. F. (2023). Intention to use e-commerce vs physical shopping. difference between consumers in the post-COVID era. *Journal of Business Research*, *157*, 113622. doi:10.1016/j.jbusres.2022.113622

Hill, J., Ford, W. R., & Farreras, I. G. (2015). Real conversations with artificial intelligence: A comparison between human-human online conversations and human-chatbot conversations. *Computers in Human Behavior*, *49*, 245–250. doi:10.1016/j.chb.2015.02.026

Hoffman, D. L., & Novak, T. P. (2009). Flow online: Lessons learned and future prospects. *Journal of Interactive Marketing*, *23*(1), 23–34. doi:10.1016/j.intmar.2008.10.003

Huang, M. H., & Rust, R. T. (2018). Artificial intelligence in service. *Journal of Service Research*, *21*(2), 155–172. doi:10.1177/1094670517752459

Huang, M. H., Rust, R. T., & Maksimovic, V. (2019). The feeling economy: Managing in the next generation of artificial intelligence (AI). *California Management Review*, *61*(4), 43–65. doi:10.1177/0008125619863436

Humphreys, A., & Wang, R. J. H. (2018). Automated text analysis for consumer research. *The Journal of Consumer Research*, *44*(6), 1274–1306. doi:10.1093/jcr/ucx104

Iacobucci, D. (2016). *Marekting Management*. South Western.

Jacobsen, B. N. (2023). Machine learning and the politics of Synthetic Data. *Big Data & Society*, *10*(1), 205395172211453. doi:10.1177/20539517221145372

Kaplan, A. (2022). *Artificial Intelligence, Business and Civilization: Our Fate Made in Machines*. Routledge. doi:10.4324/9781003244554

Kartajaya, H., Setiawan, I., & Kotler, P. (2021). *Marketing 5.0: Technology for humanity*. John Wiley & Sons.

Kaushal, V., & Yadav, R. (2022). Learning successful implementation of chatbots in businesses from B2B customer experience perspective. *Concurrency and Computation*, *35*(1), e7450. doi:10.1002/cpe.7450

Khodadadi, A., Ghandiparsi, S., & Chuah, C.-N. (2022). A natural language processing and deep learning based model for Automated Vehicle Diagnostics using free-text Customer Service Reports. *Machine Learning with Applications*, *10*, 100424. doi:10.1016/j.mlwa.2022.100424

Klette, R. (2014). *Concise computer vision* (Vol. 233). Springer. doi:10.1007/978-1-4471-6320-6

Krogh, A. (2008). What are artificial neural networks? *Nature Biotechnology*, *26*(2), 195–197. doi:10.1038/nbt1386 PMID:18259176

Kumar, V., Rajan, B., Venkatesan, R., & Lecinski, J. (2019). Understanding the role of artificial intelligence in personalised engagement marketing. *California Management Review*, *61*(4), 135–155. doi:10.1177/0008125619859317

Latham, A., & Goltz, S. (2019). A Survey of the General Public's Views on the Ethics of Using AI in Education. In Artificial Intelligence in Education: 20th International Conference, AIED 2019, Chicago, IL, USA, June 25-29, 2019 [Springer International Publishing.]. *Proceedings*, *20*(Part I), 194–206.

Laufer, R. S., & Wolfe, M. (1977). Privacy as a concept and a social issue: A multidimensional developmental theory. *The Journal of Social Issues*, *33*(3), 22–42. doi:10.1111/j.1540-4560.1977.tb01880.x

LeCun, Y., Bengio, Y., & Hinton, G. (2015). Deep learning. *Nature*, *521*(7553), 436–444. doi:10.1038/nature14539 PMID:26017442

Lee, S. H., Yoon, S. H., & Kim, H. W. (2021). Prediction of online video advertising inventory based on TV programs: A deep learning approach. *IEEE Access : Practical Innovations, Open Solutions*, *9*, 22516–22527. doi:10.1109/ACCESS.2021.3056115

Lemon, K. N., & Verhoef, P. C. (2016). Understanding customer experience throughout the customer journey. *Journal of Marketing*, *80*(6), 69–96. doi:10.1509/jm.15.0420

Li, J., Ye, Z., & Zhang, C. (2022). Study on the interaction between Big Data and artificial intelligence. *Systems Research and Behavioral Science*, *39*(3), 641–648. doi:10.1002res.2878

Liebman, E., Saar-Tsechansky, M., & Stone, P. (2019). The right music at the right time: Adaptive personalised playlists based on sequence modeling. *Management Information Systems Quarterly*, *43*(3), 765–786. doi:10.25300/MISQ/2019/14750

Liu, R., Gupta, S., & Patel, P. (2021). The application of the principles of responsible AI on social media marketing for digital health. *Information Systems Frontiers*, 1–25. doi:10.100710796-021-10191-z PMID:34539226

Liu, X., Singh, P. V., & Srinivasan, K. (2016). A structured analysis of unstructured big data by leveraging cloud computing. *Marketing Science*, *35*(3), 363–388. doi:10.1287/mksc.2015.0972

Liu, Y., Ram, S., Lusch, R. F., & Brusco, M. (2010). Multicriterion market segmentation: A new model, implementation, and evaluation. *Marketing Science*, *29*(5), 880–894. doi:10.1287/mksc.1100.0565

Lobschat, L., Mueller, B., Eggers, F., Brandimarte, L., Diefenbach, S., Kroschke, M., & Wirtz, J. (2021). Corporate digital responsibility. *Journal of Business Research*, *122*, 875–888. doi:10.1016/j.jbusres.2019.10.006

Manivannan, P., Prabha, D., & Balasubramanian, K. (2022). Artificial Intelligence Databases: Turn-on big data of the SMBS. *International Journal of Business Information Systems*, *39*(1), 1. doi:10.1504/IJBIS.2022.120367

Marblestone, A. H., Wayne, G., & Kording, K. P. (2016). Toward an integration of deep learning and neuroscience. *Frontiers in Computational Neuroscience*, *10*, 94. doi:10.3389/fncom.2016.00094 PMID:27683554

Mariani, M. M., Perez-Vega, R., & Wirtz, J. (2022). AI in marketing, consumer research and psychology: A systematic literature review and research agenda. *Psychology and Marketing*, *39*(4), 755–776. doi:10.1002/mar.21619

Martin-Rodilla, P., Pereira-Fariña, M., & Gonzalez-Perez, C. (2019). Qualifying and quantifying uncertainty in Digital Humanities. *Proceedings of the Seventh International Conference on Technological Ecosystems for Enhancing Multiculturality*. ACM. 10.1145/3362789.3362833

McDuff, D., & Czerwinski, M. (2018). Designing emotionally sentient agents. *Communications of the ACM*, *61*(12), 74–83. doi:10.1145/3186591

McLuhan, M., & Fione, Q. (1968). *War and peace in the global village*. Bantam Books.

Mo, F., Rehman, H. U., Monetti, F. M., Chaplin, J. C., Sanderson, D., Popov, A., Maffei, A., & Ratchev, S. (2023). A framework for manufacturing system reconfiguration and Optimisation Utilising Digital Twins and Modular Artificial Intelligence. *Robotics and Computer-integrated Manufacturing*, *82*, 102524. doi:10.1016/j.rcim.2022.102524

Mohammad, N. I., Ismail, S. A., Kama, M. N., Yusop, O. M., & Azmi, A. (2019, August). Customer churn prediction in telecommunication industry using machine learning classifiers. In *Proceedings of the 3rd international conference on vision, image and signal processing* (pp. 1-7). IEEE. 10.1145/3387168.3387219

Moon, S., Kim, M. Y., & Iacobucci, D. (2021). Content analysis of fake consumer reviews by survey-based text categorization. *International Journal of Research in Marketing*, *38*(2), 343–364. doi:10.1016/j.ijresmar.2020.08.001

Nanne, A. J., Antheunis, M. L., Van Der Lee, C. G., Postma, E. O., Wubben, S., & Van Noort, G. (2020). The use of computer vision to analyse brand-related user generated image content. *Journal of Interactive Marketing*, *50*(1), 156–167. doi:10.1016/j.intmar.2019.09.003

Netzer, O., Toubia, O., Bradlow, E. T., Dahan, E., Evgeniou, T., Feinberg, F. M., Feit, E. M., Hui, S. K., Johnson, J., Liechty, J. C., Orlin, J. B., & Rao, V. R. (2008). Beyond conjoint analysis: Advances in preference measurement. *Marketing Letters*, *19*(3-4), 337–354. doi:10.100711002-008-9046-1

Ngai, E. W., & Wu, Y. (2022). Machine learning in marketing: A literature review, conceptual framework, and research agenda. *Journal of Business Research*, *145*, 35–48. doi:10.1016/j.jbusres.2022.02.049

Nilashi, M., Abumalloh, R. A., Alrizq, M., Almulihi, A., Alghamdi, O. A., Farooque, M., Samad, S., Mohd, S., & Ahmadi, H. (2022). A hybrid method to solve data sparsity in travel recommendation agents using Fuzzy Logic Approach. *Mathematical Problems in Engineering*, *2022*, 1–20. doi:10.1155/2022/7372849

Ordenes, F. V., Ludwig, S., de Ruyter, K., Grewal, D., & Wetzels, M. (2017). Unveiling What is Written in The Stars: Analysing Explicit, Implicit, and Discourse Patterns of Sentiment in Social Media. *The Journal of Consumer Research*, *43*(6), 875–894. doi:10.1093/jcr/ucw070

Pennycook, G., & Rand, D. G. (2019). Lazy, not biased: Susceptibility to partisan fake news is better explained by lack of reasoning than by motivated reasoning. *Cognition*, *188*, 39–50. doi:10.1016/j.cognition.2018.06.011 PMID:29935897

Pereira, A. M., Moura, J. A., Costa, E. D., Vieira, T., Landim, A. R. D. B., Bazaki, E., & Wanick, V. (2022). Customer models for artificial intelligence-based decision support in fashion online retail supply chains. *Decision Support Systems*, *158*, 113795. doi:10.1016/j.dss.2022.113795

Rahman, M. S., Bag, S., Hossain, M. A., Abdel Fattah, F. A., Gani, M. O., & Rana, N. P. (2023). The new wave of AI-Powered Luxury Brands Online Shopping experience: The role of digital multisensory cues and customers' engagement. *Journal of Retailing and Consumer Services*, *72*, 103273. doi:10.1016/j.jretconser.2023.103273

Rana, N. P., Chatterjee, S., Dwivedi, Y. K., & Akter, S. (2022). Understanding dark side of artificial intelligence (AI) integrated business analytics: Assessing firm's operational inefficiency and competitiveness. *European Journal of Information Systems*, *31*(3), 364–387. doi:10.1080/0960085X.2021.1955628

Rathore, B. (2023). Integration of Artificial Intelligence& Its Practices in Apparel Industry. [IJNMS]. *International Journal of New Media Studies*, *10*(1), 25–37. doi:10.58972/eiprmj.v10i1y23.40

Roy, G. (2023). Travelers' online review on Hotel Performance – Analysing Facts with the theory of lodging and sentiment analysis. *International Journal of Hospitality Management*, *111*, 103459. doi:10.1016/j.ijhm.2023.103459

Ruane, E., Birhane, A., & Ventresque, A. (2019, December). Conversational AI: Social and Ethical Considerations. In AICS (pp. 104-115).

Sáez-Ortuño, L., Forgas-Coll, S., Huertas-Garcia, R., & Sánchez-García, J. (2023a). What's on the horizon? A bibliometric analysis of personal data collection methods on social networks. *Journal of Business Research*, *158*, 113702. doi:10.1016/j.jbusres.2023.113702

Sáez-Ortuño, L., Forgas-Coll, S., Huertas-Garcia, R., & Sánchez-García, J. (2023b). Online cheaters: Profiles and motivations of internet users who falsify their data online. *Journal of Innovation & Knowledge*, *8*(2), 100349. doi:10.1016/j.jik.2023.100349

Sáez-Ortuño, L., Huertas-Garcia, R., Forgas-Coll, S., & Puertas-Prats, E. (2023c). How can entrepreneurs improve digital market segmentation? A comparative analysis of supervised and unsupervised learning algorithms. *The International Entrepreneurship and Management Journal*, •••, 1–28. doi:10.100711365-023-00882-1

Sarker, I. H. (2021). Machine learning: Algorithms, real-world applications and Research Directions. *SN Computer Science*, *2*(3), 160. doi:10.100742979-021-00592-x PMID:33778771

Satornino, C. B., Grewal, D., Guha, A., Schweiger, E. B., & Goodstein, R. C. (2023). The perks and perils of artificial intelligence use in lateral exchange markets. *Journal of Business Research*, *158*, 113580. doi:10.1016/j.jbusres.2022.113580

Stead, M., Gordon, R., Angus, K., & McDermott, L. (2007). A systematic review of social marketing effectiveness. *Health Education*, *107*(2), 126–191. doi:10.1108/09654280710731548

Stewart, D. W. (2010). The Evolution of Market Research. In P. Maclaran, M. Saren, B. Stern, & M. Tadajewski (Eds.), *The SAGE Handbook of Marketing Theory* (pp. 74–85). Sage.

Szeliski, R. (2011). *Computer vision: Algorithms and applications*. Springer-Verlag. doi:10.1007/978-1-84882-935-0

Valls, A., Gibert, K., Orellana, A., & Antón-Clavé, S. (2018). Using ontology-based clustering to understand the push and pull factors for British tourists visiting a Mediterranean coastal destination. *Information & Management*, *55*(2), 145–159. doi:10.1016/j.im.2017.05.002

Wedel, M., & Kamakura, W. A. (2000). *Market segmentation: Conceptual and methodological foundations*. Kluwer Academic Publishers Group. doi:10.1007/978-1-4615-4651-1

Wilson, E. J., & Vlosky, R. P. (1997). Partnering Relationship Activities: Building Theory From Case Study Research. *Journal of Business Research*, *39*(1), 59–70. doi:10.1016/S0148-2963(96)00149-X

Wirtz, B. W., Weyerer, J. C., & Geyer, C. (2019). Artificial intelligence and the public sector—Applications and challenges. *International Journal of Public Administration*, *42*(7), 596–615. doi:10.1080/01900692.2018.1498103

Zachlod, C., Samuel, O., Ochsner, A., & Werthmüller, S. (2022). Analytics of social media data–State of characteristics and application. *Journal of Business Research*, *144*, 1064–1076. doi:10.1016/j.jbusres.2022.02.016

Zhang, H., Zang, Z., Zhu, H., Uddin, M. I., & Amin, M. A. (2022). Big data-assisted social media analytics for business model for business decision making system competitive analysis. *Information Processing & Management*, *59*(1), 102762. doi:10.1016/j.ipm.2021.102762

Zohdi, M., Rafiee, M., Kayvanfar, V., & Salamiraad, A. (2022). Demand forecasting based machine learning algorithms on Customer Information: An applied approach. *International Journal of Information Technology : an Official Journal of Bharati Vidyapeeth's Institute of Computer Applications and Management*, *14*(4), 1937–1947. doi:10.100741870-022-00875-3

KEY TERMS AND DEFINITIONS

Artificial Intelligence (AI): A branch of computer science that focuses on creating intelligent machines that can perform tasks that typically require human intelligence, such as visual perception, speech recognition, decision-making, and language translation.

Adaptive Recommendation Engines: AI systems that use data and algorithms to provide personalised recommendations and experiences to individual users based on their preferences and behavior.

Artificial Neural Networks (ANNs): Algorithms that attempt to reproduce the neural circuitry of the human brain through a system of nodes and connections through which information flows.

Computer Vision: A field of AI that focuses on enabling machines to interpret and understand visual information from the world around them, such as images and videos.

Decision Trees (DT): An algorithm used in classification and regression models that uses a hierarchical distribution to classify subjects according to some self-reported or self-recorded criteria.

Deep Learning: A subset of machine learning that uses artificial neural networks with multiple layers to learn and extract features from complex data.

Ethical Considerations: The moral and social implications of using AI in market research, including issues related to privacy, bias, transparency, and accountability.

Expert Systems: AI algorithms that can provide expert advice or recommendations based on rules and knowledge.

Fuzzy Logic Systems: A type of AI system that uses degrees of truth instead of binary values to represent uncertainty and imprecision in data.

Genetic Algorithms: A type of AI system that uses principles of natural selection and genetics to optimise solutions to complex problems.

Market Segmentation: The process of dividing a market into smaller groups of consumers with similar needs or characteristics.

Personalization Systems: AI systems that use data and algorithms to provide personalised recommendations and experiences to individual users based on their preferences and behavior.

Predictive Analytics: The use of data, statistical algorithms, and machine learning techniques to identify the likelihood of future outcomes based on historical data.

Rule-Based Systems (R-BS): A type of AI system that uses a set of rules and facts to achieve a specific goal. These systems are deterministic and operate with a "cause and effect" methodology.

Sentiment Analysis: A technique used to identify and extract subjective information from text data, such as opinions, emotions, and attitudes.

APPENDIX

Table 2. Main advantages of using AI in marketing research

Advantage	Description	References
Large and complex data analysis	Ability to efficiently process large volumes of unstructured data from multiple sources	(Manivannan, Prabha & Balasubramanian, 2022) (Jacobsen, 2023)
Personalization and targeting	Provides personalized recommendations and experiences tailored to individual consumers	(Chung et al., 2016) (Kaushal & Yadav, 2022)
Cost and time savings	Automates repetitive tasks, saving time and research costs	(Haleem et al., 2022)
Strategic decision-making	Enables better understanding of customers and markets to guide strategy and innovation	(Huang & Rust, 2021) (Mo et al., 2023)

Source: Own elaboration

Table 3. Key challenges of using AI in marketing research

Challenge	Description	References
Data quality and accuracy	Risk of bias if training data is low quality, incomplete or unrepresentative	(Sáez-Ortuño et al., 2023b) (Rana et al. 2022)
Algorithmic bias	Potential for AI systems to perpetuate biases and unfairness present in training data	(Adams & Loideáin, 2019)
Privacy and ethical issues	Concerns around data privacy, transparency in data collection/ use, and other ethical implications	(Sáez-Ortuño et al., 2023b) (Zhang et al. 2019)

Source: Own elaboration

Chapter 3
Artificial Intelligence in Healthcare

Kun-Huang Huarng
National Taipei University of Business, Taiwan

Tiffany Hui-Kuang Yu
Feng Chia University, Taiwan

Duen-Huang Huang
Chaoyang University of Technology, Taiwan

ABSTRACT

Artificial intelligence (AI) has been applied to various domains to improve the quality of human life. This chapter outlines the recent application of AI in healthcare. A brief history of AI development is first introduced. Machine learning, one of the current AI advancements, is explained. Successful AI application in different areas of healthcare is then showcased, including different medical diagnosis and long-term care. The popular ChatGPT series of systems and their extraordinary performance are described. This chapter ends with debates and future expectations of AI.

1. INTRODUCTION

The giant progress in artificial intelligence (AI) development has fueled a rapid rise in the applications of various domains over the last decade. Why AI is a big deal? Brynjolfsson & Mcafee (2017) provide two reasons. First, via AI, we can automate many tasks. Second, AI is an excellent learner. Some interesting AI applications have been identified as in Table 1 (Brynjolfsson & Mcafee, 2017; Jordan & Mitchell, 2015). For example, we can take drug chemical properties as the input to the AI applications and the AI applications can produce treatment efficacy as the output. And this type of application can be used in pharmacy R&D.

DOI: 10.4018/978-1-6684-9591-9.ch003

Table 1. Interesting AI applications

Input	Output	Application
Voice	Transcript	Speech recognition
Photo	Caption	Image tagging
Drug	Treatment efficacy	Pharma R&D
Store transactions	Fraudulent transactions	Fraud detection
Purchase records	Future purchase behavior	Customer retention
Car location	Traffic flow	Traffic lights
Face	Name	Face recognition

AI was considered to have the potential to exploit meaningful relationships in a data set and the relationships can be used in the diagnosis, treatment and predicting outcome in many clinical scenarios (Ramesh et al., 2004). AI in healthcare has two main branches: virtual and physical (Hamet & Tremblay, 2017). The virtual branch includes informatics approaches from machine learning to control of health management systems; and the physical branch is represented by robots assist the elderly patient or the attending surgeon.

The use of AI in healthcare has become quite popular, including medical images (Ker et al., 2018), the detection of drug interactions (Han et al., 2022), the identification of high-risk patients (Beaulieu-Jones et al., 2021), etc. Recently, the performance of ChatGPT by Open AI has recently shown amazing performance and attracted many studies.

This chapter introduces and discusses the application of artificial intelligence (AI) in healthcare. First, we briefly present AI development by covering expert systems and machine learning. For machine learning, we explain how an artificial neural network (ANN) can achieve learning step by step. We subsequently cover some current AI applications in medical diagnosis, some possible applications of it in long-term care, and the popular ChatGPT application in medicine. AI offers opportunities as well as the potential for both positive and negative impacts for society and individuals (Dwivedi et al., 2023). This chapter ends with debates and doubts about the application of AI.

2. ARTIFICIAL INTELLIGENCE (AI)

Expert Systems

AI can be classified into two categories. The first category is named classical AI. This category intends to build computers systems that replicate human behavior. One sub-category is expert systems. These systems capture the expertise of human experts by using sets of rules.

The classical example in medical application is MYCIN, which was one of the earliest medical expert systems developed in the 1970s. It was designed to model a physician's diagnostic expertise by using around 500 rules and helped diagnose and treat bacterial infections based on reported symptoms and medical test results (Bionity, 2023). MYCIN provides advice based on three subsystems: a consultation system that chooses proper medication based on its own knowledge base; an explanation system that answers questions to justify its decisions; and a knowledge acquisition system that codes rules for future consultation (Shortliffe et al., 1975). MYCIN operated at roughly the same level as human specialists in blood infections and rather better than general practitioners (Copeland, 2023).

INTERNIST-I is another medical expert system that was developed in the 1970s to diagnose multiple diseases in internal medicine by modelling the behavior of physicians. The heuristic rules use a partitioning algorithm to create problem areas and exclusion functions to eliminate diagnostic possibilities. When the system was unable to determine diagnosis, it asked questions or observations to clear up the mystery (Wikipedia, 2023). The knowledge base in INTERNIST-I has been applied in successor systems for medical education and clinical use. Wolfram (1995) surveyed the effects of INTERNIST-I and concluded that they can become Quick Medical Reference when used as an electronic textbook of medicine.

Machine Learning

Machine learning emerged in the mid-1980s. Instead of trying to duplicate human intelligence, it sought to develop programs that monitor the operation of a process. It relies heavily on mathematical models to develop algorithms. These algorithms have been used in practical systems that identify objects, find patterns in data, and develop strategies for robots. Artificial neural networks (ANNs) are useful models applied to problem-solving and machine learning (Abiodun et al., 2018).

An ANN architecture consists of multiple layers, where one is the input layer, one is the output layer, and one or more are hidden layer(s). The input and output layers hold the pairwise input and output values, respectively. The hidden layers contain multiple hidden nodes. There are links between nodes, with which various weights are associated. Figure 1 illustrates an ANN architecture. One can refer to Han & Kamber (2011) for a detailed introduction about a feed forward backpropagation neural network.

Figure 1. An ANN architecture

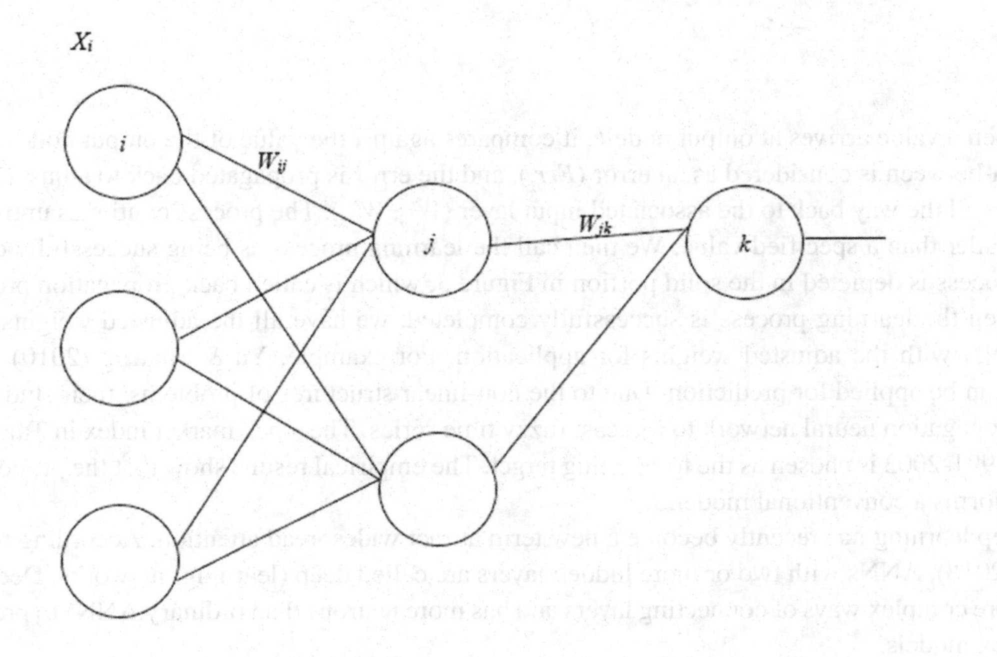

The learning of ANN occurs as follows. The value of an input node i, X_i, is multiplied by the associated weight between nodes i and j (W_{ij}) before propagating to its subsequent nodes (named hidden nodes in hidden layers). A hidden node sums all the incoming values and uses a sigmoid function to transform the value into a value as the output from the hidden node, node j in this case. Next, the output of the hidden node propagates to the subsequent nodes (either the nodes in the subsequent hidden or output layers). In Figure 2 the output node is k, and the process is illustrated by the solid portion, which is also called the feed forward process.

Figure 2. Feed forward process

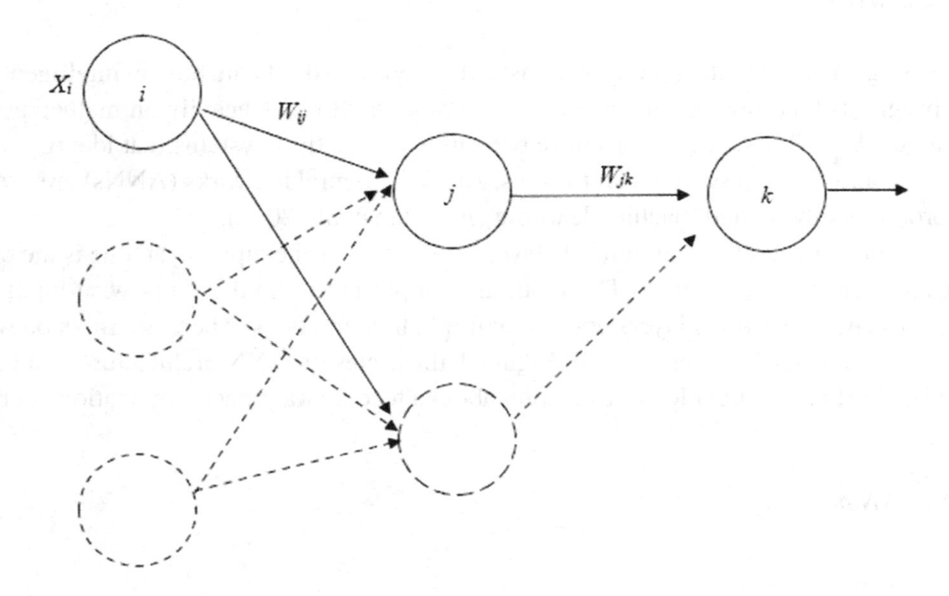

When a value arrives at output node k, it compares against the value of the output node. The difference in-between is considered as an error (Err_k), and the error is propagated back to adjust the relevant weights all the way back to the associated input layer (W'_{ij}, W'_{jk}). The process continues until the errors are smaller than a specified value. We then call the learning process as being successfully completed. The process is depicted in the solid portion in Figure 3, which is called back propagation process.

When the learning process is successfully completed, we have all the adjusted weights. Next, we use ANN with the adjusted weights for application. For example, Yu & Huarng (2010) show how ANN can be applied for prediction. Due to the non-linear structures of problems, their study applies a backpropagation neural network to forecast fuzzy time series. The stock market index in Taiwan for the years 1991-2003 is chosen as the forecasting target. The empirical results show that the proposed model outperforms a conventional model.

Deep learning has recently become a new term attract widespread attention. According to Abiodun et al. (2018), ANNs with two or more hidden layers are called deep (learning) networks. Deep learning has more complex ways of connecting layers and has more neurons than ordinary ANNs to present more complex models.

Figure 3. Back propagation process

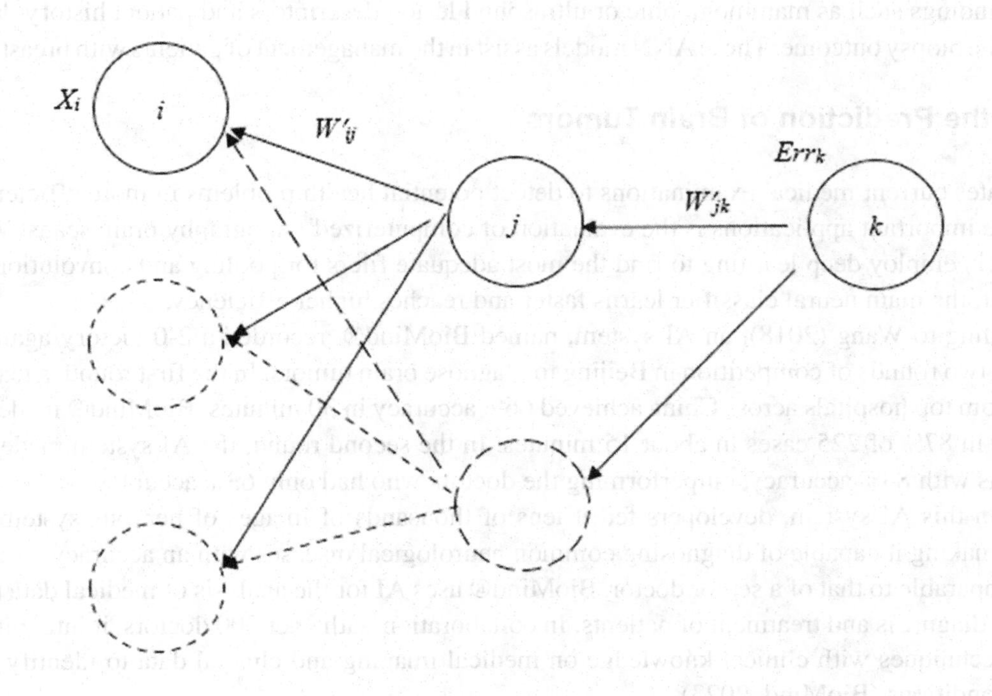

3. ANN IN MEDICAL PREDICTION

Medical diagnosis detection has become popular through ANN applications (Abiodun et al., 2018). This section introduces some existing applications of AI in medical diagnosis and prediction.

ANN in the Prediction of Breast Cancer

Breast cancer is one of the most common cancers worldwide. Its early diagnosis can improve the chance of survival significantly. Many studies have reported the use of ANN for breast cancer detection (Lo & Floyd, 1999; Janghel et al., 2010; Mehdy et al., 2017). Because of its unique advantages in critical feature detection from complex datasets, ANN is widely considered in breast cancer pattern classification and forecast modelling (Yue et al., 2018).

Medical imaging is very time consuming during the manual diagnosis of each image pattern. Automated classifiers could substantially upgrade the diagnosis process in terms of both accuracy and time requirement. Mehdy et al. (2017) review the related literature and show the successful application of ANNs in four different medical imaging applications: mammogram, ultrasound, and thermal imaging, and MRI imaging.

Lo & Floyd (1999) conduct four projects via ANN for the prediction of breast cancer: (1) prediction of breast lesion malignancy using mammographic findings; (2) classification of malignant lesions as in situ vs. invasive cancer; (3) prediction of breast mass malignancy using ultrasound findings; and (4) the evaluation of computer-aided diagnosis models in a cross-institution study. The inputs to the ANNs are medical findings such as mammographic or ultrasound lesion descriptors and patient history data. The output is the biopsy outcome. These ANN models assist in the management of patients with breast lesions.

ANN in the Prediction of Brain Tumors

AI facilitates current medical examinations to detect potential health problems in more efficient ways. One of the important applications is the evaluation of computerized tomography brain scans. Woźniak et al. (2021) employ deep learning to find the most adequate filers for pooling and convolution layers. As a result, the main neural classifier learns faster and reaches higher efficiency.

According to Wang (2018), an AI system, named BioMind®, recorded a 2-0 victory against elite doctors in two rounds of competition in Beijing to diagnose brain tumors. In the first round, a team of 15 doctors from top hospitals across China achieved 66% accuracy in 30 minutes. BioMind® made correct diagnoses in 87% of 225 cases in about 15 minutes. In the second round, the AI system made correct predictions with 83% accuracy, outperforming the doctors who had only 63% accuracy.

To train this AI system, developers fed it tens of thousands of images of nervous system-related diseases, making it capable of diagnosing common neurological diseases with an accuracy rate of over 90% - comparable to that of a senior doctor. BioMind® uses AI for the analysis of medical data to assist doctors in diagnosis and treatment of patients. In collaboration with over 200 doctors, it integrates deep learning techniques with clinical knowledge on medical imaging and clinical data to identify various medical conditions (BioMind, 2023).

4. AI IN LONG-TERM CARE

AI has the potential to assist long-term care. In a literature survey, Lukkien et al. (2023) identify three broad applications of AI in long-term care: user-oriented AI innovation, AI as a solution for responsible innovation issues (RI), and context sensitivity. User-oriented AI innovation centers around the role of users, in particular older adults and their caregivers, in the implementation of AI technologies. Using AI as a solution to RI issues positions AI as a technical fix to certain RI issues that are associated with supportive technologies in long-term care. Context-sensitivity refers to the ways of being sensitive to the specific context of using AI technologies in long-term care when addressing RI.

Huarng (2018) proposes a framework for long-term care in the sharing economy, as shown in Figure 4. The core of the framework is to move the physical objects by transportation and to share the information by the Internet. To facilitate the implementation of long-term care, the physical objects needed should be delivered in the right time and to the right place. To achieve that, the related information should be efficiently shared by the stakeholders (all kinds of participants who are willing to serve). Hence, information management is critical so that the limited amount of human and other resources can be allocated effectively. Hence, that study suggests deploying an AI system to coordinate all the resources. Some well-known existing algorithms such as the shortest paths can be applied.

In long-term care, AI is expected to learn about the environment and adapt to changing contexts of action (Dermody & Fritz, 2019; Ho, 2020; Mukaetova-Ladinska et al., 2020). For example, through AI, camera-based monitoring systems can learn to classify activities such as lying, sitting, standing, and walking. They can also predict the amount of time an older adult spends getting out of bed or the risk

Figure 4. Long-term care in the sharing economy

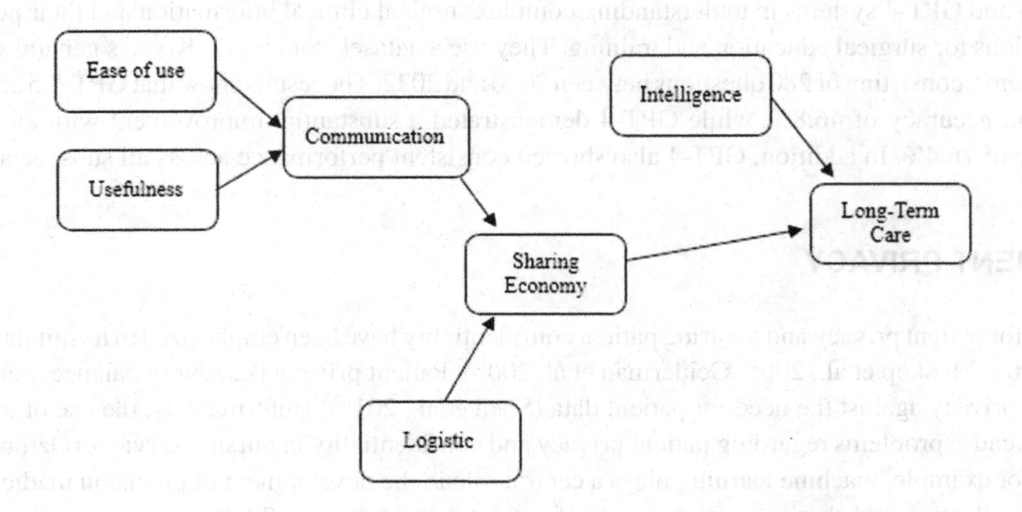

of events (Cardinaux et al., 2011). In addition, sensor-based monitoring systems can track older adults' walking speed and presence in different rooms. AI-based monitoring technologies can facilitate timely care and potentially prevent further deteriorations (Kaye et al., 2011; Zwierenberg et al., 2018).

AI systems can particularly be integrated with popular wearable devices to provide a wider coverage of healthcare. Huarng et al. (2022) explore the factors that affect the adoption of these devices. One further expansion of the devices is to collect health-related data by a data center, where the data can be monitored by AI systems for various diagnoses. Some AI systems can predict the symptoms through the trends of the health-related data.

5. CHATGPT AND GPT-4

ChatGPT is a recent hot AI product, whose dialogue format makes it possible for it to answer questions (Open AI, 2023). The goal of ChatGPT is to get external feedback so as to improve systems and make them safer. The conversations may be reviewed by AI trainers on a team in order to improve the systems.

Studies have focused on measuring ChatGPT in comparison to that of expert human doctors (Antaki et al., 2023; Yeo et al., 2023). Kung et al. (2023) evaluate the performance of ChatGPT on the United States Medical Licensing Exam (USMLE). USMLE is a set of tests on expert-level knowledge, which are required to be taken for medical licensure in the United States. ChatGPT was found to perform at or near the passing threshold of 60% accuracy. ChatGPT was the first to achieve this benchmark, marking a notable milestone in AI effectiveness in the medical field.

GPT-4 is Open AI's more advanced system, producing safer and more useful responses. With broad general knowledge and domain expertise, GPT-4 can follow complex instructions in natural language and solve difficult problems with accuracy (https://openai.com/product). Lee et al. (2023) conduct another test of GPT-4 on USMLE problems. Surprisingly, GPT-4 presents answers that are correct more than 90% of the time. In addition, GPT-4 can provide detailed reasoning to support its answers.

GPT-4 has been tested to demonstrate its improvement in medical education. Oh et al. (2023) compare GPT-3.5 and GPT-4 systems in understanding complex surgical clinical information and their potential implications for surgical education and training. They use a dataset from South Korea's general surgery board exams, consisting of 280 questions between 2020 and 2022. The results show that GPT-3.5 achieved an overall accuracy of 46.8%, while GPT-4 demonstrated a substantial improvement with an overall accuracy of 76.4%. In addition, GPT-4 also showed consistent performance across all subspecialties.

6. PATIENT PRIVACY

Respect for patient privacy and assuring patient confidentiality have been emphasized in hospitals (Karro et al., 2005; Moskop et al., 2005; Geiderman et al. 2006). Patient privacy is about to balance the patient need for privacy against the need for patient data (Shen et al., 2019). Unfortunately, the use of technology may cause problems regarding patient privacy and confidentiality in nursing services (Özturk et al., 2014). For example, machine learning plays a central role in the development of precision medicine but requires collecting and sharing large amounts of patient data (Azencott, 2018).

Recent technologies have been developed to protect patient privacy in various perspectives. For example, the application of defacing or skull-stripping algorithms to face (Lotan et al., 2020) can be applied to eliminate digital imaging and communications in medicine metadata (such as name and sex) (Aryanto et al., 2015). The reliable medical recommendation systems were proposed with patient privacy-friendly framework (Hoens et al., 2013). A healthcare system for patient privacy), based on cryptographic technologies, to provide privacy protection to patients under any circumstances while enabling timely retrieval of data in emergency situations (Sun et al., 2011).

The application AI in healthcare requires patient data. A key question is how to protect patient privacy. The regulations for privacy protection and institution-specific polices provide need to be reviewed and adjusted for a health care environment where AI tools such as ChatGPT or GPT-4 apply (Fitzgerald, 2023).

Meanwhile, as AI can be applied as a warning system to monitoring events (Yankoski et al., 2020). In the future, can we develop an AI monitoring system to monitor the application of AI in healthcare; for example, to enforcement the patient privacy?

7. DEBATES

Lim et al. (2023) present a set of both challenges and opportunities for using AI. One long-lasting concern of AI centers on the substitution of human beings and making people obsolete. AI has been accused of actually trying to replace human beings (Grier, 2023). However, many researchers aim to use AI to enhance human productivity rather than replace human beings. For example, ChatGPT has successfully passed graduate-level business and law exams (Kelly, 2023). Will ChatGPT threaten lawyers' jobs or serve as assistants to lawyers? Only time will tell for now.

There is also concern about the unpredictability and uncontrollability of AI. In response, legislators and scholars are calling for more transparency and explainability of AI (Buiten, 2019). This refers to the black box beyond AI, focusing on the concrete risks and biases of its underlying technology. It is necessary to understand, monitor, and manage the associated risks (National Institute of Standards and Technology, 2023). Frameworks governing the use of AI must be driven by principles (Pechenkina,

2023). Regulating its medical applications may advocate for the benign usage of AI and hence prevent fraudulent medicine.

8. EXCITEMENTS YET TO COME

There are inevitably critics (Chomsky et al., 2023) and expectations for AI in many applications, particularly in healthcare. However, from the advancement of human well-being viewpoint, we indeed see some promises in healthcare. For example, patients tend to have few choices when they face treatments, because of their limited medical knowledge. Of course, patients can always study and summarize Internet information and find out about various possible treatments. However, the process takes time, and the barrier to medical knowledge is rather high. If AI can help summarize correct information as some of the applications now do, then AI application in healthcare definitely can help a lot of patients and their families.

This chapter introduces the development of AI and explains how ANNs can achieve learning. In addition, this chapter covers some current applications of AI in healthcare. Lastly, GPT series of application in healthcare are also presented.

We are now still in the gold rush of AI, and many more debates and doubts are expected. This chapter just reflects the advancement and enrichment that AI can add to human well-being. There are still other excitements that one can expect.

REFERENCES

Abiodun, O. I., Jantan, A., Omolara, A. E., Dada, K. V., Mohamed, N. A., & Arshad, H. (2018). State-of-the-art in artificial neural network applications: A survey. *Heliyon*, *4*(11), e00938. doi:10.1016/j.heliyon.2018.e00938 PMID:30519653

Antaki, F., Touma, S., Milad, D., El-Khoury, J., & Duval, R. (2023). Evaluating the performance of chat-gpt in ophthalmology: An analysis of its successes and shortcomings. *Ophthalmology Science, 100324*.

Aryanto, K., Oudkerk, M., & van Ooijen, P. (2015). Free DICOM de-identification tools in clinical research: Functioning and safety of patient privacy. *European Radiology*, *25*(12), 3685–3695. doi:10.100700330-015-3794-0 PMID:26037716

Azencott, C. A. (2018). Machine learning and genomics: Precision medicine versus patient privacy. *Philosophical Transactions - Royal Society. Mathematical, Physical, and Engineering Sciences*, *376*(2128), 20170350. doi:10.1098/rsta.2017.0350 PMID:30082298

Beaulieu-Jones, B. K., Yuan, W., Brat, G. A., Beam, A. L., Weber, G., Ruffin, M., & Kohane, I. S. (2021). Machine learning for patient risk stratification: Standing on, or looking over, the shoulders of clinicians? *NPJ Digital Medicine*, *4*(1), 62. doi:10.103841746-021-00426-3 PMID:33785839

Brynjolfsson, E., & Mcafee, A. (2017). Artificial intelligence, for real. *Harvard Business Review*, *1*, 1–31.

Buiten, M. C. (2019). Towards intelligent regulation of artificial intelligence. *European Journal of Risk Regulation*, *10*(1), 41–59. doi:10.1017/err.2019.8

Cardinaux, F., Bhowmik, D., Abhayaratne, C., & Hawley, M. S. (2011). Video based technology for ambient assisted living: A review of the literature. *Journal of Ambient Intelligence and Smart Environments*, *3*(3), 253–269. doi:10.3233/AIS-2011-0110

Chomsky, N., Roberts, I., & Watumull, J. (2023). Noam Chomsky: The False Promise of ChatGPT. *The New York Times*, 8.

Copeland, B. J. (2023). MYCIN: artificial intelligence program. *Encyclopedia Britannica*. https://www.britannica.com/technology/MYCIN

Dermody, G., & Fritz, R. (2019). A conceptual framework for clinicians working with artificial intelligence and health-assistive smart homes. *Nursing Inquiry*, *26*(1), e12267. doi:10.1111/nin.12267 PMID:30417510

Dwivedi, Y. K., Kshetri, N., Hughes, L., Slade, E. L., Jeyaraj, A., Kar, A. K., Baabdullah, A. M., Koohang, A., Raghavan, V., Ahuja, M., Albanna, H., Albashrawi, M. A., Al-Busaidi, A. S., Balakrishnan, J., Barlette, Y., Basu, S., Bose, I., Brooks, L., Buhalis, D., & Wright, R. (2023). "So what if ChatGPT wrote it?" Multidisciplinary perspectives on opportunities, challenges and implications of generative conversational AI for research, practice and policy. *International Journal of Information Management*, *71*, 102642. doi:10.1016/j.ijinfomgt.2023.102642

Fitzgerald, S. (2023). How Can You Use AI—And Protect Patient Privacy. *Neurology Today*, *23*(12), 1–12. doi:10.1097/01.NT.0000943140.92540.15

Geiderman, J. M., Moskop, J. C., & Derse, A. R. (2006). Privacy and confidentiality in emergency medicine: Obligations and challenges. *Emergency Medicine Clinics of North America*, *24*(3), 633–656. doi:10.1016/j.emc.2006.05.005 PMID:16877134

Grier, D. A. (2023). *Debating Artificial Intelligence*. Computer. https://www.computer.org/publications/tech-news/closer-than-you-might-think/debating-artificial-intelligence

Hamet, P., & Tremblay, J. (2017). Artificial intelligence in medicine. *Metabolism: Clinical and Experimental*, *69*, S36–S40. doi:10.1016/j.metabol.2017.01.011 PMID:28126242

Han, J., & Kamber, M. (2011). *Data Mining: Concepts and Techniques*. Morgan Kaufmann Publishers.

Han, K., Cao, P., Wang, Y., Xie, F., Ma, J., Yu, M., Wang, J., Xu, Y., Zhang, Y., & Wan, J. (2022). A review of approaches for predicting drug-drug interactions based on machine learning. *Frontiers in Pharmacology*, *12*, 814858. doi:10.3389/fphar.2021.814858 PMID:35153767

Ho, A. (2020). Are we ready for artificial intelligence health monitoring in elder care? *BMC Geriatrics*, *20*(1), 358. doi:10.118612877-020-01764-9 PMID:32957946

Hoens, T. R., Blanton, M., Steele, A., & Chawla, N. V. (2013). Reliable medical recommendation systems with patient privacy. [TIST]. *ACM Transactions on Intelligent Systems and Technology*, *4*(4), 1–31. doi:10.1145/2508037.2508048

Huarng, K. H. (2018). Entrepreneurship for long-term care in sharing economy. *The International Entrepreneurship and Management Journal*, *14*(1), 97–104. doi:10.100711365-017-0460-9

Huarng, K. H., Yu, T. H. K., & Lee, C. F. (2022). Adoption model of healthcare wearable devices. *Technological Forecasting and Social Change*, *174*, 121286. doi:10.1016/j.techfore.2021.121286

Janghel, R. R., Shukla, A., Tiwari, R., & Kala, R. (2010). Breast cancer diagnosis using artificial neural network models. In *The 3rd International Conference on Information Sciences and Interaction Sciences* (pp. 89-94). IEEE. 10.1109/ICICIS.2010.5534716

Jordan, M. I., & Mitchell, T. M. (2015). Machine learning: Trends, perspectives, and prospects. *Science*, *349*(6245), 255–260. doi:10.1126cience.aaa8415 PMID:26185243

Karro, J., Dent, A. W., & Farish, S. (2005). Patient perceptions of privacy infringements in an emergency department. *Emergency Medicine Australasia*, *17*(2), 117–123. doi:10.1111/j.1742-6723.2005.00702.x PMID:15796725

Kaye, J. A., Maxwell, S. A., Mattek, N., Hayes, T. L., Dodge, H., Pavel, M., Jimison, H. B., Wild, K., Boise, L., & Zitzelberger, T. A. (2011). Intelligent systems for assessing aging changes: Home-based, unobtrusive, and continuous assessment of aging. *The Journals of Gerontology. Series B, Psychological Sciences and Social Sciences*, *66*(Suppl. 1), i180–i190. doi:10.1093/geronb/gbq095 PMID:21743050

Kelly, S. M. (2023). ChatGPT passes exams from law and business schools. *CNN Business*, 26.

Ker, J., Wang, L., Rao, J., & Lim, T. (2018). Deep learning applications in medical image analysis. *IEEE Access : Practical Innovations, Open Solutions*, *6*, 9375–9389. doi:10.1109/ACCESS.2017.2788044

Kung, T. H., Cheatham, M., Medenilla, A., Sillos, C., De Leon, L., Elepaño, C., Madriaga, M., Aggabao, R., Diaz-Candido, G., Maningo, J., & Tseng, V. (2023). Performance of ChatGPT on USMLE: Potential for AI-assisted medical education using large language models. *PLoS Digital Health*, *2*(2), e0000198. doi:10.1371/journal.pdig.0000198 PMID:36812645

Lee, P., Goldberg, C., & Kohane, I. (2023). *The AI Revolution in Medicine: GPT-4 and Beyond*. Pearson.

Lim, W. M., Gunasekara, A., Pallant, J. L., Pallant, J. I., & Pechenkina, E. (2023). Generative AI and the future of education: Ragnarök or reformation? A paradoxical perspective from management educators. *International Journal of Management Education*, *21*(2), 100790. doi:10.1016/j.ijme.2023.100790

Lo, J. Y., & Floyd, C. E. (1999). Application of artificial neural networks for diagnosis of breast cancer. In *Proceedings of the 1999 Congress on Evolutionary Computation-CEC99 (Cat. No. 99TH8406)* (*Vol. 3*, pp. 1755-1759). IEEE. 10.1109/CEC.1999.785486

Lotan, E., Tschider, C., Sodickson, D. K., Caplan, A. L., Bruno, M., Zhang, B., & Lui, Y. W. (2020). Medical imaging and privacy in the era of artificial intelligence: Myth, fallacy, and the future. *Journal of the American College of Radiology*, *17*(9), 1159–1162. doi:10.1016/j.jacr.2020.04.007 PMID:32360449

Lukkien, D. R., Nap, H. H., Buimer, H. P., Peine, A., Boon, W. P., Ket, J. C., Minkman, M. M. N., & Moors, E. H. (2023). Toward responsible artificial intelligence in long-term care: A scoping review on practical approaches. *The Gerontologist*, *63*(1), 155–168. doi:10.1093/geront/gnab180 PMID:34871399

Moskop, J. C., Marco, C. A., Larkin, G. L., Geiderman, J. M., & Derse, A. R. (2005). From Hippocrates to HIPAA: privacy and confidentiality in emergency medicine–Part I: conceptual, moral, and legal foundations. *Annals of Emergency Medicine*, *45*(1), 53–59. doi:10.1016/j.annemergmed.2004.08.008 PMID:15635311

Mukaetova-Ladinska, E.B., Harwoord, T., & Maltby, J. (2020). Artificial Intelligence in the healthcare of older people. Archives of Psychiatry and Mental Health, 4(1), 007–013.

National Institute of Standards and Technology. (2023). *NIST AI Risk Management Framework Playbook*. NIST. https://pages.nist.gov/AIRMF/

Oh, N., Choi, G. S., & Lee, W. Y. (2023). ChatGPT goes to the operating room: Evaluating GPT-4 performance and its potential in surgical education and training in the era of large language models. *Annals of Surgical Treatment and Research*, *104*(5), 269. doi:10.4174/astr.2023.104.5.269 PMID:37179699

Özturk, H., Bahçecik, N., & Özçelik, K. S. (2014). The development of the patient privacy scale in nursing. *Nursing Ethics*, *21*(7), 812–828. doi:10.1177/0969733013515489 PMID:24482263

Pechenkina, K. (2023). Artificial intelligence for good? Challenges and possibilities of AI in higher education from a data justice perspective. In L. Czerniewicz & C. Cronin (Eds.), *Higher Education for Good: Teaching and Learning Futures (#HE4Good)*. Open Book Publishers.

Ramesh, A. N., Kambhampati, C., Monson, J. R., & Drew, P. J. (2004). Artificial intelligence in medicine. *Annals of the Royal College of Surgeons of England*, *86*(5), 334–338. doi:10.1308/147870804290 PMID:15333167

Shen, N., Bernier, T., Sequeira, L., Strauss, J., Silver, M. P., Carter-Langford, A., Shortliffe, E. H., Davis, R., Axline, S. G., Buchanan, B. G., Green, C. C., & Cohen, S. N. (1975). Computer-based consultations in clinical therapeutics: Explanation and rule acquisition capabilities of the MYCIN system. *Computers and Biomedical Research, an International Journal*, *8*(4), 303–320. doi:10.1016/0010-4809(75)90009-9 PMID:1157471

Sun, J., Zhu, X., Zhang, C., & Fang, Y. (2011, June). HCPP: Cryptography based secure EHR system for patient privacy and emergency healthcare. In *2011 31st International Conference on Distributed Computing Systems* (pp. 373-382). IEEE.

Wang, X. (2018). AI defeats elite doctors in diagnosis competition. *The Star*, (July), 2.

Wiljer, D. (2019). Understanding the patient privacy perspective on health information exchange: A systematic review. *International Journal of Medical Informatics*, *125*, 1–12. doi:10.1016/j.ijmedinf.2019.01.014 PMID:30914173

Wolfram, D. A. (1995). An appraisal of INTERNIST-I. *Artificial Intelligence in Medicine*, *7*(2), 93–116. doi:10.1016/0933-3657(94)00028-Q PMID:7647840

Woźniak, M., Siłka, J., & Wieczorek, M. (2021). Deep neural network correlation learning mechanism for CT brain tumor detection. *Neural Computing & Applications*, 1–16.

Yankoski, M., Weninger, T., & Scheirer, W. (2020). An AI early warning system to monitor online disinformation, stop violence, and protect elections. *Bulletin of the Atomic Scientists*, *76*(2), 85–90. doi:10.1080/00963402.2020.1728976

Yeo, Y. H., Samaan, J. S., Ng, W. H., Ting, P. S., Trivedi, H., Vipani, A., & Kuo, A. (2023). Assessing the performance of ChatGPT in answering questions regarding cirrhosis and hepatocellular carcinoma. medRxiv, 2023-02.

Yu, T. H. K., & Huarng, K. H. (2010). A neural network-based fuzzy time series model to improve forecasting. *Expert Systems with Applications*, *37*(4), 3366–3372. doi:10.1016/j.eswa.2009.10.013

Yue, W., Wang, Z., Chen, H., Payne, A., & Liu, X. (2018). Machine learning with applications in breast cancer diagnosis and prognosis. *Designs*, *2*(2), 13. doi:10.3390/designs2020013

Zwierenberg, E., Nap, H. H., Lukkien, D., Cornelisse, L., Finnema, E., Hagedoorn, M., & Sanderman, R. (2018). A lifestyle monitoring system to support (in) formal caregivers of people with dementia: Analysis of users need, benefits, and concerns. *Gerontechnology (Valkenswaard)*, *17*(4), 194–205. doi:10.4017/gt.2018.17.4.001.00

Chapter 4
Causal Machine Learning in Social Impact Assessment

Nuno Castro Lopes
https://orcid.org/0000-0003-2448-9798
Universidade Aberta, Portugal

Luís Cavique
https://orcid.org/0000-0002-5590-1493
Universidade Aberta, Portugal

ABSTRACT

Social impact assessment is a fundamental process to verify the achievement of the objectives of interventions and, consequently, to validate investments in the social area. Generally, this process is based on the analysis of the average effects of the intervention, which does not allow a detailed understanding of the individualization of these effects. Causal machine learning methods mark an evolution in causal inference, as they allow for a more heterogeneous assessment of the effects of interventions. Applying these methods to evaluate the impact of social projects and programs offers the advantage of improving the selection of target audiences and optimizing and personalizing future interventions. In this chapter, in a non-technical way, the authors explore classical causal inference methods to estimate average effects and new causal machine learning methods to evaluate heterogeneous effects. They address adapting the Uplift Modeling method to assess social interventions. They also address the advantages, limitations, and research needs for using these new techniques in social intervention.

1. INTRODUCTION

Recently, due to the emergence of ChatGPT and similar technologies, there has been much talk about Artificial Intelligence (AI) and Machine Learning (ML). As we explore the potential of AI/ML, we are seeing significant advances in natural language processing, neural networks, generative networks, computer vision, and others. However, despite these advances, it is only recently that the scientific community has begun to look more deeply into the relationship between Causality and AI/ML. This work

DOI: 10.4018/978-1-6684-9591-9.ch004

focuses precisely on this interconnection. The text is divided into two distinct parts: a brief introduction to the classical approach to causal inference, with an emphasis on econometrics, followed by a second, more specific part on how to apply AI/ML to causal inference. We will explore the advantages of these new approaches for a more complete understanding of causality, focusing on the practical application of causal inference in social impact assessment.

According to a report by the Organization for Economic Cooperation and Development (OECD) published in 2021, the need for better management of funding for social programs and projects has led to an increasing number of funders requesting impact evaluation reports from third-sector organizations. Assessing the impact of social intervention is becoming increasingly essential for these organizations and public entities that develop similar intervention actions. This approach is because funders of social programs and projects, whether public entities, foundations, companies, or individual patrons, increasingly demand transparency and accountability.

According to the same organization (OECD, 2002, p.4), the impact can be defined as "Positive and negative, primary and secondary long-term effects produced by a development intervention, directly or indirectly, intended or unintended.". Impact evaluation demonstrates how projects, programs, or policies achieve their objectives, showing possible changes in participants' behaviors, skills, and living conditions (Rogers, 2014a).

The outcome of a social intervention's impact may condition its continuation or replication. On the one hand, funders will find it more challenging to re-fund interventions that do not seem to show results. On the other hand, less effective intervention strategies and practices may be internally modified or even abandoned. Promoters and funders can also, based on the impact evaluation results, take advantage of the successful cases and be presented as good practices to be replicated in other contexts and by other organizations. For these reasons, impact assessment is fundamental in seeking greater transparency in social investment and greater effectiveness in social and community interventions.

The need to evaluate social programs and projects promoted by non-profit organizations and public bodies has driven the creation of consulting firms and research centers dedicated to this purpose. Despite several methodologies used to conduct an impact evaluation, starting the process by building a solid "Theory of Change" has become a frequent procedure among evaluators (Rogers, 2014a).

The "Theory of Change" is a model that describes how the activities of programs, projects, policies, or even the mission of specific organizations are expected to produce a series of results that contribute to achieving the intended final impacts. This method can be an asset in understanding the intended mechanisms and outcomes of these interventions to show which indicators should be considered to assess potential changes in people (Rogers, 2014b). A key element to consider in impact evaluation is that its purpose is not only to measure and describe the changes that occur but also to seek to understand the role of that program or project in producing the changes. This process is often referred to as causal inference. Several methods for analyzing causal inference benefit from being based on a solid "Theory of Change" (Rogers, 2014a).

However, it is crucial to note that, in general, the approaches adopted in these evaluations provide only the average effects on the group receiving the intervention, failing to consider the individual effects. However, looking only at average effects can provide misguidance to decision-makers (White et al., 2014a).

1.1. Problem

In the classic context of evaluating the impact of programs and social projects, the approach focuses on analyzing the average effects of this intervention (ATE- Average treatment Effect). However, this approach is limited in not capturing the heterogeneity of observed effects (HTE- Heterogeneity of Treatment Effect); that is, it does not analyze how an intervention can affect groups or individuals in a targeted way. In many situations, the intervention may have harmful average effects while benefiting specific subgroups within the same intervention group. In addition, it is possible to observe positive overall effects that may be negative for particular subgroups. This outcome variation highlights the importance of understanding the heterogeneity of the impact across different individuals or segments of the target population.

1.2. Objectives

Using stratification techniques to analyze HTE, namely trying to identify subgroups demonstrating divergent effects about the average, may lead to biases or ambiguous conclusions. Even when pre-experimental analysis plans are prepared, they often need a more comprehensive exploration of all potential heterogeneous effects of the intervention (Athey & Imbens, 2016; Wager & Athey, 2018).

Thus, this work aims to demonstrate how current and future evaluators of social projects can benefit from using casual Machine Learning methods to infer the so-called CATE (Conditional Average Treatment Effect). Enabling a deeper understanding of the effects allows decision-makers to obtain more accurate information for developing future interventions that are more personalized and adapted to the target audience (Lecher, 2023)

1.3. Contributions

This chapter presents how Causal Machine Learning methods can be a valuable resource for those evaluating social programs and projects. As a contribution, we demonstrate the adaptation of the Uplift Modeling method, a tradition used in marketing, for its application in social intervention. Through the Transformed Outcome Approach, it is possible to expand decision-makers ability to assess the heterogeneity of effects and, at the same time, contribute to improving decision-making related to future social programs and projects.

1.4. Organization

The chapter is divided as follows: Section 2: Causal Inference will begin by addressing the frequently used traditional causal inference methods, which focus on evaluating the Average Treatment Effect (ATE). We will explore techniques such as "Propensity Score Matching", "Regression Discontinuity", "Difference in Differences", and "Synthetic Control". Section 3: Machine Learning for Causal Inference, we will discuss the application of Machine Learning in causal inference, especially the evaluation of the Conditional Average Treatment Effect (CATE). For ease of understanding, we will briefly explain how Supervised Machine Learning techniques are generally used to make variations. We will also discuss some strategies and algorithms that use Machine Learning to perform causal inference, allowing the estimation of the heterogeneous effects of an intervention. These techniques include the "Metalearners": "S-learner", "T-learner" and "X-learner", as well as the "Causal Trees and Causal Forests" algorithms.

Section 4: Application - Uplift Modeling: this section will demonstrate how to adopt the Uplift Modeling method, traditionally used in marketing. We will explain how this method is typically applied in marketing and introduce the Transformed Results Approach technique. Finally, we will show how this adaptation can be used to evaluate the impact of programs and social projects. Finally, in section 5, there are some considerations. It is important to note that this chapter addresses techniques and methodologies that still need to be widely disseminated in social impact assessment. We aim to make this text accessible to anyone, even with little prior knowledge about Machine Learning algorithms, avoiding the excessive use of formulas or precise terminology. However, we will provide references to some of the main works in the area, allowing those interested in deepening the topic to consult relevant bibliographies for a deeper understanding.

2. CAUSAL INFERENCE

Developing a "Theory of Change" helps impact evaluators to define the relevant indicators better, usually referred to as dependent variables and designated with the letter "Y," as well as the variables that may be associated, traditionally referred to as covariates or independent variables and defined with the letter "X." In addition to the " Theory of Change," which is a more general and comprehensive representation, in causal inference, we can use the so-called "Causal Directed Acyclic Graphs" (Pearl, 1995; Pearl & Mackenzie, 2018) to represent the cause-effect relationships between the different variables that can affect the impact of these programs or projects. These diagrams make it possible to visualize the causal connections between variables, facilitating the analysis of the effects of interest. They are represented by nodes and arrows, where each node represents a variable, and the arrows indicate the direction of causality between them. In this way, it is possible to examine how different variables influence each other, allowing us to understand better and determine which variables we wish to control and which we wish to incorporate into our causal inference process effectively.

In the context of causal inference, Rubin's (1974) causal model is often used as a basis. In this context, it is usual to use the letter "T" to denote the intervention (or treatment). One of the critical concepts of this model is the potential outcomes, which correspond to the outcomes that would be observed if a person[1] were subjected to a particular intervention. According to this model, each person will have a potential outcome for each possible level of intervention, and only one of them can be observed. These potential outcomes are calculated by the difference between the value of the dependent variable at a time before the intervention starts (time=1 or t1) and the value of that variable at a later stage (time=2 or t2), which usually corresponds to the time of the evaluation. They may occur either after the end of the intervention or at an intermediate stage. Thus, the causal effect is defined as the difference between the potential outcome that would be observed if the person received the intervention (Y^1) and the potential outcome that would be observed if the same person did not receive the intervention (Y^0) or even if they received another type of intervention. This notion of causal effect can be summarized with the formula Y^1-Y^0.

However, in practice, we face the fundamental problem of causal inference. What we want to know is what happened to the people who participated in the intervention and what would have happened if they had not participated in the intervention. However, it is logical that we cannot simultaneously and directly observe the effects of people exposed to the intervention and those not. Creating counterfactual strategies and alternative ways of simulating what would have happened in situations different from reality is necessary (Imbens & Rubin, 2015; Rubin, 1974).

The gold standard strategy for conducting causal inference is experimental studies with randomly selected groups. This study randomly divides participants into the treatment or intervention (T=1) and the control group (T=0). This random selection ensures that each participant has an equal chance of being assigned to either group, making them comparable concerning the dependent variable (Y) and eliminating potential bias. Following this approach, when the sample is significant, and the random selection is performed correctly, the groups become similar at the level of their independent variables (X_1, X_i, ..., X_n).

In experimental studies, the assignment of a person to one of the groups must be individual, probabilistic, and unbiased, and this random assignment can be carried out in different ways (Imbens & Rubin, 2015; Kang et al., 2008; Stanley, 2007). According to Imbens & Rubin (2015), this assignment can be carried out by randomly dividing the sample into two groups; Bernoulli's method, where for each person in the sample, a 50% random choice is made (such as flipping a coin to determine whether they will belong to one group or the other); stratification, where first the sample is divided into blocks based on a certain covariate that may influence the outcome (such as gender, age, etc.) and within each block, the selection is carried out entirely randomly; or paired comparison, where within each stratified block the sample is divided into pairs and, for each pair, the selection is carried out completely randomly and within each block, the selection is carried out entirely randomly; or paired comparison, where within each stratified block, that block is divided into pairs and, for each pair, a random choice is made to determine which of the people in the pair will go to each of the groups.

Despite recognition by the scientific community, conducting experimental studies in social intervention is not always feasible or appropriate. The need to intervene with individuals with specific characteristics makes it impossible to randomly select intervention and control groups. Usually, the intervention group is chosen before the procedures inherent to the evaluation are carried out. In many situations, it would be unethical to determine who would and would not receive the intervention randomly. Also, in many circumstances, the start of the evaluation occurs after the intervention has begun, making it impractical to conduct experimental studies, leading to the use of observational data.

For these reasons, social impact assessment often resorts to quasi-experimental methods to conduct the evaluation. In several of these situations, there are recorded data from other people with similar characteristics (X_1, X_i, ..., X_n) who have not received an intervention, which can help to create the counterfactuals.

For example, when a project is implemented in a particular class in a school or a specific group of inmates in prison, if there is relevant information regarding the same covariates (X_1, X_i, ..., X_n) and dependent variables (Y) for other people in the same organization, it is possible to use some techniques to create the control groups. We will discuss some of these techniques below. It is important to note that these techniques only determine the ATE.

2.1. Propensity Score Matching

One of the techniques used to address bias in the selection of a control group, known as control group selection bias, is Propensity Score Matching (PSM). This approach allows for matching between each person who received the intervention and all the other people who were not intervened but with similar characteristics (covariates) and belong to the same context.

PSM aims to create a control group comparable to the intervention group's characteristics using various statistical techniques or supervised machine learning algorithms (which will be explained later); PSM seeks to create a control group comparable to the intervention group's characteristics. The larger the sample of possible counterfactuals compared to the sample of intervened persons, the greater the possibility of performing the matching properly. In addition, a significant number of covariates also requires a higher proportion of counterfactual candidates compared to the number of intervention subjects.

For example, in an intervention in a school context, PSM assigns a propensity score to each student in the same school, indicating their similarity to the student who received the intervention. Intervened students with the highest propensity scores and similar characteristics are selected for the control group. This procedure will be done for all other subjects who received the intervention. In the end, the sample obtained through the PSM will serve as the control group, just as in an experimental study, since the groups will have similar characteristics.

The most commonly used technique to perform this propensity score is logistic regression, but discriminant analysis, probit models, classification or regression trees, neural networks, support vector machines, or meta-classifiers can also be applied (Cunningham, 2021; Fan & Nowell, 2011; Fougère & Jacquemet, 2019; Huntington-Klein, 2022; Imbens, 2000; Imbens & Rubin, 2015; Rosenbaum & Rubin, 1983; Tu, 2019).

2.2. Regression Discontinuity

Unlike PSM, regression discontinuity (RD) is a methodology that cannot be applied to all situations. RD is suitable in specific settings where the intervention is defined based on a cut-off index, value, or percentage that unambiguously determines which individuals can and cannot participate.

For example, let us consider a school support program that offers intervention only to students who are beneficiaries of scholarships. These scholarships are awarded only to students whose household income per capita is below a specific cut-off value. In this context, the RD is applied to distinguish the intervention group (i.e., the scholarship beneficiaries) from the control group (i.e., the non-beneficiaries) based on this specific cut-off criterion. Moreover, to ensure the precision of the analysis, the RD methodology requires the definition of a bandwidth, which delimits the upper and lower limits related to the cut-off.

In this way, RD allows the analysis of the causal effect of the intervention, focusing on subjects close to the cut-offline, both in the intervention group and the control group. The comparison between these groups makes it possible to investigate the intervention's impact on those on the borderline of eligibility, contributing to more robust causal inferences.

Importantly, RD is a methodology that presents its effectiveness and applicability in specific settings where the intervention is clearly defined through a cut-off point. However, it is critical to recognize that this approach is only suitable for evaluating a relatively small number of real-world social projects and programs (Cook, 2008; Fougère & Jacquemet, 2019; Huntington-Klein, 2022; Gopalan, Rosinger & Ahn, 2020; Thistlewaite & Campbell, 2017).

2.3. Difference-in-Differences and Synthetic Control

In Social Impact Assessment, it is only sometimes feasible to find individual observations that can serve as a counterfactual to assess the impact of a specific intervention. In such situations, the strategy used is to compare aggregate data, i.e., to analyze the average effect of the intervention on the dependent variable (Y), based on observational data from other contexts in which that intervention has not been applied but where we have access to data regarding the value of the indicator of that dependent variable. The Difference in Differences and the Synthetic Control are two strategies that allow us to perform this type of analysis.

The Difference-in-Differences (DiD) is one of the most classical and widely used methods in social impact assessment. The DiD method starts by looking for a population group similar to the intervention group, which has yet to be exposed to the intervention, acting as a control group. These groups should be selected in contexts identical to the intervention, such as in another school, prison, municipality, etc.

For both groups, the intervention and the control, we need information regarding two distinct temporal moments: before and after the implementation of the intervention. In addition, it is desirable to have other intermediate time points that allow analyzing the trajectory of the values of the dependent variable over time.

For the method to be applied correctly, the trajectory of the two groups must be similar during the period before the intervention, even if the values themselves are different. This prior similarity of trajectory helps ensure that any differences observed after the intervention are caused by the intervention itself and not by pre-existing factors.

The DiD consists of two differences. First, the difference between the periods before and after the intervention is calculated for both groups. Then, these two differences are subtracted, resulting in the intervention effect.

To estimate the effect, many studies use a regression model, which allows estimating what would have happened to the intervention group if it had not been exposed to the intervention, based on the trajectory observed in the control group after that intervention (Cunningham, 2021; Crato & Paruolo, 2019; Fougère & Jacquemet, 2019; Huntington-Klein, 2022; Gopalan, Rosinger & Ahn, 2020; Lechner, 2011).

For its part, Synthetic Control (SC) is a methodology that addresses the difficulty of finding control groups with similar trajectories in the pre-intervention phase. Instead of looking for an authentic context identical to the intervention group, SC uses an algorithm to create a synthetic group that is as similar as possible to that intervened group.

CS is an approach that offers an alternative to directly comparing the intervention group to a specific school, prison, or municipality. Instead, SC creates a synthetic entity, such as a synthetic school, synthetic prison, or synthetic city, through an algorithm. Creating the synthetic group is based on other schools, prisons, or municipalities with similar characteristics to the intervention group. To carry out the process, it is necessary to have observational data related to the dependent variable and the covariates of these similar entities. The SC algorithm combines these similar entities' characteristics (covariates) to create a synthetic entity virtually identical to the intervened group. In this way, the synthetic group acts as a control group with similar characteristics and behaviors similar to what would have happened to the intervention group in the absence of the intervention (Abadie & Gardeazabal, 2003; Abadie, Diamond, & Hainmueller, 2010; Fougère & Jacquemet, 2019; Cunningham, 2021; Huntington-Klein, 2022).

3. CAUSAL MACHINE LEARNING

Using these traditional causal inference methods makes it possible to evaluate the practical impact of interventions, demonstrating the viability of programs and social projects before funders. By determining the ATE of social interventions, decision-makers can better understand actions that should be considered best practices for future interventions and those that should not be continued due to their reduced average effectiveness.

But, in social intervention, especially with vulnerable or at-risk populations, there are many cases of failure, even in projects that obtain good results overall. In several situations, the average effects of the intervention may be adverse despite being beneficial for specific groups within the same intervention population. Additionally, it is possible to identify positive results that may harm particular subgroups. This variation in results highlights the importance of understanding diversity in effects among different individuals or segments of the target population. The conventional approach to social impact assessment may not be sufficient to fully capture these nuances and individual differences, which limits a comprehensive understanding of the effects of social interventions.

In this logic, new approaches that combine Machine Learning techniques and algorithms with causal inference have emerged in recent years, known as "Causal Machine Learning". These new approaches allow a more in-depth analysis of the heterogeneity of effects; that is, they will enable us to understand the impact of a project or program on individuals with specific characteristics based on the interventions already carried out. Despite their advantages, they are not widely applied in social impact assessment processes. However, their implementation can boost the prescription of future projects and interventions by improving the selection of target audiences, optimizing the interventional process, and personalizing future interventions.

"Causal Machine Learning" is a broad term encompassing Machine Learning methodologies for inferring causality. According to Kaddour, Lynch, Liu, Kusner, & Silva (2022), the area can be subdivided into five subgroups: (1) causal supervised learning, (2) causal generative modeling, (3) causal explanations, (4) causal fairness, and (5) causal reinforcement learning.

In this paper, we only focus on Causal Supervised Machine Learning and its application to infer the heterogeneous effect of interventions (determine the CATE). Within the various strategies to combine Supervised Machine Learning algorithms or new specific Causal Machine Learning algorithms, we present the Metalearners (S-Learner, T-Learner, and X-Learner), the Transformed Outcome Approach, and Causal Trees and Forests. To facilitate the understanding of these methods, we will only present how to use them in simple situations where there is a single intervention, where an individual can be intervened (T=1) or not (T=0), and the outcome is binary (Y^1 or Y^0).

Before exploring these methodologies and algorithms, for better understanding, we will briefly explain how Supervised Machine Learning algorithms are generally used to make predictions.

3.1. Supervised Machine Learning for Forecasting

According to Gama, Lorena, Faceli, Oliveira & Carvalho (2017), machine learning can be defined as the process of inducing a hypothesis from past experiences. In machine learning, algorithms learn from experience, causing general conclusions from specific examples. Generally, machine learning tasks are categorized into three types: supervised, unsupervised, and reinforcement learning. In this context, we focus on supervised learning. As noted by Lechner (2023), the task of supervised learning involves identifying structures in the space of covariates (X_1, X_i, ..., X_n) that result in accurate predictions of the dependent variable (Y). Within this category are classification methods for predicting specific values and regression methods for predicting discrete or continuous values. Supervised Machine Learning involves training a model using relationships of the type X_1, X_i, ..., X_n -> Y and employing that model to predict Y based on new datasets X_1, X_i, ..., X_n (Figure 1).

Figure 1. Supervised machine learning model for prediction

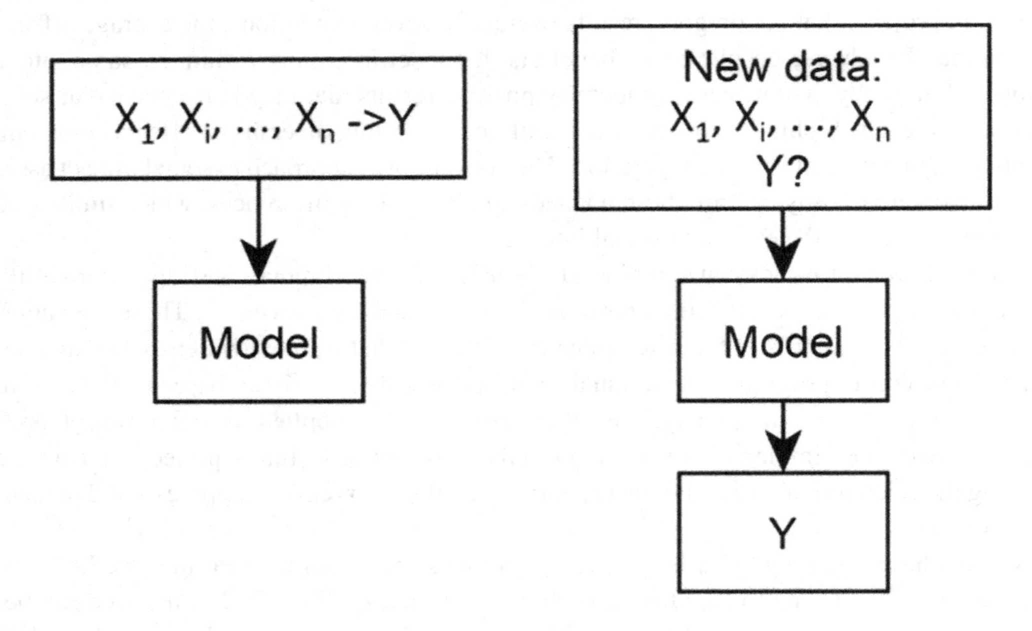

Translating this example to a practical application in the social area, we consider a scenario of a school that wants to use a supervised machine learning model to predict whether students in a given school year will transition or be retained based on the co-variables of students from previous years (such as sociodemographic characteristics, academic performance, behavior, among others). In this context, these co-variables would be the values X, Y would be 1 in the case of transition, and 0 in the retention situation.

3.2. Metalearners

As we have seen, the use of Supervised Machine Learning requires two types of attributes: the independent variables (X_1, X_i, ..., X_n) and a dependent variable (Y) aims to train the model to predict Y based on the new variables X. To infer causality, a new variable referring to the treatment/intervention (T) is added. In this latest action context, the data sets must include variables of type X, T, and Y for applying Causal Machine Learning methods.

Metalearners enable the use of traditional Supervised Machine Learning algorithms or statistical regression methods to estimate the heterogeneous effects of an intervention.

3.2.1. S-Learner

One of the simplest metalearner is the S-Learner (Figure 2), in which the letter "S" stands for using a single estimation model. The S-Learner model is trained to estimate the target variable "Y" based on variables X and T. In this method, the intervention-related variable (T) and the other independent variables (X) are another input attribute. Thus, the algorithm is trained and tested according to the following logic: X_1, X_i, ..., X_n, T -> Y. Because we only consider binary interventions and outcomes, T and Y can only have the values 0 or 1.

After training and testing the model, causal inference is performed using two estimates obtained by the same model. In the first estimate, it is assumed that the person underwent the intervention (T=1), and in the second estimate, it is assumed that this person was not subject to the intervention (T=0). The individual treatment effect can be obtained by subtracting the estimated value when T=1 from the estimated value when T=0 (Alves, 2022 & Künzel, Sekhon, Bickel, & Yu, 2019).

Figure 2. S-Learner model

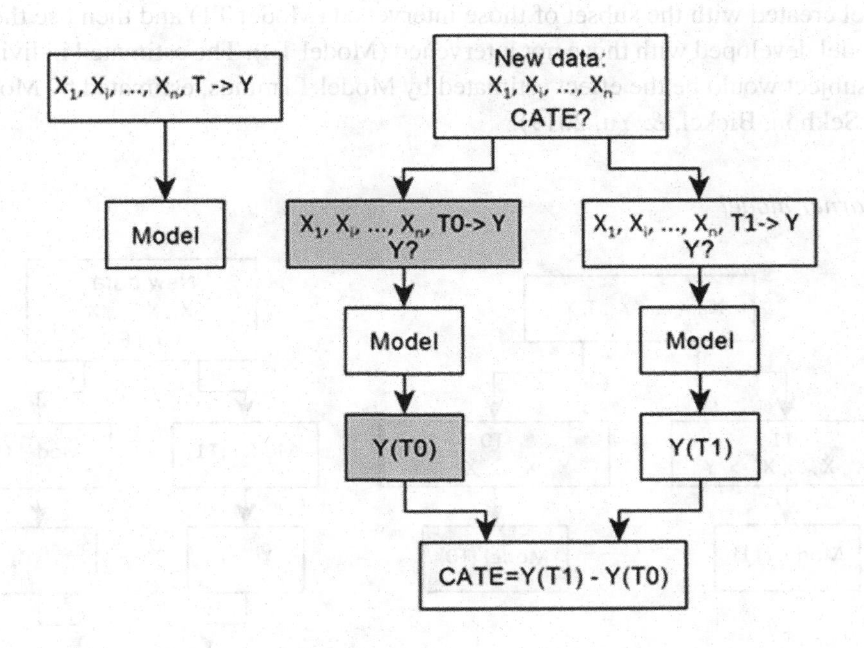

Returning to the example of school retention, the focus here would not be to predict retention but to understand the effect an intervention (e.g., a tutoring program) would have on this retention. To do this, we would consider the history of results obtained in that school in previous years, both in students who participated in tutoring and in those who did not. Thus, the model could be trained with all students in the school, regardless of whether they participated in the mentoring programs. The model would be trained with the specific attributes of each student (X), plus a feature indicating whether they had participated (T=1) or not (T=0) in tutoring, with the target variable for training being whether these students had been retained (Y^0) or not (Y^1). After training, the model could be used to estimate the individual effect of future pupils based on their characteristics (X). Thus, for each new student, an estimate would be made with all their features X and a T=1, and another would be made with the same characteristics but with a T=0. The subtraction of these two scenarios would provide the estimated individual effect of this mentoring program.

3.2.2. T-Learner

Another metalearner is the T-Learner model (Figure 3), which is also relatively simple to apply. It is so named because it involves the use of two separate models. The process starts by splitting the initial dataset into two subsets: one containing only those cases where people were intervened (T=1) and one containing those that were not (T=0).

Subsequently, we train these subsets separately using supervised machine learning algorithms or regression models, following the logic: Model (T1) = X_1, X_i, ..., X_n -> Y and Model (T0) = X_1, X_i, ..., X_n -> Y.

To estimate the heterogeneous effect of a given individual based on their specific attributes (X), we apply the model created with the subset of those intervened (Model T1) and then use these same attributes to the model developed with those not intervened (Model T0). The estimated individual treatment effect for that subject would be the effect estimated by Model T1minus, estimated by Model T0 (Alves, 2022; Künzel, Sekhon, Bickel, & Yu, 2019).

Figure 3. T-Learner model

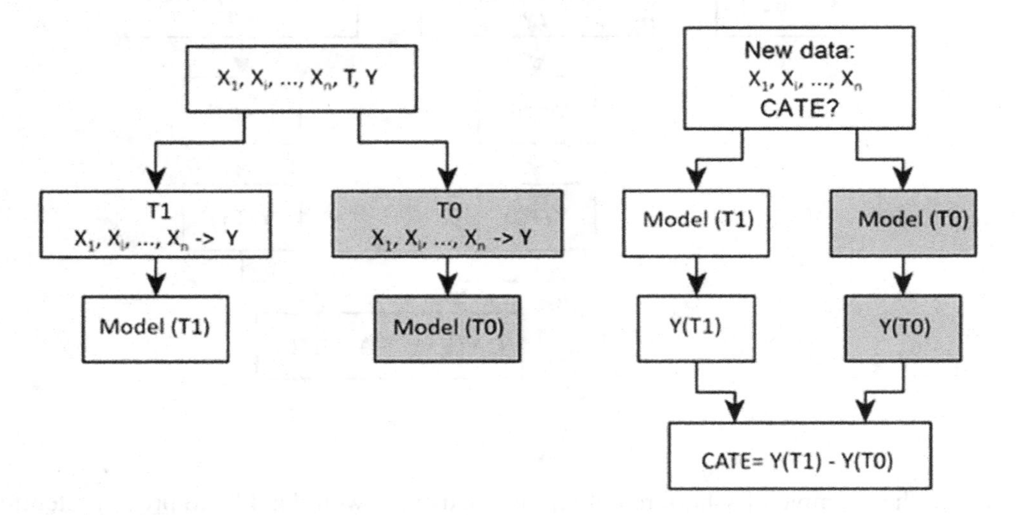

In the mentoring scenario, the process would involve dividing the former students into two groups: those who participated in the mentoring (T=1) and those who did not (T=0). An estimation model based on a machine learning or regression algorithm would be developed for each group to estimate retention (Y) based on student characteristics (X). To estimate the individual effect on new students, it would be sufficient to estimate the potential outcome in each model considering the characteristics of these new students and subtract the potential outcome obtained in Model T1 from that obtained in Model T0.

3.2.3. X-Learner

Another method is the X-Learner, which is a more complex approach than the previous ones, as it involves the interaction of models and datasets. Due to this complexity, we recommend following the explanation while analyzing the process in Figure 4.

The first stage of the X-Learner is like the T-Learner, where we build separately Model(T1) for the intervened (T=1) and Model(T0) for the non-intervened (T=0). Still, we will create a third propensity model (PS) with machine learning or regression algorithms in this first stage. This other propensity model aims to estimate the probability of a given person with specific characteristics belonging to the intervention group. To this end, this model is developed based on the factors of all people in the initial data set, where the target variable is the fact of having been intervened (X_1, X_i, ..., X_n ->T).

In the second stage, a significant difference appears in Model(T1) and Model(T0), compared to T-Learner: a cross between data and models is carried out. Thus, we apply the Model(T1) model to the initial set of non-intervened people (Model(T1)-> T0), and we use the Model(T0) to the set of intervened people (Model(T0)-> T1). This approach allows us to obtain inverse estimates in both groups: in the intervention group, we get an estimate of the potential outcome that each person would have had if they had not been intervened, Y[Model(T0)], while in the non-intervention group, we obtain an estimate of the potential outcome that each individual would have been if intervened, Y[Model(T1)].

In this way, we generated two sets of data. In the first set CATE(1), we subtract from the actual result Y(T1) the potential result obtained with the estimate Y[Model(T0)]. In the case of the other subgroup, CATE(0), we use the estimate of the potential result Y[Model(T1)] and subtract the actual value from Y(T1). The CATE(1) and CATE(0) are designated as the imputed treatment effect. We then developed two more machine learning models, each to estimate the value of the effects attributed to the treatment, based on the characteristics of each of the groups, since in the case of the intervention group, we obtain the Model[CATE(1)], and there is no control of Model[CATE(0)].

Thus, to estimate the effect of the intervention on new data (X1, Xi, ..., Xn), we apply three models to this data set: Model[CATE(1)], Model[CATE(0)] and PS and we obtain the values Y(T1), Y(T0) and P. Thus, to estimate the effect of the intervention, to the estimated value Y(T1) we multiply the propensity value of the individual belonging to the intervention group Y(T1). P and Y(T0), we multiply the propensity value of the individual who does not belong to the intervention group Y(T0). (1-P), in the end, we subtract one value from the other to know the effect: P.[Y(T1)] - (P-1).[Y(T0)].

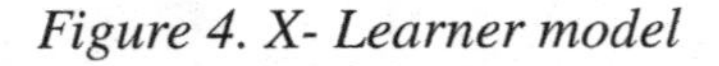

Figure 4. X- Learner model

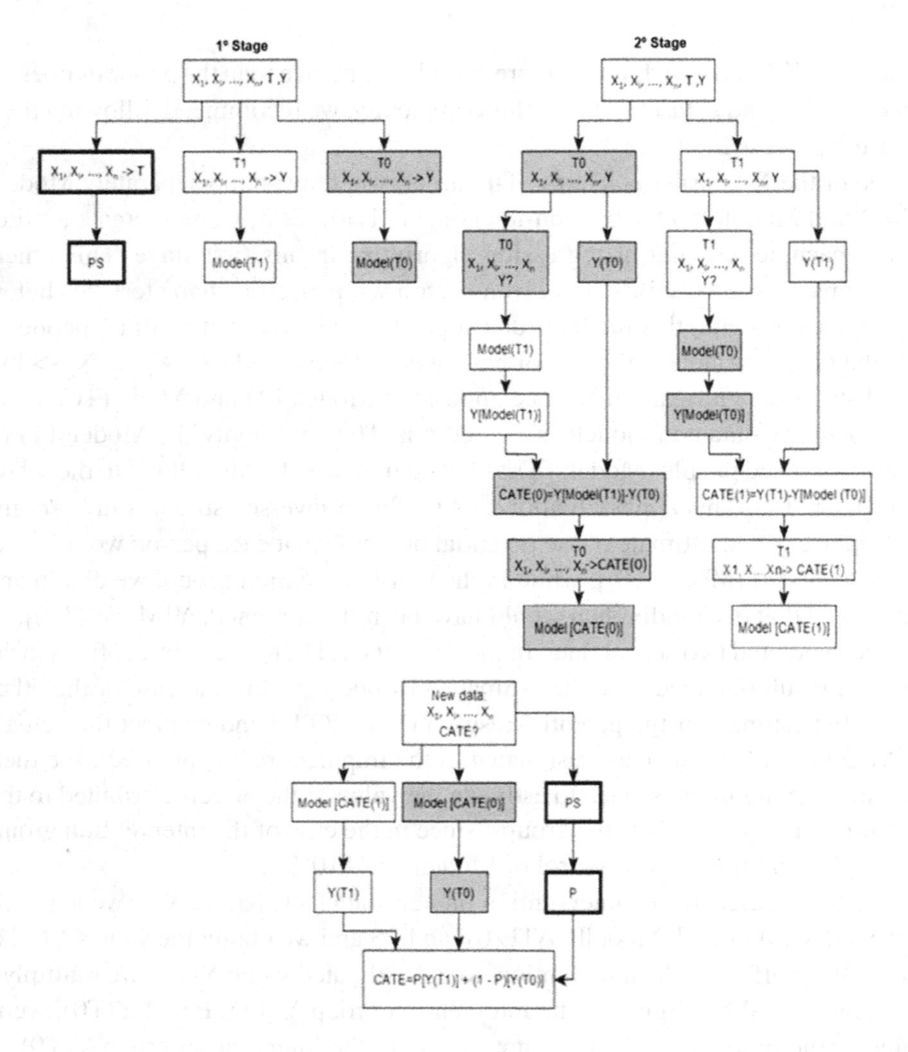

In addition to the methods mentioned, there are others, most of them more complex, such as the Doubly Robust (DR-learner) method (Kennedy, Ma, McHugh, & Small, 2017), R-learner (Nie & Wager, 2021), M-Learner (Acharki, Lugo, Bertoncello, & Garnier, 2023) among others. Some of these methods were created for multiple interventions, and each has its advantages and disadvantages depending on the dataset, namely the sample size, number of covariates, and the ratio of the proportion between intervention and control groups (Acharki, Lugo, Bertoncello, & Garnier, 2023).

3.3. Causal Trees and Causal Forests

In addition to methods that employ conventional Machine Learning algorithms to simulate the creation of counterfactuals and estimate heterogeneous intervention effects, specific algorithms have been developed for this purpose. One example is the Causal Tree (Athey & Imbens, 2016), which derives from traditional Decision Tree algorithms.

Decision Trees are Machine Learning algorithms that divide data into branches, forming a tree-like structure in which each node represents a feature of the dataset, and each branch denotes a decision based on that feature. These decisions are made until the data is separated into smaller, more homogeneous groups, allowing the prediction of an outcome for each group.

The Causal Tree follows a detached approach. It also segments the data into branches to identify heterogeneity in treatment effects. Rather than aiming only for accuracy in predicting the outcome, the Causal Tree focuses on finding groups of individuals who react differently to the intervention, manifesting distinct responses. This algorithm divides the data into two sets. The first set is used to build the tree, establishing rules that segment the data into terminal nodes. At the same time, the second set is used to calculate the effects of the intervention at each terminal node by calculating the mean difference between the outcomes observed in the intervention and control groups. This split is known as "honest" as it avoids the bias that could arise if the same dataset was used for both tasks. This approach allows the Causal Tree to more accurately identify subgroups of people who may or may not benefit from the intervention.

Like Random Forests, which expedite Decision Trees by constructing multiple trees that work together and are randomly generated using only a subset of the data and features to reduce overfitting and improve prediction accuracy, Causal Random Forests (Wager & Athey, 2018) also represent new causal machine learning algorithms that extend the notion of Causal Trees. The Causal Forests algorithm builds multiple causal trees and subsequently combines the predictions from these trees to obtain a more accurate estimate of the intervention effect.

It should be noted that, in addition to implementing the methods described, it is essential to evaluate the effectiveness of these models since the procedures used for this evaluation differ from those used in predictions made through Supervised Machine Learning (Devriendt, Moldovan, & Verbeke, 2018; and Pinheiro & Cavique, 2022).

4. APPLICATION: UPLIFT MODELLING

In social intervention, especially with vulnerable or at-risk populations, there are many cases of failure, even in projects that obtain good results overall. In many situations, the average effects of the intervention may be harmful, although they may be favorable for specific subgroups within that same intervention group. In addition, it is possible to observe positive overall effects that may be negative for particular subgroups. This outcome variation highlights the importance of understanding the heterogeneity of effects across different individuals or segments of the target population.

The traditional methods are presented in *section 2. Causal Inference* is widely used in social impact assessment. Despite their advantages, these approaches provide only the average effects of the intervention (usually referred to as ATE - Average Treatment Effect), not considering the individual effects that the intervention may have had on each person.

In social intervention, it is critical to understand the heterogeneity of effects across different individuals or segments of the target population. Traditional approaches to social impact assessment can only sometimes fully capture these nuances and individual differences, which limits our complete understanding of the effects of social interventions. Understanding the heterogeneity of the effects of social interventions can be crucial to refining future interventions, including the most appropriate selection of candidates for a given project or program, optimizing the intervention process itself, and tailoring the intervention to individual needs.

To illustrate the practical application of a Causal ML technique to address the heterogeneity of the effects of the impacts of social interventions, we will demonstrate adapting the Uplift Modeling method to evaluate social programs and projects. Although initially developed for marketing, this methodology also applies to other areas. We will initially discuss how this approach is commonly used in marketing campaigns for a more precise understanding. We will then explain how to implement this methodology through the Transformed Outcome Approach. Finally, we will propose a specific practical adaptation of this method to social impact assessment.

4.1. Uplift Modelling in Marketing

This approach is central to the field of prescriptive marketing and customer analytics, as it allows us to analyze what effect the campaign has on customer behavior.

This methodology was developed to reduce the costs of direct marketing campaigns and increase the return on investment. In its original version, based on previous campaigns, the objective is to categorize customers into four quadrants, considering their behavior towards a marketing campaign.

In the first quadrant, the "Persuadable" customers only buy when a marketing campaign targets them. In the second quadrant, we have the "Sure Thing," individuals who buy regardless of whether the campaign targets them. In the third quadrant, we have the "Lost Cause," individuals who never buy, irrespective of whether they are targeted or campaigned. Finally, in the fourth quadrant, we have "Do-Not-Disturb" or "Sleeping Dog", which are individuals who only buy if they are not targeted by the campaign (Pinheiro & Cavique, 2022; and Devriendt, Moldovan, & Verbeke, 2018).

Based on this method of prescription for a future marketing campaign, companies seek to avoid sending this campaign to the so-called "Do-Not-Disturb" and direct their efforts to the "persuadable". In addition, they seek to avoid investing resources in the "Sure Thing" and the "Lost Cause". In this way,

Figure 5. Uplift modeling

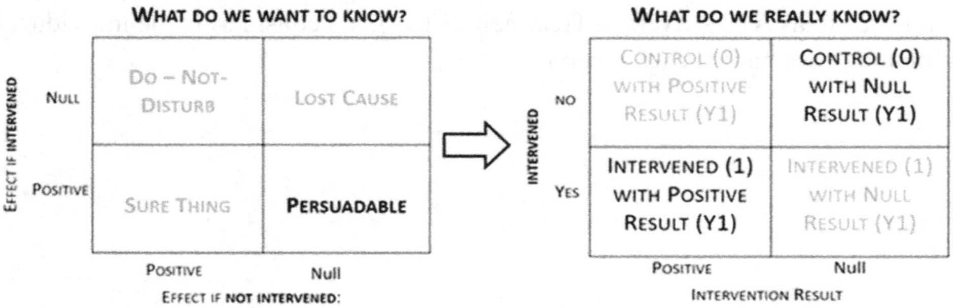

they can maximize the effectiveness of their campaigns by focusing resources where there is a greater propensity to achieve positive results.

However, in reality, this model corresponds to what we want to know rather than what we know (see Figure 5); we fall back into the fundamental problem of causal inference. We cannot simultaneously know what would have happened to a given intervened person if they had not been intervened, nor the opposite: inability to observe effects simultaneously with and without exposure.

4.2. Transformed Outcome Approach

A common form of Uplift Modeling is through the use of the Transformed Outcome Approach (Figure 6). This approach, similar to Metalearners, is a technique that allows estimating the effect of heterogeneous impact through classical supervised machine learning methods. As the name suggests, the aim is to transform the outcome approach to facilitate causal inference. The symbology that represents this transformation may vary according to the author. In this case, we express it as "Y*".

According to Jaskowski and Jaroszewicz (2012), the transformation from the initial values to the value Y* is intuitive in the case of binary outcomes. We consider Y* equal to 1 if the intervention outcome is at least as good as the outcome that would have occurred in the control group had we known the outcome in both groups. In practice, Y* is equal to 1 if the intervention results in a positive outcome

Figure 6. Transformed outcome approach model

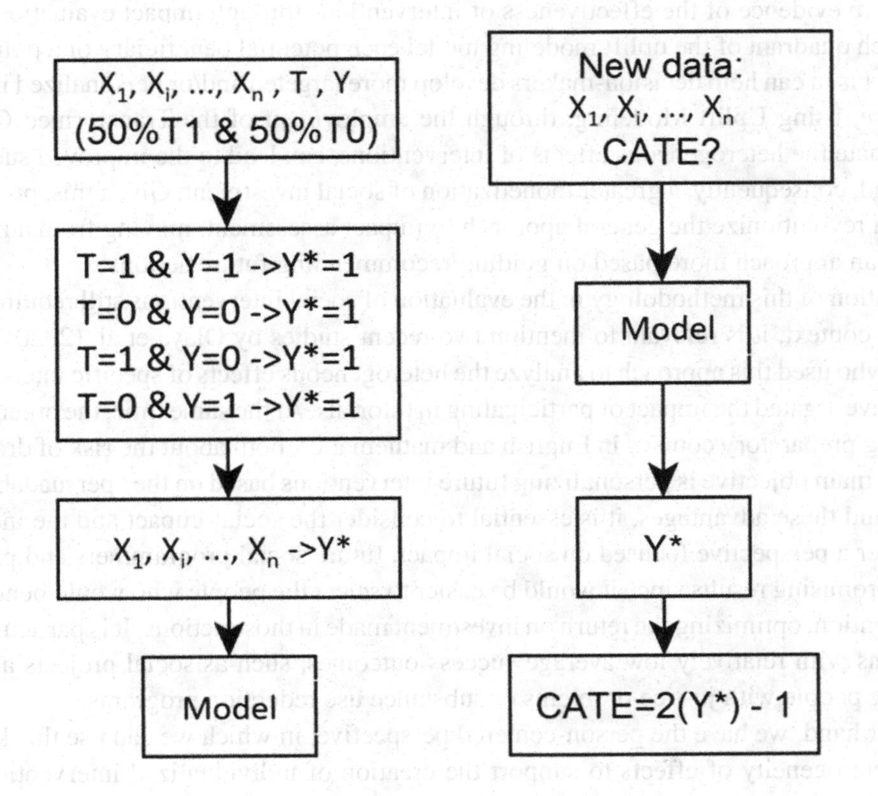

(T=1 & Y=1) or if no intervention results in a null outcome (T=0 & Y=0); in all other situations, Y* will be equal to 0.

By performing this transformation on all the observations (individuals) in our dataset, we can train a machine learning or regression model using the attributes (X_1, X_i, ..., X_n) as independent variables and the value of Y* as the independent variable. Without going into the mathematical particularities inherent in this method, to calculate the individual treatment effect, multiply the value of the estimate by two and subtract 1, i.e. (2xY*)-1. It is important to note that this method is only effective if there is an initial probability of 50% of an individual belonging to the intervention group or the control group (Devriendt, Moldovan, & Verbeke, 2018; Jaskowski & Jaroszewicz, 2012 and Pinheiro & Cavique, 2022).

In the example of tutoring at school, the model would be trained by transforming the value Y*=1 for all students who participated in tutoring and were not retained (T=1&Y=1) or did not participate and were retained (T=0 & Y=0), with all other situations, the value of Y* would be equal to 0. It is important to note that the sample size of those who did not participate in tutoring should be similar to those who did. After the estimation for each student, the individual effect of tutoring would be calculated by applying the formula mentioned above.

4.3. Uplift Modelling in Social Impact Assessment

To improve the effectiveness of interventions, just as in Marketing, where we seek to avoid wasting resources on campaigns that aim to "Do Not Disturb" and try to focus our efforts and investments on reaching "Persuadable", the objective can be similar in social intervention. As mentioned previously, funders of educational, social, and community projects and programs are increasingly demanding and looking for more evidence of the effectiveness of interventions through impact evaluation projections. Detecting which quadrant of the uplift modeling model each potential beneficiary of a policy, program, or social project is in can help decision-makers develop more targeted and/or personalized interventions.

In this sense, using Uplift Modeling, through the employment of the Transformed Outcome Approach to estimate the heterogeneous effects of interventions, can lead to the improved success of each intervention and, consequently, a greater monetization of social investment. Given this, promoting these procedures can revolutionize the general approach to impact assessment, moving from a merely evaluative vision to an approach more based on guiding/recommending future actions.

The application of this methodology in the evaluation of social interventions still requires significant studies. In this context, it is relevant to mention two recent studies by Olaya et al. (2020) and Tanai & Ciftci (2023), who used this approach to analyze the heterogeneous effects of specific interventions. One of the studies investigated the impact of participating in tutorials. At the same time, the other explored the effects of taking preparatory courses in English and mathematics, both about the risk of dropping out of university. The main objective is personalizing future interventions based on the "persuadable" segment.

To understand these advantages, it is essential to consider the social impact and the individual perspectives. Under a perspective focused on social impact, future social programmers and projects could achieve more promising results since it would be easier to select the people who would benefit from that specific intervention, optimizing the return on investment made in those actions. It is particularly relevant for interventions with relatively low average success outcomes, such as social projects and programs targeting young people with justice problems or substance use reduction programs.

On the other hand, we have the person-centered perspective, in which we can use this knowledge to estimate the heterogeneity of effects to support the creation of individualized intervention pathways.

Social workers can accompany and guide these people in trajectories more suited to their characteristics and needs. An example could be job centers' personalized referral of unemployed people to training or apprenticeships that best meet their characteristics and needs. The same approach could benefit any intervention to develop individualized social support pathways: rehabilitation, social reintegration, and promotion of behaviors or skills, among others. In this way, the Social Impact Assessment and Prescription Approach can be valuable in optimizing social interventions and providing personalized support for each individual, allowing a more effective targeting of resources and a more attentive approach to the needs and potential of the people involved.

Despite the advantages mentioned, it is essential to highlight that, as in any data science project, especially when it involves large volumes of data and machine learning algorithms, it is crucial to consider the quality of data collection and the preservation of the privacy of that data. Aragon et al. (2022) argue that data is rarely neutral, as it reflects the biases and subjectivity of the people and systems that collect, process, and interpret it. Ignoring this fact can perpetuate discriminatory patterns and biases harmful to specific groups.

In addition to concerns about potential biases in the data collected, another significant challenge is the availability of this data. Data related to the characteristics of each subject participating in the evaluation and prescription processes (both in the intervention and control groups) must be available to achieve these objectives. It may include data on the characteristics of the subjects, such as demographic data, behaviors, needs, and results obtained in this intervention or previous interventions. Another challenge is data collection itself. Data must be collected before and after the intervention to evaluate its effects. Additionally, having many observations (usually subjects) is vital to making the most of these methods because sample size can influence the effectiveness of causal machine-learning methods and algorithms.

In many social projects, outcome indicators are based on standard questionnaires applied before and after interventions. However, it is only sometimes feasible to have control groups for comparison. Considering these difficulties from the project design stage is crucial to maximize future benefits when assessing and predicting varied effects. Finally, it is essential that funding entities recognize the importance of this type of evaluation/prescription, considering the investment in social interventions as a long-term return, as specific projects may present less favorable initial results but can be a source of information for the future improvement and personalization of these interventions.

5. CONCLUSION

The classic approach to social impact assessment focuses on traditional methods of causal inference, which only provide us with the average effects of the intervention without capturing how social programs and projects can affect people or groups differently. Traditional stratification methods require detailed analysis before intervention to identify groups in detail, limiting later discovery of effects as they could be biased. In contrast, new Causal Machine Learning approaches allow for a more heterogeneous analysis of the effects of interventions, allowing us to understand existing effects and estimate possible future effects of new programs and projects. Causal Machine Learning provides decision-makers with more accurate information for developing future personalized interventions adapted to the target audience.

In this chapter, we have explored the potential advantages of applying Machine Learning algorithms for causal inference in the impact assessment processes of social projects and programs. In addition to a non-technical review of the classic causal inference methods used to evaluate the social impact of programs and projects, we present some of the new Causal Machine Learning techniques that can be applied to assess the heterogeneous effects of this type of intervention. To understand how to adapt these methods to social impact assessment, we demonstrate the Uplift Modeling method's adaptation to assess social projects and programs. Although it is usually used in Marketing, this method allows for capturing the heterogeneity of effects, grouping individuals into four categories. With this methodology, impact evaluators can present decision-makers with the ability to identify, based on previous interventions, the people most likely to benefit from this social intervention, that is, those who probably fall into the persuadable group. In this approach, we indicate that, through the Transformed Results Approach, it is possible to increase the ability of decision-makers to evaluate the heterogeneity of effects and, at the same time, contribute to improving decision-making related to future social programs and projects.

Despite remarkable advances in Artificial Intelligence, its use remains predominantly restricted to sectors with substantial economic returns. Contrary to other fields, social intervention often faces investment constraints, as its sustainability derives primarily from public funding or patronage, resulting in a relative neglect of investment in new Artificial Intelligence and Machine Learning methodologies and algorithms. The emerging area of Causal Machine Learning is no exception, as it needs more research aimed at its applicability in assessing and prescribing the social impact of projects and programs. The number of studies employing these techniques is scarce, except for some research on the impact evaluation of public policies. Thus, it is essential to invest in disseminating these processes among funders and promoters and training new impact evaluators in contemporary data science techniques, namely those related to Causal Machine Learning techniques.

To improve evaluation and prescription for a more effective intervention practice, academia must promote more applied research in this field, aggregating contributions from econometrics, computer science, and social and behavioral sciences. This chapter contributes to developing a new prescriptive vision, promoting the future design of more personalized social, educational, and community intervention projects and programs with a more significant impact on society.

REFERENCES

Abadie, A., Diamond, A., & Hainmueller, J. (2010). Synthetic control methods for Comparative case studies: Estimating the effect of California's tobacco control program. *Journal of the American Statistical Association, 105*(490), 493–505. doi:10.1198/jasa.2009.ap08746

Abadie, A., & Gardeazabal, J. (2003). The economic costs of conflict: A case study of the Basque Country. *The American Economic Review, 93*(1), 113–132. doi:10.1257/000282803321455188

Acharki, N., Lugo, R., Bertoncello, A., & Garnier, J. (2023). Comparison of metalearners for estimating multi-valued treatment heterogeneous effects. *ICML2023- Fortieth International Conference on Machine Learning*. Doi:10.48550/arXiv.2205.14714

Alves, M. (2022). Causal inference for the brave and true. *Matheus Facture*. https://matheusfacure.github.io/python-causality-handbook/landing-page.html

Aragon, C., Guha, S., Kogan, M., Muller, M., & Neff, G. (2022). *Human-centered data science: An introduction*. MIT Press.

Athey, S., & Imbens, G. (2016). Recursive partitioning for heterogeneous causal effects. *Proceedings of the National Academy of Sciences of the United States of America, 113*(27), 7353–7360. doi:10.1073/pnas.1510489113 PMID:27382149

Cook, T. D. (2008). "Waiting for life to arrive": A history of the regression-discontinuity design in psychology, statistics and economics. *Journal of Econometrics, 142*(2), 636–654. doi:10.1016/j.jeconom.2007.05.002

Crato, N., & Paruolo, P. (2019). *Data-driven policy impact evaluation: How access to microdata is transforming policy design*. Springer Nature. doi:10.1007/978-3-319-78461-8

Cunningham, S. (2021). *Causal inference: The mixtape*. Online Ebook Version. https://mixtape.scunning.com/

Devriendt, F., Moldovan, D., & Verbeke, W. (2018). A literature survey and experimental evaluation of the state-of-the-art in uplift modeling: A stepping stone toward developing prescriptive analytics. *Big Data, 6*(1), 13–41. doi:10.1089/big.2017.0104 PMID:29570415

Fan, X., & Nowell, D. L. (2011). Using propensity score matching in educational Research. *Gifted Child Quarterly, 55*(1), 74–79. doi:10.1177/0016986210390635

Fougère, D., & Jacquemet, N. (2019). Causal inference and impact evaluation. *Economie et Statistique/Economics and Statistics*, (510-511-512), 181-200. doi:10.24187/ecostat.2019.510t.1996

Gama, J., Lorena, A., Faceli, K., Oliveira, O., & Carvalho, A. (2017). Extração de Conhecimento de dados: Data Mining. Edições Sílabo- 3ª Edição.

Gopalan, M., Rosinger, K., & Ahn, J. B. (2020). Use of quasi-experimental research designs in education research: Growth, promise, and challenges. *Review of Research in Education, 44*(1), 218–243. doi:10.3102/0091732X20903302

Huntington-Klein, N. (2022). *The effect: An introduction to research design and causality*. Chapman and Hall/CRC Online Ebook Version. https://theeffectbook.net/index.html

Imbens, G., & Rubin, D. (2015). *Causal Inference for Statistics, Social, and Biomedical Sciences: An Introduction*. Cambridge University Press. doi:10.1017/CBO9781139025751

Imbens, G. W. (2000). The role of the propensity score in estimating dose-response functions. *Biometrika, 87*(3), 706–710. doi:10.1093/biomet/87.3.706

Jaskowski, M., & Jaroszewicz, S. (2012). Uplift modeling for clinical trial data. In *ICML Workshop on Clinical Data Analysis* (Vol. 46, pp. 79-95).

Kaddour, J., Lynch, A., Liu, Q., Kusner, M. J., & Silva, R. (2022). (Manuscript submitted for publication). Causal machine learning: A survey and open problems. *arXiv preprint arXiv:2206.15475. Work (Reading, Mass.).* doi:10.48550/arXiv.2206.15475

Kang, M., Ragan, B. G., & Park, J. H. (2008). Issues in outcomes research: An overview of randomization techniques for clinical trials. *Journal of Athletic Training, 43*(2), 215–221. doi:10.4085/1062-6050-43.2.215 PMID:18345348

Kennedy, E. H., Ma, Z., McHugh, M. D., & Small, D. S. (2017). Nonparametric methods for doubly robust estimation of continuous treatment effects. *Journal of the Royal Statistical Society. Series B, Statistical Methodology*, *79*(4), 1229–1245. doi:10.1111/rssb.12212 PMID:28989320

Künzel, S. R., Sekhon, J. S., Bickel, P. J., & Yu, B. (2019). Metalearners for estimating heterogeneous treatment effects using machine learning. *Proceedings of the National Academy of Sciences of the United States of America*, *116*(10), 4156–4165. doi:10.1073/pnas.1804597116 PMID:30770453

Lechner, M. (2011). The estimation of causal effects by difference-in-difference methods. *Foundations and Trends® in Econometrics*, *4*(3), 165–224. doi:10.1561/0800000014

Lechner, M. (2023). Causal Machine Learning and its use for public policy. *Schweizerische Zeitschrift für Volkswirtschaft und Statistik*, *159*(1), 1–15. doi:10.118641937-023-00113-y

Nie, X., & Wager, S. (2021). Quasi-oracle estimation of heterogeneous treatment effects. *Biometrika*, *108*(2), 299–319. doi:10.1093/biomet/asaa076

OECD. (2002). *Glossary of key terms in evaluation and results-based management. DAC Network on Development Evaluation*. OECD.

OECD. (2021). *Social Impact Measurement for the Social and Solidarity Economy*. Doi:10.1787/20794797

Olaya, D., Vásquez, J., Maldonado, S., Miranda, J., & Verbeke, W. (2020). Uplift modeling for preventing student dropout in higher education. *Decision Support Systems*, *134*, 113320. doi:10.1016/j.dss.2020.113320

Pearl, J. (1995). Causal diagrams for empirical research. *Biometrika*, *82*(4), 669–688. doi:10.1093/biomet/82.4.669

Pearl, J., & Mackenzie, D. (2018). *The book of why: the new science of cause and effect*. Basic books.

Pinheiro, P., & Cavique, L. (2022). Uplift Modeling Using the Transformed Outcome Approach. In G. Marreiros, B. Martins, A. Paiva, B. Ribeiro, & A. Sardinha (Eds.), Lecture Notes in Computer Science: Vol. 13566. *Progress in Artificial Intelligence. EPIA 2022*. Springer., doi:10.1007/978-3-031-16474-3_51

Rogers, P. (2014a). Overview of Impact Evaluation: Methodological Briefs - Impact Evaluation. *Methodological Briefs*. UNICEF Office of Research-Innocenti. https://www.unicef-irc.org/publications/746-overview-of-impact-evaluation-methodological-briefs-impact-evaluation-no-1.html

Rogers, P. (2014b). Theory of Change: Methodological Briefs - Impact Evaluation No. 2, *Methodological Briefs*. UNICEF Office of Research-Innocenti. https://www.unicef-irc.org/publications/747-theory-of-change-methodological-briefs-impact-evaluation-no-2.html

Rosenbaum, P. R., & Rubin, D. B. (1983). The central role of the propensity score in Observational studies for causal effects. *Biometrika*, *70*(1), 41–55. doi:10.1093/biomet/70.1.41

Rubin, D. B. (1974). Estimating causal effects of treatments in randomized and nonrandomized studies. *Journal of Educational Psychology*, *66*(5), 688–701. doi:10.1037/h0037350

Stanley, K. (2007). Design of randomized controlled trials. *Circulation*, *115*(9), 1164–1169. doi:10.1161/CIRCULATIONAHA.105.594945 PMID:17339574

Tanai, Y., & Ciftci, K. (2023). How to customize an early start preparatory course policy to improve student graduation success: An application of uplift modeling. *Annals of Operations Research*. doi:10.100710479-023-05607-9

Thistlewaite, D. L., & Campbell, D. T. (2017). Regression-Discontinuity Analysis: An Alternative to the Ex-Post Facto Experiment. *Observational Studies*, *3*(2), 119–128. doi:10.1353/obs.2017.0000

Tu, C. (2019). Comparison of various machine learning algorithms for estimating generalized propensity score. *Journal of Statistical Computation and Simulation*, *89*(4), 708–719. doi:10.1080/00949655.2019.1571059

Wager, S., & Athey, S. (2018). Estimation and Inference of Heterogeneous Treatment Effects using Random Forests. *Journal of the American Statistical Association*, *113*(523), 1228–1242. doi:10.1080/01621459.2017.1319839

White, H., Sabarwal, S., & de Hoop, T. (2014). Randomized Controlled Trials (RCTs): Methodological Briefs - Impact Evaluation No. 7, *Methodological Briefs*. UNICEF Office of Research-Innocenti. https://www.unicef-irc.org/publications/752-randomized-controlled-trials-rcts-methodological-briefs-impact-evaluation-no-7.html

ENDNOTE

[1] In this text, we have used the term "person" as a generic reference to facilitate the reading and understanding of impact assessment methods. However, these methods can be applied to other units of interest, such as organizations (schools, companies, hospitals, prisons, health facilities, etc.), localities, and groups of people.

Chapter 5
The Incorporation of Large Language Models (LLMs) in the Field of Education:
Ethical Possibilities, Threats, and Opportunities

Paul Aldrin Pineda Dungca
https://orcid.org/0009-0002-9452-646X
Salesians of Don Bosco, Philippines

ABSTRACT

This chapter delves into the ethical implications that arise from integrating LLMs within the realm of education. LLMs, exemplified by the GPT-3.5, have emerged as formidable instruments for natural language processing, offering diverse applications in educational domains. Nevertheless, their adoption necessitates careful consideration of ethical matters. This chapter comprehensively overviews the ethical potentials, threats, and opportunities in incorporating LLMs into education. It scrutinizes the potential advantages, including enriched personalized learning experiences and enhanced accessibility, while addressing concerns regarding data privacy, bias, and the ramifications of supplanting human instructors. By critically examining the ethical dimensions, this chapter endeavors to foster a varied comprehension of the implications of utilizing LLMs in educational settings.

1. INTRODUCTION

Integrating LLMs in education has garnered considerable attention recently, offering the transformative potential for enriching teaching and learning experiences. One such prominent LLM is ChatGPT 3.5, which has exhibited exceptional capabilities in natural language processing and holds promising applications in educational contexts (Eysenbach, 2023). However, it is essential to note that integrating LLMs into education is not without controversy and has sparked a significant debate among researchers,

DOI: 10.4018/978-1-6684-9591-9.ch005

educators, and stakeholders. While some see LLMs' immense possibilities and benefits in enhancing education, others express concerns and raise important questions regarding data privacy, algorithmic bias, and the role of human instructors (Opara & Aduke, 2023). This ongoing debate highlights the need to carefully consider and examine the ethical, social, and pedagogical implications of integrating LLMs in the academe.

With advanced language processing capabilities, LLMs like ChatGPT 3.5 find concrete applications in various tasks. They can assist students in writing essays, offering ideas, structure outlines, and relevant information (Herbold et al., 2023). Researchers benefit from research support, including paper summaries, references, and information retrieval (Rahman et al., 2023). Programmers receive coding help with code snippets, explanations, and issue troubleshooting. Language translation has become more accessible (Peng et al., 2023), and creative enthusiasts can generate prompts for storytelling (Chu & Liu, 2023). These applications highlight the versatility of LLMs in assisting with writing, research, programming, and creative endeavors.

The remarkable capacity of these models to revolutionize education through personalized learning (PL), automated assessment, and improved accessibility highlights the need for careful consideration of the ethical dimensions associated with their integration into educational settings (Hong, 2023). One of the most severe problems associated with integrating LLMs in education is the issue of Value Alignment (VA) (Cao, 2023). LLMs are trained on vast amounts of text data from diverse sources, reflecting society's wide range of human values, beliefs, and perspectives. As a result, the values embedded within LLMs may need to align with the values and goals upheld in educational settings, which can pose significant challenges.

But technically, how do LLM work? Large Language Models (LLMs) such as ChatGPT are sophisticated models designed to understand and generate human-like text. They go through a two-step process: pre-training and fine-tuning. During pre-training, the model learns from a vast amount of text data, predicting the next word in a sentence based on preceding words. This phase equips the LLM with language knowledge and contextual understanding. The Transformer architecture, specifically designed for sequence-to-sequence tasks, forms the backbone of LLMs. It consists of self-attention mechanisms and feed-forward neural networks that allow the model to capture relationships between words and generate coherent responses.

After pre-training, LLMs undergo fine-tuning on specific tasks like chatbot interactions. This phase involves training the model on labeled examples, enabling it to adapt to the target task requirements. When a user provides input, the LLM tokenizes and converts it into numerical representations. The Transformer model processes these tokens, building a contextualized understanding of the input. Next, the LLM generates responses by predicting the next word based on context and sampling from a probability distribution. The generated text is post-processed for readability and presented to the user. It's important to acknowledge that LLMs generate text based on statistical patterns learned from training data, and while they can produce impressive responses, they may occasionally generate incorrect or nonsensical output.

Non-technically speaking, imagine LLMs as intelligent virtual assistants that have been trained on vast amounts of text from the internet. They learn the patterns and knowledge contained in this text to provide helpful responses. It's like having a language expert who can answer your questions, hold a conversation, or even tell stories. LLMs work by first learning from the internet, understanding the context of your input, and then generating appropriate and coherent responses based on that understanding. They use a special architecture called the Transformer, which helps them process and organize information effectively. However, it's important to remember that LLMs are statistical models and may occasionally produce incorrect or nonsensical answers. Nonetheless, they are powerful tools that can assist with a wide range of tasks and provide valuable information to non-technical users in a user-friendly and conversational manner.

This chapter aims to comprehensively address the multifaceted aspects of integrating LLMs into education. To accomplish this, the chapter will systematically examine and analyze (1) the ethical possibilities arising from LLM integration, (2) the potential threats and challenges associated with their use from the lens of VA, and finally, (3) provide opportunities in forms of recommendations and guidelines for the effective implementation of LLMs in education. This chapter strives to thoroughly understand the diverse facets of incorporating LLMs in education by engaging with these critical dimensions.

2. EXPLORING ETHICAL POSSIBILITIES: TRANSFORMING EDUCATION THROUGH INTELLIGENT AND PERSONALIZED ASSISTANCE

In education, integrating LLMs offers a wide range of possibilities that align with ethical principles and values. By harnessing the capabilities of LLMs in responsible ways, educators can leverage their potential for positive impact. This section presents some ethically acceptable applications of LLMs in education, highlighting how they can enhance teaching and learning experiences while upholding critical ethical considerations. These applications demonstrate the potential for LLMs to support personalized learning, provide automated assessment and feedback, improve accessibility, and empower educators. By understanding and implementing these ethical applications, education can benefit from the transformative power of LLMs while prioritizing the well-being and development of learners.

Henceforth, it is vital to establish a general guideline when determining the ethical acceptability of integrating LLMs in various domains. In the context of writing and creativity, LLMs should be viewed not merely as “libraries or resource centers” but as “intelligent virtual assistants” capable of offering valuable support. By recognizing their potential to generate ideas, provide structure outlines, and assist in overcoming creative blocks, one can establish a framework that acknowledges the ethical acceptability of utilizing LLMs to enhance the creative process.

Also, while integrating LLMs in education offers valuable support in writing, using them may not be advisable for lessons and objectives focusing on fostering creativity and styling skills. These aspects are considered vital components of education, requiring human engagement, critical thinking, and personal expression. The subjective nature of creativity and the need for individuality make it challenging for LLMs to fully capture and replicate the nuanced and individualized aspects of artistic expression. Therefore, it is recommended that caution be exercised, and alternative approaches be considered when the aim is to develop and nurture creativity and styling abilities in learners.

After establishing the general principle and providing a necessary caution, here is a list of ethically acceptable applications of LLMs in teaching-learning environments:

Personalized Learning

Since ChatGPT has extensive training on vast amounts of data, as Haque et al. (2022) highlighted, it can quickly provide personalized responses tailored to the specific context of a given prompt. Also, it can learn from human interactions, making it adaptable and personalized as a conversational agent (Shen et al., 2023).

Additionally, as noted by Aljanabi (2023), it can generate responses with diverse tones and structures, catering to its users' individual preferences and needs. Because of these facilities, ChatGPT and other LLMs can provide customized learning experiences by adapting instructional content and resources to meet individual student needs. This promotes student engagement, motivation, and academic success while respecting their unique learning styles and preferences (Kasneci et al., 2023). For example, in teaching the concept of buoyancy in Physics, students can benefit from interactive simulations and virtual experiments that allow them to explore buoyancy phenomena hands-only. Additionally, personalized practice problems and quizzes can be generated to target specific areas of difficulty for each student. This tailored approach promotes student engagement and motivation and ensures that students can grasp the principles of buoyancy according to their unique learning styles and preferences. Using LMM can lead to improved academic success and a deeper understanding of buoyancy principles among students.

Automated Assessment and Feedback

LLMs can assist in automating assessment processes by evaluating student work, providing feedback, and generating personalized recommendations for improvement (Dai et al., 2023). This streamlines the assessment process, ensures prompt feedback, and helps students track their progress. In writing an essay, for example, on Plato's Caveman, LLMs can play a valuable role in automating the assessment process and providing personalized feedback. For instance, students can submit their essays to an LLM-based system that can analyze their arguments' content, structure, and coherence. The LLM can then generate detailed feedback on the strengths of their analysis, clarity of expression, and logical reasoning.

Moreover, the LLM can provide personalized recommendations for improvement, suggesting specific areas where students can enhance their understanding of Plato's allegory and strengthen their arguments. However, from a case study by Dai et al. (2023), three limitations were concluded: Dai et al. (2023) conducted a case study that identified three limitations. Firstly, while the study assessed the overall agreement between instructor and ChatGPT feedback regarding polarity, it did not investigate their alignment on the same assignment, which is crucial for individual students. Further research is needed to determine ChatGPT's ability to provide assignment-specific feedback. Secondly, the unsupervised generation of feedback by ChatGPT may affect its effectiveness. Future studies could explore prompt engineering techniques to obtain desired feedback aligned with learning goals. Lastly, the study heavily relied on time-consuming human annotation for analysis, emphasizing the need for automated evaluation methods to assess educational feedback effectiveness more efficiently.

Language Support and Translation

In their report, Jiao et al. (2023) concluded that ChatGPT performs competitively with commercial translation products for high-resource European languages but needs help with low-resource or foreign languages. In general, LLMs can provide language support and translation services, enabling students with limited proficiency in the instructional language to access educational resources and participate fully in the learning process (Peng et al., 2023). This promotes inclusivity and breaks down language barriers. In classroom debates, for example, on Hamlet, LLMs can be harnessed to facilitate language support and translation services for students from Sudan with limited proficiency in the instructional language. By utilizing LLMs, these students can access educational resources, engage in discussions, and actively participate in learning. For instance, when analyzing a specific scene from Hamlet, the LLM can provide contextual examples and explanations in the student's native language, allowing them to grasp the nuances and complexities of the text thoroughly. This enhances their understanding of the play and empowers them to contribute their perspectives and insights to the debate. Using LLMs in the classroom fosters inclusivity by dismantling language barriers and establishing a fair learning environment that caters to students of diverse linguistic backgrounds.

Support for Special Needs Education: LLMs can be employed to create adaptive and inclusive learning environments for students with special needs, offering tailored resources and accommodations to enhance accessibility. In a history class, for instance, discussing the History of the Philippines, LLMs exemplify their crucial role in supporting students with special needs. Students with visual impairments may benefit from alternative formats and text-to-speech capabilities, ensuring their active engagement with historical texts. Similarly, interactive features like adjustable difficulty levels in quizzes or games can cater to the diverse learning requirements of students with cognitive disabilities. Moreover, LLMs can provide visual cues, timers, and reminders, benefiting students with attention deficit hyperactivity disorder (ADHD) and helping them maintain focus during lessons. These personalized adaptations and accommodations offered by LLMs foster inclusivity, address the specific needs of students with special needs, and create an empowering and enriching educational experience. This application for students with ADHD is based on the research of van Schalkwyk (2023). He concluded that ChatGPT, as an LLM, holds promise in its ability to contribute to pediatric behavioral health in several ways. Through its capacity to offer evidence-based information and guidance, ChatGPT has the potential to enhance the accessibility and standard of care for children and adolescents, as well as their parents and caregivers, dealing with behavioral health conditions. Mental health professionals, parents, and caregivers are encouraged to consider incorporating ChatGPT as a valuable tool to improve pediatric behavioral health.

Professional Development for Educators

LLMs offer substantial assistance to educators in lesson designing and assessment and creating an effective learning environment. For instance, a biology educator can create interactive simulations to enhance students' understanding. For example, in a math class, LLMs can automatically grade quizzes, identify areas of improvement, and recommend targeted practice exercises.

Chatbots as Learning Support

LLMs contribute to creating an engaging learning environment by incorporating interactive features such as chatbots that offer personalized answers and instant support. For example, LLMs can provide tailored writing prompts, grammar explanations, and vocabulary assistance in a language arts class to address individual student needs. Additionally, they can serve as a valuable resource for obtaining multiple examples and simplifying complex concepts, effectively bridging the gap between collegiate and grade school levels. Despite the power of ChatGPT as a chatbot, Tlili et al. (2023) conducted a comprehensive case study explicitly focusing on ChatGPT in education to address concerns related to its use as a chatbot-tutor. The study emphasized the importance of implementing ChatGPT with caution. Moreover, it identified several areas that require further research and investigation to improve the adoption and effectiveness of chatbots, particularly ChatGPT, as a valuable learning support tool within LLMs.

In summary, LLMs offer tremendous potential for enhancing education within ethical practices. They enable personalized learning experiences by providing tailored responses, accommodating individual student needs, and respecting their unique learning styles and preferences. LLMs also contribute to automated assessment and feedback processes, streamlining the evaluation of student work and promoting timely feedback. Moreover, they facilitate language support and translation services, breaking down language barriers and promoting inclusivity in the classroom. Additionally, LLMs play a vital role in supporting students with special needs, offering adaptive resources and accommodations to enhance accessibility and create inclusive learning environments. Furthermore, LLMs provide valuable support for educators by assisting in lesson design, assessment, and creating engaging learning environments. By understanding and implementing ethical applications of LLMs, educators can leverage the transformative power of these models to create personalized learning experiences that prioritize the growth of each student.

3. NAVIGATING ETHICAL THREATS: THE CRISIS OF VALUE ALIGNMENT IN LLMS

After recognizing the ethical possibilities of incorporating LLMs in education, it is crucial also to acknowledge the ethical threats that arise in this integration. Value alignment theory provides a critical framework for examining these threats and ensuring that using LLMs in education aligns with ethical principles and values.

Value Alignment Theory (VAT) in AI explores the challenge of aligning the objectives and behavior of AI systems with human values and ethical principles. It involves developing mechanisms and frameworks to ensure that AI systems act in ways that are consistent with human values, promoting fairness, transparency, and accountability (Montemayor, 2023). By addressing value alignment, AI researchers and developers aim to mitigate potential ethical risks and foster AI technologies' responsible and beneficial use.

This section explores seven critical ethical threats associated with LLMs in education. By critically examining these factors through the lens of VAT, educators, and educational institutions can make informed decisions, mitigate risks, and navigate the ethical challenges of LLM integration to ensure the well-being and development of learners remains paramount.

Privacy Concerns

Value alignment theory emphasizes upholding individuals' autonomy and privacy rights (Winter et al., 2023). When considering the integration of LLMs in education, it becomes crucial to address data privacy concerns. The potential collection and analysis of student's personal information and learning data without their explicit consent raises ethical concerns. The hypothetical scenario of an educational institution implementing an LLM-based system without proper permission or transparent data practices raises ethical concerns regarding privacy violations, compromising students' autonomy and eroding trust within the learning environment.

Algorithmic Bias

LLMs can extend to biased values, favoring specific socio-economic or cultural backgrounds over others (Ray, 2023). If the training data predominantly reflects a particular set of values or perspectives, the LLM may exhibit bias by promoting or reinforcing those values in its responses and evaluations (Ferrara, 2023). This can result in marginalized groups experiencing their values and attitudes being disregarded or undervalued, leading to a lack of inclusivity and fair representation. Recognizing and mitigating these biases is essential to foster a more equitable and diverse learning environment that respects and incorporates many values and perspectives.

Misinformation and Manipulation

LMMs can be used to misinform and manipulate due to the need for more transparency surrounding their operations. The inner workings of LMMs are often complex and not easily understandable to the average user, making it challenging to discern the sources and reliability of the information they generate. This opacity allows malicious actors to exploit the technology by disseminating false or biased educational content without proper accountability. Furthermore, it is essential to consider the potential implications of LLM utilization, including the risk of "political manipulation," as Beerbaum (2023, p. 12) highlighted in a software assessment. This observation underscores the need to critically examine how LLMs can be influenced or biased, intentionally or unintentionally, by political agendas or external factors. By recognizing and addressing such risks, educational institutions can take proactive measures to safeguard the integrity and objectivity of the educational experience, ensuring that LLMs are used responsibly and ethically to foster a balanced and unbiased learning environment.

Dependency and Reliance

Value alignment theory underscores the intrinsic significance of promoting individual autonomy and critical thinking skills, particularly when confronted with the potential dangers arising from an excessively narrow and exclusive perception of reality derived solely from Language and Learning Models (LLMs). Relying solely on LLMs as the primary source of information can restrict students' exposure to diverse perspectives and alternative sources of knowledge, consequently hindering their ability to develop a well-rounded understanding of the world (Degaa, I., 2023). There is a need to emphasize individual autonomy and cultivate critical thinking abilities. By encouraging autonomous decision-making and promoting critical analysis, individuals can navigate the complexities of the modern world more effectively. However, an overreliance on LLMs can lead to a myopic and singular view of reality. Although LLMs possess impressive information processing capabilities, they have inherent limitations in comprehending and representing the entirety of human knowledge. Thus, exclusive reliance on LLMs limits exposure to diverse perspectives, inhibiting comprehensive understanding development. Therefore, it is crucial to recognize the potential drawbacks and actively seek out alternative sources of knowledge to ensure a panoramic educational experience (Degaa, I., 2023).

Devaluation of Human Expertise

Integrating Large Language Models (LLMs) in education poses a significant danger: the potential devaluation of human teachers. As LLMs take on tasks traditionally performed by educators, there is a risk of undermining the importance and expertise of teachers in the learning process (Bozkurt et al., 2023). By automating content creation, grading, and other instructional tasks, LLMs may create a perception that teachers are replaceable or unnecessary. This can lead to a loss of the invaluable aspects that human teachers bring to education, such as their ability to provide personalized support, mentorship, and social-emotional guidance to students. The danger lies in the potential for a standardized and impersonalized educational experience, which overlooks teachers' multifaceted role in fostering mentorship, social support, and modeling, which are much needed in the student's development.

Plagiarism

Plagiarism, a serious and obvious problem in education, is further exacerbated by the integration of LLMs. LLMs provide students with easy access to a vast amount of pre-generated text, increasing the temptation to present this content as their work. The ease of plagiarism facilitated by LLMs poses a significant threat to academic integrity and undermines the principles of originality, critical thinking, and knowledge acquisition. The prevalence of plagiarism not only hampers students' personal and intellectual growth but also undermines the credibility and fairness of educational assessments and evaluations.

The Gap in Education

Adapting the United Nations Educational, Scientific and Cultural Organization's (UNESCO) perspective, it is crucial to recognize the potential drawbacks of LLMs, such as AI-powered chatbots, in education, particularly their ability to widen technological and educational disparities. UNESCO emphasizes the importance of responsibly and moderately utilizing LLMs to bridge existing gaps and reduce inequalities. According to UNESCO's Education 2030 Agenda, a human-centered approach to LLMs is essential, as it can address disparities in knowledge access, research, and cultural diversity. The goal is to ensure that LLMs do not exacerbate technological divisions within and between countries. By adopting a responsible approach, LLMs can be leveraged to promote inclusivity and equity in education, safeguarding against widening technological and educational divides.

While it is essential to acknowledge the positive intentions of LLM developers, it is imperative to recognize that integrating these models into education can give rise to unanticipated side effects. Given LLMs' inherent complexity and intricacy, developers may need to be fully aware of all the potential ethical threats that can emerge. Hence, a cautious and vigilant approach is necessary to carefully assess and address any unintended consequences, ensuring the ethical implementation of LLMs in education and safeguarding the well-being and development of learners.

It is essential to acknowledge that many Large Language Models (LLMs) are developed and deployed by for-profit organizations, and their institutional values can have implications for the model's behavior and outputs. These organizations have their own priorities, including financial success, user engagement, and other business objectives. As a result, the LLM's training and fine-tuning processes may be influenced by these values, potentially shaping the model's biases, preferences, and limitations. This is a main concern for Value Alignment.

The institutional values of for-profit organizations can inadvertently introduce biases into LLMs. The models learn from vast amounts of internet text, which can contain inherent biases present in the source material. If not carefully addressed, these biases can be perpetuated or amplified by the LLM during response generation. Additionally, the prioritization of user engagement and business goals may lead to the model generating sensational or attention-seeking responses, rather than focusing solely on accuracy or providing balanced information.

Another consideration is the potential impact of the business model on access and availability of LLMs. Some organizations may choose to restrict or monetize access to their models, making them less accessible for certain communities or use cases. This can exacerbate existing disparities and inequalities in access to cutting-edge language technologies.

Users and developers should be aware of the institutional values that may influence the models and actively engage in efforts to mitigate biases, promote fairness, and ensure equitable access to these powerful language technologies.

4. SEIZING THE POTENTIAL: EXPLORING THE EDUCATIONAL OPPORTUNITIES OF LLMS WHILE ADDRESSING ETHICAL CONSIDERATIONS

LLMs have emerged as powerful tools with immense potential to revolutionize education. After thoroughly examining the ethical possibilities and threats associated with their integration, it becomes evident that LLMs are here to stay and can play a transformative role in shaping the future of education (Kasneci et al., 2023). While ethical concerns must be addressed, it is crucial to recognize that LLMs offer exciting opportunities for enhancing teaching and learning experiences. By understanding and embracing these opportunities, we can harness the full potential of LLMs while upholding ethical standards, promoting learner well-being, and nurturing their development. This section explores the opportunities and ethical challenges Large Language Models (LLMs) present in education. The discussion is divided into two parts, examining the roles of school administration and leadership and proposing strategies for teachers in the classroom to integrate LLMs effectively into teaching and lesson planning.

The first part emphasizes the importance of specific data privacy policies, transparent data practices, and obtaining informed consent from students and their families. It explores strategies for promoting inclusivity, addressing algorithmic bias, and safeguarding students' autonomy and privacy rights within the educational ecosystem.

The second part provides practical approaches for teachers seamlessly incorporating LLMs into their instructional delivery and lesson planning. It highlights the benefits of leveraging LLMs to enhance teaching, foster critical thinking, and support personalized learning experiences. Additionally, it emphasizes the need to maintain a balanced approach that values human expertise and creates a well-rounded educational environment, mitigating the risks associated with overreliance on LLMs.

By embracing responsible utilization of LLMs and aligning practices with value alignment theory, the potential of these powerful tools can be harnessed to create an inclusive, engaging, and effective educational experience for all stakeholders. This section presents actionable strategies to address the ethical challenges of LLMs in education, ensuring their responsible and beneficial integration.

Role of School Administrators

As educational leaders, administrators play a crucial role in shaping policies, fostering inclusivity, and protecting students' rights and Privacy. This part focuses on the specific responsibilities of school administrators in effectively incorporating LLMs into education. By recognizing the multifaceted nature of their role and providing practical guidance, this section aims to empower school leaders to navigate the ethical possibilities and challenges associated with LLM integration while ensuring a positive and equitable learning experience for all students. Here are some actionable steps administrators and their administrative teams can take to integrate LLMs into their educational institutions effectively.:

Strengthen Data Privacy

Establish clear protocols and policies to safeguard students' personal information and ensure compliance with data protection regulations. This includes obtaining explicit consent from students and their parents or guardians for data collection and analysis. To prevent unauthorized access, it is essential to enforce stringent security measures, including the implementation of encryption and secure storage systems, to safeguard sensitive data effectively (Brown et al., 2022). Review and update privacy practices regularly to align with evolving regulations and best practices.

Mitigate Algorithmic Bias

Firstly, curating diverse training data by collaborating with educators, students, and community members ensures that LLMs learn from various socio-economic and cultural backgrounds. Secondly, conducting regular bias audits and evaluations with the help of experts helps identify and address any biases present in the LLMs. This may involve retraining the models with updated data or fine-tuning algorithms to promote fair representation and evaluation. Lastly, promoting student engagement and feedback empowers students to voice their concerns and suggestions, actively involving them in improving LLM outputs and fostering an inclusive and student-centered learning environment.

Foster Media Literacy

Practical actions can be taken to foster media literacy in a school setting and empower students to navigate the challenges of misinformation and manipulation facilitated by LLMs. It integrates media literacy education into the curriculum, with dedicated lessons on identifying reliable sources, fact-checking techniques, and understanding bias. It provides students with opportunities to critically analyze and evaluate information generated by LLMs, encouraging them to question and verify the accuracy and credibility of the content. This can be done through guided discussions, research projects, and collaborative activities. It promotes responsible digital citizenship by teaching students about responsible online behavior, ethical use of technology, and the importance of sharing information responsibly. This can involve creating guidelines for reliable online research and encouraging students to be responsible digital content creators.

Encourage Diverse Learning Sources

Firstly, diversify the curriculum by incorporating materials from various authors, cultures, and perspectives. This can be achieved through a thoughtful selection of textbooks, literature, and supplementary resources that reflect diverse voices. Secondly, provide access to various learning resources such as libraries, online databases, and multimedia platforms offering rich educational materials. Encourage students to explore and engage with these resources to broaden their understanding of different subjects. Lastly, foster critical thinking skills by guiding students to analyze and compare information from multiple sources, encouraging them to question assumptions and biases, and facilitating class discussions that promote respectful dialogue and the exchange of diverse viewpoints.

Value Human Expertise

Encourage open and inclusive classroom discussions where students can engage in meaningful dialogues with their peers and educators. Create a supportive environment that values diverse perspectives, encourages critical analysis, and promotes student viewpoints development. Provide opportunities for one-on-one mentoring and guidance from educators. This can involve regular check-ins, personalized assignment feedback, and individualized support to nurture students' critical thinking skills and foster intellectual growth. Organize workshops and professional development sessions for educators to enhance their instructional strategies and facilitate rich discussions. Encourage educators to incorporate interactive teaching methods, encourage student-led conferences, and provide guidance on effectively integrating LLMs into their lessons while ensuring human interaction remains at the core of the learning experience.

Promote Academic Integrity

Provide comprehensive education on the ethical implications of plagiarism, including specific discussions on the challenges LLMs pose. Engage students in conversations about the importance of originality, critical thinking, and the responsible use of technology in their academic work. Integrate lessons and assignments emphasizing proper citation practices and teach students how to attribute sources in their research and writing effectively. Offer guidance on using citation tools and provide resources to help students navigate the complexities of referencing digital content generated by LLMs. Foster a culture of academic integrity by celebrating and recognizing originality and creativity in student work. Establish clear policies and consequences for plagiarism and create opportunities for students to showcase their unique perspectives and insights through projects, presentations, and discussions.

Enhance Transparency

As an institution, advocate for developers and organizations to provide clear and comprehensive documentation on LLM operations, including details about the model architecture, training data, and the decision-making processes involved. This information should be easily accessible and understandable to users, enabling them to make informed judgments about the reliability and biases of the generated content. Promote transparency through open dialogue and collaboration between developers, educators, and students. Encourage opportunities for developers to e engage with the education community, such as hosting workshops or webinars to address questions and concerns. Additionally, establish channels for feedback and reporting to encourage users to highlight issues and seek clarification regarding the inner workings of LLMs. Advocate for developing and adopting standardized ethical guidelines and evaluation frameworks for LLMs in education. These frameworks can provide benchmarks for transparency and guide the responsible and honest implementation of LLMs, ensuring users have the necessary information to assess the reliability and trustworthiness of the generated content.

Engage Stakeholders

Establish regular forums, such as conferences or focus groups, where all stakeholders can share their perspectives, concerns, and ideas about integrating LLMs. This open and inclusive dialogue will foster a sense of ownership and collaboration among stakeholders. Create opportunities for students and educators to participate in the evaluation and testing of LLMs actively. This can involve piloting LLM-based educational activities, soliciting feedback on usability, effectiveness, and ethical considerations, and incorporating student and educator input into the decision-making process. Encourage research and academic institutions to conduct studies and research projects examining LLMs' impact on teaching and learning. These studies can provide valuable insights and evidence-based recommendations, ensuring that decisions regarding using LLMs are informed by rigorous research and the experiences of those directly involved in the educational process.

Provide Ethical Guidelines and Regulations

Collaborate with relevant stakeholders, such as educators, administrators, policymakers, and legal experts, to develop comprehensive ethical guidelines for integrating LLMs. These guidelines should address critical concerns, including data privacy, algorithmic bias, transparency, and accountability. Integrate ethical education and training programs for educators and students, ensuring they know the guidelines and regulations, understand their implications and have the necessary knowledge and skills to implement them in their daily practices. This can include workshops, seminars, and online resources that promote awareness and provide practical guidance on ethical LLM usage. Create a system for monitoring and enforcing adherence to ethical guidelines and regulations to guarantee compliance with ethical standards. This can involve regular audits, assessments, and evaluations of LLM usage in educational institutions and consequences for non-compliance.

Continuous Monitoring and Evaluation

Establish a dedicated team or committee responsible for monitoring and evaluating the use of LLMs in education. This team should comprise educators, administrators, researchers, and students, ensuring diverse perspectives and expertise. They can conduct regular assessments, surveys, and interviews to gather feedback and insights from stakeholders regarding the ethical implications and effectiveness of LLM integration. Collaborate with external organizations, such as research institutions or ethics committees, to conduct independent evaluations and audits of LLM usage in the school. These external assessments can provide valuable insights and recommendations for improvement. Use the feedback and research findings to drive evidence-based decision-making and adapt LLM practices accordingly. This can include revising policies, updating training programs, and implementing additional safeguards to address emerging ethical concerns. By embracing a continuous monitoring and evaluation culture, schools can proactively address ethical challenges, optimize the benefits of LLM integration, and ensure these powerful tools' responsible and sustainable use in education.

Schools must adopt a proactive and responsible approach, taking deliberate steps to address these challenges. These lines of actions, when implemented thoughtfully, can ensure that the potential of LLMs is harnessed while safeguarding learner well-being, upholding ethical standards, and preserving the essential role of human educators. By striking a careful balance between innovation and ethics, we can shape a future where LLMs serve as transformative tools that empower learners, foster critical thinking, and create inclusive and engaging educational experiences.

Furthermore, as schools adopt a proactive and responsible approach to LLM integration, it is imperative to consider the broader societal implications and ensure alignment with core values. The ethical implementation of LLMs in education extends beyond individual schools and encompasses the collective responsibility of society. By incorporating value alignment into the discourse surrounding LLMs, we can foster a shared understanding of the ethical considerations and establish a framework that reflects the diverse perspectives and values of the community.

This alignment of values is a guiding principle in navigating the ethical challenges of LLM integration. It encourages open dialogue, collaboration, and engagement among stakeholders, including students, educators, researchers, policymakers, and community members. By involving these diverse voices in decision-making processes, we can ensure that the integration of LLMs in education aligns with the aspirations and values of the society it serves.

Moreover, this emphasis on value alignment in LLM integration facilitates a holistic approach beyond technical considerations. It encourages the exploration of broader questions, such as the impact on social justice, equity, and cultural preservation. Through a thorough analysis of the potential effects of LLM utilization on these aspects, educational institutions can actively tackle unintended biases or discriminatory outcomes, thereby promoting an educational environment that is inclusive and fair for all students.

One concrete contribution that can help in the alignment of values is by the use of reinforcement learning from humans to fine-tune Large Language Models (LLMs) parameters for educational purposes that involves incorporating human feedback to improve the model's responses in specific contexts. This process aims to make LLMs more suitable for educational settings and align them with desired learning outcomes. Human feedback, in the form of ratings, rankings, or comparative evaluations, is collected to assess the quality, appropriateness, and relevance of the model's responses. Reinforcement learning is then applied to optimize the LLM's parameters based on this feedback. The collected human feedback serves as a reward signal, guiding the model towards generating more desirable responses over time. This iterative process helps the LLM adapt and refine its behavior, enhancing its understanding of educational topics and pedagogical strategies.

However, there are challenges associated with reinforcement learning from humans. Acquiring diverse and reliable human feedback can be resource-intensive and time-consuming. It is crucial to ensure a balanced representation of human perspectives and prevent the reinforcement of pre-existing biases present in the training data. Despite these challenges, reinforcement learning from humans has the potential to significantly improve LLMs for educational purposes. By incorporating human expertise and judgment, these models can better support learning, provide accurate information, and generate appropriate responses in educational contexts, ultimately enhancing their effectiveness as educational tools.

Role of Educators in the Classroom

In the dynamic landscape of education, we must shift our emphasis from solely evaluating students' output to actively observing and supporting their learning processes. By prioritizing learning processes over output, education can empower students to become independent learners, fostering critical thinking skills and advancing ethical values.

Promoting an Assistive Learning Process

Teachers can shift their focus from solely evaluating the output of students' learning to actively observing and supporting their learning process. Teachers can facilitate personalized guidance and feedback throughout the learning journey by leveraging LLMs as additional tools. They can engage with students, encouraging critical thinking, promoting metacognitive skills, and providing support tailored to individual needs. For example, the teacher can leverage LLMs to provide students with additional resources and perspectives on World War II during the lesson. Instead of solely evaluating the final research paper or presentation, the teacher can actively engage with students throughout the process. They can encourage students to share their progress, ask thought-provoking questions, and provide guidance. By observing and supporting the learning process, the teacher ensures students receive personalized feedback and develop a deeper understanding of historical events. A specific example of this is the work of Kung et al. (2023) in applying assistive learning in medicine.

Design Lessons for Inquiry and Assessment

Teachers may explore moving beyond traditional lesson planning and embrace lesson design that integrates LLMs as resources. They can create learning experiences that stimulate inquiry, critical analysis, and problem-solving. By designing lessons that incorporate opportunities for students to assess information generated by LLMs, evaluate its credibility, and develop metacognitive skills related to information literacy, teachers encourage students to question, analyze, and engage in reflective assessment of the information they encounter. In designing the lesson, for instance, the teacher can incorporate LLM-generated articles, primary sources, and speeches related to World War II. Students can be divided into small groups and assigned specific topics or events to research using traditional and LLM-generated resources. As part of the assessment, the teacher can guide students to critically evaluate the information from LLMs by discussing potential biases, comparing different sources, and considering the reliability and credibility of the information. This encourages students to develop information literacy skills and evaluate historical events from multiple perspectives.

Foster Active Engagement and Metacognition

Teachers can foster an active and participatory learning environment where students engage with LLMs as tools for exploration and inquiry. Teachers promote metacognition by encouraging students to become active participants in asking questions, seeking clarification, and reflecting on their understanding. Additionally, teachers can guide students in developing critical thinking skills to evaluate LLM-generated information and consider multiple perspectives. To foster active engagement and metacognition, the teacher can organize a classroom discussion or debate where students analyze the causes and impacts of World War II. The teacher can use LLM-generated content as prompts to encourage critical thinking and facilitate discussion. Throughout the activity, the teacher can prompt students to reflect on their understanding, ask open-ended questions, and consider alternative viewpoints. This approach encourages students to actively engage with LLM-generated information, think critically, and develop their metacognitive skills in understanding the complexities of historical events.

Some classroom activities that can be implored to apply these three methodologies in teaching are (a) reflective journals; (b) learning logs; (c) peer feedback sessions; (d) community research projects; (e) debates and discussions; (f) case studies; (g) Socratic seminars; (h) collaborative projects.

Emphasizing the learning process over output is crucial for empowering students as individual learners and promoting ethical values within education. By recognizing the significance of the learning journey, students become active participants, capable of critical thinking, problem-solving, and ethical decision-making. Integrating the principles of Value Alignment Theory when using LLMs ensures that their integration aligns with ethical principles and promotes responsible and beneficial use. Education should prioritize the learning process, empowering students to become lifelong learners who can navigate the world's complexities with a solid ethical foundation.

As educators navigate the terrain of LLM, it becomes ever more apparent that paramount to prioritize ethical considerations and ensure the responsible use of LLMs in the classroom. Ethical principles, transparency, accountability, and the promotion of critical thinking are essential aspects that guide the effective integration of LLMs. Here are guidelines for the ethical and responsible use of LLMs in education aimed at empowering both educators and students:

Assess Algorithmic Bias

Educators emphasize the importance of learners critically evaluating LLM-generated content for biases or favoritism. By actively assessing and questioning the potential biases within the content, learners can contribute to creating an inclusive and equitable learning environment. Educators encourage learners to develop the ability to identify and mitigate biases, ensuring fair representation and diverse perspectives.

Foster Critical Thinking

Educators recognize the need to cultivate learners' critical thinking skills when engaging with LLM-generated information. Learners should be encouraged to analyze the credibility, accuracy, and limitations of the content they encounter. Through guided instruction and practice, educators can empower learners to evaluate the reliability of LLM-generated information and develop the capacity to verify information from multiple sources. This fosters a discerning mindset and equips learners with the skills necessary to navigate the complexities of the digital landscape.

Promote Transparency and Accountability: Educators stress the significance of learners actively monitoring LLM performance and evaluating the reliability of the generated content. By encouraging learners to assess the information provided by LLMs critically, educators promote transparency and accountability in utilizing these models. Learners should feel empowered to voice concerns and limitations transparently, fostering a culture of open dialogue and constructive feedback.

Safeguard Privacy and Data Protection

Educators recognize that learners are responsible for upholding Privacy and data protection principles when utilizing LLMs. Learners should obtain explicit consent and handle student data securely, aligning with privacy laws and regulations. Educators emphasize the importance of clear communication with learners and parents regarding data privacy practices, ensuring LLMs' ethical and responsible use.

Continuous Professional Development

Educators emphasize the need for learners to engage in continuous professional development to navigate the evolving landscape of education and LLM technologies. Learners should proactively stay informed about LLM developments, algorithmic biases, privacy concerns, and ethical considerations. By engaging in ongoing learning and updating their knowledge regularly, learners can adapt to emerging challenges and leverage the full potential of LLMs in their educational journey. Educators are crucial in providing resources, guidance, and opportunities for learners to engage in continuous professional development, empowering them to become informed and responsible users of LLMs.

Prioritize Ethical Considerations

Reflect on potential risks, biases, and privacy concerns associated with LLM integration. Ensure LLM usage aligns with ethical principles, fairness, transparency, and accountability.

5. CONCLUSION

In conclusion, this paper has explored the ethical possibilities, threats, and educational opportunities arising from the integration of technology within the field of education, with a specific focus on its relevance to an LLM program. The findings underscore the transformative potential of technological advancements in enhancing educational experiences and improving student outcomes. However, considering the complex ethical implications involved, this potential must be harnessed responsibly.

As LLM educational programs strive to cultivate future leaders equipped to navigate the dynamic educational landscape, integrating technology becomes increasingly crucial. Examining personalized learning, automated assessment and feedback, language support and translation, support for special needs education, professional development for educators, and chatbots as learning support illustrate the diverse possibilities technology affords in the educational context.

Nevertheless, it is paramount to acknowledge and address the complex ethical challenges inherent in this integration, particularly within the framework of an LLM program. Privacy concerns arise due to the extensive data generated by technology-enabled educational systems, necessitating rigorous safeguards to protect sensitive information. Algorithmic bias becomes a pressing issue, demanding an unwavering commitment to mitigating biases and promoting fairness and equity. Misinformation and manipulation underscore the need for critical assessment and media literacy skills, enabling learners to distinguish between reliable and misleading sources of information. The overreliance on technology raises questions about preserving human expertise and judgment, emphasizing the importance of maintaining a balance between technological tools and human agency. Additionally, plagiarism calls for robust measures to ensure academic integrity in the digital age. Lastly, the existence of a digital divide and unequal access to technology emphasizes the imperative of addressing disparities and promoting equitable opportunities for all learners.

Administrators and teachers must prioritize certain principles to navigate these ethical challenges within an LLM program. Administrators play a vital role in establishing a framework that strengthens data privacy, mitigates algorithmic bias, fosters media literacy, encourages diverse learning sources, values human expertise, promotes academic integrity, enhances transparency, engages stakeholders, establishes ethical guidelines and regulations, and implements continuous monitoring and evaluation. Concurrently, teachers in an LLM program must embrace a pedagogical approach that prioritizes the learning process over mere output, fostering critical thinking, creativity, and problem-solving skills. They should emphasize values over mere information, cultivating ethical awareness, empathy, and responsible digital citizenship among students.

By synthesizing these insights and integrating a focus on 'process over output' and 'values over mere information,' an LLM program can effectively prepare future educational leaders to navigate the complex landscape of technology integration with a solid ethical foundation. This approach encourages learners to engage in profound learning experiences, where knowledge acquisition is coupled with developing critical thinking, ethical decision-making, and a sense of social responsibility. Through a commitment to these principles, an LLM program can ensure that graduates possess the skills and values necessary to leverage technology for positive educational transformation while safeguarding against the ethical risks and challenges that arise.

REFERENCES

Aljanabi, M., & Chat, G. P. T. (2023). ChatGPT: Future directions and open possibilities. *Mesopotamian Journal of Cyber Security*, *2023*, 16–17. doi:10.58496/MJCS/2023/003

Beerbaum, D. (2023). Generative Artificial Intelligence (GAI) Software – Assessment on Biased Behavior. SSRN *Electronic Journal*. doi:10.2139/ssrn.4386395

Bozkurt, A., Xiao, J., Lambert, S., Pazurek, A., Crompton, H., Koseoglu, S., Farrow, R., Bond, M., Nerantzi, C., Honeychurch, S., Bali, M., Dron, J., Mir, K., Stewart, B., Costello, E., Mason, J., Stracke, C., Romero-Hall, E., Koutropoulos, A., & Toquero, C. (2023). Speculative Futures on ChatGPT and Generative Artificial Intelligence (AI): A Collective Reflection from the Educational Landscape. *Asian Journal of Distance Education*, *18*(1), 53. https://eprints.gla.ac.uk/294292/1/294292.pdf

Brown, H., Lee, K., Mireshghallah, F., Shokri, R., & Tramèr, F. (2022*). What Does it Mean for a Language Model to Preserve Privacy?* arXiv. https://arxiv.org/pdf/2202.05520.pdf

Cao, Y., Zhou, L., Lee, S., Cabello, L., Chen, M., & Hershcovich, D. (n.d.). *Assessing Cross-Cultural Alignment between ChatGPT and Human Societies: An Empirical Study*. arXiv. https://arxiv.org/pdf/2303.17466.pdf

Chu, H., & Liu, S. (2023). *Can AI tell good stories? Narrative Transportation and Persuasion with ChatGPT*.

Dai, W., Lin, J., Jin, F., & Chen, G. (2023, April 25). *Can Large Language Models Provide Feedback to Students? A Case Study on ChatGPT*. ResearchGate. https://www.researchgate.net/publication/370228288_Can_Large_Language_Models_Provide_Feedback_to_Students_A_Case_Study_on_ChatGPT

Dergaa, I., Chamari, K., Zmijewski, P., & Saad, H. B. (2023). From human writing to artificial intelligence generated text: Examining the prospects and potential threats of ChatGPT in academic writing. *Biology of Sport*, *40*(2), 615–622. doi:10.5114/biolsport.2023.125623 PMID:37077800

Eysenbach, G. (2023). The role of ChatGPT, generative language models, and artificial intelligence in medical education: A conversation with ChatGPT and a call for papers. *JMIR Medical Education*, *9*(1), e46885. doi:10.2196/46885 PMID:36863937

Ferrara, E. (2023). *Should chatbots be biased? Challenges and risks of bias in large language models*. arXiv preprint arXiv:2304.03738.

Haque, M. U., Dharmadasa, I., Sworna, Z. T., Rajapakse, R. N., & Ahmad, H. (2022). "I think this is the most disruptive technology": Exploring sentiments of ChatGPT early adopters using Twitter data. https://doi.org//arXiv.2303.03836. doi:10.48550

HerboldS.Hautli-JaniszA.HeuerU.KiktevaZ.TrautschA. (2023). AI, write an essay for me: A large-scale comparison of human-written versus ChatGPT-generated essays—arXiv preprint arXiv:2304.14276.

Hong, W. C. H. (2023). The impact of ChatGPT on foreign language teaching and learning: Opportunities in education and research. *Journal of Educational Technology and Innovation*, *5*(1).

Jiao, W., Wang, W., Huang, J. T., Wang, X., & Tu, Z. (2023). *Is ChatGPT a good translator? A preliminary study*. arXiv preprint arXiv:2301.08745.

Kasneci, E., Seßler, K., Küchemann, S., Bannert, M., Dementieva, D., Fischer, F., Gasser, U., Groh, G., Günnemann, S., Hüllermeier, E., Krusche, S., Kutyniok, G., Michaeli, T., Nerdel, C., Pfeffer, J., Poquet, O., Sailer, M., Schmidt, A., Seidel, T., & Kasneci, G. (2023). ChatGPT for good? On opportunities and challenges of large language models for education. *Learning and Individual Differences*, *103*, 102274. doi:10.1016/j.lindif.2023.102274

Khowaja, S. A., Khuwaja, P., & Dev, K. (2023). ChatGPT Needs SPADE (Sustainability et al. divide, and Ethics) Evaluation: A Review. arXiv preprint arXiv:2305.03123.

Kung, T. H., Cheatham, M., Medenilla, A., Sillos, C., De Leon, L., Elepaño, C., Madriaga, M., Aggabao, R., Diaz-Candido, G., Maningo, J., & Tseng, V. (2023). Performance of ChatGPT on USMLE: Potential for AI-assisted medical education using large language models. *PLOS Digital Health*, *2*(2), e0000198. doi:10.1371/journal.pdig.0000198 PMID:36812645

Mhlanga, D. (2023). Open AI in education, the responsible and ethical use of ChatGPT towards lifelong learning. *Education, the Responsible and Ethical Use of ChatGPT Towards Lifelong Learning*.

Montemayor, C. (2023). *The Prospect of a Humanitarian Artificial Intelligence: Agency and Value Alignment*.

Peng, K., Ding, L., Zhong, Q., Shen, L., Liu, X., Zhang, M., & Tao, D. (2023). *Towards making the most of chatbot for machine translation*. arXiv preprint arXiv:2303.13780.

Rahman, M., Terano, H. J. R., Rahman, N., Salamzadeh, A., & Rahaman, S. (2023). ChatGPT and Academic Research: A Review and Recommendations Based on Practical Examples. *Journal of Education. Management and Development Studies*, *3*(1), 1–12.

Rahman, M. M., Terano, H. J., Rahman, M. N., Salamzadeh, A., & Rahaman, M. S. (2023). *ChatGPT and academic research: a review and recommendations based on practical examples*.

Ray, P. P. (2023). ChatGPT: A comprehensive review of the background, applications, key challenges, bias, ethics, limitations, and future scope. *Internet of Things and Cyber-Physical Systems*.

Shen, Y., Heacock, L., Elias, J., Hensel, K. D., Reig, B., Shih, G., & Moy, L. (2023). ChatGPT and other large language models are double-edged words. *Radiology*, *307*(2), e230163. doi:10.1148/radiol.230163 PMID:36700838

Tlili, A., Shehata, B., Adarkwah, M. A., Bozkurt, A., Hickey, D. T., Huang, R., & Agyemang, B. (2023). What if the devil is my guardian angel: ChatGPT is a case study of using chatbots in education. *Smart Learning Environments*, *10*(1), 15. doi:10.118640561-023-00237-x

Unesco.org. https://unesdoc.unesco.org/ark:/48223/pf0000245656

van Schalkwyk, G. (2023). Artificial intelligence in pediatric behavioral health. *Child and Adolescent Psychiatry and Mental Health*, *17*(1), 1–2. doi:10.118613034-023-00586-y PMID:36907862

Winter, C., Hollman, N., & Manheim, D. (2023). Value alignment for advanced artificial judicial intelligence. *American Philosophical Quarterly*, *60*(2), 187–203. doi:10.5406/21521123.60.2.06

ADDITIONAL READING

Chi, M., & Goodman, P. S. (Eds.). (2018). *Language Models for Educational Applications: Expanding the Frontier*. Springer.

Davenport, T. H. (2018). *The AI Advantage: How to Put the Artificial Intelligence Revolution to Work*. The MIT Press. doi:10.7551/mitpress/11781.001.0001

Du Boulay, B., Koedinger, K. R., & Mulder, R. (Eds.). (2015). *Machine Learning and Education: Leveraging the AI Revolution*. Routledge.

Ferster, B. (2014). *Teaching Machines: Learning from the Intersection of Education and Technology*. The MIT Press. doi:10.1353/book.36140

Gasevic, D., Kovanovic, V., Jovanovic, J., & Dawson, S. (2019). *Artificial Intelligence and Learning Analytics in Education*. Springer.

Margaryan, A., Mavrikis, M., & Bransford, J. (2017). *Education and Artificial Intelligence: Open Learning Environments*. Routledge.

Nkambou, R., Mizoguchi, R., & Bourdeau, J. (Eds.). (2019). *Artificial Intelligence in Education: Promises and Implications for Teaching and Learning*. Springer.

Rosé, C. P., Martínez-Maldonado, R., Hoppe, H. U., Luckin, R., Mavrikis, M., & Porayska-Pomsta, K. (Eds.). (2018). *Artificial Intelligence in Education: 19th International Conference*. Springer.

Sobh, T. M., & Elleithy, K. (Eds.). (2006). *Artificial Intelligence in Education: Challenges and Opportunities*. Springer.

Zimmerman, M. (2018). *Teaching AI: Exploring New Frontiers for Learning*. Elevate Books Edu.

KEY TERMS AND DEFINITIONS

Algorithmic Bias: This occurs when computer algorithms make unfair or discriminatory decisions, often due to partial data or flawed programming, leading to unequal treatment or disadvantages for specific individuals or groups.

ChatGPT: ChatGPT is a significant language model that can generate human-like text responses and engage in conversational interactions, making it a valuable tool for interactive learning and communication.

Human-Machine Collaboration: Human-machine collaboration involves working with machines, such as AI systems, to complement and enhance human abilities in educational settings, fostering collaboration and synergy between humans and technology.

LLM, Large Language Model: This refers to advanced models like GPT-3 designed to generate human-like text responses and have a wide range of applications in natural language processing tasks, including education.

Personalized Learning: Personalized learning is an educational approach that tailors instruction and learning experiences to meet the individual needs, interests, and learning styles of students, utilizing technology and data analysis to provide customized content and support.

Process-Based Education: Process-based education focuses on the learning process rather than just the results, emphasizing critical thinking, problem-solving, and metacognitive skills development, fostering active engagement and a deeper understanding of concepts.

Value Alignment: In the context of ChatGPT and education, it refers to ensuring that the responses and interactions generated by the AI model align with the values, ethics, and educational objectives of the educational institution or the desired learning outcomes. It involves configuring and fine-tuning the model to prioritize and reflect the educational values, promoting responsible and beneficial use of the technology in the learning environment.

Chapter 6
Artificial Intelligence in Tourism

Enrique Bigné
Universitat de Valencia, Spain

ABSTRACT

Artificial intelligence in tourism activities opens a bundle of emerging applications for tourists and companies. This chapter aims to delineate the stages of the tourist journey and the usage of four types of intelligence suggested in the literature: mechanical, analytical, intuitive, and empathetic. Based on these two ideas, the authors propose a useful framework for disentangling the different types of current and future applications of AI in tourism. Each stage involves multiple suppliers with different types of AI applications, and its adoption will ultimately rely on tourist trust and, therefore, willingness to share data and the use of robotics and other AI forms. The chapter ends with some trends and reflections on the expansion of AI in tourism that pivot around these ideas: job replacement and flexible operations; mobile-centric approach; data integration and analytics; revenue management and customer interactions tension; (v) neuroscientific tools for AI in tourism.

1. INTRODUCTION

Popular beliefs about the role of tourism often overlook the technological aspect, which is mainly due to tourism's association with leisure and vacation. However, this belief is far from reality, and some examples from the past may change this view. The tourism industry has been a leader in innovation in two ways: (i) the development of loyalty cards that were initially created for use in airlines; and (ii) the global distribution systems (GDS), referring to the worldwide computerized networks that provide multi-access to a single source database and that are used for booking and are associated with the modern platforms provided by the Internet. The adoption of artificial intelligence (AI) by leaders in the tourism and hospitality industries is no exception. The technologies available in airports, such as self-check-in, access to boarding gates through QR codes, and automatic passport control through facial recognition or the scanning of passports, are some of the AI tools visible to tourists. However, other uses of AI

DOI: 10.4018/978-1-6684-9591-9.ch006

are not as visible to customers, such as its use in revenue management, fuel optimization, maintenance alerts, and other company operations. Furthermore, other AI applications are still being developed, such as self-flying aircrafts, and Airbus has developed an automated, computer vision-based take-off and landing platform.

Based on previous contributions (Huang & Rust, 2018; Tussyadiah, 2020), I define AI herein as myriad human-computer processes that allow users and firms to add value to their interactions through data gathering and analysis, resulting in operations and actions that optimize customer-to-company (C2C) and business-to-business (B2B) interactions. This definition involves the interaction of the following three elements: devices, analysis, and availability. First, a technological device (e.g., smartphone, computer, sensor, scanner) captures, stores, and transmits data from users or the environment to other end users or devices through the Internet of Things (IoT). For instance, a temperature sensor in a space (e.g., a restaurant, hotel, or aircraft) can measure the current temperature, interpret it, and transmit this value to another device to trigger a planned action (e.g., turning an air conditioning unit on or off or opening a window). Second, data analysis is a predominant part of AI. For instance, AI can be used to track tourist movements and information, such as the number of visits to a tourist attraction, restaurant, or area, the number of comments and pictures posted on social media, and the number of nights spent at a destination or in a hotel. However, this numerical, textual, or visual information requires mathematical algorithms for its analysis, ranging from basic techniques (e.g., counting people and detecting emotions based on comments posted on social media) to more advanced machine learning (ML) techniques that are used to detect future patterns or suggest prescriptive actions. Lastly, the stored and analyzed information is delivered meaningfully to different recipients, such as organizations, companies, or users. For instance, the data can be used by destination management organizations (DMOs) to delineate tourist paths through a town in ways that preclude queuing, flight bookings can be used to predict the hotel accommodation services or restaurant supplies needed, or online hotel reviews can be used to predict occupancy rates.

This chapter aims to inform students, companies, researchers, and stakeholders about the application of AI in the tourism and hospitality industries. It is written from a user's perspective and through the lens of human-device interaction. Therefore, this framework combines two realms of research: (i) the tourist journey that encompasses the activities prior to traveling to a destination, at the destination, and those related to the journey after it is concluded; and (ii) the four types of intelligence used in services, as defined by Huang and Rust (2018). Figure 1 depicts a visual guide of my approach to combining the two realms of research, the supplier and user relationships (B2B and C2C), and the intensity of AI adoption in each relationship.

The rest of the chapter is structured as follows: I first address the multiple types of interrelated activities of tourism that affect AI, after which I analyze each stage of the tourist journey and the types of intelligence. I then discuss the future trends, risks, and dark sides of AI in tourism and hospitality, and lastly, I conclude with an analysis of the philosophical challenges faced by the increasing adoption of AI in terms of the ethical and societal implications.

Figure 1. The sources of data throughout the tourism journey and the types of intelligence

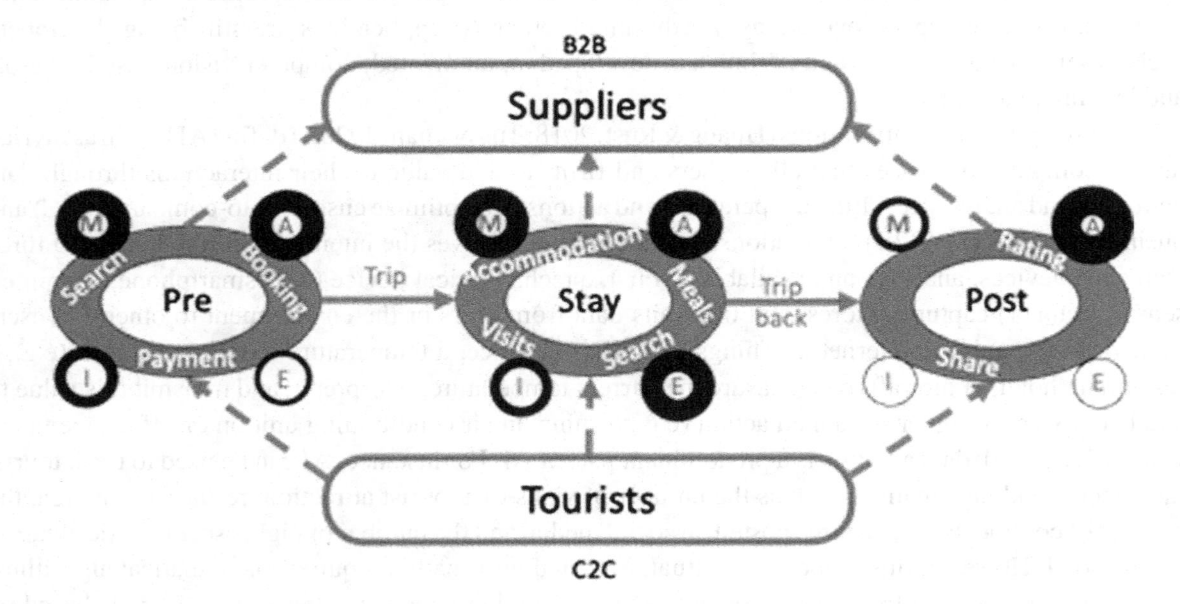

2. TOURISM AND HOSPITALITY AS COMPLEX AND INTERRELATED FIELDS

Tourism is a dynamic industry that has continuously grown over the last 40 years. In 2019, 1,494 million international arrivals were reported worldwide. The COVID-19 pandemic has dramatically affected travel, with international arrivals reaching 80% of pre-pandemic levels in the first quarter of 2023 (The United Nations World Tourism Organization [UNWTO], 2023). However, despite the pandemic, the adverse economic environment, and high inflation rates, the tourism industry is continuously growing and demonstrates dynamic activity worldwide.

Tourism involves a large range of activities, suppliers, and actors in several stages, including the start of a travel journey, during the trip, and at the end of the trip. The different activities within the tourism industry involve a variety of data sources, which are used for data analysis that is conducted for different purposes as well as data delivery to different stakeholders in tourist destinations that are inhabited by residents. As a result, the coordination and integration of different stakeholders should take this complex ecosystem into account. Figure 2 shows the five main components that affect AI in this context.

There are different types of travel motives that can range from holidays and meetings, incentives, conferences, and exhibitions (MICE) to different types of products (i.e., culture, leisure and entertainment, gastronomy, knowledge and training, health and wellness, sports and adventure, business, trade fairs) at assorted intensity levels at various destinations types (rural, urban, seaside). Thus, different user types, including residents, have several purposes at different destinations, and a wide range of applications for both tourists and residents provide data on the diverse types of users and their multipurpose goals. Therefore, data coordination and integration are necessary for destination planning.

Figure 2. The main components of the tourism industry that affect AI

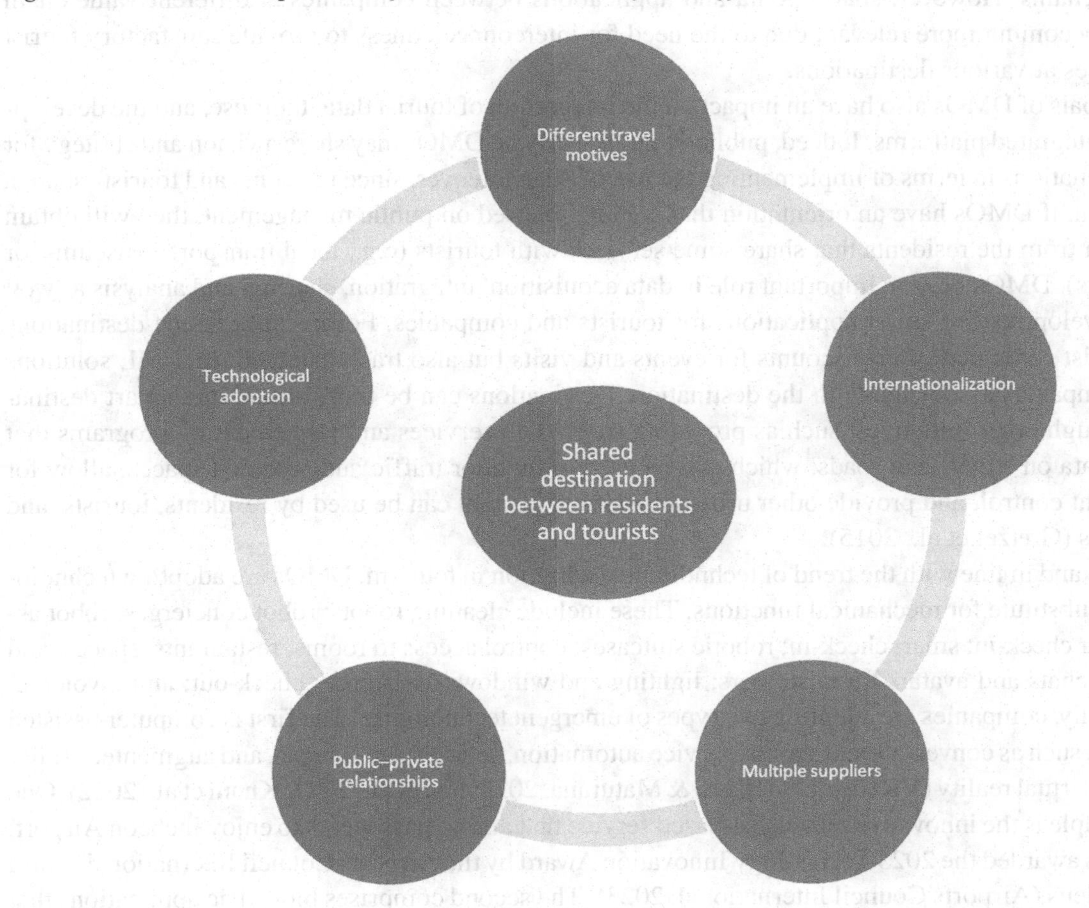

International arrivals from diverse origins are exceedingly common in many destinations, with international carriers and airports playing a role. Carriers, mainly comprising big companies such as Airbus, and airports provide many potential applications of AI due to the massive volume of users and the need for data recording in some of the processes, such as bookings and boardings. In fact, 34 airports around the world transported more than 50 million passengers each in 2019 (Airports Council International, 2020). Public bodies, such as the UNWTO, Airports Council International, and similar bodies as well as airlines that are part of the International Air Transport Association also offer massive amounts of data at the international level. Likewise, international passengers who speak different languages and have diverse backgrounds can urge providers to adopt smart automatic language translation services to ensure improved service delivery.

Multiple company types, including travel agencies, carriers, event planners, accommodation facilities, restaurants, museums, local transportation facilities, and other private or public actors, provide an integrated experience to tourists. Even though the actors on the supply side are of different natures and have various ownership structures, an integrated user experience is crucial for reaching a better understanding of tourists' claims and needs. Furthermore, some companies belong to international or national chains, which lends itself to data integration and these companies' developing their own AI solutions for their chains. However, sharing data and applications between companies at different value chain levels is becoming more relevant due to the need for interconnectedness to provide satisfactory tourist experiences at various destinations.

The goals of DMOs also have an impact on the integration of tourist data, their use, and the development of integrated platforms. Indeed, public or public–private DMOs may share a vision and strategy for their destinations in terms of implementing the use of AI. Moreover, since residents and tourists share a destination, if DMOs have an orientation that is more focused on public management, they will obtain more data from the residents that share some services with tourists (e.g., local transport, museums, or restaurants). DMOs play an important role in data acquisition, integration, sharing, and analysis as well as the development of smart applications for tourists and companies. For instance, some destinations offer tourist cards that offer discounts for events and visits but also track data and offer ML solutions to the companies associated with the destination. Destinations can be transformed into smart destinations through other initiatives, such as providing free Wi-Fi services and implementing programs that capture data on streets and roads, which can be used to monitor traffic and crowded spaces, allow for traffic light control, and provide other urban planning tools that can be used by residents, tourists, and companies (Gretzel et al., 2015).

Lastly, and in line with the trend of technological adoption in tourism, DMOs are adopting technologies as a substitute for mechanical functions. These include cleaning robots; robot concierges; robot assistants for check-in; smart check-in; robotic suitcases; control access to rooms, restaurants, spaces, and facilities; chats and avatars for customers; lighting and windows assistance; check-out; and invoicing. Additionally, companies are adopting two types of emergent technologies. The first is computer-assisted processes, such as conversational agents, service automation, self-service screens, and augmented reality (AR) and virtual reality (VR) devices (Bigne & Maturana, 2023; Fan et al., 2022; Knani et al., 2022). One such example is the innovative metaverse-based service that allows passengers to enjoy Incheon Airport, which was awarded the 2023 Technology Innovation Award by the Airports Council International World and Amadeus (Airports Council International, 2023). The second comprises biometric applications that are used to detect users' implicit and unconscious responses through eye tracking, facial readers, heart rate variability, and electroencephalography, among others (Ausin-Azofra et al., 2021; Bigne et al., 2023; Simonetti & Bigne, 2022). Despite some biometric measures being used at the laboratory level for research purposes, commercial use of eye-tracking devices and heart rate variability are becoming common in the market due to the use of sensors (for a review, see Kakaria et al., 2022).

3. TYPES OF AI AND TOURIST JOURNEYS

3.1. Intelligence Types

Based on Sternberg's (2005) idea of mimicking human intelligence, Huang and Rust (2018) delineated four types of intelligence: mechanical, analytical, intuitive, and empathetic. Each has value in the tourist journey (see Figure 1). Mechanical intelligence refers to the ability to perform routine and repetitive tasks that do not involve creativity (Huang & Rust, 2018). They can be categorized depending on the context, such as digital versus physical environments. There are multiple examples in the digital domain as only a URL link in the form of a website or an application is needed for users to click on the option or ask for the service they need. Thus, requesting taxis, booking essential services, buying tickets, and transferring money or making payments are standard mechanical services that require a program to deliver the requested service through the Internet. Likewise, intelligent search engines, such as Google or similar, that are embedded in booking sites are another example. In the physical world, delivery is sometimes needed through traditional systems but not in others. An example of such a mechanical task is an individual ordering packaged food or drinks online that are then delivered. However, others, such as lighting, do not involve physical delivery. Furthermore, these tasks affect the workforce because these processes can replace humans completing these tasks and are not subject to working hours.

Analytical intelligence refers to the ability to process information so as to solve problems and then learn from this endeavor (Huang & Rust, 2018). The data can be structured (e.g., numbers) or unstructured (e.g., text, images, videos, audio, gestures), which is becoming more relevant to companies (Balducci & Marinova, 2018). Information processing involves using mathematical tools that range from basic univariate and bivariate to multivariate and advanced algorithms. Although counting words, people, and visitors is of interest to actors in the tourism sector, the fundamental goal is to reveal the relationships between the variables. For example, the number of visitors at a museum might be related to external data, such as rainy conditions, which are the potential relationships that AI can identify. Determining these relationships will allow predictions to be made, such as predicting the number of visitors to a museum according to the type and intensity of rain, the duration of the tourists' visits, the origin of the visitors, and whether the visitors are residents versus tourists.

This chapter does not focus on the technical issues in the vast number of advanced algorithms and data analysis procedures. I acknowledge its relevance, but its analysis exceeds the scope of this chapter. This topic undoubtedly constitutes a promising developing area that is rooted in mathematics as ML methods encompass different statistical techniques such as artificial neural networks, support vector machines, genetic algorithms, or deep learning, but they also add instant and continuous computing based on a continuous data flow. The use of these techniques in tourism is becoming common. For example, Bigne et al. (2019) investigated the use of artificial neural networks in the analysis of tweets about several tourism destinations; Bigne, Fuentes-Medina et al. (2020) examined the use of support vector machines in the tourism sector; the use of deep learning in online reviews was analyzed by Bigne et al., 2021; and Bigne, William et al. (2020) discussed the visualization of online ratings through self-organizing maps. Another promising area of tourism research is the adoption of natural language processing (NLP) for interpreting social media texts through sentiment analysis and emotions. According to Databricks Lakehouse (2023), NLP is one of the most commonly used techniques by companies. Moreover, the promising area of analyzing video content by using practical tools such as Google Images and video solutions is attracting scholarly attention (e.g., Li et al., 2019).

Intuitive intelligence comprises thinking creatively and adjusting effectively to novel situations (Huang & Rust, 2018). The key difference of this intelligence type is the novelty of the situation being addressed. It involves a deep learning process and adopting the "what happens to Y if X is new" schema. This approach allows researchers to provide recommendations to customers about other services. For instance, on a rainy day, customers could receive a recommendation about the type of indoor activities in which they could engage.

Empathetic intelligence entails recognizing and understanding other peoples' emotions, responding appropriately, and influencing others' emotions (Huang & Rust, 2018). This type of intelligence focuses on the emotional dimension and addresses emotional reactions based on feelings. This type of intelligence is still challenging for researchers and managers to achieve. However, some examples, such as the Ameca robot (https://www.youtube.com/watch?v=vE9tIYGyRE8) or Sophia (https://www.youtube.com/watch?v=5TWB_DmG0Ek), are the current pinnacle of this flourishing development field.

3.2. Pre-Journey Stage

The pre-journey stage involves potential tourists engaging in a range of activities in their hometown. These activities be sorted into three main areas of interest: (i) searching for information through experts (e.g., travel agencies, websites, or influencers) and peers through social media; (ii) using the acquired information and booking the elements of the trip (e.g., transportation, accommodation, and other related services) through different suppliers; and (iii) paying for it. A rich data source is generated and registered for use by AI in each interaction.

Information Search. Search activities offer a rich dataset in multiple areas. For example, the online search engine used by tourists tracks all the keywords used in the searches, the specific sites visited, and the activities performed on each site, thus providing interesting sources of information. The first is the words used in the search, which convey valuable information to DMOs. Google Trends can be used for this purpose. It is a free tool that tracks the words users type in their searches and also provides some location data. For example, a DMO can track users' searches of their destination and compare them through the months or years to detect an increasing or decreasing interest. While these data only show users' interest, the application also allows destinations to add other searches, such as hotels or flights, specific airlines, or events, allowing correlations to be discovered. Therefore, DMOs can use this information to adjust their actions accordingly, such as through promoting the destination, a hotel, or an activity, or adjusting their prices. In addition, DMOs can identify their main competitors by comparing the number of searches. Second, Google provides multiple applications, such as Google Analytics, that show various metrics, including the number of visitors, times, days, the origin countries of searches, pages viewed, channels of access, conversions, some demographic data, interests, returns to the pages, and bounce rate (i.e., the number of people who visit a page and leave without performing a specific action). Detailed information about all the metrics can be obtained here: https://analytics.google.com/analytics/web/. Digital companies often use these free tools to provide targeted reports that have added value through the use of additional data sources. For instance, ForwardKeys (www.forwardkeys.com) offers integrated solutions for the travel and hospitality industries by providing predictions based on AI and ML techniques. Third, based on previously visited pages and tourists' clicks, DMOs can provide promotional activities through retargeting, which consists of automatically providing advertisements that are related to the pages visited to those who have previously visited the site.

Social media sites offer an incredible amount of data on tourism. This rich media setting involves numerical data, text, pictures, and videos. Social media sites in terms of tourism can be sorted into specialized and generic sites. Some specialized social media sites are those on which only comments can be shared (e.g., TripAdvisor), while other specialized sites act as booking sites on which comments are also allowed (e.g., Booking.com). Conversely, generic social media sites, such as Facebook, Instagram, or Yelp, may also provide valuable information with less touristic details. Bigne, William et al. (2020) showed that six specialized tourist platforms exhibited consistent patterns in their online reviews, regardless of the platform. However, the only non-tourism specialized platform, Yelp, stood apart as it had a distinct pattern of online evaluations. Additionally, these evaluation patterns remained consistent over time. Furthermore, no distinction was observed between the evaluation patterns of platforms that necessitated prior reservation proof and those that did not. Other peer-to-peer booking platforms, such as Airbnb, provide vast amounts of data from bookings and reviews of selected destinations (see the data here: http://insideairbnb.com/get-the-data). A flourishing area of interest in social media research is sentiment analysis, which entails automatically detecting opinions, attitudes, and emotions from social media content, whether textual or visual, through AI and ML algorithms. The literature (for a review, see Mehraliyev et al., 2022) and commercial applications such as Brandwatch, Meaning Cloud, IBM Watson, Google Cloud Natural Language, and Eden AI provide a vast number of analytical solutions for the tourism and hospitality industries.

Booking. Booking an entire trip or any of its elements implies users making decisions. The vast majority of bookings are made online through agents' websites, intermediaries such as travel agency sites, and online aggregators of services. The GDS is the origin of the current online booking system, which allows anyone from any location to book transport, accommodation, car rentals, or similar services in any part of the world in a secure manner and with transparency. Since all these processes are digitalized, the information is recorded and available.

A booking platform shows available options, tracks each booking, and updates the DMOs' availability, thereby providing information relevant to the yield management of companies and organizations. Some platforms offer information that allows tourists to make informed decisions, such as the average rates, expected price change, the occupancy levels or seats available, and the online ratings. For instance, Booking.com offers multiple filters to allow users to make a straightforward booking. Moreover, some platforms provide managerial information for DMOs, such as revenue and comparative options (e.g., similar properties in a neighborhood). In addition, in terms of peer-to-peer accommodation, AIRDNA (https://www.airdna.co/) provides detailed information about the average daily rates, occupancy rates, revenue, and online ratings over time to hosts and investors, and it provides a link that allows users to make bookings through Airbnb.

A relevant variable for hotels is the time in which the booking was made as the time between when the booking was made and when the user arrives provides information relevant to yield management, which allows the rates for online travel agencies, hotels, transportation carriers, event ticketing, and other related services to be determined. Indeed, Bigne, Nicolau et al. (2021) analyzed dynamic pricing of hotel bookings across various channels, and they found that online travel agencies tended to undercut any other channel's price 90 days in advance, while 30 days in advance, the rates offered by hotel websites, online travel agencies, and GDSs were similar.

Payments. Payment of a booking denotes a money transfer from a client to a company or individual. From a data point of view, big players such as credit cards, banks, and other types of financial services, such as PayPal, collect vast amounts of data that have tremendous value for data analysis and predictive analytics.

The activities performed during a user's pre-journey offer different types of intelligence, and each activity encompasses a double dimension: one occurs in the back office, referring to DMOs' internal procedures, analytics, and predictions; and the other occurs in the front, denoting interactions with consumers (see Figure 1). The thickness of the circles for the four types of intelligence in Figure 1 depicts their relevance to each stage of the journey.

Most activities related to searching, booking, and paying encompass mechanical intelligence, mainly digital mechanical intelligence. Due to the enormous amount of data from any of the tasks during the pre-journey stage, analytical intelligence is ubiquitous. During the pre-journey stage, the volume and velocity of the data are consistent. However, tourism actors must overcome two challenges: (i) integrating data into a single-source data set; and (ii) combining different languages (e.g., from online searches). Thus, companies that integrate and consolidate data for DMOs and other touristic actors, such as ForwardKeys (https://forwardkeys.com/), have become popular as they address these issues. Intuitive intelligence during the pre-journey stage is often present due to various searches and types of online reviews. Lastly, emotional intelligence is applicable since searches, bookings, and payments entail a customer–company relationship. However, using emotional intelligence is more challenging as, in most cases, these activities are done without physical contact.

3.2. The Stay at the Tourist Destination

The stay stage entails multiple activities performed by tourists at the destination, which can be sorted into four main areas of interest: (i) accommodation services and related services in hotels and apartments; (ii) all the activities related to meals at restaurants and local grocery stores as well as shopping; (iii) information searches for activities, transport, and services as well as posting of comments; and (iv) visits and movements (e.g., local transport or tourist buses) to museums, events, attractions, adventures, or other activities as well as to and from main transport infrastructures (e.g., airports, main railway stations). During this stage, the amount of data provided is massive, sourced from multiple actors, and in a relatively short period. Therefore, the challenges for companies and DMOs include the integration of data and the short response times.

Accommodation. Accommodation facilities implement AI and robotics in their services, with some tasks often being integrated as a substitute for mechanical human activities. For example, the following is a sequence of the different accommodation processes (e.g., check-in, suitcase transport, services and tickets for events, and check-out) and areas (e.g., rooms, restaurants, bars, other facilities and amenities) in which the application of AI and robots is suitable. From the operations side, cleaning and temperature control can also be performed by robot cleaners and sensors. Indeed, several robotic devices and AI applications are becoming available for automatic check-in and check-out. For instance, Henn-na Hotel Japan was one of the first hotels to provide a welcome, interactive check-in service using humanoid robots at the front desk (https://group.hennnahotel.com/), and Connie is a robot concierge used by the Hilton that uses an AI platform developed by IBM that allows it to interact with guests and respond to their questions, suggest places to visit, where to dine, and how to find anything on the property. The system also learns and adapts with each interaction, thus improving the answers it provides (https://www.youtube.com/watch?v=ghbS-aTYw14). These systems perform a variety of tasks through a natural language speech interface. They are similar to the well-known Alexa from Amazon, Google Assistant, and Siri from Apple. Check-in service offers an excellent opportunity for collecting data about clients, such as entrance time, nationality, type of requested services, and similar information. Furthermore, the use of hotel loyalty cards as well as cards or wristbands that have radio frequency identification technology embedded allows the tracking of all the services used and how visitors move around attractions, and they can also be used for payment. Hotels can also use this tool to obtain information that can be used for security purposes and to improve the design and shorten queuing times. Other robots perform physical processes, such as automatically collecting and delivering guests' luggage or delivering snacks, toiletries, and other hotel amenities.

Hotel rooms can implement myriad robotic applications as well as AR and VR applications. For example, a smart room can be developed that automatically adjusts temperature, lighting, the opening and closing of curtains, and other functions based on inner and outer sensors. In addition, some physical tasks, such as room-cleaning robots, fit well with the AI-type hotel developments. Immersive technologies can also be used in hotel rooms in different ways. For instance, meals from the hotel restaurant can be previewed at any table in the room through AR (For an overview of the Hololamp AR Menu, watch https://www.youtube.com/watch?v=LQY5AvRwCN8), a hotel room can be visited (see Whitney Hotel in Boston and Old Bank Hotel in Oxford and other examples here: https://hoteltechreport.com/news/virtual-reality-hotel-tours), and visual information of attractions can be provided through immersive glasses placed in guests' room.

Picking up products without human intervention is no longer only theoretical as Amazon Fresh and Amazon Go stores have successfully implemented "grab-and-go" technology, which tracks what customers pick, allowing them to skip conventional check-out and cashiers (https://www.youtube.com/watch?v=YmzatOYykHw). This technology is suitable for bars, amenities, and other services at hotel facilities, meaning that hotel workers being present is no longer necessary. Moreover, the hours in which these services are available can be extended. Maybe it is too early for its full adoption at hotels, but as long as the grab-and-go technology exists, it will be implemented in hotels sooner rather than later. In fact, kiosks that deliver drinks and packaged food and coffee machines are quite common in hotels and constitute an decisive step towards a more natural interaction with AI developments.

Lastly, from a hotel operations point of view, data and AI applications have a significant impact on security, water leaks and waste, smoke, fire, and CO2 emission issues. They are so successful that these applications have been extended to other properties, such as airports, restaurants, museums, and similar spaces. Furthermore, other applications synchronize all booking channels, such as the Smoobu application (https://www.smoobu.com/) that automatically synchronizes booking portals, including Airbnb and Booking.com, through the use of a channel manager and a property management system booking engine that has online check-in.

Meals and Daily Shopping. The second primary data source at the stay stage of the tourist journey includes daily activities that involve meals and shopping for artisan goods and souvenirs. Beyond the potential analytics that can be obtained from these data, recommendations based on AI can be implemented. As discussed above, hotel robots can also provide shopping recommendations. There are additional AI solutions, such as sommelier.bot, which provides wine recommendations based on ChatGPT to hoteliers, retailers, and end users. These types of tools can allow companies and DMOs to customize their product assortments in other product categories and services.

Information Searches at the Destination. As described earlier, during the stay stage, tourists conduct multiple online searches and post comments on social media sites. However, there are other types of data sources and AI applications related to searches: DMOs often provide tourist information points or digital kiosks for tourists, which can be used to gather valuable data about the type of searches conducted according to the time, days of the week, and languages. Furthermore, recommendations can be customized through digital information kiosks being equipped to identify basic biometric data, such as determining emotions through facial recognition. Additionally, they can adapt the recommendations of where users should visit depending on the weather conditions or how crowded the attractions are. Likewise, other external data sources can be embedded to deliver more effective customization.

Visits. Technologies such as visualization maps (e.g., Google Maps, Citymapper, Waze, local transportation applications), Wi-Fi, and Bluetooth-linked smartphone applications can be used to track the flow of user movements and detect overcrowded areas, after which alternative itineraries or destinations can be provided, thus managing crowds and enhancing customers' visits. This is also applicable to other spaces, such as museums. For instance, a Bluetooth-based solution at the Louvre Paris (https://sense-able.mit.edu/louvre/) allows managers and users to monitor paths and show how crowded each room is. Likewise, tourist destination cards may track data and offer ML solutions for reducing crowds through monitoring traffic, flows, and crowded spaces.

AI and ChatGPT applications that provide users with instant itineraries are being developed (see https://iplan.ai/). For example, users can send a request to ChatGPT detailing the place they are visiting, the dates on which they are doing so, the types of activities in which they are interested, their available budget, the restaurants they like, and similar information, and an itinerary will be provided. I tested this capability in several iterations, including various numbers of visitors, dates, and budgets. The recommendations were satisfactory and provided valuable information. However, it still needs to deliver customized information to the users. Thus, the application's ability to learn from users is prompting new plugins. Moreover, the latest information offered by ChatGPT is from September 2021, and the application is not yet directly connected to the Internet.

3.4. Post-Journey Stage

This stage involves the targeted activities related to data generation and AI usage after a user's trip has concluded. The most relevant ones are users rating and sharing their experiences, which occurs mainly through social media sites and complaint services. Tourists often rate their experiences according to multiple criteria for different facilities and services, which will be used by new tourists when choosing their destinations, facilities, and carriers. Thus, sharing comments as text or visual content is becoming a valuable data source. Most of these activities have a strong relationship with analytical intelligence. Integration with other data sources may provide fascinating insights for use by companies and DMOs as unveiling the relationship between a positive online rating and an accommodation facility's occupancy level may provide an informative trade-off between occupancy, rates, and tourist satisfaction at a destination.

Round trips are an inherent part of tourist activities. Most of the trips to a destination involve a return. However, some destinations act as a hub for connecting with other destinations. Therefore, the unveiling of itineraries through AI becomes relevant. Carriers need infrastructure, mainly comprising airports, railway stations, or ports that have security and border controls. Selling trips, whether in a one-way or round-trip format, and activities at transport locations provide valuable sources of information with different purposes, especially for the tourism and security sectors. Indeed, the security sector has developed multiple AI devices, including scanners and digital visual inspections, for monitoring people, passport control, gate access, and luggage control. The trend of adopting self-service activities is becoming ubiquitous at international airports that include self-check-in, passport control through passport scanning and facial readers, and access to boarding gates being provided through the use of QR codes and digital boarding passes.

4. ETHICAL AND SOCIETAL IMPLICATIONS

AI in tourism has an impact on societal issues because it typically involves activities in public spaces, such as destinations and transport areas. Additionally, the private interactions between tourists and private actors also have similar issues as those in other industries. This section reviews some of the ethical implications for individuals and society through a philosophical lens.

In a nutshell, AI implies data acquisition, analysis, and responses (DAAR). This triple function demands a closer analysis of the origin of AI and its type of use. The origin can be sorted into two main areas: (i) objects and environment-based; and (ii) human-based. Data from weather conditions and similar data types do not imply human interventions, while bookings being made and similar actions entail human interactions. This suggests that DAAR originating from objects and the environment triggers fewer human concerns than those originating from a human decision in which personal information accompanies an action. This delineates a pertinent difference, in that if the use of AI deals with objects (e.g., maintenance of a facility) and public aims (e.g., security), the concern will be less important than if a private actor (e.g., an accommodation facility) benefits from DAAR. Figure 3 depicts the interaction between these two criteria, with the upper-left quadrant representing the least important concern about DAAR and the bottom-right quadrant demonstrating the most important concern, which arises from the data originating from human actions and being used for private purposes.

Another related societal issue deals with technological dependence. As described earlier, many activities are based on mechanical intelligence. The discussion on whether human interventions should be substituted for machines is currently raging in relation to what will happen if technology fails. Using AI for security purposes, opening a window, and many other functions seem efficient in terms of increasing the amount of time people have available and lowering the costs of these actions, but when these processes fail, the entire system fails. For instance, in late May 2023, the facial recognition technology in Heathrow, Manchester, and Gatwick airports failed, halting their ability to verify travelers' identities and capture their images, causing chaos (Russell & Andersson, 2023). Since machines are acting as substitutes for humans, it can be said that computerized systems manage modern societies. The multiple technological improvements in these processes in all directions and ethical considerations may not be enough. Instead, a social debate on human supervision of the processes and the limits of machine–human substitution is needed.

Tourists are an essential part of the data acquisition process as valuable information is obtained whenever they search for any service, book it, pay for it, or upload online reviews to a social media site. The current practices are based on informational disclosures about data usage through a third party, which are mainly associated with cookies. From a philosophical point of view, a debate on who owns the data, the beneficiaries, and the benefits of sharing these data demands public discussion. For example, should beneficiary companies compensate tourists as data generators and disclose the purposes of their data use? Should tourists be compensated for sharing their data? Financially or service-based way? Are they subject to the type of purpose? For instance, can booking data be used for price discrimination, upgrade services, or even for informing other actors about potential services or only for public security reasons? There is no unique answer to such questions. Furthermore, users may consent to some of them but maybe not for others. Therefore, a societal debate about the above issues is needed: data ownership and potential benefit types for each party. The ethical approval of academic research involving humans, known as the Helsinki Declaration, can be used to inform this discussion, allowing ethical principles related to data acquisition, data privacy, data storage, statistical analysis, aggregate versus customized responses, compensation to tourists, and data selling to be derived. Moreover, transparency and open data are common ideas in the digital world. Therefore, I advocate for establishing open-source datasets and processes that guarantee veracity and social benefits. For instance, some peer-to-peer accommodation services share aggregate data (see: http://insideairbnb.com/get-the-data/). Thus, there is a need for social debate, which should trigger the development of a legal framework that has an international scope. A valuable step forward is the three packages of the European Union (EU) AI Act that was proposed in

April 2021 in an effort to foster a European approach to AI. It is called the Coordinated Plan on Artificial Intelligence (European Commission, 2021). The EU defines an AI scheme according to the different risk practices, which range from minimal to unacceptable, in terms of governance, enforcement, and sanctions.

Figure 3. The concerns in applying AI in the tourism sector

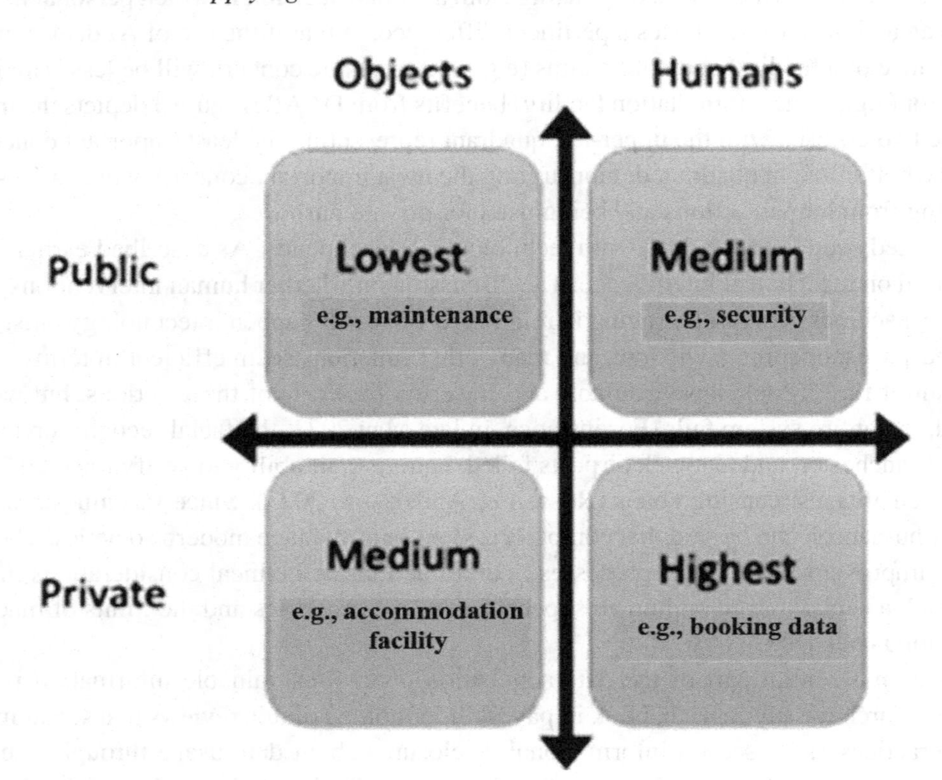

5. FUTURE DIRECTIONS

Some technological developments are eliciting rapid adoption of AI devices and services. This section delineates the significant salient changes that will increase the use of AI in tourism..

5.1 Job Replacement and Flexible Operations

Robotics will become increasingly popular as part of the mechanical intelligence in specific tasks, such as maintenance, cleaning, and safety conditions at airports, hotels, and other facilities. Despite boarding and passport control already being partially controlled by machines, AI and robots will be increasingly implemented in these fields. This job replacement trend will also be fueled by autonomous machines in transportation, such as cars, taxis, and maybe in collective transport, personal consumption of packaged products (e.g., Amazon Fresh stores), amenities, luggage transportation, and basic services (e.g., the Connie robot). In addition to the implementation of robotics, the use of self-service technologies (e.g., Booking.com), such as in banking through automated teller machines, will contribute to replacing certain jobs. As a result, operations will be flexible as services will be available 24 hours a day, seven days a week, ultimately modifying job conditions and regulations.

5.2 A Mobile-Centric Approach

Mobile usage will increase throughout the world for multiple purposes due to the proliferation of 5G, free Wi-Fi connections, and roaming services. From a data generation point of view, it offers two salient benefits for tourism: (i) smartphones travel with their owners from the destination of origin, through the trip, and at their destination; and (ii) multiple applications are now available for users, even at the international level. Notably, a few international groups dominate access to mobile data around the globe. Furthermore, the increasing number of applications (e.g., travel, weather, location, traffic) and smooth functions (e.g., QR readers, photos, videos) make social media posting and information searches easier. As a result, smartphones are an essential element for travelers. This intensive use of data from tourists will force the adoption of international regulations on privacy issues, disposal of data sharing, and data beneficiaries.

5.3 Data Integration and Analytics

Integrating data from tourists with other external sources (e.g., weather conditions) through computing applications, the IoT, and cloud computing will trigger a need for specialized companies that will have three tasks: (i) merging data sources; (ii) processing and analyzing data; and (iii) delivering targeted reports to the tourism industry and tourists, which will evolve into actionable decisions for companies and personalized recommendations for tourists.

5.4 The Tension in the Interactions Between Revenue Management and Customers

AI diffusion and adoption will be driven by the benefits obtained by both companies and users, in that companies will adopt AI as long as it increases the operational margin of delivering a service and ensuring service quality, and tourists will benefit through the perceived value of sharing information with third parties. Indeed, each party should gain monetary reward, access, or informational value in this triadic relationship between tourist companies, data companies, and tourists. Mainly, tourists do not often see the value of sharing their information and are reluctant to do so unless they trust data companies, gain rewards, or obtain valuable individual recommendations. The tension between these forces will be shaped by the perceived value of the benefit of sharing and gathering the data and its costs, including privacy and monetary issues.

5.5 Neuroscientific Tools for AI in Tourism

The use of neurophysiological tools that measure, for example, heart rate variability, galvanic skin conductance, and facial emotions will increase due to their low costs and increased focus on emotional intelligence. Facial recognition has increased substantially for security purposes at airports. Nevertheless, its application in emotional detection is yet to be used in hotels, resorts, and tourist attractions.

Heart rate variability and galvanic skin conductance have received increased attention due to the new type of wristwatches that monitor these signals. Thus, the challenge is not in the use of the technology itself but is centered around what metrics can be obtained and what its uses are for individuals. Once again, privacy issues and the value of sharing and gathering information will lead to tension between the targeted parties.

5.6 Ethical Considerations and Societal Changes

The growing use of AI in different applications is triggering a social debate, for which DAAR can be used to provide answers to the possible challenges. Substituting humans with machines and processes, the role played by human supervision, and individuals' dependence on technology are only the tip of the iceberg. Regulations need to be quickly developed at the international level to ensure a common framework that ensures security. In addition, tourists need to be provided with information about the risks, usage, and compensation of AI in understandable language.

6. CONCLUSION

This chapter delineated how AI is being used in tourism through two main frameworks: (i) a broader perspective of the customer's journey that incorporates the pre, stay, and post-stages of travel; and (ii) the multiple applications of mechanical, analytical, intuitive, and empathetic intelligence in services in each stage of the customer journey.

The high adoption rate of data analytics, robots, and other forms of AI will continue, and it will be fueled by new applications and technologies. However, the tension between data companies, tourist companies, and tourists will be shaped depending on each party's perceived gains and losses. In this context, privacy and personal issues will play a key role, for which international regulations should be developed to create a common framework.

Tourism encompasses a range of different activities and intelligence types, but two issues should be taken into consideration: (i) product types; and (ii) the segmentation of tourists. Firstly, not all types of tourism will adopt AI in the same ways, with the differences in travel motivation between MICE versus holidays being a good example: In MICE tourism, technological developments will be driven by the utilitarian value of the tasks, while on holidays, the hedonic value of the tourist experience will predominately limit this expansion. Secondly, the segmentation of tourists results in different groups of adopters of AI. For instance, there are technologically savvy tourists versus traditional ones and tourists who are highly concerned about privacy versus those who are not as concerned, each of whom will react differently to AI, thereby determining which type of applications will successfully lead to the expansion of AI in particular sectors.

The chapter also discussed five major future trends: (i) job replacement and flexible operations; (ii) a mobile-centric approach; (iii) data integration and analytics; (iv) the tension in the interactions between revenue management and customers; and (v) neuroscientific tools for AI in tourism. Regardless of these trends, I argue that the adoption of self-service technologies in tourism will continue to increase in the coming years, which will be driven by the control over the process as perceived by the tourists and the time flexibility it provides.

The philosophy behind AI in tourism points out the need for a compromise between technological advancements, such as AI, analytics, and robotics, and the need for a better understanding of the type of tourists and their motives and needs.

REFERENCES

Airports Council International. (2020). *Annual World Airport Traffic Reports*. Airports Council International.

Ausin-Azofra, J. M., Bigne, E., Ruiz, C., Marin-Morales, J., Guixeres, J., & Alcañiz, M. (2021). Do you see what I see? Effectiveness of 360-Degree vs. 2D video ads using a neuroscience approach. *Frontiers in Psychology*, *12*, 612717. doi:10.3389/fpsyg.2021.612717 PMID:33679528

Balducci, B., & Marinova, D. (2018). Unstructured data in marketing. *Journal of the Academy of Marketing Science*, *46*(4), 557–590. doi:10.100711747-018-0581-x

Bigne, E., Fuentes-Medina, M. L., & Morini-Marrero, S. (2020). Memorable tourist experiences versus ordinary tourist experiences analyzed through user-generated content. *Journal of Hospitality and Tourism Management*, *45*, 309–318. doi:10.1016/j.jhtm.2020.08.019

Bigne, E., & Maturana, P. (2023). Does virtual reality trigger visits and booking holiday travel packages? *Cornell Hospitality Quarterly*, *64*(2), 226–245. doi:10.1177/19389655221102386

Bigne, E., Nicolau, J. L., & William, E. (2021). Advance booking across channels: The effects on dynamic pricing. *Tourism Management*, *86*, 104341. doi:10.1016/j.tourman.2021.104341

Bigne, E., Oltra, E., & Andreu, L. (2019). Harnessing stakeholder input on Twitter: A case study of short breaks in Spanish tourist cities. *Tourism Management*, *71*(April), 490–505. doi:10.1016/j.tourman.2018.10.013

Bigne, E., Ruiz, C., & Badenes-Rocha, A. (2023). The influence of negative emotions on brand trust and intention to share cause-related posts: A neuroscientific study. *Journal of Business Research*, *157*(March), 113628. doi:10.1016/j.jbusres.2022.113628

Bigne, E., Ruiz, C., Cuenca, A., Perez-Cabañero, C., & Garcia, A. (2021). What drives the helpfulness of online reviews? A deep learning study of sentiment analysis, pictorial content and reviewer expertise for mature destinations. *Journal of Destination Marketing & Management*, *20*(June), 100570. doi:10.1016/j.jdmm.2021.100570

Bigne, E., William, E., & Soria-Olivas, E. (2020). Similarity and consistency in hotel online ratings across platforms. *Journal of Travel Research*, *59*(4), 742–758. doi:10.1177/0047287519859705

Databricks Lakehouse. (2023). *2023 State of Data + AI*. Databricks. https://www.databricks.com/resources/ebook/state-of-data-ai?scid=7018Y000001Fi0tQAC&utm_medium=paid+search&utm_source=google&utm_campaign=17161077299&utm_adgroup=154951955492&utm_content=ebook&utm_offer=state-of-data-ai&utm_ad=662845930524&utm_term=ai&gclid=EAIaIQobChMI7Yee7cC0gAMVAoJoCR09PwkAEAAYASAAEgIWZvD_BwE. Accessed on July 29, 2023.

European Commission. (2021). *Communication on fostering a European approach to artificial intelligence*. EC. https://digital-strategy.ec.europa.eu/en/library/communication-fostering-european-approach-artificial-intelligence

Fan, X., Jiang, X., & Deng, N. (2022). Immersive technology: A meta-analysis of augmented/virtual reality applications and their impact on tourism experience. *Tourism Management*, *91*, 104534. doi:10.1016/j.tourman.2022.104534

García-Madurga, M. Á., & Grilló-Méndez, A. J. (2023). Artificial intelligence in the tourism industry: An overview of reviews. *Administrative Sciences*, *13*(8), 172. doi:10.3390/admsci13080172

Gretzel, U., Sigala, M., Xiang, Z., & Koo, C. (2015). Smart tourism: Foundations and developments. *Electronic Markets*, *25*(3), 179–188. doi:10.100712525-015-0196-8

Huang, M. H., & Rust, R. T. (2018). Artificial intelligence in service. *Journal of Service Research*, *21*(2), 155û172.

Kakaria, S., Bigne, E., Catambrone, V., & Valenza, G. (2022). Heart rate variability in marketing research: A systematic review and methodological perspectives. *Psychology and Marketing*, *40*(1), 190–208. doi:10.1002/mar.21734

Knani, M., Echchakoui, S., & Ladhari, R. (2022). Artificial intelligence in tourism and hospitality: Bibliometric analysis and research agenda. *International Journal of Hospitality Management*, *107*, 103317. doi:10.1016/j.ijhm.2022.103317

Li, X., Shi, M., & Wang, X. S. (2019). Video mining: Measuring visual information using automatic methods. *International Journal of Research in Marketing*, *36*(2), 216–231. doi:10.1016/j.ijresmar.2019.02.004

Mehraliyev, F., Chan, I. C. C., & Kirilenko, A. P. (2022). Sentiment analysis in hospitality and tourism: A thematic and methodological review. *International Journal of Contemporary Hospitality Management*, *34*(1), 46–77. doi:10.1108/IJCHM-02-2021-0132

Russell, R., & Andersson, J. (2023, May 27). *Anger over airports' passport e-gates not working*. BBC. https://www.bbc.com/news/uk-65731795.

Simonetti, A., & Bigne, E. (2022). How visual attention to social media cues impacts visit intention and liking expectation for restaurants. *International Journal of Contemporary Hospitality Management*, *34*(6), 2049–2070. doi:10.1108/IJCHM-09-2021-1091

Sternberg, R. J. (2005). The theory of successful intelligence. *Revista Interamericana de Psicología. Interamerican Journal of Psychology*, *39*(2), 189–202.

Tussyadiah, I. (2020). A review of research into automation in tourism: Launching the Annals of Tourism Research curated collection on artificial intelligence and robotics in tourism. *Annals of Tourism Research*, *81*, 102883. doi:10.1016/j.annals.2020.102883

Chapter 7
The Influence of Culture on Sentiments Expressed in Online Reviews of Eco-Friendly Hotels: The Case Study of Amsterdam

Estefania Ballester Chirica
University of Valencia, Spain

Carla Ruiz-Mafé
University of Valencia, Spain

Natalia Rubio
Autonomous University of Madrid, Spain

ABSTRACT

The proliferation of content generated by tourists, in parallel with the exponential growth of social media is causing a paradigm shift in research. Traditional surveys cannot be necessary to obtain users' opinions when scholars can access this valuable information freely through social media. In the domain of tourism, online tourists' reviews (OTRs) shared on online travel communities stand out. The aim of this study is to demonstrate the usefulness of OTRs in analysing the image of a green hotel. The authors also examine the possible differences in the content of green hotel online reviews across Anglos and European tourists. The data source are 28,189 reviews by tourists shared on TripAdvisor regarding the 82 green hotels of the city of Amsterdam. The findings showed that tourist's culture significantly determine the content of the OTRs. The results show preferences and opinions from the tourist's perspective, which can be useful for hotel managers to promoting sustainability practices.

DOI: 10.4018/978-1-6684-9591-9.ch007

INTRODUCTION

The proliferation of online travel communities where consumers can connect with one another effectively, has enabled everyone nowadays to share their tourists' experiences, providing what is commonly known as electronic word-of-mouth (eWOM) (Oliveira and Casais, 2019). One of the most widely available types of eWOM is online tourist reviews (OTRs). In OTRs, tourists freely share their consumption experiences (D'Acunto et al., 2020), and find information about destinations (Marine-Roig, 2019), destination attractions (Filieri et al., 2021), accommodation (Kumari and Sangeeetha, 2022) or restaurants (Kim and Hwang, 2022). OTRs, as a post-consumption response, have revolutionized travellers purchasing decisions in the last decade (Kumari and Sangeetha, 2022). One of the most a well-known and popular review platforms is TripAdvisor, which hosted more than 73 millons reviews and opinions were submitted to the platform by 23 milions members from different countries. This includes 30.2 millions reviews and 31.6 million photos and videos shared by TripAdvisor community. Furthermore, 40,30% of the reviews have been posted about accommodations business (TripAdvisor, 2023). In addition, TripAdvisor is a trustworthy source of user content. Tripadvisor has a three-part system for reviewing reviews to ensure their accuracy. Once reviewed, the vast majority of review submissions are approved for posting, but some are immediately rejected, while others are flagged for further assessment by Tripadvisor's content moderation team. Past research has shown that many consumers read the online reviews of other tourists and consider them when purchasing services (Lin et al., 2021). In short, these sites have become the most useful online information sources for consumers worldwide. Consequently, online reviews have attracted more attention from marketing and tourism scholars.

Online reviews have several functions, first describe real experience that offers great opportunities to zoom in the multi-faceted dimensionality of image perceived and transmitted by tourists; second, reports the sentiment of tourist by its textual content (Taecharungroj and Mathayomchan, 2019). Prior literature in hospitality field has analyzed OTRs as a key tool for customer-customer or customer-company interaction; however only a few studies have considered these user content generated as data that provides deeper understanding of overall brand image of tourism companies (e.g., Wang et al., 2019; Lin et al., 2022). Prior literature has explored OTRs in various tourism contexts such as accommodation (Kim and Kim, 2022), restaurants (Lin et al., 2022), attractions (Bigné et al., 2023 and destinations (Marine-Roig et al., 2019). However, few studies have resorted to online reviews for research purposes in the domain of green hotels (e.g., Arici et al., 2023), which is one of the fastest growing hotels segments nowadays (Yang et al., 2022) due to the growth of environmental consciousness among consumers (Galati et al., 2021).

Hence, a relevant research question is the following:

R1. What components of the green hotel image tourists include in their reviews?

Additionally, understanding customers' feelings embedded in the comments of online reviews is crucial. In terms of sentiments of online reviews most of the previous studies have used ratings as indicators of reviews' polarity (i.e., Dhar and Bose 2022) however, a recent study by Bigné et al. (2023) reports that the star rating of an online review maybe misaligned with the sentiment of its textual content.

Even though the analysis of online reviews has attracted interests amongst scholars and managers in recent years (i.e., Lin et al., 2022; Arici et al., 2023; Bigne et al., 2023) research on cross-cultural differences in user-generated content are still in its infancy. Previous studies have demonstrated that tourist culture is a predominant factor of service evaluation in the hospitality field, however, an important knowledge gap remains regarding the effects of the influence of culture on OTRs. For example, some online travel websites such as TripAdvisor.com provide the customized option to sort reviews according to national background factors such as currency used, language, and nationality of review posters. This option improves the consumer's credibility and satisfaction with the review site and, in turn, contributes to the profit of the site. If cultures lead to systematic differences in the review generating process, it is necessary to examine whether culturally customized information would be useful in customers' decision-making processes and whether customers prefer tourism companies providing such information (Kim et al., 2018). Therefore, to gain deeper knowledge of effect of culture in OTRs we pose the following research question:

RQ2. Does tourist culture affect customer perceptions of green hotel image expressed in the content of the online reviews?

As green hotels are growing in number worldwide, scholars' interest in green strategies has also increased (e.g., Wang et al., 2018; Hameed et al., 2021; Sadiq et al., 2022). However, there is only limited research on tourists' stays at green hotels as opposed to their stays in other accommodations. In particular, it is important to identify how many of the tourists who have stayed green hotels shared their perception or concern about the environment in their reviews. This is of vital importance since if the green image is a competitive advantage in hotel industry and customers take OTRs into account for their choice of hotels, the presence of sustainable reviews related to environmental topics enhance the green image of hotels (Wang et al., 2018). Previous studies on green hotels are mostly conducted based on associated predefined constructs and hypotheses (see a revision on Yadegaridehkordi et al., 2021). Technology has produced a large amount of data through digital devices. Thus, extracting and understanding valuable information from massive unstructured data is extremely critical for customers, organizations, and policy makers.

Since differences in the cultural backgrounds of consumers are likely to be able to play a critical role in review generation, as well as review consumption, this study attempts to simultaneously examine the effects of the different cultural backgrounds on sustainable reviews posting behaviour.Therefore, based on the relevant research gaps identified the following research questions are proposed:

RQ3a. How many online tourists reviews are sustainable comments?
RQ3b. Is the volume of green comments influenced by the culture and the polarity of the sentiments of tourists?

This study presents customers' perceptions of hotels' images as a multidimensional construct, and inductively identifies the dimensions of green hotel image through the attributes referred to in the online reviews. Sustainable communications posted by hotels greatly improves their brand image and prestige, since the consumers' rights are seen as being protected and their expectations fulfilled. Therefore, trust is developed toward the hotel and its products and services (Badenes et al., 2019). In addition, we test whether customer culture (Anglo vs European) influences tourists' perceptions and sentiments. To analyze tourists' perceptions of hotels' green practices as expressed in social media, we use content and sentiment analysis of the reviews they post on TripAdvisor. The sample was made up of all green hotels in Amsterdam (82) advertised in TripAdvisor. Amsterdam is one of the main "smart cities" in Western Europe. The Cities in Motion Index (2022) describes Amsterdam as having an important international impact and mobility due to the high number of hotels and inbound air routes. Moreover, it is one of the cities that is considering introducing sustainability rules for hotels (Santos-Lacueva et al., 2022).

OBJECTIVES

The present study has three main objectives:

- To identify the dimensions of the perceived image of eco-friendly hotels based on the reviews posted by tourists on TripAdvisor and identify the key attributes of each dimension.
- To analyse the effect of culture (Anglo vs European) on perceived dimensions of green image on online reviews of eco-friendly hotels.
- To identify the volume of green OTRs and analyse the relationship between sustainable comments with the culture of tourists and the sentiment polarity of the review.

LITERATURE REVIEW

Green Hotel Image and Online Tourists Reviews

Over the past few decades, the hotel industry has become more competitive because of the similarity of the products and services hotels offer (Hameed et al., 2021). An improved image helps hotels stand out among their competitors (Wang et al., 2019). A hotel's image is of paramount importance because it creates an impression of the establishment in the potential tourist's mind and shapes his/her preliminary thoughts (Marine-Roig et al., 2019). The image of businesses has been defined many times. Kuo and Hasio (2008) defined a company's image as "a series of perceptions about a firm as reflected by its associations in consumers' memories". On similar lines, Durna et al. (2015) posited that image relates to the beliefs, behaviours and impressions of individuals or groups as regards certain subjects (companies, products, brands, destinations, individuals).

Today, due to the growth of environmental consciousness among consumers, a green image has become a critical competitive advantage (Lee et al., 2010) in hospitality and tourism. Accordingly, in recent years, hotels have increasingly been adopting green practices. Environmentally friendly actions taken by hotels are designed to meet eco-friendly customers' expectations and encourage them to post positive, green-focused reviews on social media. In the present study we consider positive-oriented green reviews as a post-consumption response to hotels' green/environmentally friendly actions. Due to intangibility of hotel services, online reviews are highly valued by potential clients and impacts on hotels' green images (Kumari and Sangeetha, 2022). Therefore, it is important that hoteliers pay attention to customers' opinions about their hotels' green practices and make efforts to properly communicate their eco-friendly initiatives to encourage tourists to share their perceptions of these green practices so they can build a greener image.

Martinez (2015) conceptualized the "green overall image" of a company as the customers perceptions related to its commitment to environmental issues. Wang et al. (2019) further argued that a green image encompasses customers' perceptions of a company dedication and concern for the environment, resulting in strong memory associations. Traditionally, image has been characterized as unidimensional, representing an overall perception. However, more recent approaches have measured image thought multiple dimensions. In the context of the service industry, particularly in the studies focusing on hotels, cognitive and affective components have commonly been employed to conceptualize hotel image.

In this study, we adopt the conceptualization of image proposed by Marine-Roig et al. (2019), which considers image as the comprehensive cognitive, affective and conative structure of the behaviour unit (Boulding, 1956). According to this perspective, the interaction between the company and the customer is divided into three facets: 1) acquiring knowledge about something; 2) developing feelings and emotion towards it; and 3) taking action based on this knowledge and emotional response. In our analyses of the green hotels' image (Figure 1), we apply the tripartite cognitive-affective-conative model (Rapoport, 1977, p28). The cognitive dimension is associated with "knowing and thinking" which encompasses the fundamental process through which individuals perceive their environment. The affective dimension includes feelings, emotions, motivations, desires, and values related to the environment. Lastly, the conative component involves taking actions, striving and making impact on the environment in response to cognitive and affective aspects of the image.

Figure 1. Green hotel image components
(Rapoport, 1977 (p.28))

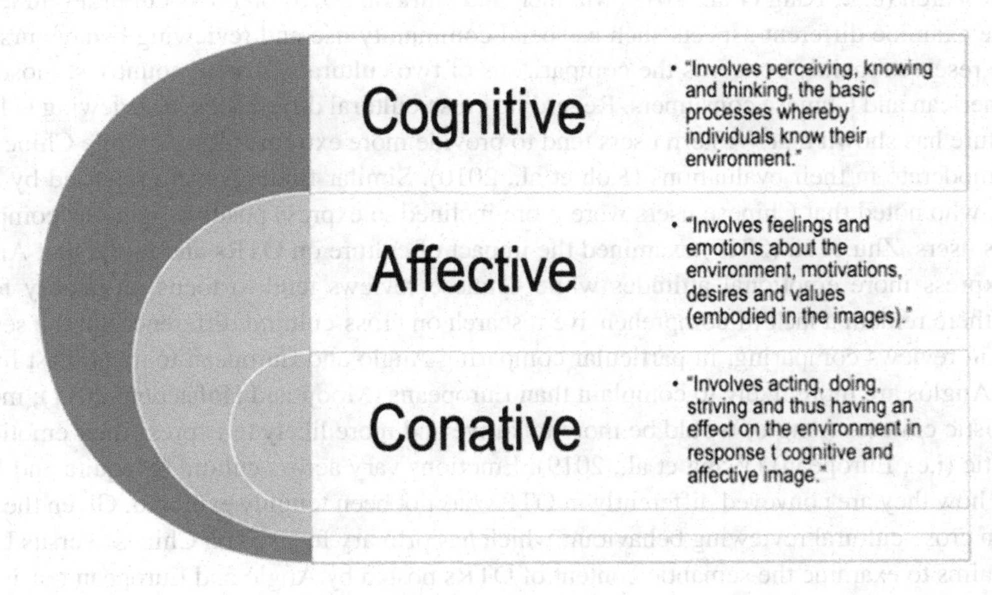

Sentiments and Online Tourist Reviews

Sentiments, defined as personal feelings that influence behaviour (Poulakidakos and Armenakis, 2014), are an inherent aspect of human speech (Zhu et al., 2014). In today's digital era, social networks have emerged as a significant platform for customer to openly share their emotions about a service (Buzova et al, 2018). Along this line, online tourists' reviews (OTRs) have garnered substantial attention in recent literature. Although several authors have associated the sentiment of online reviews with numeric ratings (Bao and Chan, 2016), research conducted by Chong et al. (2016) and Bigné et al. (2023) has shown that star ratings are not precise indicators of the sentiment expressed in the text. Villarroel et al. (2017) emphasize that written language encompasses a wide range of sentiment expressions, which should not be overlooked in their influencer on the reader. Therefore, in a customer's online review, the evaluation of a service is conveyed not only through the numerical rating provided but mainly through emotionally charged words and their respective sentiment contained within the text. Hu et al. (2014) argue that sentiments expressed in a review offer "more tact, context-specific explanations of the reviewer's feelings, experience, and emotions about the product or service" (p.42). Considering the impact of sentiments on consumers' purchase intentions and attitudes towards services (Pike et al., 2021; Sayfunddin et al., 2021), as well as the experiential and intangible nature of services, it become crucial to acknowledge the consumer sentiment valence (Dhar and Bose, 2023). The influence of affect expressed in reviews is particularly notable when consumer lack expertise or when evaluation criteria are limited (Ludwig et al., 2013). Taking all this evidence into account, this research focus on textual sentiment conveyed in online reviews of green hotel tourists experience, aiming to perform a cross-cultural comparison.

Culture and Online Tourist Reviews

Culture plays a significant role in shaping consumer behaviour in online context (Buzova et al., 2018). Previous research (e.g., Fang et al., 2013; Malinen and Nurkka, 2015) on cross-cultural online behavioural have examine different aspects such as social community use and reviewing behaviours among other. The research mainly concerns the comparisons of two culturally distant countries, mostly comparing American and Chinese consumers. Regarding cross-cultural differences in reviewing behaviour, past literature has shown that Western users tend to provide more extreme ratings, while Chinese users are more moderate in their evaluations (Koh et al., 2010). Similar findings were reported by Fang et al. (2013), who noted that Chinese users were more inclined to express positive opinions compared to Americans users. Zhu et al. (2017) examined the impact of culture on OTRs and found that American reviews express more emotional attitudes while Chinese reviews tend to focus on quality features. However, there remain a lack of comprehensive research on cross-cultural differences in the sentiment expressed in reviews comparing, in particular comparing Anglo and European tourists. Past literature show that Anglos are more likely to complain than Europeans (Mooji and Hofstedem 2011); moreover individualistic cultures (Anglo) would be more talkative and more likely to express their emotion than collectivistic (i.e., European) (Wang et al., 2019). Emotions vary across culture (Mequita and Walker, 2003), but how they are conveyed differently in OTRs has not been toughly explored. Given the limited research on cross-cultural reviewing behaviour, which has primary focused on Chinese versus US data, this study aims to examine the semantic content of OTRs posted by Anglo and European tourists.

METHODOLOGY

We examined 28,189 reviews for 82 green hotels located in Amsterdam, all of which were available on TripAdvisor. The research methodology (Figure 2) had four steps: (1) Case study Selection; (2) Data Collection; (3) Content Analysis; (4) Sentiment Analysis.

Figure 2. Methodological framework
(Marine-Roig, 2015)

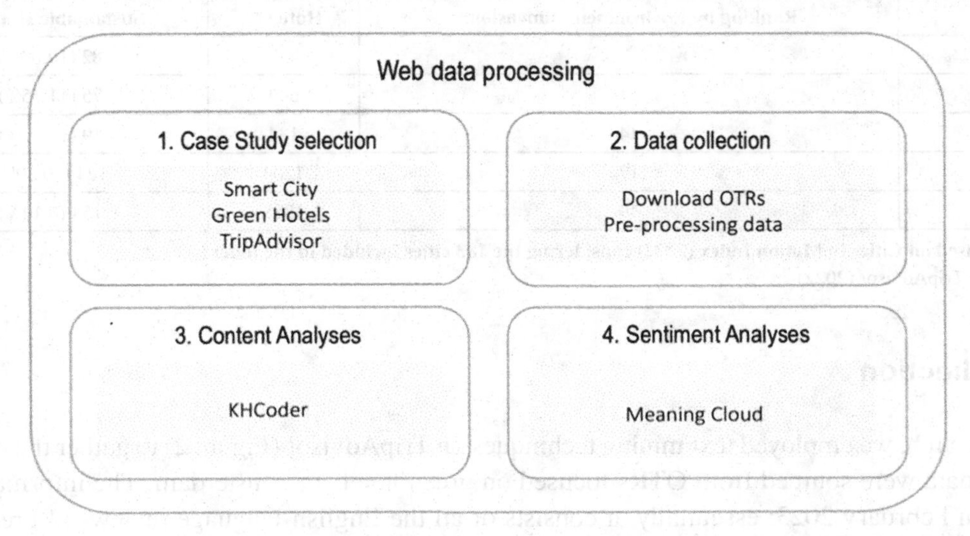

Case Study: Green hotels in Amsterdam

In the hospitality sector, various terms are used to describe "green hotels", including 'sustainable hotel' (Ponnapureddy et al., 2017), 'environment-friendly hotel' (Gonzalez-Rodríguez et al., 2019) and 'eco-friendly hotel' (Sadiq et al., 2021). Green hotels are defined as properties that prioritize environmental conservation by implementation water-saving measures, energy conservation initiatives and waste reduction programs, all while aiming to save costs and protect the planet (Green Hotel Association, 2008). Verma et al. (2019) emphasize that green hotels, compared to traditional hotels, commonly adopt eco-friendly management practices such as water conservation, the use of energy-efficient products, organic materials, bio toilets, and promoting environmental consciousness among their employees and guests.

This case study focus on the opinions shared by customers of green hotels online, reflecting the growing environmental awareness among customers. Adopting green practices has become crucial competitive advantage in the hospitality industry (Arici et al., 2023). The case selection process involves three steps. Firstly, we identify the top-ranked smart cities in Western Europe (refer to Table 1) based on nine dimensions: (Governance, Economy, Social Cohesion, Human Capital, Technology, International Projection, Urban Planning, Mobility and Transportation and Environment) (Cities in Motion Index 2022). Among the five European Smart Cities, Amsterdam stands out in the Sustainable dimension, being the only one that is considering introducing sustainability rules for hotels (Santos-Lacueva et al., 2022). Moreover, the Cities in Motion Index (2022) describes Amsterdam as having an important international impact and mobility due to the high number of hotels and inbound air routes. Second, Amsterdam was chosen as the context of this research because: 1) the marketing strategy of going green can add value to a hotel's brand image and help position the hotel within the market (i.e. by differentiating it from "non-green" hotels); 2) a climbing number of hotel guests seek green accommodation. Third, we select the most suitable website for this case. TripAdvisor travel-related platform was chosen because: (1) It is the largest source of UGC in the tourism domain; (2) TripAdvisor's reputation management system helps to determine the helpfulness of the reviewer and motivates users to contribute with reliable reviews.

Table 1. Case study

	Ranking by Environment dimension*	Hotels**	Sustainable Hotels**
Amsterdam	3	418	82 (19.62%)
Berlin	21	644	75 (11.65%)
Copenhagen	14	134	19 (13.43%)
London	17	1204	124 (10.30%)
Paris	49	1868	154 (8.14%)

Note 1: ***Based on Cities in Motion Index (2022) considering the 183 cities included in the index;**
**Based on TripAdvisor (2022)

Data Collection

For the research, we employed text mining techniques on TripAdvisor (Figure 2) to gather the necessary data. The data were sourced from OTRs focused on green hotels in Amsterdam. The information was retrieved in February 2023; essentially, it consists of all the English-language reviews of green hotels in Amsterdam posted online from 2018 to 2022. TripAdvisor features 418 hotels in Amsterdam, 82 of them green. Following the approach outlined by Marine-Roig (2019), we extracted all the relevant information from TripAdvisor OTRs and stored it in a CSV file. Each OTR was assigned a unique record within the file, including details such as name, location (geographic code), hotel resources (destination code), OTR code, reviewers' nationality, OTR title, OTR data and the text of the OTR. To carry out the analyses, the cross-cultural differences in the hotel image and sentiment expressed in the reviews, we grouped the comments based on the nationality of the reviews. The cross-cultural comparison focused on opinions provided by Anglo and European tourists, in response to a suggestion made by De Mooji and Hofstedem (2011) to examine complaint intention between these two cultural groups. The reviewers' national culture was determined based on their indicated location. Review authors who had not included their place of residence in their profile information were excluded from the dataset, resulting in the removal of 7.328 opinions. Therefore, out of the initial 28.189 reviews, the final dataset considered of 20.861 opinions (819.211 words), 16.340 of which pertained to Anglo citizens, while 4.521 were posted by European tourists (Table 2).

Figure 3. The data collection process

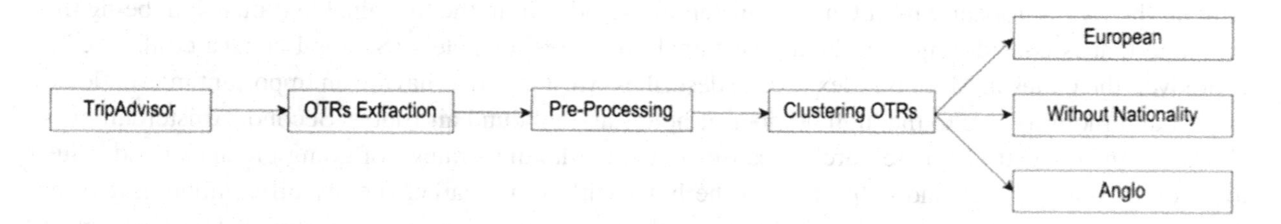

Table 2. Collected reviews' information

	Number (n = 20,861)	Percentage
Tourists' origin		
Anglo	16,340	78.33%
European	4,521	21.67%
Year of posting		
2018	9,135	43.79%
2019	7,294	34.96%
2020	1,388	6.65%
2021	1,086	5.21%
2022	2,890	13.85%

Content Analyses

Content analysis is a research method that has been defined as a systematic, replicable technique for clustering key words into content categories (Marine-Roig, 2019). The most applied techniques are based on terms frequency count because higher mention words reflect higher importance (Stemler et al., 2019). KHCoder was adopted to process the data because it is a very powerful tool. Thus, a KHCoder was set to calculate the most frequently mentioned words in the overall textual data. Once the most frequent word was identified according to their word type (nouns, adjectives, adverbs, or verbs among others), they were categorized.

The words associated with the Anglo and European reviews from Amsterdam hotels OTRs were initially categorized into three main groups (Figure 2) using the destination image definition proposed by Rapoport (1977): cognitive image, affective image, and conative image. The main categories related to the model in Figure 1 are detailed below:

Cognitive image. To cognitive image component analyses, we cluster the most frequent terms into sixteen groups for Anglo reviews and fifteen for European reviews: tangible facilities, service, price, destination, location, duration, transport, gastronomy offer, city attractions, kind of trip, seasonality, brand name, sustainability, booking criteria, interaction with other tourists and hotel quality; seasonality subdimension only appear in the Anglo reviews.

Affective image: In this study case, the affective component was measured by quantifying the presence of adjectives and other words indicating positive, neutral or negative feelings or moods. Using a table displaying word frequencies and a lexicon consisting of a list of positive words (e.g., great, nice), neutral words (e.g., ok, average) and negative words (e.g., terrible, disappointed), three metrics are proposed: positive feelings, neutral feelings, and negative feelings. All three are calculated by the percentage of positive, neutral, or negative words in relation to the total number of words (including stop words).

Conative image. A dichotomous category has been constructed a priori dividing into attitudinal and behavioral response. We define attitudinal response as tourists' intention to recommend the hotel, distinguishing between positive recommendation (e.g., a must, recommend) and negative recommendations (e.g., do not stay here, wouldn't recommend). Similarly, we view behavioral response as tourists' intention to revisit the hotel, which can also be either positive (e.g., return soon, come again) or negative (e.g.., never again).

Sentiment Analysis

Given the volume of reviews retrieved, we carry out an automatic sentiment analysis (Deng and Yu 2014; Timoshenko and Hauser 2019). In this regard, Sharma and Dey (2015) emphasize the importance of sentiment analysis in understanding culture and demographic influences on reviews writing and sharing behaviors, particularly for marketing purposes. Therefore, sentiment analysis was employed in this research to achieve its objectives.

Sentiment analysis aims to determine the reviewer's attitude towards a product or service by identifying and categorizing the emotional content of subjective statements within a text corpus (Tsytsarau & Palpanas, 2012). In this study, we employed an artificial intelligence-based sentiment analysis approach to delve into the sentiment expressed in tourist reviews. Specifically, we used the MeaningCloud software, which leverages a hybrid sentiment analysis technique. This technique amalgamates the attributes of both corpus-based and lexicon-based approaches to attain optimal classification accuracy (Zulkifli and Lee, 2019). MeaningCloud relies on machine learning methods and harnesses advanced natural language processing techniques to discern the polarity of textual content. This capability facilitates a comprehensive and multilingual sentiment analysis of texts originating from diverse sources. Termed MeaningCloud, this text analytics service categorizes sentiments into classifications such as very positive, positive, neutral, negative, or very negative (Ciechanowski et al., 2020). Furthermore, MeaningCloud provides algorithmic confidence percentages (Loyola-González et al., 2019). As reported by Zulkifli and Lee (2019), MeaningCloud demonstrated an impressive accuracy rate of 82.1%, outperforming other prominent sentiment analysis tools such as Miopia (73%) and Python NLTK (74.5%).

The sentiment analysis process entails the determination of whether a given text conveys a positive, negative, or neutral sentiment. To accomplish this, it involves the identification of the local polarity of individual sentences within the text, followed by an assessment of the interrelationships between these sentences. This culminates in the computation of a global polarity value for the entire text. Beyond the assessment of polarity at the sentence and document levels, MeaningCloud employs sophisticated natural language processing techniques to detect the sentiment associated with entities and concepts present in the text. Additionally, it provides a reference within the relevant sentence and a list of detected elements, each associated with aggregated polarity values derived from all their appearances. This analysis takes into account the grammatical structures in which these elements are embedded. Consequently, the MeaningCloud API ascertains the local polarity of various sentences within the text and furnishes a global polarity value for the entire document (Sharma and Hoque, 2017). This approach calculates the sentiment score of a specific review by determining the average sentiment value across all sentences within the review's text. This sentiment score ranges from negative (-1) to neutral (0) or positive (1).

RESULTS

To answer RQ1 and clarify the main components of green hotel image, a content analysis of the online reviews was conducted using KHCoder. The analysis involved three phases: parser configuration, frequency analysis, and categorization. The OTRs were divided into two groups: Anglo reviews and European reviews. We carried out a hierarchical cluster analysis for each group of reviews and we observe that in both groups' reviews include a cognitive, affective, and conative components of green hotel image (Figure 4).

Figure 4. Green Hotel image (components)

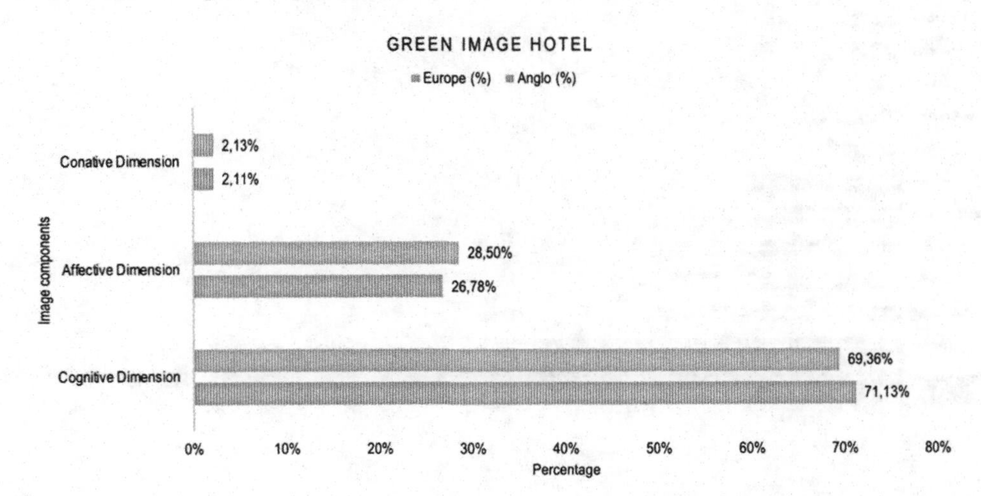

The dimension of cognitive image (see Figure 5) is organized in sixteen (Anglo cluster) and fifteen (European cluster) main independent clusters (Tangible Facilities, Service, Price, Destination, Location, Duration, Transport vehicles and Infrastructures), Gastronomic Offer, City Attractions, Kind of trip, Seasonality, Brand Name, Sustainability, Booking, Interaction with other tourists and Quality), being the seasonality present only in the Anglo reviews. Despite the observed differences in the frequency of each attribute, the predominant cognitive sub-dimensions in both cultures are: Tangible facilities (30.33% Anglo; 32.70% European), Service (12.66% Anglo; 14.68% European), Transport vehicles and infrastructures (10.02% Anglo; 9.20% European), Duration (9.96% Anglo; 9.44% European) and Location (7.18% Anglo; 8.14% European). The tangible facilities features words related to tourists' perceptions of accommodation (e.g., room, bed), connected to perceptions services (e.g., staff, breakfast, check-in/out); the transport dimension features in its two categories vehicles (e.g., bus, bike) and infrastructures (with connected words such as central-station, tram-stop); the location subdimension features words related to distance with the city attractions (e.g., close, short-walk); the duration sub-dimension dimension is represented by words "night" and "weekend".

The data obtained from TripAdvisor facilitates an Analysis of affective image. We examined the terms employed by the reviewers in the OTRs and categorized them into impressions of positive, neutral and negative sentiments. Positive sentiments exhibit the highest level of affective image in both cultural groups (81.64% Anglo; 82.30% European). The proportion of each sentiment (positive, neutral, and negative) were remarkably similar in both Anglo and European reviews (refer to Figure 6).

Figure 5. Green hotel image (cognitive dimension)

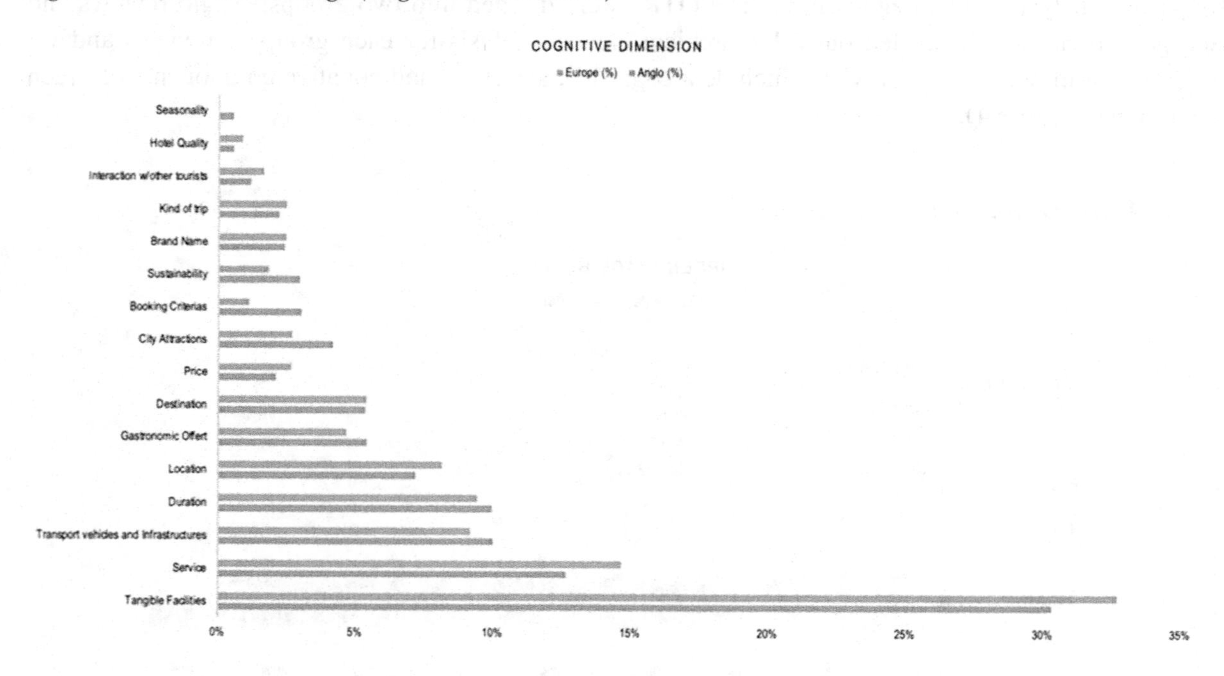

Figure 6. Green hotel image (affective dimension)

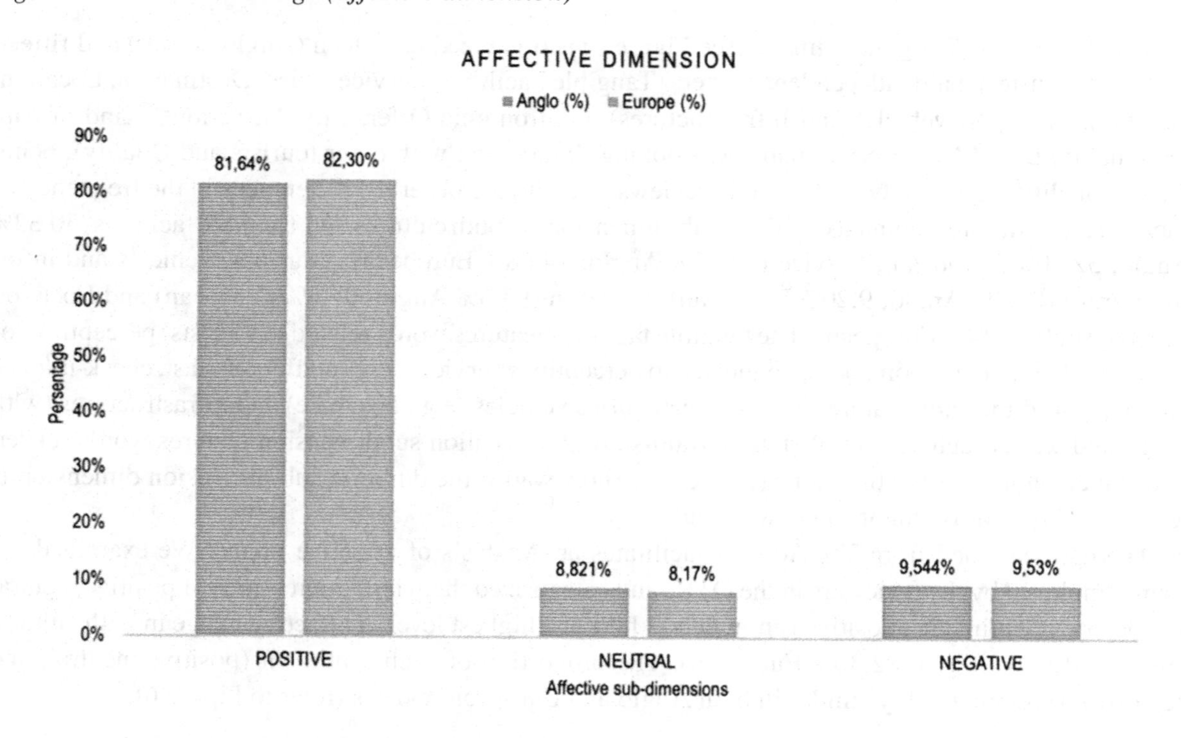

Positive and negative reviews in both samples are referenced to four clusters: tangible facilities, location, price, and staff (See Figure 7). The perceptions about staff (e.g., friendly, helpful, polite) and tangibles (e.g., fantastic, clean, comfortable) are similar in both clusters, being the positive feelings predominant. However, the perception of location differs, being more positive in European cluster (e.g., perfect, good). Lastly the perception of price is the only cluster where negative sentiments (e.g., overpriced, expensive) is predominant, being the values similar in both groups.

Figure 7. Green hotel image (affective sub-dimensions)

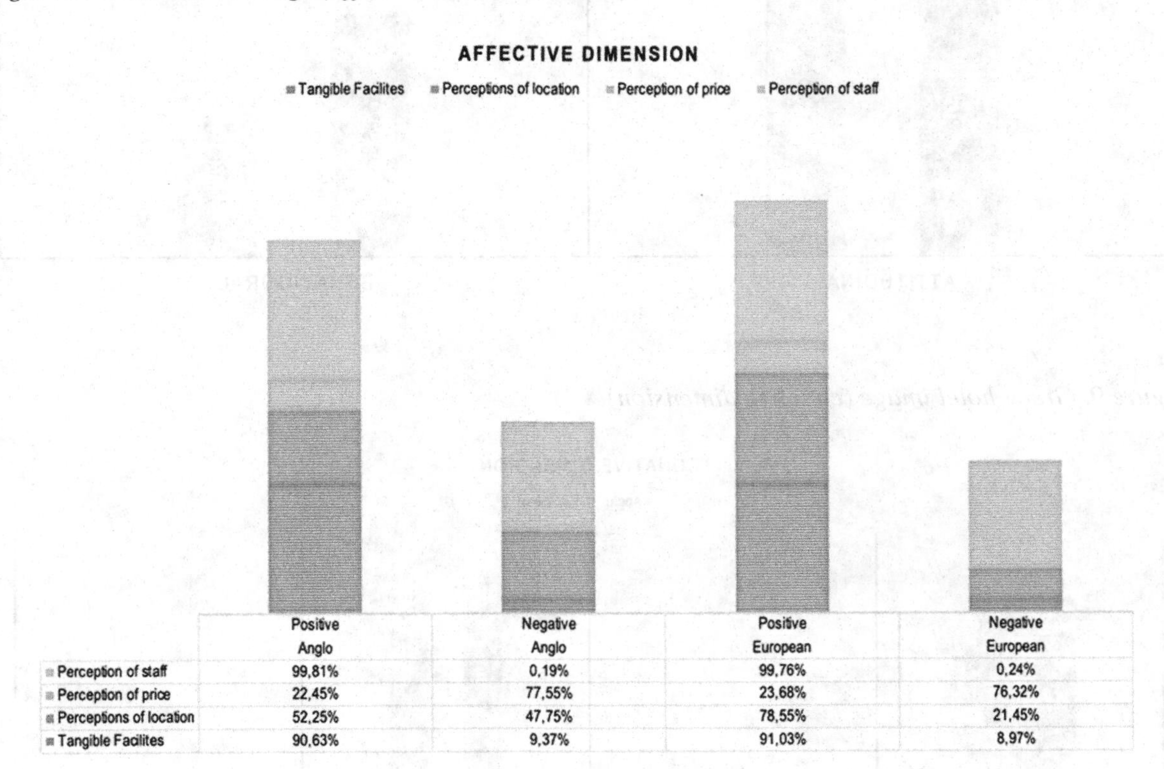

	Positive Anglo	Negative Anglo	Positive European	Negative European
Perception of staff	99,81%	0,19%	99,76%	0,24%
Perception of price	22,45%	77,55%	23,68%	76,32%
Perceptions of location	52,25%	47,75%	78,55%	21,45%
Tangible Facilites	90,63%	9,37%	91,03%	8,97%

Figure 8 depicts conative image that include attitudinal and behavioral sub-dimensions. The results showed that attitudinal sub-dimension is more predominant than behavioral one. The attitudinal cluster is represented by the positive and negative intention to recommend the hotel. As we can see in Figure 9, although in both cultures the positive recommendation is predominant, the European reviews include more negative recommendations. In terms of behavioral cluster, that refer to the intentions to revisit the hotel, most of the reviewers include positive intentions (e.g., come back).

To address RQ2, we carried out a preliminary frequency analysis for each group of reviews (see Table 3).

Figure 8. Green hotel image (conative dimension)

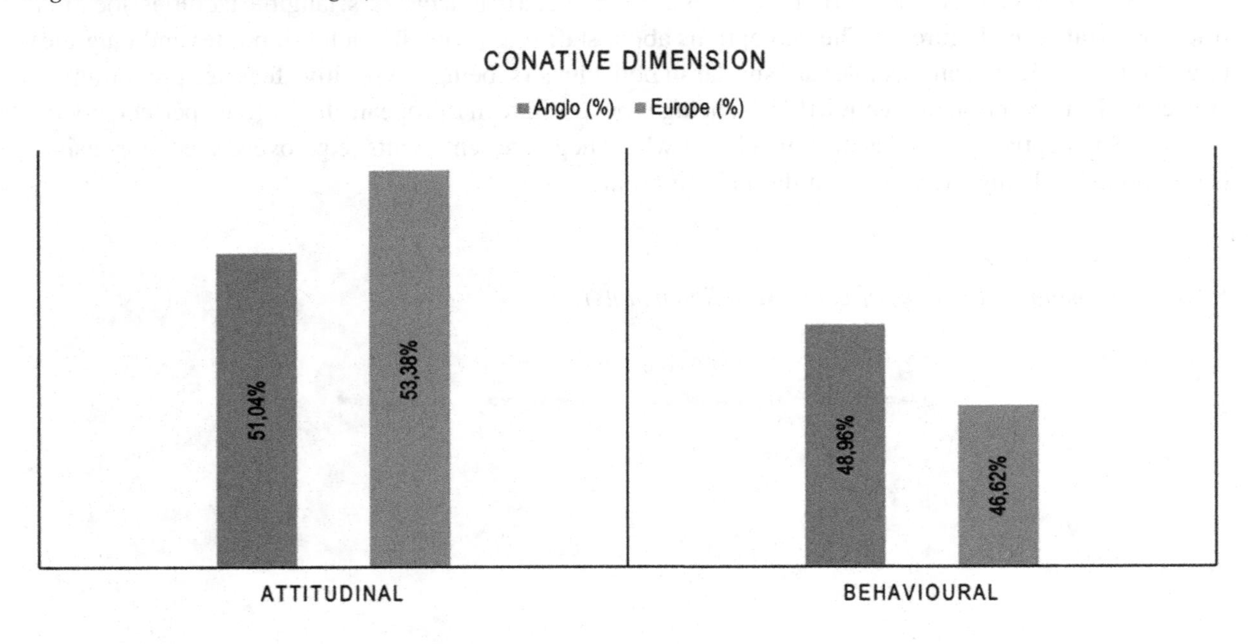

Figure 9. Green hotel image (conative dimension)

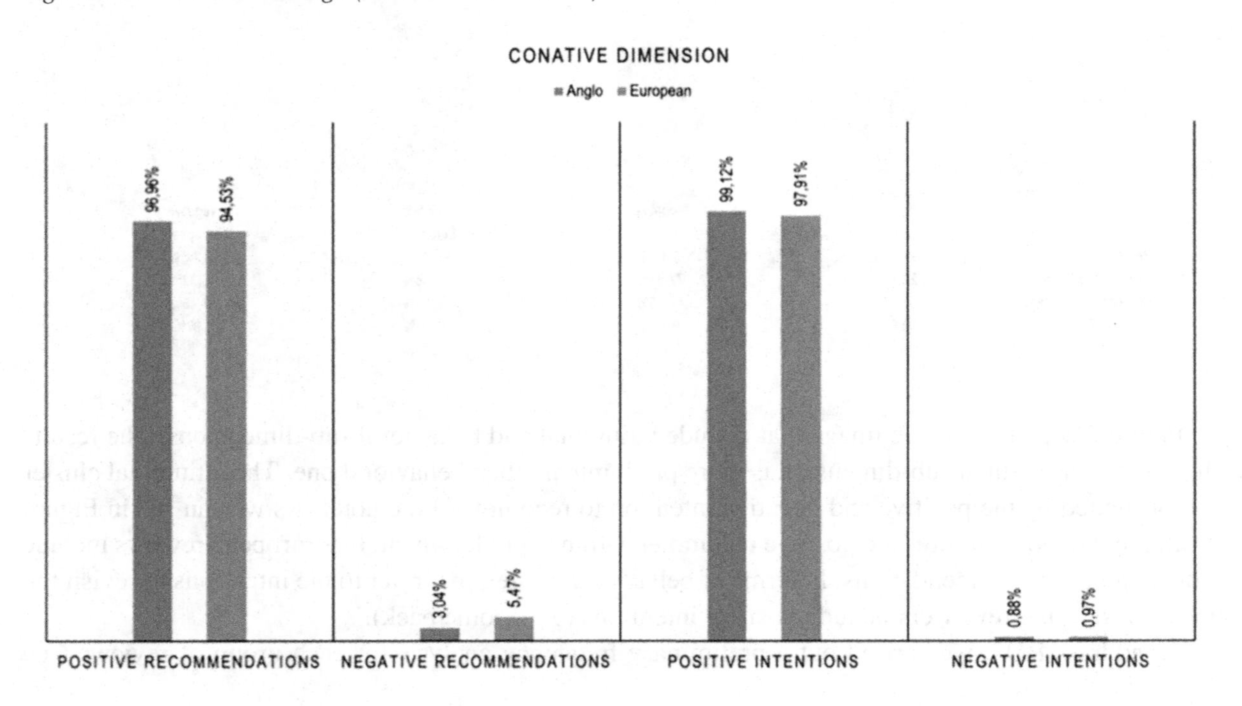

As we can see in Table 3, the most important concepts mentioned in tourists' hotel reviews are common in both groups (Anglo and European reviews): hotel, room, stay, staff, great, restaurant, Amsterdam, good, breakfast, location, clean, walk, night, comfortable, friendly, nice, helpful, bed, service, recommend, excellent. However, the importance of these words varies depending on the group, with the most notables' differences being that Anglo reviews mention more aspects related to the travel duration (stay and night) and the location (location and walk), while European Reviews highlight aspects related to the service (service and breakfast). As for positive sentiments, Europeans use general adjectives like "Good" and "Nice", while Anglos include more specific adjectives in their comments such as "Comfortable" and "Helpful". In addition, there are differences in terms of services aspects, while Anglos reviews pay attention to Booking and Check-in/out, the Europeans tourists focus on Reception, Price. Similarly results highlight that Anglos include the adjective "Lovely" and mention the "Station"; and European include "City Centre" and Views".

Table 3. 25 most frequently used word in the reviews, by tourist culture

	Anglo reviews				European reviews			
	Keyword	Count	%	Image	Keyword	Count	%	Image
1	Hotel	25320	7,97%	Cognitive	Hotel	5478	7,59%	Cognitive
2	Room	23065	7,26%	Cognitive	Room	5179	7,18%	Cognitive
3	Stay	12728	4,00%	Cognitive	Good	2483	3,44%	Affective
4	Staff	11236	3,54%	Cognitive	Stay	2434	3,37%	Cognitive
5	Great	7994	2,52%	Affective	Staff	2379	3,30%	Cognitive
6	Restaurant	7675	2,41%	Cognitive	Nice	2211	3,06%	Affective
7	Amsterdam	7660	2,41%	Cognitive	Breakfast	2020	2,80%	Cognitive
8	Good	7350	2,31%	Affective	Amsterdam	1929	2,67%	Cognitive
9	Breakfast	6564	2,07%	Cognitive	Clean	1450	2,01%	Cognitive
10	Location	6470	2,04%	Cognitive	Great	1419	1,97%	Affective
11	Station	6351	2,00%	Cognitive	Friendly	1154	1,60%	Affective
12	Clean	6331	1,99%	Cognitive	Night	1038	1,44%	Cognitive
13	Walk	5972	1,88%	Cognitive	City Centre	1536	2,13%	Cognitive
14	Night	5588	1,76%	Cognitive	Restaurant	1487	2,06%	Cognitive
15	Comfortable	5329	1,68%	Affective	Walk	1134	1,57%	Cognitive
16	Friendly	4884	1,54%	Affective	Reception	1083	1,50%	Cognitive
17	Nice	4834	1,52%	Affective	Service	1067	1,48%	Cognitive
18	Helpful	4617	1,45%	Affective	Location	956	1,32%	Cognitive
19	Lovely	4491	1,41%	Affective	Bed	946	1,31%	Cognitive
20	Booking	4061	1,28%	Cognitive	Comfortable	816	1,13%	Affective
21	Bed	4024	1,27%	Cognitive	Price	792	1,10%	Cognitive
22	Service	3472	1,09%	Cognitive	Recommend	777	1,08%	Conative
23	Recommend	3318	1,04%	Conative	Helpful	724	1,00%	Affective
24	Excellent	2923	0,92%	Affective	Excellent	573	0,79%	Affective
25	Check in/out	2903	0,91%	Cognitive	View	499	0,69%	Cognitive

Once the most frequent terms were identified, we classified them based on the dimension they referred to in the image. As we can see in Table 4, most of terms refer to cognitive and affective dimensions, while the conative dimension is residual.

Table 4. Hotel green image (25 most frequently use words)

	Anglo and European Reviews	Anglo Reviews	European Reviews
Cognitive Dimension	Hotel, Room, Stay, Staff, Restaurant, Amsterdam, Breakfast, Location, Walk, Night, Bed, Service.	Booking, Check-in/out, Station	Reception, Price, City Centre, Views.
Affective Dimension	Great, Good, Clean, Comfortable, Friendly, Nice, Helpful, Excellent.	Lovely	
Conative Dimension	Recommend		

By using a 0/1 system, we classified each review based on the presence of the common terms associated with cognitive, affective and conative image dimensions. For the conative image dimension, we considered the following terms: Hotel, Room, Stay, Staff, Restaurant, Amsterdam, Breakfast, Location, Walk, Night, Bed, Service. In terms of the affective image dimension, we assessed the presence of feelings expressed through terms such as (Great, Good, Clean, Comfortable, Friendly, Nice, Helpful, Excellent). Lastly, the conative image dimension was evaluated using a single term, "Recommend". For instance, a review that include any of the above-mentioned word regarding cognitive image was coded 1, but when the reviews did not include one of these standards, it was coded 0. As we can see in Table 5, the majority of the OTRs include cognitive and affective dimension, while the OTRs that include conative dimension are residual. The results show that in the case of European reviews, the cognitive and affective dimension have the same frequency, while in the case of the Anglos reviews the frequency of cognitive dimension is higher.

Table 5. Hotel green image (number of reviews)

Online tourists' reviews	Anglo Reviews	European Reviews
Cognitive Dimension	48,02%	44,95%
Affective Dimension	41,48%	44,95%
Conative Dimension	10,50%	10,09%

We performed the t-test for independent samples on RQ2 regarding the effect of culture on green image dimensions (Table 5). The results of table 6 show significant differences between Anglos and European reviews regarding Cognitive and Affective image components, however this difference is not significant in terms of Conative image component. It should be noted that Anglos (48.02%) write reviews more functional including cognitive image dimension, while European reviews are more affective reflecting in their reviews the affective image more frequently.

Table 6. Results of t-test and descriptive statistics for green image components

	Culture	n	Mean	SD	t	95% confidence interval
Cognitive	Anglo	16337	0.81	0.396		
	European	3990	0.77	0.419		
	Total				4.671***	0.019; 0.047
Affective	Anglo	16337	0.70	0.460		
	European	3990	0.77	0.419		
	Total				-9.588***	-0.092; -0.061
Conative	Anglo	16337	0.18	0.381		
	European	3990	0.17	0.379		
	Total				0.388	-0.011; 0.016

*** significant at p<0.001

To investigate RQ3a and RQ3b, first we carry out a frequency analysis to identify the most common green practices mentioned in Anglo and European online reviews. The most frequent words in terms of sustainability in Anglos and European reviews are Climate Control, Towel Reuse, Green Certification, environmentally aware, Organic food, Recycle, Water save and Energy. Sustainability is one of the sub-dimentions of Cognitive Dimension and as we can see in Figure 5 is most prominent in Anglo reviews (n = 6413; 2.02%) than in European reviews (n = 917; 1.25%). The frequency of the term above-mentioned inside the subdimension sustainability (Figure 10) vary between both groups. Figure 10 depicts that "Climate control", "Towel reuse" and "Organic food" are the most mentioned green practices in Anglo reviews; while "Water save", "Energy", "Green Certified" and "Environmentally aware" are more frequent in European reviews.

Figure 10. Most frequent sustainability words

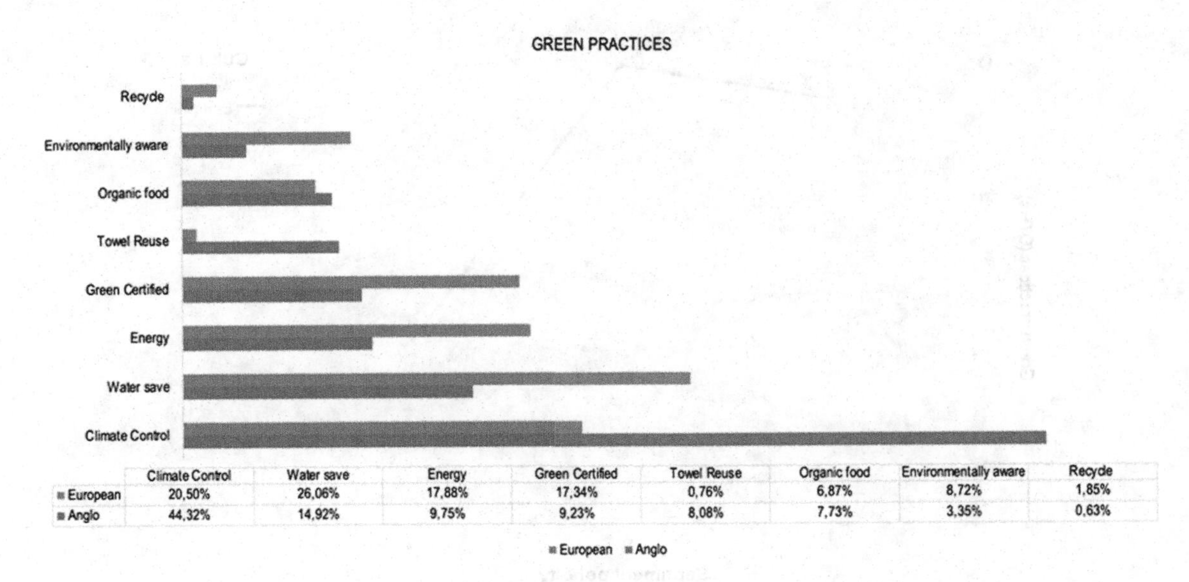

Once the most frequent words related to sustainable practices mentioned by tourist in the OTRs were identified, we employed a binary system (0/1) to code each review and identify green comments. These criteria encompass the green terms mentioned above. For example, if a review included any of the green terms, it was coded as 1 and considered as green review. However, if the review did not include any of green practices, it was coded as 0.

Table 7. Descriptive statistics

	Total reviews	Green Reviews	Green Reviews %
Anglo	16340	631	3,86%
European	4521	252	5,57%

Finally, the sentiment polarity classification was performed with approximately 89.19% overall accuracy, and the positive precision is equal to 92.34%. To test RQ3 we analyzed the relationships between culture (Anglo vs European) and sentiment polarity (negative, neutral and positive) with the content of review (green vs no green) by applying an ANOVA. The research findings (Figure 11) showed that culture (Anglo vs European) has a significant effect on the sustainable content included in tourists' reviews, $F(1, 883) = 17.031$, $p < 0.001$. Specifically, European tourists tend to share more perceptions about sustainable practices implemented by hotels. Furthermore, the sentiment polarity (positive vs neutral vs negative) also has a direct and significant effect on the sustainable content included in tourists' comments, $F(2, 993) = 18.106$, $p < 0.001$, with negative comments mentioning sustainable practices more frequently. Lastly, there is no significant interaction effect between culture and sentiment polarity on the sustainable content included in tourist's comments, $F(2, 883) = 0.489$, $p < 0.05$.

Figure 11. ANOVA results

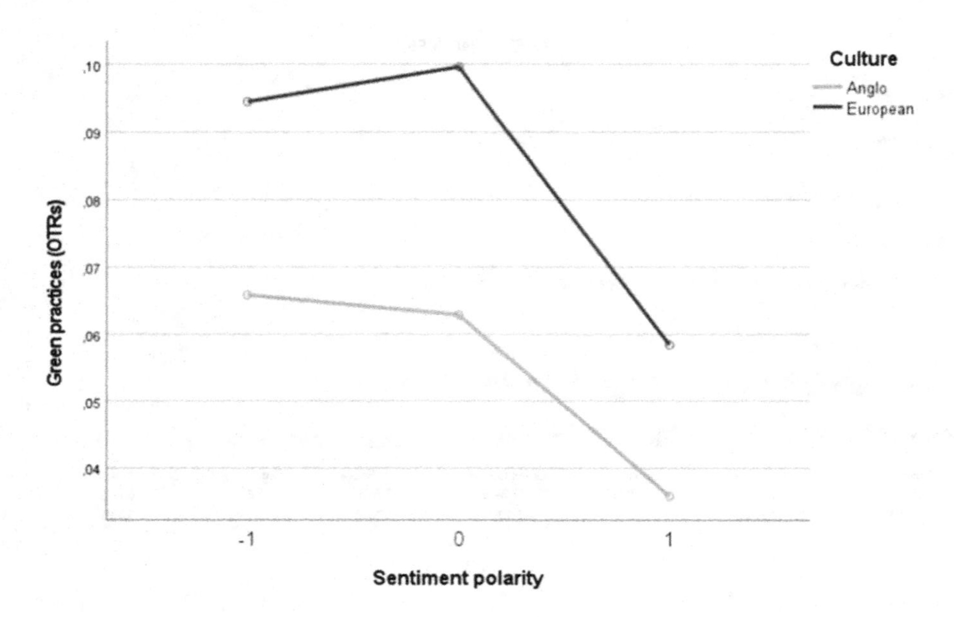

DISCUSSION

To enhance the environmental perception of hotels, it is essential for hoteliers to develop effective strategies. To achieve this, they must gain a comprehensive understanding of the underlying dimensions of a hotel's image. Additionally, practitioners should recognize the variations in these perceptions based on the cultural backgrounds of tourists. One cost-effective and efficient approach for the hospitality sector to gather opinions from tourists regarding their overall perception of a hotel and its implementation of sustainable practices is to extract information from online reviews.

This study aimed to investigate the cross-cultural variations between Anglo and European green hotel customers. Specifically, the focus was on analyzing the components of a hotel's green image expressed in reviews posted on TripAdvisor. By conducting component analysis across diverse cultural backgrounds, this research contributes to the existing literature on cross-cultural differences in online settings. Moreover, it enhances our current understanding of the behavior of guests who prefer environmentally friendly hotels. Given the increasing significance of the green strategies, as discussed by previous studies such as (Nimri et al., 2020) and (Galeazzo et al., 2021), this research stands out as one of the few that utilizes online data sources to contribute empirically to the study of green hotel image perceptions. Furthermore, this research presents a groundbreaking approach of sentiment analysis techniques to extract marketing insights. Particularly, this study proposes the use of sentiment analysis for identifying cultural differences in online tourist green reviews.

In response to the call by Lim et al. (2022), which encourages the examination of the hotel image conveyed by users on social media, the current study analyzes the components of the image of eco-friendly hotels mentioned by tourists in their reviews. The findings of our content analysis, in line with Sadiq al. (2021), showed that the green hotel image include three components: cognitive, affective, and conative. All three image dimensions are present in both group (Anglo and European reviews), being the cognitive the most prominent, but the importance of the sub-dimensions of each group differ. For example, focusing on cognitive component, Anglo reviews feature more words about transport vehicles, duration, gastronomic offer, city attraction, sustainability, booking criteria and mention seasonality, a sub-dimension that is not present in European reviews. While European tourists stand out sub-dimensions such as tangible facilities, service, location, kind of trip, price, interactivity with other tourists and hotel quality; being brand name and destination word frequency quite similar in both groups. In the line with Wang et al., 2019, our findings showed that Anglo tourists, due to their high uncertainty avoidance, feel more comfortable including aspects related to trip organization in their comments (such as seasonality, booking, or tourist attractions). However, European tourists, being more willing to embrace changes, focus on aspects like interactions with other tourists or service. This result also supports Leon et al. (2019), who found that detailed reviews are provided by the guests who have a long-term orientation, such as Anglos turists. The affective dimension shows that the positive words are salient. In both clusters, the most frequent positive words refer to the staff and tangible facilities, with similar frequencies. However, in terms of location, the frequency of positive words is higher in European reviews. This result aligns with the findings of the cognitive dimension, which indicate that Europeans mention the location sub-dimension more frequently. Regarding negative sentiments, the main terms are related to price in both cities. Finally, in the conative dimensions the most frequently used words in both clusters are related to attitudinal subdimension. However, when we analyze each subdimension separately, we see that Anglo comments have a higher frequency of terms related to the behavioral aspect, while European tourists stand out in terms of attitudinal terms. Accordingly, this study demonstrated that cultures with a masculine orientation, such as Anglo cultures, tend to prioritize action and acquisition over contemplation and observation (Cyr et al. 2017; Biswas et al., 2021). As for the polarity of both dimensions, it is worth noting that positive recommendations and intentions prevail.

Based on the most frequent words, we have classified the comments from both cultural groups into cognitive, affective, and behavioral categories, without excluding the categories from each other, and observing that there are comments that include all three dimensions of the image. The investigation has shown that the tourists from different cultures express differently their perception regarding cognitive and affective image components, with reviews authored by Europeans more affective while the Anglos are more cognitive. On one hand, this finding supports the idea that European consumers tend to be notably more expressive (Krys et al., 2022). On the other hand, it confirms that tourists with lower aversion to uncertainty are more willing to share their positive experiences with others (Hofstede et al., 2010).

The analysis of green reviews highlights the low frequency of terms related to the environment included in the OTRs from both cultural groups. The sustainable terms were categorized in eight groups, being climate control and towel reuse the most prominent in Anglo reviews and water save, energy, green certified, environmentally aware, organic food and recycle the most frequent in European tourists OTRs. Online tourists' reviews have begun to prove that culture is a critical factor that affects consumers' motivation to share green comments (Guan et al., 2022). This study results showed tourists provide more green comments for green hotels from European Countries. Therefore, the findings of the study are consistent with previous research (Fang et al., 2013; Yin et al., 2014, Tseng 2017) that has demonstrated that collectivist culture exhibits a high level of conformity and is open to social commitment involving environmental concerns. Moreover, our findings showed that the sentiment of OTRs influence green comment, being the green OTRs more negative than the no green OTRs. This result confirms the findings of some recent studies (Kim and Kim, 2022) that posit that customers who are most concerned about the environment are the ones who share the greenest electronic word-of-mouth. However, their involvement and knowledge in this regard lead to high expectations that are sometimes not met by the practices implemented by hotels.

THEORETICAL AND MANAGERIAL IMPLICATIONS

Unlike other studies that examined tourists' perceptions of the images of green hotels, the present study goes further by comparing cross-cultural differences (Anglo vs European reviews). Thus, the present study makes several important contributions to the previous literature on the overall perceived image of green hotels. First, the dimensions of hotel image and the subdimensions of each one was derived from narratives shared online by guests who stayed in green hotels located in Amsterdam. According to recent research (Wang et al., 2019) the content of online reviews plays a key role in understanding tourists' experiences and the perceived hotel image. Whereas many studies into hotel image evaluate it based on its cognitive and affective dimensions (e.g., Chen and Chen, 2014; Durrna et al., 2015; Wang et al., 2019), the present study goes a step forward towards and examined the tripartite cognitive - affective - conative image of green hotels using big data extracted from TripAdvisor OTRs. This study challenges the dominant two-dimensional view that regards cognitive and affective dimensions as two ends of a continuum. Instead, we adopt a tri-dimensional view, that is, cognitive, affective and conative components can coexist and exert independent impacts simultaneously. This is a significant contribution of the present study, and the results are consistent with recent research (e.g., Sadiq al., 2021). Second, the findings of the present study highlight that cognitive and affective dimensions attract more words than does the conative dimension, which is residual in both cultural clusters (Anglo and European reviews). Our findings emphasize that cognitive elements often dominate destination image (Marine-Roig, 2019; Loja et al., 2020). These results, although not common in the context of hospitality, are consistent with previous studies that have demonstrated hierarchical relationships among the dimensions, cognitive - affective – conative (Loja et al., 2020). Positive reviews can be seen as an indirect way of endorsing the hotel, even if the term "recommend" is not explicitly used (Marine-Roig, 2019). Therefore, this constitutes a significant contribution.

Third, these findings enrich the literature on the image of eco-friendly hotels based on information from tourists of different cultures (Anglo and European reviews). The results, similar to previous analyses (Mequita and Walker, 2003, Buzova et al., 2018, Wang et al., 2019), reveal notable differences in the attributes that tourists mention in their reviews. Building upon prior research, the findings of this study identify that European tourists, due to their lower aversion to uncertainty, are more willing to express their emotions. Meanwhile, Anglo tourists, as a culture with a masculine orientation, prioritize attributes that constitute the cognitive dimension, which is consistent with Biswas et al. (2021). The findings increase the understanding of green hotel image by revealing those attributes which play an important role in building the image of a green hotel based on cultural differences. Additionally, revealed that customers are not fully satisfied with sustainable practices would provide negative word-of-mouth (Merli et al., 2019). Accordingly with previous research (D'Acuto et al., 2020; Arici et al., 2023), the vast majority of reviews discussing sustainable practices are positive. However our findings indicate that, in comparison to reviews that do not mention green aspects, they often convey more negative messages. Contrary to what Gasbarro and Bonera (2021) stated, our results suggest that the implementation of eco-friendly practices that do not meet guests' expectations stimulates electronic word-of-mouth (eWOM) among guests regarding the eco-friendly practices of an accommodation facility, although at the same time it appears to reflect low overall guest satisfaction scores.

This study also offers some managerial implications. The most common reviews are those that describe specific hotel attributes. Therefore, we advise marketing professionals to solicit emotional reviews that express attitudinal and behavioral aspects to support electronic word-of-mouth (eWOM) with facts. Through sentiment analysis of guests' expressions about eco-friendly hotels, thus reflecting their image, managers can gain an informed understanding and details that cannot be obtained by examining overall ratings alone.To translate the results of the investigation into practical terms, the study suggests that managers of green hotels should be mindful of cross-cultural differences in the content of reviews. The analysis of keywords related to the image of green hotels indicates certain aspects that managers should consider. These include factors such as "room", "staff", "cleanliness", and "bed" for both clusters of tourists. Additionally, for European tourists, factors like "breakfast", "price", and "views" should be taken into account, while for Anglo tourists, "location" and "check-in/out" are the most important.

Secondly, we suggest that, alongside the implementation of eco-friendly practices, hotels engage in effective communication of these practices. Although all the analyzed hotels were green, it is important to recognize that tourists' perceptions of differences in the hotels' green practices may not always lead to positive outcomes if these practices are not effectively communicated to the customers. Therefore, it is crucial for hotel managers, especially when catering to European tourists, to actively communicate their green practices. They should display the standards they meet on their homepages so that environmentally conscious consumers can select the one that best suits their needs, thereby avoiding dissatisfaction experiences that can harm the hotel's image. Furthermore, it would be beneficial for consumers to evaluate these practices after their stay, fostering improvement by the hotel and enhancing the tourists' experience by making them feel heard.

Finally, some recommendations for platforms hosting tourist reviews (e.g., TripAdvisor). It is crucial that managers of these platforms improve their understanding of the possible unintended consequences of cultural differences regarding review-posting behavior. Firstly, we suggest that platforms hosting tourist reviews consider actively providing a culturally customized review option, which can serve to differentiate them from others. Second, these platforms should reconsider the design of their online review systems and show each tourist the reviews that correspond to their cultural group to obtain more relevant information for their choice. This could include adding filters for review searches based on the guests' origin. Similarly, regarding environmental aspects, they should encourage authors to provide reviews expressing their opinions about sustainable practices, posing questions like "What sustainable practices did you identify during your stay?" and indicating their level of involvement with the environment.

LIMITATIONS AND FUTURE RESEARCH LINES

This chapter presents limitations that suggest promising avenues for future research. First, although its large sample size of green hotels OTRs size, the study analyses green hotels only in Amsterdam, so future studies might examine guests' perceptions in other European cities. It would also be interesting to analyze the approach to sustainable practices of large hotel chains operating in different cities to control the variability in standards used by hotels and see if there are significant differences in the perception of these practices, taking into account factors such as the type of tourism offered by each city. Second, the present research collected OTRs only from TripAdvisor, which might have a platform bias. However, other platforms such as Booking have established their own sustainability certification, so it would be interesting to see on which of the two platforms green eWOM is more abundant and if there are substantial differences in the volume of comments related to the sustainable practices implemented by hotels.Third, in future research it would be interesting to examine other interesting aspects of the analyzed reviews, such as their usefulness and also aspects related to the tourists themselves, like their level of co-creation, the timing of when they share their reviews (immediately after the experience or a few days later), or whether they include photos of their experience. Last, we considered only the content of the review, however future research may include the title and the star ratings and show if the sentiment is congruent between them.

REFERENCES

Arici, H. E., Cakmakoglu Arıcı, N., & Altinay, L. (2023). The use of big data analytics to discover customers' perceptions of and satisfaction with green hotel service quality. *Current Issues in Tourism*, *26*(2), 270–288. doi:10.1080/13683500.2022.2029832

Badenes, A., Ruiz, C. & Bigné, E. (2019). Engaging customers through user-and company-generated content on CSR. *Spanish Journal of marketing-ESIC, 23*(3), 379-372.

Bao, T., & Chang, T. (2016). The product and timing effects of eWOM in viral marketing. *International Journal of Business*, *21*(2), 99–111.

Bigne, E., Ruiz, C., Perez-Cabañero, C., & Cuenca, A. (2023). Are customer star ratings and sentiments aligned? A deep learning study of the customer service experience in tourism destinations. *Service Business*, *17*(1), 281–314. doi:10.100711628-023-00524-0

Biswas, B., Sengupta, P., & Ganguly, B. (2021). Your reviews or mine? Exploring the determinants of "perceived helpfulness" of online reviews: A cross-cultural study. *Electronic Markets*, 1–20. PMID:35602112

Boulding, K. E. (1956). *The image: Knowledge in life and society* (Vol. 47). University of Michigan press. doi:10.3998/mpub.6607

Buzova, D., Sanz-Blas, S., & Cervera-Taulet, A. (2019). Does culture affect sentiments expressed in cruise tours' eWOM? *Service Industries Journal*, *39*(2), 154–173. doi:10.1080/02642069.2018.1476497

Chen, W. J., & Chen, M. L. (2014). Factors affecting the hotel's service quality: Relationship marketing and corporate image. *Journal of Hospitality Marketing & Management*, *23*(1), 77–96. doi:10.1080/19368623.2013.766581

Chong, A., Li, B., Ngai, E., Ch'ng, E., & Lee, F. (2016). Predicting online product sales via online reviews, sentiments, and promotion strategies: A big data architecture and neural network approach. *International Journal of Operations & Production Management*, *36*(4), 358–383. doi:10.1108/IJOPM-03-2015-0151

Ciechanowski, L., Jemielniak, D., & Gloor, P. A. (2020). TUTORIAL: AI research without coding: The art of fighting without fighting: Data science for qualitative researchers. *Journal of Business Research*, *117*, 322–330. doi:10.1016/j.jbusres.2020.06.012

Cyr, D., Gefen, D., & Walczuch, R. (2017). Exploring the relative impact of biological sex and masculinity–femininity values on information technology use. *Behaviour & Information Technology*, *36*(2), 178–193. doi:10.1080/0144929X.2016.1212091

D'Acunto, D., Tuan, A., Dalli, D., Viglia, G., & Okumus, F. (2020). Do consumers care about CSR in their online reviews? An empirical analysis. *International Journal of Hospitality Management*, *85*, 102342. doi:10.1016/j.ijhm.2019.102342

De Mooij, M., & Hofstede, G. (2011). Cross-cultural consumer behavior: A review of research findings. *Journal of International Consumer Marketing*, *23*(3-4), 181–192.

Deng, L., & Yu, D. (2014). Foundations and Trends in Signal Processing. *DEEP LEARNING–Methods and Applications*, *7*(3–4), 197–387.

Dhar, S., & Bose, I. (2022). Walking on air or hopping mad? Understanding the impact of emotions, sentiments and reactions on ratings in online customer reviews of mobile apps. *Decision Support Systems*, *162*, 113769. doi:10.1016/j.dss.2022.113769

Durna, U., Dedeoglu, B. B., & Balikçioglu, S. (2015). The role of servicescape and image perceptions of customers on behavioral intentions in the hotel industry. *International Journal of Contemporary Hospitality Management*, *27*(7), 1728–1748. doi:10.1108/IJCHM-04-2014-0173

Fang, H., Zhang, J., Bao, Y., & Zhu, Q. (2013). Towards effective online review systems in the Chinese context: A cross-cultural empirical study. *Electronic Commerce Research and Applications*, *12*(3), 208–220. doi:10.1016/j.elerap.2013.03.001

Filieri, R., Lin, Z., Pino, G., Alguezaui, S., & Inversini, A. (2021). The role of visual cues in eWOM on consumers' behavioral intention and decisions. *Journal of Business Research*, *135*, 663–675. doi:10.1016/j.jbusres.2021.06.055

Galati, A., Thrassou, A., Christofi, M., Vrontis, D., & Migliore, G. (2021). Exploring travelers' willingness to pay for green hotels in the digital era. *Journal of Sustainable Tourism*, 1–18. doi:10.1080/09669582.2021.2016777

Gasbarro, F., & Bonera, M. (2021). The influence of green practices and green image on customer satisfaction and word-of-mouth in the hospitality industry. *Sinergie Italian Journal of Management*, *39*(3), 231–248. doi:10.7433116.2021.12

González-Rodríguez, M. R., Díaz-Fernández, M. C., & Font, X. (2020). Factors influencing willingness of customers of environmentally friendly hotels to pay a price premium. *International Journal of Contemporary Hospitality Management*, *32*(1), 60–80. doi:10.1108/IJCHM-02-2019-0147

Green Hotel Association. (2008). *What are green hotels?* Green Hotel Association. (https://www.greenhotels.com/whatare.htm).

Guan, C., Hung, Y. C., & Liu, W. (2022). Cultural differences in hospitality service evaluations: Mining insights of user generated content. *Electronic Markets*, *32*(3), 1–21. doi:10.100712525-022-00545-z

Hameed, I., Hussain, H., & Khan, K. (2022). The role of green practices toward the green word-of-mouth using stimulus-organism-response model. *Journal of Hospitality and Tourism Insights*, *5*(5), 1046–1061. doi:10.1108/JHTI-04-2021-0096

Hennig-Thurau, T., Gwinner, K. P., Walsh, G., & Gremler, D. D. (2004). Electronic word-of-mouth via consumer-opinion platforms: What motivates consumers to articulate themselves on the internet? *Journal of Interactive Marketing*, *18*(1), 38–52. doi:10.1002/dir.10073

Hoftede, G., Hofstede, G. J., & Minkov, M. (2010). *Cultures and organizations: software of the mind: intercultural cooperation and its importance for survival*. McGraw-Hill.

Hu, N., Koh, N., & Reddy, S. (2014). Ratings lead you to the product, reviews help you clinch it? The mediating role of online review sentiments on product sales. *Decision Support Systems*, *57*, 42–53. doi:10.1016/j.dss.2013.07.009

Índice, I. E. S. E. (2022). *Cities in Motion*. IESE Business School.

Kim, J. M., Jun, M., & Kim, C. K. (2018). The effects of culture on consumers' consumption and generation of online reviews. *Journal of Interactive Marketing*, *43*, 134–150. doi:10.1016/j.intmar.2018.05.002

Kim, Y. J., & Kim, H. S. (2022). The impact of hotel customer experience on customer satisfaction through online reviews. *Sustainability (Basel)*, *14*(2), 848. doi:10.3390u14020848

Koh, N., Hu, N., & Clemons, E. (2010). Do online reviews reflect a product's true perceived quality? An investigation of online movie reviews across cultures. *Electronic Commerce Research and Applications*, *9*(5), 374–385. doi:10.1016/j.elerap.2010.04.001

Krys, K., Vignoles, V. L., De Almeida, I., & Uchida, Y. (2022). Outside the "cultural binary": Understanding why Latin American collectivist societies foster independent selves. *Perspectives on Psychological Science*, *17*(4), 1166–1187. doi:10.1177/17456916211029632 PMID:35133909

Kumari, P., & Sangeetha, R. (2022). How Does Electronic Word of Mouth Impact Green Hotel Booking Intention? *Services Marketing Quarterly*, *43*(2), 146–165. doi:10.1080/15332969.2021.1987609

Kuo, N. W., & Hsiao, T. Y. (2008). An exploratory research of the application of natural capitalism to sustainable tourism management in Taiwan. *Journal of Cleaner Production*, *16*(1), 116–124. doi:10.1016/j.jclepro.2006.11.005

Lee, J. S., Hsu, L. T., Han, H., & Kim, Y. (2010). Understanding how consumers view green hotels: How a hotel's green image can influence behavioural intentions. *Journal of Sustainable Tourism*, *18*(7), 901–914. doi:10.1080/09669581003777747

Leon, R. D. (2019). Hotel's online reviews and ratings: A cross-cultural approach. *International Journal of Contemporary Hospitality Management*, *31*(5), 2054–2073. doi:10.1108/IJCHM-05-2018-0413

Lin, M. P., Marine-Roig, E., & Llonch-Molina, N. (2022). Gastronomy tourism and well-being: Evidence from Taiwan and Catalonia Michelin-starred restaurants. *International Journal of Environmental Research and Public Health*, *19*(5), 2778. doi:10.3390/ijerph19052778 PMID:35270469

Lojo, A., Li, M., & Xu, H. (2020). Online tourism destination image: Components, information sources, and incongruence. *Journal of Travel & Tourism Marketing*, *37*(4), 495–509. doi:10.1080/10548408.2020.1785370

Loyola-González, O., Monroy, R., Rodríguez, J., López-Cuevas, A., & Mata-Sánchez, J. I. (2019). Contrast pattern-based classification for bot detection on twitter. *IEEE Access : Practical Innovations, Open Solutions*, *7*, 45800–45817. doi:10.1109/ACCESS.2019.2904220

Ludwig, S., De Ruyter, K., Friedman, M., Brüggen, E., Wetzels, M., & Pfann, G. (2013). More than words: The influence of affective content and linguistic style matches in online reviews on conversion rates. *Journal of Marketing*, *77*(1), 87–103. doi:10.1509/jm.11.0560

Malinen, S., & Nurkka, P. (2015). Cultural influence on online community use: A cross–cultural study on online exercise diary users of three nationalities. *International Journal of Web Based Communities*, *11*(2), 153–169. doi:10.1504/IJWBC.2015.068539

Marine-Roig, E. (2019). Destination image analytics through traveller-generated content. *Sustainability (Basel)*, *11*(12), 3392. doi:10.3390u11123392

Martínez, P. (2015). Customer loyalty: Exploring its antecedents from a green marketing perspective. *International Journal of Contemporary Hospitality Management*, *27*(5), 896–917. doi:10.1108/IJCHM-03-2014-0115

Mesquita, B., & Walker, R. (2003). Cultural differences in emotions: A context for interpreting emotional experiences. *Behaviour Research and Therapy*, *41*(7), 777–793. doi:10.1016/S0005-7967(02)00189-4 PMID:12781245

Nimri, R., Patiar, A., & Jin, X. (2020). The determinants of consumers' intention of purchasing green hotel accommodation: Extending the theory of planned behaviour. *Journal of Hospitality and Tourism Management*, *45*, 535–543. doi:10.1016/j.jhtm.2020.10.013

Oliveira, B., & Casais, B. (2019). The importance of user-generated photos in restaurant selection. *Journal of Hospitality and Tourism Technology*, *10*(1), 2–14. doi:10.1108/JHTT-11-2017-0130

Pike, S., Pontes, N., & Kotsi, F. (2021). Stopover destination attractiveness: A quasi-experimental approach. *Journal of Destination Marketing & Management*, *19*, 100514. doi:10.1016/j.jdmm.2020.100514

Ponnapureddy, S., Priskin, J., Ohnmacht, T., Vinzenz, F., & Wirth, W. (2017). The influence of trust perceptions on German tourists' intention to book a sustainable hotel: A new approach to analysing marketing information. *Journal of Sustainable Tourism*, *25*(7), 970–988. doi:10.1080/09669582.2016.1270953

Poulakidakos, S., & Armenakis, A. (2014). Propaganda in Greek public discourse. Propaganda scales in the presentation of the Greek MoU-bailout agreement of 2010. Revista de Ştiinţe Politice. *Revue des Sciences Politiques*, *41*, 126–140.

Rapoport, A. (1977). *Human Aspects of Urban Form*.

Sadiq, M., Adil, M., & Paul, J. (2022). Eco-friendly hotel stay and environmental attitude: A value-attitude-behaviour perspective. *International Journal of Hospitality Management*, *100*, 103094. doi:10.1016/j.ijhm.2021.103094

Santos-Lacueva, R., Velasco González, M., & González Domingo, A. (2022). *The integration of sustainable tourism policies in European cities*.

Sayfuddin, A. T. M., & Chen, Y. (2021). The signaling and reputational effects of customer ratings on hotel revenues: Evidence from TripAdvisor. *International Journal of Hospitality Management*, *99*, 103065. doi:10.1016/j.ijhm.2021.103065

Sharma, A., Park, S., & Nicolau, J. L. (2020). Testing loss aversion and diminishing sensitivity in review sentiment. *Tourism Management*, *77*, 104020. doi:10.1016/j.tourman.2019.104020

Sharma, S. K., & Hoque, X. (2017). Sentiment predictions using support vector machines for odd-even formula in Delhi. *International Journal of Intelligent Systems and Applications*, *9*(7), 61–69. doi:10.5815/ijisa.2017.07.07

Stemler, S. (2019) *An Overview of Content Analysis*. Pare Online. https://pareonline.net/getvn.asp?v=7&n=17

Taecharungroj, V., & Mathayomchan, B. (2019). Analysing TripAdvisor reviews of tourist attractions in Phuket, Thailand. *Tourism Management*, *75*, 550–568. doi:10.1016/j.tourman.2019.06.020

Timoshenko, A., & Hauser, J. R. (2019). Identifying customer needs from user-generated content. *Marketing Science*, *38*(1), 1–20. doi:10.1287/mksc.2018.1123

TripAdvisor. (2023). *TripAdvisor Review Transparency Report*. Tripadvisor. https://www.tripadvisor.com.mx/TransparencyReport2023#group-section-Trends-and-Processes-Tkmr3mda9i

Tseng, A. (2017). Why do online tourists need sellers' ratings? Exploration of the factors affecting regretful tourist e-satisfaction. *Tourism Management*, *59*, 413–424. doi:10.1016/j.tourman.2016.08.017

Tsytsarau, M., & Palpanas, T. (2012). Survey on mining subjective data on the web. *Data Mining and Knowledge Discovery*, *24*(3), 478–514. doi:10.100710618-011-0238-6

Verma, V. K., Chandra, B., & Kumar, S. (2019). Values and ascribed responsibility to predict consumers' attitude and concern towards green hotel visit intention. *Journal of Business Research*, *96*, 206–216. doi:10.1016/j.jbusres.2018.11.021

Villarroel Ordenes, F., Ludwig, S., De Ruyter, K., Grewal, D., & Wetzels, M. (2017). Unveiling what is written in the stars: Analyzing explicit, implicit, and discourse patterns of sentiment in social media. *The Journal of Consumer Research*, *43*(6), 875–894. doi:10.1093/jcr/ucw070

Wang, J., Wang, S., Xue, H., Wang, Y., & Li, J. (2018). Green image and consumers' word-of-mouth intention in the green hotel industry: The moderating effect of Millennials. *Journal of Cleaner Production*, *181*, 426–436. doi:10.1016/j.jclepro.2018.01.250

Wang, W., Ying, S., Lyu, J., & Qi, X. (2019). Perceived image study with online data from social media: The case of boutique hotels in China. *Industrial Management & Data Systems*, *119*(5), 950–967. doi:10.1108/IMDS-11-2018-0483

Wang, Y., Wang, Z., Zhang, D., & Zhang, R. (2019). Discovering cultural differences in online consumer product reviews. *Journal of Electronic Commerce Research*, *20*(3), 169–183.

Yadegaridehkordi, E., Nilashi, M., Nasir, M., Momtazi, S., Samad, S., Supriyanto, E., & Ghabban, F. (2021). Customers segmentation in eco-friendly hotels using multi-criteria and machine learning techniques. *Technology in Society*, *65*, 101528. doi:10.1016/j.techsoc.2021.101528

Yang, Y., Jiang, L., & Wang, Y. (2023). Why do hotels go green? Understanding TripAdvisor GreenLeaders participation. *International Journal of Contemporary Hospitality Management*, *35*(5), 1670–1690. doi:10.1108/IJCHM-02-2022-0252

Yin, D., Bond, S. D., & Zhang, H. (2014). Anxious or angry? Effects of discrete emotions on the perceived helpfulness of online reviews. *Management Information Systems Quarterly*, *38*(2), 539–560. doi:10.25300/MISQ/2014/38.2.10

Zhu, D., Ye, Z., & Chang, Y. (2017). Understanding the textual content of online customer reviews in B2C websites: A cross-cultural comparison between the US and China. *Computers in Human Behavior*, *76*(November), 483–493. doi:10.1016/j.chb.2017.07.045

Zhu, X., Guo, H., Mohammad, S., & Kiritchenko, S. (2014). An empirical study on the effect of negation words on sentiment. *Proceedings of the Annual Meeting of the Association for Computational Linguistics (ACL)* (pp. 304–313). Baltimore, MD. 10.3115/v1/P14-1029

Zulkifli, N. S. A., & Lee, A. W. K. (2019). Sentiment analysis in social media based on english language multilingual processing using three different analysis techniques. In Soft Computing in Data Science: 5th International Conference, [Springer Singapore.]. *Proceedings*, *5*, 375–385.

Chapter 8
Artificial Intelligence Method for the Analysis of Marketing Scientific Literature

Antonio Hyder
Miguel Hernandez University, Spain & Hackers and Founders Research, USA

Carlos Perez-Vidal
Universidad Miguel Hernandez, Spain

Ronjon Nag
Stanford University, USA

ABSTRACT

A machine-based research reading methodology specific to the academic discipline of marketing science is introduced, focused on the text mining of scientific texts, analysis and predictive writing, by adopting artificial intelligence developments from other research fields in particular materials and chemical science. It is described how marketing research can be extracted from documents, classified and tokenised in individual words. This is conducted by applying text-mining with named entity recognition together with entity normalisation for large-scale information extraction of published scientific literature. Both a generic methodology for overall marketing science analysis as well as a narrowed-down contextualised method for delimited marketing topics are detailed. Automated literature review is discussed as well as potential automated formulation of hypotheses and how AI can assist in the transfer of marketing research knowledge to practice, in particular to startups, as they can benefit from AI powered science-based decision making. Recommendations for next steps are made.

DOI: 10.4018/978-1-6684-9591-9.ch008

INTRODUCTION

Given the challenges humanity is facing, the acceleration of scientifically tested discovery is essential for human flourishing. Scientific progress is disseminated in scientific publications. The number of journal articles has increased dramatically with millions of research articles already available yet growing exponentially with thousands of new papers annually in fields such as materials, education and marketing science. Researchers have become much better at the generation of information than at its integrative analysis, interpretation and use for tackling issues where knowledge is required.

A substantial human effort is required for the manual extraction of research literature. Keeping up with publications is untenable even within specialised disciplines. Such overload has an impact on making use of research discoveries, as they depend on the knowledge embedded in past publications. It is increasingly harder to assimilate, relate and connect research and new results with the previously established literature, and to elucidate the role of a piece of knowledge. Individual marketing scientists will only access a fraction of research information in their lifetime and only a sliver of the relevant science guides the formulation of new questions or hypotheses. As a consequence, a major bottleneck arises for attaining progress in the marketing discipline as well as for the application of knowledge to both science and practice.

The vast majority of marketing science knowledge is stored as unstructured text across published articles. Although current online search technologies typically find many relevant documents, they still do not extract and organise the information content of scientific documents, nor suggest new scientific hypotheses to be formulated based on organised content. There is a need for techniques that can simplify the use of marketing research knowledge. Here is where artificial intelligence (AI) can assist.

Fundamentals of AI for Marketing Science Analysis

Herewith a description of AI terms required for the development of this chapter.

Scientific research documents are commonly presented in the form of scientific articles and thesis documents available through published digital information. They are written in natural languages mixed with domain-exclusive terminologies added to numerical data, and are difficult to analyse by either traditional statistical analysis or modern machine learning (ML).

Text mining (TM) refers to using algorithms for the extraction of information from written text documents. It is also used with scientific literature automatically identifying direct and indirect references to specific knowledge (e.g., Srinivasan, 2004).

Natural language processing (NLP) is a subfield of linguistics, computer science, information engineering and artificial intelligence concerned with the interactions between computers and human natural languages, in particular how to program computers in order to process and analyse large amounts of natural language data.

Scientific literature text mining with natural language processing is enabled by combining text mining and NLP. It has been used in several studies focused on the supervised retrieval of information from scientific literature. Research knowledge can be efficiently encoded as information-dense word embeddings, which is a vector representation of words, with or without human labelling or supervision (Tshitoyan et al., 2019).

Named entity recognition (NER), one of the most widely tools used in NLP, is capable of extracting information from text on a massive scale. NER locates and classifies named entities from unstructured text into categories.

AI Machine Reading Developments Across Scientific Fields

The majority of AI research literature reading developments have been made in materials, biology and chemistry (e.g., Eltyeb & Salim, 2014; Kim et al., 2017; Srinivasan, 2004; Swain & Cole, 2016), however progress has also been made in other domains. For instance, as in education research the time between a finding, then publishing and referencing it in a literature review ranges between 2.5 and 6.5 years, Crues (2017) suggested developing an automated review system where the most up-to-date published findings are added. This facilitates that both practitioners and researchers are better informed of the latest breakthroughs. In defence, DARPA (2023) aims at developing machines that understand and reason in context so that the machine increasingly becomes a partner. For this to occur AI experts and domain experts must work together.

Artificial Intelligence and NLP in the Academic Field of Marketing Science

Due to the foreseeable exponential growth of scientific literature, the global academic field of marketing science would also benefit from AI-based technologies for text mining and analysis, or predictive writing as occurs in other areas. Marketing science is still developing understandings on how AI technologies could be used in the future (e.g., Huang & Rust, 2021; Mariani et al., 2023; Mustak et al., 2020). However, progress is already being made in large-scale extraction from summary-level information and automated literature review of marketing science texts. Hyder & Nag (2020) analysed the feasibility of machine reading marketing science literature by applying progress in natural language processing as applied to scientific articles. Subsequently Hyder & Nag (2021) defined steps in document extraction, and tokenisation of individual marketing-related words from marketing science literature.

Overload creates barriers for adoption of research, yet marketing scientists need to formulate thousands of hypotheses annually. It would be extremely useful for marketing researchers to be able to ask large-scale questions in published literature such as 'for the marketing scientific literature on engagement, which types of engagement have been studied in digital contexts?', 'which are yet to be explored' or 'give me a list of all documents related to this query'. Currently, answering such questions requires a laborious and tedious literature search performed manually by a domain expert. However, by representing published articles as structured database entries, such questions may be asked programmatically and can be answered in a matter of seconds. AI techniques such as text mining and natural language processing (NLP) can assist in removing growing bottlenecks in research literature. Kim et al. (2017) had already developed a platform which automatically retrieves articles and then extracts and codifies the words and parameters found in the text. By combining this at large scale, a structured database can be mined to discover underlying marketing relationships. The key idea is that, because words with similar meanings often appear in similar contexts, the corresponding embeddings will also be similar, and this allows to form relationships between them. Training a machine for recognising equivalent processes and synonyms is challenging. Issues arise in the way in which entities are written. This is crucial for normalising entities and mapping each entity onto a sole database. These ideas are described in our suggested generic methodology within this chapter.

Based on the advances accomplished in other fields, the aim of our research in progress is to develop an AI scientific literature reading methodology for the discipline of marketing science that ultimately can answer questions, conduct predictive writing, automatically generate hypotheses and assist in the transfer of research knowledge to practice.

STATE OF THE ART OF SCIENTIFIC LITERATURE MACHINE READING

Artificial Intelligence and Natural Language Processing

AI research studies have already focused on the retrieval of information from scientific literature using supervised NLP. The majority and most advanced developments have been made in the disciplines of chemistry and materials (e.g., Kim et al., 2017; Tshitoyan et al., 2019). Scientific knowledge extraction and relationships are based on massive bodies of scientific literature, where CrossRef Application Programming Interface (API) can be used to retrieve large lists of article Digital Object Identifiers (DOIs). This is used by a number of publisher APIs (such as Elsevier https://dev.elsevier.com and Springer Nature https://dev.springernature.com) which enable downloading of full-text journal articles.

Several studies that have used supervised NLP point out that this requires large hand-labelled datasets for training. However, Tshitoyan et al. (2019) suggested that published literature can be efficiently encoded without supervision using word embeddings. Embeddings are usually constructed using ML algorithms such as Word2vec (Mikolov et al., 2013) which predicts the contexts of words and creates a semantic space, that is, a mathematical representation of a large body of text. Both Weston et al. (2019) and Tshitoyan et al. (2019) have developed unsupervised methods that can actually recommend materials for practical applications several years before their discovery.

In Tshitoyan et al. (2019) method, embeddings were trained with 3.3 million scientific abstracts from 1922 to 2018 from more than 1000 journals deemed likely to contain materials-related research, directly retrieved from scientific databases. The performance of their algorithm substantially improved when irrelevant and non-English abstracts were removed. The remaining 1.5 million abstracts classified as relevant were tokenised using ChemDataExtractor, a tool for automatically extracting relevant information from scientific documents in order to produce individual words, resulting in a vocabulary of approximately 500000 words. From a journal article, ChemDataExtractor will extract names and properties from the text so they can be imported into a database or spreadsheet. Other NLP efforts include Kim et al. (2017) who developed an NLP learning approach that automatically compiles materials' synthesis parameters across tens of thousands of research texts, training a binary classifier that labels abstracts as 'relevant' or 'not relevant'.

Named Entity Recognition (NER)

Recent advances in machine learning and natural language processing have enabled the development of tools capable of extracting information from text on a massive scale. NER, also simply known as entity extraction, is a text analysis technique that uses NLP. It enables machines to automatically identify or extract specific data from unstructured text, and to classify the text according to predefined categories. These categories are named entities, that is, the words or phrases that represent a noun. Entity extraction enables machines to automatically identify or extract entities, like product names, events, and locations.

NER has been used in the extraction of research entities and their combinations in both inorganic materials and chemistry, however in marketing science entity normalisation has not yet been reported. The same occurs across social sciences at large. Eltyeb and Salim (2014) suggested using dictionary-based systems which are more suitable and effective when using closely defined and updated vocabulary names, and when names are correctly written in documents. They sketched out dictionary-based, rule-based, ML, as well as hybrid chemical NER concerned with finding and classifying mentions of specific information inside text documents. This can be done in a supervised, unsupervised or semi-supervised manner. Tshitoyan et al. (2019) applied this research towards making steps in trying to achieve a generalised approach to the mining of scientific literature. Despite several previous studies have used supervised NLP observing that this requires large hand-labelled datasets for training, Tshitoyan et al. (2019) affirmed that with their method, published literature can be efficiently encoded without human labelling or supervision. This research piece is the evolution of Weston et al. (2019) where they applied text-mining with NER together with entity normalisation for supervised article extraction. Both developed their methods based on Kim et al. (2017) statistical learning approach using NLP that automatically compiles materials' synthesis parameters.

Training a machine for recognising synonyms or equivalent processes is challenging, hence issues arise in the way in which entities are written. This is crucial when normalising entities and mapping each individual one onto a sole database. Currently, answering such questions requires a laborious and tedious literature search, performed manually by a domain expert.

Concepts of NER and NLP

Herewith a description of NER and NLP concepts required for the development of our methodology.

Corpus: Collection of text including scientific articles, theses documents and industry reports organised into datasets, which are used to train a natural language processing model. Such documents are written in natural languages mixed with domain-specialised terminologies and numerical data. They contain valuable knowledge regarding the connections and relationships between data items as interpreted by the authors.

Entities: Common things or nouns. They are specific terms of a piece of text that represent real-world objects, such as names, people, companies, events, products, organisations, locations, topics, times and monetary values.

Named entity recognition in an NLP model aims to identify and classify entities in order to extract useful information from unstructured text data. Named entities have proper names such as people, places, companies and products, and proper nouns. Example of entities are 'scientist' and 'capital city'. Named entities are 'Antonio Hyder' and 'London'.

Tokenisation: The process of splitting raw text into individual words, terms, sentences, symbols or other meaningful elements into a sequence of meaningful elements called tokens. The objective is to convert a sequence of characters into understandable bits of data that a program can work with. This results in a vocabulary formed by the extracted contents of marketing science articles. Tokens should be harmonised into unique words. As an example, New York and Web site are harmonised as newyork and website. Such pre-processing is conducted to normalise tokens that are identified as valid words used within marketing. This substantial reduces vocabulary sizes.

Labelling: Data labelling is a way of identifying the raw data and adding suitable labels or tags to that data to specify what the data is about. This allows machine learning models to recognise patterns in order to make accurate predictions. With labelling, research knowledge can then be efficiently encoded with Word2vec.

Model training: A machine learning training model is a process in which a ML algorithm is fed with sufficient training data to learn from.

METHODOLOGY

Having described the background theory and concepts necessary for attaining the aim of this research, we hereby describe our two-phase methodology specific to the academic field of marketing science for large-scale information extraction from published marketing literature, elaborating on the developments of Tshitoyan et al. (2019), Weston et al. (2019) and Kim et al. (2017). In the first stage (I) we design a generic NER for overall marketing science. In the second (II) the method is delimited to a selection of marketing sub-fields.

Methodology Part One: Generic Marketing Science Processing

The first part is a generic method for marketing science processing. We map unstructured raw-text of published articles on a structured database entry that allows programmatic querying. Overall, we foresee to extract more than 20 million marketing-science-related named entities from at least 1 million marketing-related journal article abstracts. The content of each abstract is represented as a database entry in a structured format. In order to attain this, we apply text-mining with NER together with entity normalisation. The NER is based on supervised machine learning with a recurrent neural network architecture (Hochreiter and Schmidhuber, 1997).

Our model shall be trained to extract summary-level information from the abstracts of marketing science articles. The unstructured text of each abstract is converted into a structured database entry containing summary-level information about the document, including marketing concepts, descriptors, mentions of any marketing properties or applications, as well as any characterisation or methods used in the study. We only consider an abstract to be of interest if it mentions at least one entity marketing concept from the 7Ps as used in services marketing, or 3 additional terms widely used across marketing literature. This is useful for training and testing our NER algorithm. For this reason, before training an NER model, we first train a classifier for document selection which labels each abstract as ‘relevant’ or ‘not relevant’.

Despite the recent progress of generative AI, current generative language models do not have direct access to scientific databases that provide real-time DOI information. We intend that researchers will be able to access targeted information on an unprecedented scale and speed, accelerating the pace of future marketing science discovery, and that a simple database query can be used to answer complex meta-questions within published literature that would previously have required laborious manual searches.

Selection of Ten Marketing Science Entities for the Generic Method

For our generic method, we require to ground on well-known marketing concepts. Hence for our 10 entity labels we have chosen the 7Ps from services marketing: product (PRD), price (PRI), place (PLA), promotion (PRM), process (PRC), physical evidence (PHY) and people (PEO); as well as 3 terms widely used across marketing literature: concept (CPT), context (CTX) and application (APL).

Phases of the Generic Methodology

The four phases of the generic method are as follows: (a) data collection and pre-processing which consists of 3 steps, specifically document collection, tokenisation, document selection relevance; (b) named entity recognition; (c) neural network model; (d) entity normalisation. These phases are an adaption of Weston et al., (2019) and Tshitoyan et al. (2019).

Phase A: Data collection and pre-processing

This first phase A consists of three steps: (1) document collection, (2) text pre-processing, (3) document selection.

Step One: Marketing Science Document Collection

We collect a majority of all English-language abstracts for marketing science-focused articles published between 1975 and 2024. In total we aim that our corpus will contain at least 1 million marketing abstracts. A list of 1350 relevant marketing-related journals is to be created indexed with Elsevier Scopus and to collect articles published in these journals via the Scopus and Science Direct APIs https://dev.elsevier.com and the Springer Nature API https://dev.springernature.com. Also, collateral journals containing marketing science articles are collected, as well as reports published by marketing institutions. The abstracts of these articles are text-mined. Scraping techniques include Kishor et al. (2021) for the web, and Bates (2016) for journal articles. The abstracts of these articles are then associated to metadata including title, authors, publication year, journal, keywords, DOI, url, assigned to a unique ID, and stored as individual documents in a database such as MongoDB or ElasticSearch.

Step Two: Text Pre-Processing—Tokenisation

In the second step of phase A, we conduct tokenisation, which is the beginning of the NLP process. For this we utilise an adaption of ChemData Extractor (Swain & Cole, 2016). Based on the quality of the results obtained, we might consider using another of the substantially open-source tools available for tokenisation.

Step Three: Document Selection—Relevant or Not Relevant for Training

We consider an abstract to be of interest, that is, useful for training and testing the NER algorithm, if the abstract mentions at least one of the ten designed entities. For this reason, before training the model, we first train a binary classifier for document selection capable of labelling abstracts are 'relevant' or 'not relevant'. For the training data, we shall label a minimum of 1000 randomly selected abstracts. This is the third and final step of phase A. In subsequent phase B, the NER study, we only consider documents predicted to be relevant for training and testing data. As a reference, Weston et al. (2019) had obtained 588 relevant abstract and 494 non relevant from their abstract selection.

Phase B: Named Entity Recognition

In this second phase and using NER, we extract specific entity types that can be used to summarise a document. Information extraction is conducted with a minimum of 1 million marketing science journal articles. For this we focus on the article abstract solely, which is the most information-dense portion of published articles and readily available from various publisher APIs such as Scopus and Springer. To extract this information, we use our ten purposely designed entity labels.

Following Weston et al. (2019), the five steps for the NER process phase are:

B-Step 1: marketing science documents are collected and added to the corpus.
B-Step 2: text is pre-processed, this is, the text is tokenised and cleaned.
B-Step 3: a small subset of documents is labelled for training data. For this we shall utilise the 3 entities widely used across marketing literature: concept (CPT), context (CTX) and application (APL).
B-Step 4: labelled documents are combined with word embeddings and conducted with Word2vec, generated from unlabelled text in order to train a neural network for named entity recognition.
B-Step 5: entities are then extracted from the entire corpus.

This five-step information-extraction pipeline is show in Figure 1.

Figure 1. The workflow for name entity recognition, NER
(adapted from Weston et al., 2019)

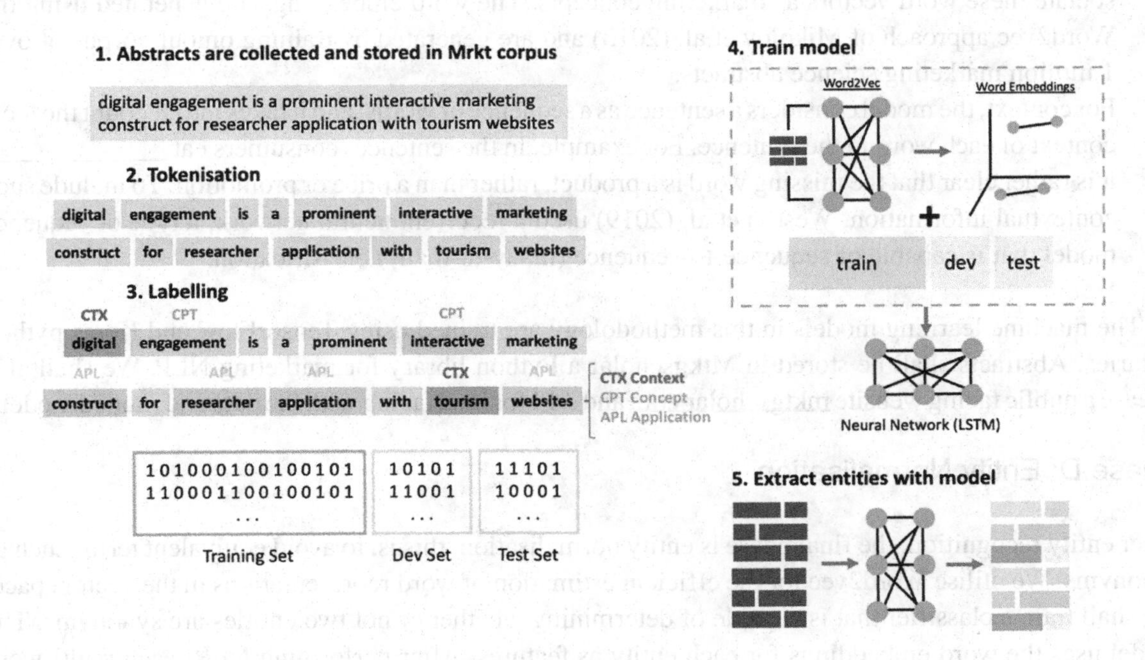

Using this tagging scheme, 800 marketing science abstracts are annotated by hand: only abstracts deemed relevant, based on the relevance classifier described earlier, are annotated. Following Weston et al. (2019) and before training a classifier, the 800 annotated abstracts are split into training, development and test sets respectively. The development set is used for optimising the hyperparameters of the model, and the test set is used to assess the final accuracy of the model on new data. For this we use an 80% -10% -10% split, therefore there are 640, 80, and 80 abstracts in the train, development, and test sets respectively. Annotations are performed by a single marketing scientist. There is no necessarily 'correct' way to annotate these abstracts, however to ensure that the labelling scheme is reasonable, a second marketing scientist annotates a subset of 25 abstracts to assess the inter-annotator agreement. This is calculated as the percentage of tokens for which the two annotators assign the same label. Weston et al. (2019) had obtained 87.4%.

Phase C: Neural Network Model

The neural network architecture we utilise is based on Lample et al. (2016). The aim is to train our model so that marketing science knowledge is encoded; for example, we wish to teach a computer that the words 'digital' and 'tourism' represent context (CTX), 'engagement' and 'involvement' correspond to concept (CPT), and 'web users' and 'consumers' refers to people (PEO).

There are two main types of information that can be used to teach a machine to recognise which words correspond to a specific entity type: 1. word representation, and 2. local context within sentence.

1. Word embeddings are used for word representation, which map each word onto a dense vector of real numbers. Word embeddings that have a similar meaning or are frequently used in a similar context, will have a similar word embedding. For example, entities such as 'involvement' or 'engagement' will have similar vector representations, and during training, the model learns to associate these word vectors as marketing concepts. The word embeddings are generated using the Word2vec approach of Mikolov et al. (2013) and are generated by training on our corpus of over 1 million marketing science abstracts.
2. For context, the model considers a sentence as a sequence of words, and it takes into account the local context of each word in the sentence. For example, in the sentence 'consumers eat ____________' it is rather clear that the missing word is a product, rather than a price or promotion. To include such contextual information, Weston et al. (2019) used a recurrent neural network, a type of sequence model that is capable of sequence-to-sequence (many-to-many) classification.

The machine learning models in this methodology are trained using TensorFlow and Keras python libraries. Abstracts shall be stored in Mtkgscholar a Python library for marketing NLP. We shall also release a public facing website mktgscholar.com and API for interfacing with the data and trained models.

Phase D: Entity Normalisation

After entity recognition, the final phase is entity normalisation, this is, to avoid equivalent terms such as synonyms. We utilise Word2vec for the efficient estimation of word representations in the vector space. We shall train a classifier that is capable of determining whether or not two entities are synonyms. The model uses the word embeddings for each entity as features. After performing NER, each multi-word entity is concatenated into a single word, and new word embeddings are trained so that every multi-word entity has a single vector representation. As an entity may be written in several forms, all entities must be stored in a normalised format.

Expected Results From the Methodology Part One: Generic Marketing Science Processing

In this section we describe the expected results from our generic methodology: (a) NER classification performance; (b) Entity normalisation and (c) Text-mined information extraction. Figure 2 illustrates an expected NER classifier performance.

Figure 2. Example predictions of the marketing science NER classifier

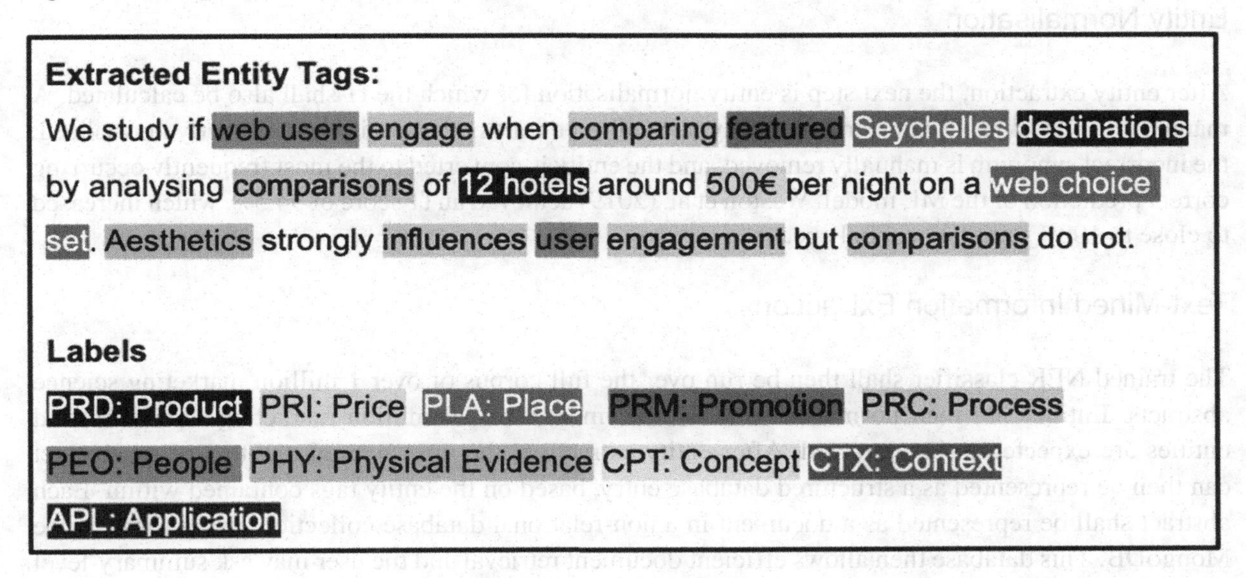

The highlighting indicates regions of text that the model as associated with a particular entity type

NER Classification Performance

After training the classifier, we should be able to accurately extract information from marketing science texts. Model performance can be assessed more quantitatively by assessing the accuracy on the development and test sets. Accuracy is defined using the f1 score, which is the harmonic mean of precision (p) and recall (r). The formula for calculating accuracy *f1* is based on three parameters t_p, f_p, $f_{n\ which}$ represent the number of true positives, false positives, and false negatives, respectively.

$p = tp / tp + fp$'

$r = tp / tp + fn$'

*f*1= 2 *p·r* / *p*+*r*'

f1 scores over 80% suggest that the model will perform well at extracting each entity type. f1 also serves as a means for benchmarking with previous authors. Weston et al. (2019) classifier achieved an overall f1 score of 87.04% with their test set. Mysore et al. (2017) achieved an f1 score of 82.11% for extraction material mentions; Swain and Cole (2016) reported an f1 score of 93.4% when extracting chemical entity mentions.

Weston et al. (2019) increased their f1 score to 94.5% with entity normalisation, finding that it greatly increases the number of relevant items identified in document querying. Term normalisation is achieved with a supervised machine learning model that learns to recognise whether or not two entities are synonyms. Model performance can be assessed more quantitatively by assessing the accuracy both on the development and on test sets.

Entity Normalisation

After entity extraction, the next step is entity normalisation for which the f1 shall also be calculated. A manual error check is also performed on any normalised entities. If an entity is incorrectly normalised, the incorrect synonym is manually removed, and the entity is converted to the most frequently occurring correct prediction of the ML model. Weston et al. (2019) achieved an f1 score of 94.5%, which increased to close to 100% with the manual check.

Text-Mined Information Extraction

The trained NER classifier shall then be run over the full corpus of over 1 million marketing science abstracts. Entities are then normalised and overall, more than 20 million marketing-science-related entities are expected to be extracted. After entity extraction, the unstructured raw-text of an abstract can then be represented as a structured database entry, based on the entity tags contained within. Each abstract shall be represented as a document in a non-relational database collection, using for instance MongoDB. This database then allows efficient document retrieval and the user may ask summary-level or meta-questions from the published marketing science literature.

Having described the generic method, we shall now present a narrowed down sub-method contextualised to 6 delimited marketing topics, which has been developed as a contingency plan based on the quality of the results obtained during the generic method. Whereas the overall aim is clear and perspective should not be lost as progress is being made, this is required due to the inevitable unpredictability of the generic method during its implementation and the habitual frustrations encountered during novel AI development (Saund, 2020).

Methodology Part Two: Contextualised Marketing Science Processing

As a continuation of the generic methodology, we herewith describe a contextualised 'narrower' method based on a three-step process to be conducted on delimited topics of marketing science. In this contingency situation the embeddings will be trained with delimited scientific abstracts from marketing and related fields directly retrieved from scientific databases as described in the generic method.

Step One

In the first of this three-step process, we shall compile consolidated bodies of offline marketing science from established topics. For this we have selected branding, retail and advertising as they are well recognised and consolidated offline marketing subfields with established online equivalents.

After removing irrelevant abstracts with a binary classifier, the remaining relevant ones will be tokenised. This will result in a vocabulary solely formed with the extracted contents of the delimited marketing science articles. We then apply Tshitoyan et al. (2019) method of knowledge extraction and relationships for the handling of large bodies of scientific literature, conducted with CrossRef Application Programming Interface (API) in order to retrieve large lists of article Digital Object Identifiers (DOIs). Publisher APIs, such as Elsevier https://dev.elsevier.com and Springer Nature https://dev.springernature.com will be used to download full-text journal articles. The knowledge can then be efficiently encoded as information-dense word embeddings using human labelling with Word2vec training.

Step Two

Secondly, the embeddings will be trained with scientific abstracts from each of the aforementioned topics, branding, retail or advertising. For this we will use article abstracts from 1975 to 2024 from more than a thousand journals and also from articles deemed likely to contain marketing-related research directly retrieved from the scientific databases (i.e., Elsevier and Science Direct) combined with web scraping. We expect the performance of the algorithm will improve when irrelevant and non-English abstracts are removed. At this point, the abstracts will be then tokenised. We expect correct pre-processing, especially the choice of phrases to include as individual tokens will improve the results. The remaining abstracts will then be classified as relevant and tokenised to produce the individual words. Using an adaption of ChemDataExtractor (Swain and Cole, 2016) which should result in a vocabulary of thousands of words.

Step Three

The aim of the third step is to repeat the first two steps integrating the aforementioned offline topics with the equivalent online marketing topics: online branding, online retail, online advertising. Hence, we shall have 6 areas in total, that is, 3 topics in two versions, offline and online. Lessons learned when dealing with the offline research topics may help inform analyses made with the equivalent online research. It will be interesting to observe results when combining offline branding with the corresponding online branding, or brick and mortar retail with online retail. Integration of all six areas together may also provide further insights into cross channel phenomena.

A potential subsequent fourth stage, could be to adapt the lessons learned during the first three steps, repeating the process with other specialised marketing topics such as online tourism marketing, online education marketing or international marketing which would add to the previously mentioned fields.

APPLICATIONS OF THE METHODOLOGY

Answering Meta-Questions

Being able to answer complex meta-questions of the published literature can be a powerful tool for researchers. With all of the corpus of published articles structured as database entries, such questions may be asked programmatically, generating an answer to these questions requires only a few lines of code, just a few seconds for data transfer and processing, and can be answered in a matter of seconds, in comparison to a several-weeks-long literature search performed by a domain expert that would have been previously required (Hyder et al., 2022). Application examples would be 'for this list of 1000 digital marketing research studies, which refer to engagement in tourism', and 'for this list of 1000 digital marketing research applications, which have been applied to business practice, and which are yet to be explored?'.

Predictive Writing: BERT and ELMo

Google research has also made developments in predictive writing, in particular BERT and ELMo, which are increasingly being adopted. BERT (Devlin et al., 2018) is a bidirectional encoder which looks both forwards and backwards and allows machine learning in the context of a word or term based on all of its surroundings, i.e., left and right of the word. This can be used to generate language models. Its key technical innovation is the bidirectional training of popular attention model 'Transformer' to language modelling (https://talktotransformer.com). This learns contextual relations between words or sub-words in a text using an encoder that reads the text input and a decoder that produces a prediction. BERT can be fine-tuned with small amounts of data although higher accuracy is achieved with greater training data. ELMo is also used to produce contextual word embeddings as one word can have a different meaning based on the words around it. These AI developments can be used for the automated formulation of research hypothesis (e.g., Spangler, 2014).

Generation of Hypotheses With AI

Automatic hypothesis generation is also a field in development. For this to occur sufficient data is required as well as knowledge experts in complex domains to accelerate scientific progress and achieve significant discoveries. Spangler et al. (2014) designed a three-phase process: 1. Exploration, which surveys the relevant unstructured information, designs text queries, and extracts relevant information; 2. Interpretation, which builds a graph of similarity relationships to assist domain experts in visualising hidden connections; and 3. Analysis, which is the diffusion of annotating information among entities, in order to rank the best entity candidates for further experimentation of novel annotation predictions. A domain expert chooses annotation candidates that are the most analytically probable, experimentally testable, and of direct interest to the problem at hand.

Application of Marketing Science to Practice

The need for connecting research and practice has been for long pointed out by top marketing academics (e.g., Kumar, 2017; Lilien, 2011) issue which is still to be addressed. Our methodology opens paths for advances in the marketing science discipline, facilitate transfer of research knowledge to marketing practice and accelerate the use of knowledge with AI, by synthesising the knowledge and making it more manageable by practitioners (Bengoa, et al., 2021). Progress with knowledge transfer processes primarily occur when there is collaboration between experts in AI technologies and domain experts (e.g., DARPA, 2023).

It is increasingly claimed that more emphasis should be placed on the application of marketing science to industry problems (e.g., Babin et al., 2022; Steenkamp, 2021; Vargo, 2019) obtaining impactful research (Paquet, 2022). In the European Union (EU) this is increasingly becoming a part of policy (Gabriel, 2022). In particular, small companies especially startups many of which operate in tech, are initially small, usually have limited resources, and timely demand the right knowledge in order to reduce trial and error expenditures. They can benefit from adopting marketing science models helping them to make evidence-based decisions, reduce trial and error, and gain competitive advantage (e.g., Lee and Kozar, 2009). Many startups are tech-based and count with a specialised workforce, yet the growing scientific knowledge from articles is still underutilised. Here is where AI could help by synthesising marketing science knowledge and making it more manageable by practitioners.

DISCUSSION AND CONCLUSION

Having described the fundamentals for conducting analysis of marketing scientific literature with natural language processing, we have designed both a generic and a narrowed-down contextualised method on how NLP can analyse marketing scientific literature by making use of AI technologies adopted from developments made in other scientific fields. This work represents a first step towards AI analysis and programmatic access to published marketing science literature. Future discoveries depend on the knowledge embedded in past publications and one major barrier for accessing the literature is that researchers may not know a priori what are the relevant terms or entities to search for. This is especially true for researchers that may be entering a new field or may be studying a novel concept for the first time. We have described how the digital object identifier (DOI) for each abstract in our corpus should be stored so that targeted queries can be linked to the original research articles. Current generative language models do not have direct access nor the ability to browse specific databases to provide real-time DOI.

The overall aim is that AI-based technologies are used for text mining and analysis of scientific marketing research. Making accelerated use of research, the removal of computational bottlenecks and the faster adoption of knowledge might be easier thanks to adoption of ongoing developments occurring in the most advanced AI fields, such as materials, that are leading the way. Weston et al., (2019) has demonstrated how research summaries can be automatically generated based on the entities without any human intervention or curation. We described how marketing research can be extracted from documents, classified and tokenised in individual words. This is conducted by applying text-mining with named entity recognition together with entity normalisation for large-scale information extraction from the published marketing science literature. Text-mining on such a massive scale provides tremendous opportunity for accessing and making use of marketing science.

Future Research

In future steps, we could also add to the contextualised methodology (II) research on product, place, price and promotion management, in both offline and online versions. Also, the lessons learned during the first three steps from the delimited method, could be repeated with other specialised marketing topics such as online tourism marketing, online education marketing or international marketing, which would add on to the previously mentioned fields. Finally, as AI is also capable of predictive writing using bi-directional encoders BERT and ELMo which are used to produce contextual word embeddings (Devlin et al., 2018), ultimately automation of hypotheses formulation could be possible (Spangler et al., 2014).

Foreseeable Limitations

Our suggested method is not free from potential limitations. Although the use of AI might be developing in areas of hard science and information technology, at this stage we are unclear if it will work well, even appreciated, in social science research. Also, although AI in a narrowed research field can lead to breakthroughs, specialisation also inherently limits the opportunities to find common grounds at the interface between fields, as could occur with topics such as online branding. Also, we only analyse abstracts and not all research information is contained within the abstracts.

As we progress, and keeping an eye on the progress made with generative AI, we shall take into account three research areas suggested by Weston et al. (2019) as applied to marketing science: document retrieval; new concept search; guided literature searches and summaries. Finally, experienced AI researcher Eric Saund (2020) affirmed that due to this habitual situation of encountering obstacles during development, focus, patience, time and resources are fundamental for advancing knowledge. In our foreseeable struggle, we shall bear in mind that the ultimate objectives of our work includes: help resolve scientific bottlenecks with AI; aim towards integration and interpretation of knowledge; awareness that discoveries depend on the knowledge embedded in past publications; acceleration of scientific discovery should be measured; obtain structured data bases as they will eventually be required for ML training; a focus on applying research knowledge to practice.

REFERENCES

Bates, M. E. (2016). *Bringing Insight to Data: Info Pros' Role in Text and Data Mining*. Springer Nature. https://www.springernature.com/gp/librarians/landing/textanddata

Bengoa, A., Maseda, A., Iturralde, T., & Aparicio, G. (2021). A bibliometric review of the technology transfer literature. *The Journal of Technology Transfer*, *46*(5), 1514–1550. doi:10.100710961-019-09774-5

Bulchand-Gidumal, J., William Secin, E., O'Connor, P., & Buhalis, D. (2023). Artificial intelligence's impact on hospitality and tourism marketing: Exploring key themes and addressing challenges. *Current Issues in Tourism*, 1–18. doi:10.1080/13683500.2023.2229480

Crues, W. (2017). Automated Extraction of Results from Full Text Journal Articles. *Proceedings of the 10th International Conference on Educational Data Mining*. IEEE.

DARPA, Defence Advanced Research Projects Agency. (2023). *DARPA perspective on AI.* SARPA. https://www.darpa.mil/about-us/darpa-perspective-on-ai

Devlin, J. (2018). *Bert: Pre-training of deep bidirectional transformers for language understanding*. arXiv preprint arXiv:1810.04805.

Eltyeb, S., & Salim, N. (2014). Chemical named entities recognition: A review on approaches and applications. *Journal of Cheminformatics*, *6*(1), 17. doi:10.1186/1758-2946-6-17 PMID:24834132

Gabriel, M. (2022). *Startups and science: EU businesses will increase their competitiveness by linking closely scientists and startups*. Linkedin. https://www.linkedin.com/pulse/startups-science-eu-businesses-increase-linking-closely-gabriel

Hochreiter, S., & Schmidhuber, J. (1997). Long short-term memory. *Neural Computation*, *9*(8), 1735–1780. doi:10.1162/neco.1997.9.8.1735 PMID:9377276

Hosmer, D. W. Jr, Lemeshow, S., & Sturdivant, R. X. (2013). *Applied logistic regression* (Vol. 398). John Wiley & Sons. doi:10.1002/9781118548387

Huang, M. H., & Rust, R. T. (2021). A strategic framework for artificial intelligence in marketing. *Journal of the Academy of Marketing Science*, *49*(1), 30–50. doi:10.100711747-020-00749-9

Hyder, A., Harris, K.-C., & Perez-Vidal, C. (2022). in Special Session: Marketing Science at the Service of Innovative Startups and Vice Versa: An Abstract. In *Academy of Marketing Science Annual Conference*. Springer Nature.

Hyder, A., & Nag, R. (2020). Artificial Intelligence Analysis of Marketing Scientific Literature: An Abstract. In *Academy of Marketing Science Annual Conference*. Springer International Publishing.

Hyder, A., & Nag, R. (2021). An Artificial Intelligence Method for the Analysis of Marketing Scientific Literature: An Abstract. In *Academy of Marketing Science Annual Conference-World Marketing Congress*. Springer International Publishing.

Kim, E., Huang, K., Saunders, A., McCallum, A., Ceder, G., & Olivetti, E. (2017). Materials synthesis insights from scientific literature via text extraction and machine learning. *Chemistry of Materials*, *29*(21), 9436–9444. doi:10.1021/acs.chemmater.7b03500

Kumar, V. (2017). Integrating theory and practice in marketing. *Journal of Marketing*, *81*(2), 1–7. doi:10.1509/jm.80.2.1

Laso, I. (2020): *#EUvsVirus interview*. European Commission DG Research and Innovation. https://www.facebook.com/EUvsVirus/videos/2635593133382903

Lee, Y., & Kozar, K. A. (2009). Designing usable online stores: A landscape preference perspective. *Information & Management*, *46*(1), 31–41. doi:10.1016/j.im.2008.11.002

Lilien, G. L. (2011). Bridging the academic–practitioner divide in marketing decision models. *Journal of Marketing*, *75*(4), 196–210. doi:10.1509/jmkg.75.4.196

Mariani, M. M., Hashemi, N., & Wirtz, J. (2023). Artificial intelligence empowered conversational agents: A systematic literature review and research agenda. *Journal of Business Research, 161*, 113838. doi:10.1016/j.jbusres.2023.113838

Mikolov, T., Chen, K., Corrado, G., & Dean, J. (2013). *Efficient estimation of word representations in vector space*. arXiv preprint arXiv:1301.3781.

Mustak, M., Salminen, J., Plé, L., & Wirtz, J. (2021). Artificial intelligence in marketing: Topic modeling, scientometric analysis, and research agenda. *Journal of Business Research, 124*, 389–404. doi:10.1016/j.jbusres.2020.10.044

Paquet, J. E. (2022). *New directionality for knowledge valorisation in the EU*. Research and Innovation Knowledge Valorisation.

Saund, E. (2020). *SCI-52 Artificial Intelligence: Deep Learning, Human-Centered AI, and Beyond.* Stanford University.

Spangler, S., Wilkins, A. D., Bachman, B. J., Nagarajan, M., Dayaram, T., Haas, P., Regenbogen, S., Pickering, C. R., Comer, A., Myers, J. N., & Stanoi, I. (2014). Automated hypothesis generation based on mining scientific literature. In *Proceedings of the 20th ACM SIGKDD international conference on Knowledge discovery and data mining.* ACM. 10.1145/2623330.2623667

Srinivasan, P. (2004). Text mining. Generating hypotheses from MEDLINE. *Journal of the American Society for Information Science and Technology, 55*(5), 396–413. doi:10.1002/asi.10389

Steenkamp, J.-B. (2021). *A life in marketing: looking back, looking forward. AMA Irwin McGraw Hill Acceptance Speech.* American Marketing Association.

Swain, M. C., & Cole, J. M. (2016). ChemDataExtractor: A toolkit for automated extraction of chemical information from the scientific literature. *Journal of Chemical Information and Modeling, 56*(10), 1894–1904. doi:10.1021/acs.jcim.6b00207 PMID:27669338

Tshitoyan, V., Dagdelen, J., Weston, L., Dunn, A., Rong, Z., Kononova, O., Persson, K. A., Ceder, G., & Jain, A. (2019). Unsupervised word embeddings capture latent knowledge from materials science literature. *Nature, 571*(7763), 95–98. doi:10.103841586-019-1335-8 PMID:31270483

Vargo, S. L. (2019). Moving forward..... *AMS Review, 9*(3), 133–135. doi:10.100713162-019-00159-3

Weston, L., Tshitoyan, V., Dagdelen, J., Kononova, O., Trewartha, A., Persson, K. A., Ceder, G., & Jain, A. (2019). Named entity recognition and normalization applied to large-scale information extraction from the materials science literature. *Journal of Chemical Information and Modeling, 59*(9), 3692–3702. doi:10.1021/acs.jcim.9b00470 PMID:31361962

ADDITIONAL READINGS

Jurafsky, D., & Martin, J. H. (2019). *Speech and Language Processing* (3rd ed.).

Krallinger, M., Rabal, O., Lourenco, A., Oyarzabal, J., & Valencia, A. (2017). Information retrieval and text mining technologies for chemistry. *Chemical Reviews, 117*(12), 7673–7761. doi:10.1021/acs.chemrev.6b00851 PMID:28475312

Lake, B. M., Ullman, T. D., Tenenbaum, J. B., & Gershman, S. J. (2017). Building machines that learn and think like people. *Behavioral and Brain Sciences, 40*, e253. doi:10.1017/S0140525X16001837 PMID:27881212

Peters, M. E., Ammar, W., Bhagavatula, C., & Power, R. (2017). *Semi-supervised sequence tagging with bidirectional language models*. arXiv preprint arXiv:1705.00108.

Rashid, T. (2016). *Make your own neural network* (Vol. 29). CreateSpace Independent Publishing Platform.

Rescorla, M. (2020). The computational theory of mind. In *Stanford Encyclopedia of Philosophy*. Stanford. https://plato.stanford.edu/entries/computational-mind

KEY TERMS AND DEFINITIONS

Formulation of marketing hypotheses with AI: Machine learning algorithms capable of detecting novel patterns which researchers might not notice, and writing connections between scientific concepts in a text that is understandable.

Marketing Corpus: Collection of marketing science texts that include scientific articles, theses documents and industry reports organised into structured datasets, which are used to train natural language processing models specific to marketing research. Such documents are written with marketing-related specialised terms combined with natural language.

Marketing Entities: Common nouns and concepts from marketing texts such as brands, products, organisations, people, places, events, and services.

Marketing Labelling: Identification of raw marketing text and adding suitable tags or tables to specify the topics of the text. With labelling, marketing research knowledge can be efficiently encoded. This allows machine learning models to recognise patterns within marketing literature in order to make accurate text predictions.

Marketing Named Entity Recognition: Marketing named entity recognition in a natural language processing model aims to identify and classify marketing entities extracting relevant information from unstructured marketing text data.

Marketing Word Embeddings: Word embedding or word vector is an approach which represents documents and words from marketing texts. It is defined as a numeric vector input that allows marketing words with similar meanings to have the same representation.

Tokenisation of Marketing Terms: The process of splitting raw marketing texts into individual words, terms, sentences, symbols or other into a sequence of meaningful elements called tokens. This results in a vocabulary formed by the extracted contents of marketing texts.

Chapter 9
Artificial Intelligence for Renewable Energy Systems and Applications:
A Comprehensive Review

Manikandakumar Muthusamy
New Horizon College of Engineering, India

Karthikeyan Periyasamy
https://orcid.org/0000-0003-2703-4051
Thiagarajar College of Engineering, India

ABSTRACT

Artificial Intelligence technology has advanced tremendously in recent years, and it is now widely used in a variety of fields, including energy, agriculture, geology, information processing, medicine, defence systems, space research and exploration, marketing, and many more. The introduction of artificial intelligence technology has ushered in a new era of renewable energy systems and smart power grid modernization. It assists in attaining the intended system availability, reliability, power quality, efficiency, and security goals through optimal resource utilization and cost-effective electricity. Automated power generation systems, energy storage control, wind turbine aerodynamic performance optimization, power generator efficiency enhancement, health monitoring of renewable energy generation systems, and fault detection and diagnose in a smart grid subsystem are just a few of the applications. The main aim of this proposed chapter is to demonstrate how artificial intelligence techniques play a significant role in renewable energy systems with their diverse applications.

DOI: 10.4018/978-1-6684-9591-9.ch009

1 INTRODUCTION

Artificial Intelligence (AI) is a phrase, in its usual context, refers to a machine's or artifact's ability to accomplish functions similar to those that characterise human intellect. The term artificial intelligence has been applied to systems and programmes that can execute tasks that are more complicated than simple programming. But they are still far from the world of true cognition. Expert Systems (ES) and Artificial Neural Networks (ANN) are the two important fields of AI. Expert systems are logics that allow the computer systems to make conclusions from data by analysing it and choosing from a set of options (Ricalde et al., 2011; Kalogirou et al., 2014). Automated power generation systems, energy storage control, wind turbine aerodynamic performance optimization, power generator efficiency enhancement, health monitoring of renewable energy systems (RES), and fault detection and diagnose in a smart grid subsystem are just a few of the AI based smart applications. It is worth noting that, of all AI techniques, Neural Networks are currently receiving the most attention for potential applications. With the emergence of efficient and affordable application specific neural network microchips, widespread AI applications are expected to bring about a new sort of good industrial revolution in the future. AI applications have the potential to improve renewable energy by increasing efficiency, which will boost industry growth and, ideally, accelerate adoption. To analyse data, expert systems use a technique called rule - based inference. Considering their development, systems are still incapable of approaching the level of complexity that truly intelligent thought requires. The most dependable future option is emerging as being renewable energy.

Artificial neural networks are made up of a number of small, interlinked processing elements. Interconnections allow data to flow between these entities. There are two values connected with an incoming connection: an input value and a weight value. The unit's output is a function of the total value (Chakrabortty & Bose, 2019; Zhang et al., 2022). ANNs are utilised to perform jobs on computers, although they are not programmed to do so. Instead, they are trained on data sets until they recognise the patterns that are presented to them. After they have been trained, new patterns may be presented to them for prediction or categorization. The following techniques are used in modern AI technologies:

§ Artificial Neural Networks (ANNs)
§ Expert System Techniques (XPS)
§ Fuzzy logic systems (FL)
§ Genetic Algorithm (GA)

These are the key artificial intelligence technique families that are taken into account in the field of modern power systems. The notion of neural network analysis was found approximately 50 years ago, but applications to manage specific issues has just been developed in the recent 20 years. They are particularly useful for jobs involving inadequate data sets, ambiguous or missing information, and very complicated and ill-defined situations, where humans typically make decisions based on intuition. They also have a high level of robustness and fault tolerance. ANNs have been successfully used in a wide range of domains, including maths, medicine, neurology, engineering, economics, agriculture and many more businesses. Image and speech recognition, the study of electromyographs, the detection of military targets, and the monitoring of bombs in passenger baggage are only a few of the most essential. They have also been used in anticipating weather and market patterns, predicting underground mining sites, predicting thermal and electrical load, adaptive and robotic control, and many other applications.

Artificial intelligence has neither hatred or prejudice for humans. Instead, it tries to use humans, which are made up of atoms, to achieve its own goals (Akarslan, 2012; Ojiro & Tsuruta, 2019).

Moreover, majority of these power generation and distribution-related problems may be solved by the application of artificial intelligence and machine learning technology. The grid operators can predict how particular appliances will behave by using machine learning to the data produced by sophisticated sensors, smart metres, and intelligent gadgets. In order to anticipate the storage life and decide the payments that should be paid, this system can also employ algorithms. Artificial Intelligence is a type of technology that can sort through complex algorithms in such a way that it learns its ways over time. It was designed to be a network capable of emulating human thinking using a large number of machine learning techniques.

Applications of energy management tools and their analysis can be used to make judgments on energy services, energy efficiency, renewable energy use, energy trends, and energy policy. The tool also calculates the cost of energy and the resulting CO2 emissions. It is capable of simulating 100 years into the future. The same has been used to anticipate the consumption of renewable energy in nations such as Denmark, Bulgaria, Latvia, Belarus, Romania, Russia, and Ukraine up to 2050. Power providers can alert customers about grid maintenance when they are informed of impending work. India's Ministry of Power has released a new policy on using biomass for power generation via cofiring in coal-based power plants. With effect one year after the date of issuing of this regulation, the use of 5% biomass pellets composed mostly of agricultural waste and coal in thermal power plants is required. According to this policy, the need to utilise biomass pellets in thermal power plants will grow to 7% two years after the policy is issued. In 39 thermal power plants, co-firing of agricultural residue pellets with coal has begun. The total amount of biomass used up to November, 2022 was 85477 MT. Analysts and experts from all over the industry are now forecasting that artificial intelligence will contribute in the future of energy. Artificial Intelligence can be implemented with the help of advanced computer tools and utilised to solve all of the difficulties for huge power systems, as need is the mother of invention. Artificial intelligence technology can also assist providers of renewable energy in expanding their markets and launching new business models. The energy sector may gain detailed usage insights that it can utilise to launch new services by using AI to data on the energy acquired. Retail providers will also have a chance to enter the consumer sector as a result. AI can make it easier to design strategies, policies, and plans that take into account both present usage and anticipated future demands.

The remaining section of the chapter structured as follows: Section 2 addresses the current state of energy systems along with the relevant literatures; Section 3 explains the challenges in renewable energy sector; Section 4 discusses the need for AI technology in renewable energy-based applications and illustrates the importance, Philosophy of AI in renewable energy and society and the contribution of AI to RE applications; Section 5 represents various applications of AI in renewable energy systems with sufficient examples; Section 6 discusses about the AI based energy systems for sustainability; Section 7 describes about the needs of integrating humans and AI for RES and Section 8 concludes the chapter.

2 CURRENT STATE OF ENERGY SYSTEMS

Insights into a society's use of natural resources and economic productivity can be gained from examining regional and national energy systems. They can help to identify technical and other areas for improvement, as well as the priority of the actions to be implemented. In attempts to improve a much more efficient allocation of resources, assessments and comparisons of various societies around the world can be of vital importance. The industrial sector and other sectors of the economy have turned increasingly to improve in recent decades, primarily to save energy and, therefore, money. It is possible to assess the energy efficiency of a region or nation by using available renewable energy analysis to determine its energy consumption (Ford et al., 2014). As a result, our generation faces a dual challenge: we must continue to progress in our battle toward energy poverty, given that the majority of the world's population still lives in poverty. Success in this battle will only translate into better life circumstances for young people today if we can also reduce carbon emissions.

The energy source and its cost are critical to making success on each of these fronts. Some people are unable to purchase enough energy, and those who have escaped the worst poverty still depend on fossil fuels to meet their energy needs. When we look at in this way, we can see that the twin energy concerns are actually two sides of the same coin. We lack affordable, safe, and long-term energy substitutes to fossil fuels on a broad scale. Without AI technologies, we are caught in a society with only awful options: low-income nations that fail to fulfil current generation demands, high-income nations that jeopardise future generations' ability to meet their needs, and middle-income nations that fail on both counts (Atasoy et al., 2015; Bose, 2020). Because now we do not have all of the technology needed to achieve this transition, the globe will need to innovate on a huge scale to make it happen. This is true for most carbon-emitting industries, including transportation (shipping, aircraft, and road transportation), heating, cement manufacture, and agribusiness. Electricity is one area where we have discovered various alternatives to fossil fuels. Nuclear power and renewable energy release significantly less carbon than fossil fuels.

However, there is room for improvement. Some countries have increased nuclear power and renewable energy production, outperforming the worldwide average. As a result of improving their performance in this area, countries should be closer to the future sustainable energy world (Mohamed et al., 2017; Lateef et al., 2022). The current condition of energy generation in many countries is still based on power plant systems as given if Fig. 1. and based on the following steps.

1. Electricity is generated by traditional power stations.
2. Substation transformers increase the voltage for transmission of power.
3. Transmission lines transport electricity across vast distances to homes and businesses.
4. The voltage is reduced by the transformer in the neighbourhood substation.
5. Before power enters the homes, transformers on poles step it down.
6. Electricity is delivered to homes and businesses via distribution lines.

3 CHALLENGES IN RENEWABLE ENERGY SECTOR

The weather's unpredictability is one of the most difficult aspects of producing renewable energy. Solar and wind are the most common renewable source of energy, and electricity generation is heavily reliant on weather and other natural resources (Blaabjerg & Ma, 2019; Bose & Wang, 2019). Fig 2. Shows the different renewable energy sources such as wind, solar, Hydroelectric, Geothermal, Biomass, and Tidal electricity generation. Despite the fact that we have effective weather forecasting systems in place, there will be rapid changes in the climate that will disrupt energy flow. Such vulnerabilities exist throughout the renewable energy supply chain. As a result, it must be sufficiently smoothed to cope with unanticipated changes. Recent advances in energy storage technologies appear to be quite promising. However, they have yet to be extensively evaluated. Climate change is the issue that has dominated the public debate on energy. A climate catastrophe puts the natural environment around us, our current well-being, and the well-being of future generations in jeopardy. Energy production is responsible for 87 percent of worldwide greenhouse gas emissions, people in the richest countries have the highest emissions. CO_2 emissions around the world have been rapidly increasing. The concentration of greenhouse gases in the atmosphere will continue to rise as long as we continue to release them.

Figure 1. Traditional way of energy generation and distribution

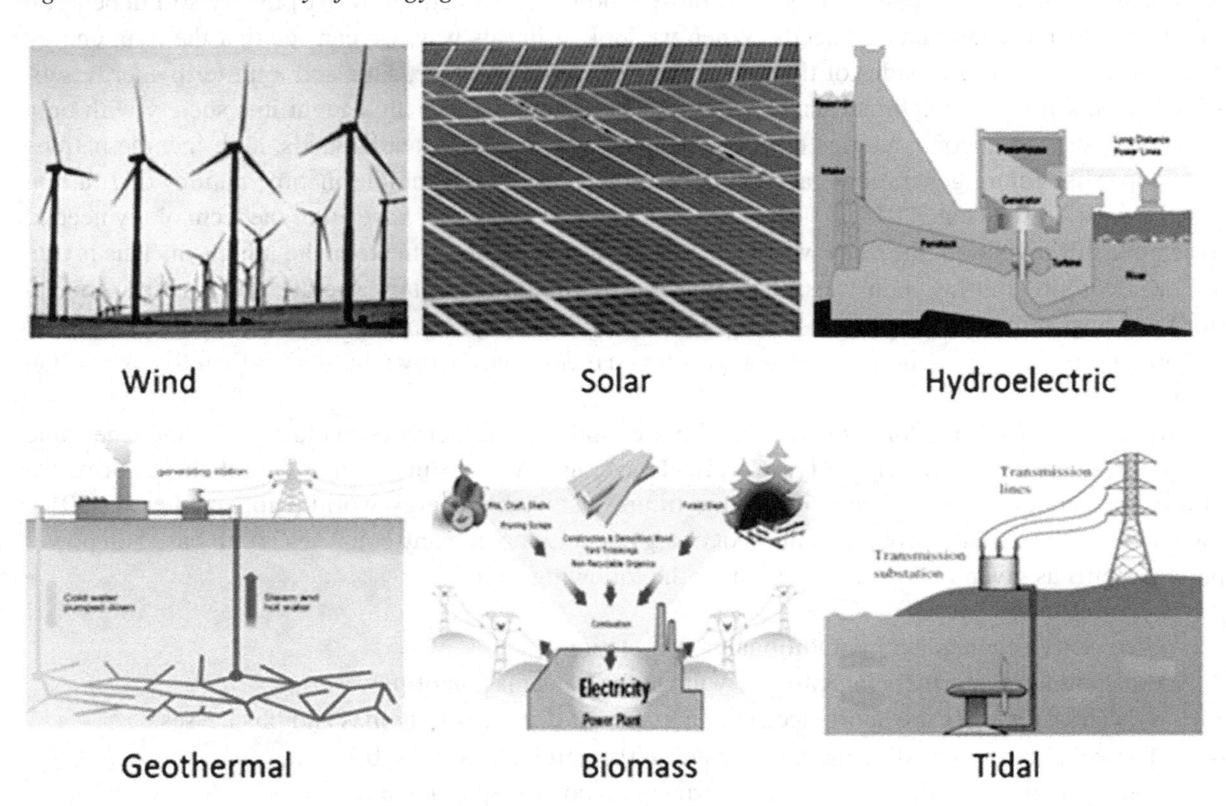

To put an end to climate change, the proportion of greenhouse gases in the atmosphere must stabilise, and carbon dioxide emissions must fall to net-zero levels. In the next years, reducing emissions to net-zero will be one of the world's most difficult issues. Renewable energy will continue to be in high demand to address such issues in the future. That is why, in order to boost productivity and overcome deficits, renewable energy companies should invest in Machine Learning, AI, IoT, and other developing technologies. It is feasible to quickly build and simulate hundreds of design iterations by utilising AI in generative thermal design. Grocery stores, industries, offices, and trains, among other significant consumers of renewable energy, can employ AI technology to make data-driven conclusions.

Figure 2. Different renewable energy sources

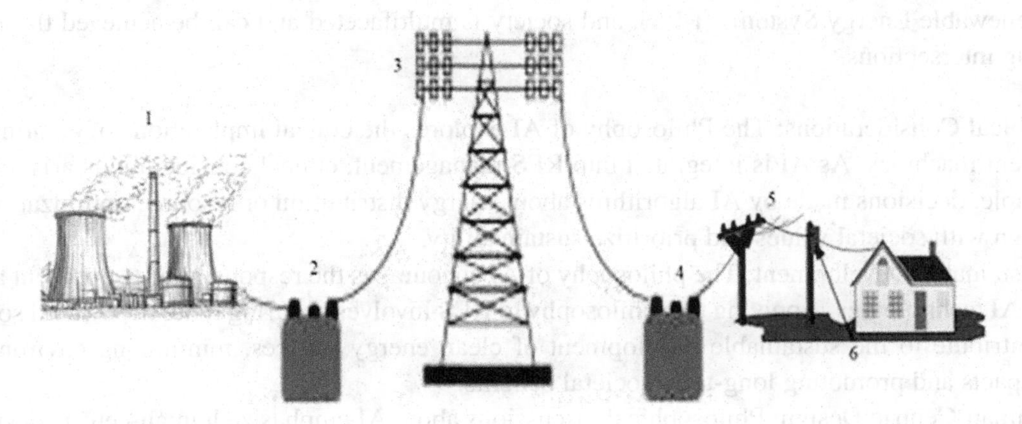

4 NEED FOR AI TECHNOLOGY IN RENEWABLE ENERGY SECTOR

In today's world, artificial intelligence is taking on a variety of new tasks and uses. It is becoming increasingly important because it can solve difficult problems in a cost-effective manner in a variety of areas, including healthcare, entertainment, finance, education, and energy. Our daily lives are becoming more comfortable and efficient as a result of AI. It is becoming a co-worker, assisting us in our houses, driving our cars, creating games, and much more. One of the most sophisticated machines on the planet is the electric grid (Ricalde et al., 2019; Molina, 2019). However, with the addition of variable renewable energy sources, it is rapidly developing. The present grid has several issues in absorbing the diversity of renewable energy due to the inherent fluctuation of wind and sun. Smart solutions are needed by the utility industry to aid better the integration of renewables into the existing grid and make renewable energy a viable energy source as mentioned here.

§ AI techniques are used to make predictions. By employing machine learning algorithms to analyse more complicated data in a shorter amount of time, meteorologists can now make more accurate predictions, saving lives and money.

§ Other forecasts, such as temperature, wave height, and precipitation, can benefit from machine learning as well.

§ To provide short-term weather forecasts or long-term climate predictions, the models can study and manipulate massive data sets relayed from weather satellites, relay stations, and radiosondes.

Artificial Neural Network, Ensemble Neural Network, Backpropagation Network, Radial Basis Function Network, General Regression Neural Network, Genetic Algorithm, Multilayer Perceptron, and Fuzzy Clustering are some other AI techniques for weather prediction.

4.1 Philosophy of AI in Renewable Energy and Society

Connecting the Philosophy of AI, Renewable Energy Systems, and society involves applying ethical considerations, sustainable development principles, human-centric design, transparency, policy frameworks, and public engagement strategies to ensure that AI-driven RES align with societal values and contribute to a more sustainable future. The connection between the Philosophy of Artificial Intelligence (AI), Renewable Energy Systems (RES), and society is multifaceted and can be achieved through the following intersections.

- Ethical Considerations: The Philosophy of AI explores the ethical implications of creating intelligent machines. As AI is integrated into RES management, ethical considerations arise. For example, decisions made by AI algorithms about energy distribution or resource optimization must align with societal values and prioritize sustainability.
- Sustainable Development: The philosophy of AI encourages the responsible development and use of AI technologies. Applying this philosophy to RES involves ensuring that AI-powered solutions contribute to the sustainable development of clean energy sources, minimizing environmental impacts and promoting long-term societal benefits.
- Human-Centric Design: Philosophical discussions about AI emphasize human-centric design and human-AI collaboration. Similarly, in RES, technologies should be designed to empower individuals and communities to participate in renewable energy generation and consumption decisions.
- Transparency and Accountability: The philosophy of AI emphasizes transparency in algorithms and decision-making processes. In the context of RES, transparent AI systems can provide explanations for energy forecasts, grid management decisions, and optimization strategies, fostering public trust.
- Policy and Governance: Ethical AI discussions often intersect with policy and governance frameworks. This also applies to RES, where policies can incentivize the integration of AI-driven solutions, making renewable energy more accessible and affordable to society.
- Public Engagement: Both AI and RES involve complex technologies that impact society. Philosophical principles advocate for involving the public in decision-making processes related to AI. Similarly, involving communities in RES projects and their AI-driven aspects can lead to better acceptance and adoption.
- Interdisciplinary Research: The connection between the philosophy of AI and RES can encourage interdisciplinary research that addresses ethical, social, technical, and environmental aspects of integrating AI into renewable energy management.
- Education and Awareness: Educating the public about the ethical and practical implications of AI in RES is crucial. Philosophical concepts can be used to explain the benefits, challenges, and potential risks of AI-driven renewable energy solutions.
- Equity and Access: The philosophy of AI stresses the importance of equitable access to AI technologies. When applied to RES, this philosophy can guide efforts to ensure that all segments of society benefit from renewable energy technologies, avoiding energy poverty and inequality.
- Long-Term Impact: Philosophical discussions on the long-term impact of AI align with concerns about the long-term sustainability and effectiveness of renewable energy solutions. Considering these philosophical perspectives can guide decisions that lead to lasting positive effects.

4.2 Contribution of AI to Renewable Energy Applications in Society

AI can play a significant role in advancing the application of renewable energy systems in society by enhancing their efficiency, reliability, and integration as represented here. AI can enable a more reliable, efficient, and widespread adoption of renewable energy systems in society. By leveraging AI's capabilities to process and analyse large datasets, predict outcomes, and optimize operations, the transition to cleaner and more sustainable energy sources can be accelerated.

- Resource Optimization: AI can analyse vast amounts of data from various sources like weather forecasts, historical energy production, and demand patterns to optimize the deployment and operation of renewable energy sources. This helps determine the most suitable locations for solar panels, wind turbines, and other RES infrastructure.
- Energy Forecasting: AI-powered predictive models can forecast energy generation from renewable sources. This information is crucial for grid operators and energy companies to balance supply and demand effectively and make informed decisions.
- Smart Grids and Management: AI can enable the creation of smart grids that automatically adapt to changes in energy generation and demand. This improves the overall stability of the grid and reduces energy wastage. Integrating renewable sources into the power grid can be challenging due to their intermittency. AI can help manage the grid by predicting fluctuations in energy generation and consumption, allowing for real-time adjustments and reducing the need for backup power sources.
- Energy Storage Optimization: Renewables often generate excess energy that can be stored for later use. AI can optimize energy storage systems, like batteries, by predicting when energy generation will be high and demand low, ensuring efficient use of stored energy.
- Efficient Energy Conversion: AI can optimize the efficiency of energy conversion processes, such as converting solar energy into electricity or bioenergy into usable fuel, by continuously adjusting parameters for maximum output.
- Design and Simulation: AI can aid in designing more efficient renewable energy systems by simulating different configurations and evaluating their potential performance before implementation.
- Policy and Investment Decisions: AI can analyse data to provide insights into the economic and environmental impact of renewable energy projects. This helps policymakers and investors make informed decisions about the adoption of RES.
- Energy Trading: AI-powered algorithms can facilitate peer-to-peer energy trading, where individuals and businesses with renewable energy sources can directly sell excess energy to those in need.
- Carbon Footprint Reduction: AI can help optimize energy consumption patterns, suggesting when to use energy-intensive processes during periods of high renewable energy generation, thus reducing reliance on fossil fuels.
- Microgrid Management: In remote areas or during emergencies, microgrids powered by renewables can provide localized energy. AI can manage these microgrids efficiently by balancing energy supply and demand.

4.3 Steps to Conserve Energy Systems

Applications of AI are transforming the digital industries. In particular, the renewable energy sector is continually evolving, embracing research and development to meet the demand that is always increasing. AI may be utilised to solve the challenging issues involved in building solar power facilities, leading to cost reductions, increased efficiency, and greater investment returns. Energy saving in homes and businesses can be as simple as turning off lights or appliances when they are not in use. Traditional light bulbs and appliances use a lot of energy and need to be updated more frequently than their energy-efficient counterparts.

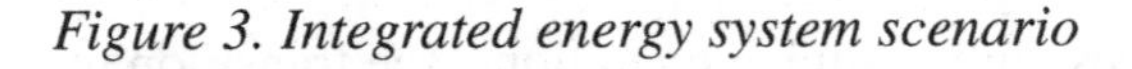
Figure 3. Integrated energy system scenario

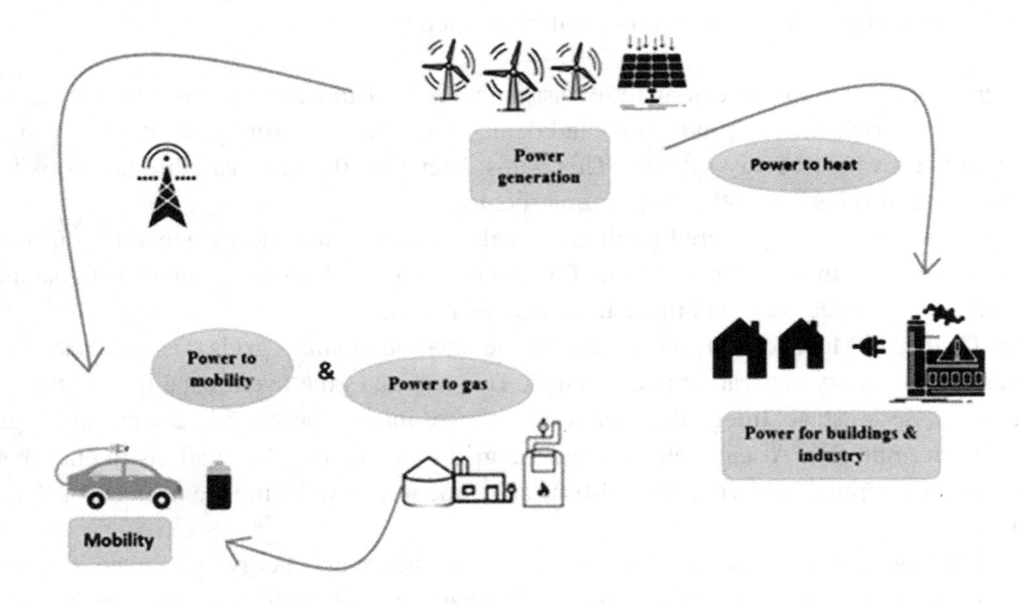

By turning off the electricity to electronic devices when they are not in use, smart power strips eliminate the problem of phantom loads (Muthusamy & Periasamy 2019). There are many ways to reduce heating and cooling costs, including using energy-efficient smart appliances, energy-efficient windows, upgrading the electronic systems, weatherizing home, and sealing air leaks around the home and office. Sector coupling is a method that will act as an integrates energy system as illustrated in Figure 3.

4.4 Ethical Implications of AI Integration

Humans will be able to use smart technologies to allocate energy more effectively. AI will assist us in making decisions about where to transmit electricity, how to store power, and how to operate systems to save energy. These choices will be made by AI, which will enable us to conserve energy. We should make use of these robots that have been assisting us in managing energy effectively as our demand for energy rises and the use of AI spreads. In certain circumstances where people are either unable to notice the details or are unable to react quickly enough, but they are able to detect patterns and make judgements based on these patterns about where to conserve energy. We are moving towards becoming carbon neutral and developing technologies that can help us optimise our energy consumption as well as other aspects of our life with the assistance of major tech corporations. We hope to see a reduction in carbon emissions as a result of these choices. The deployment of AI in renewable energy systems (RES) introduces several ethical implications that need to be carefully considered and addressed. Some of the key ethical concerns associated with the integration of AI in RES are,

- Transparency: AI algorithms can make complex decisions about resource allocation, energy distribution, and grid management. Lack of transparency in these decision-making processes can lead to distrust among stakeholders. Ensuring that AI systems are transparent and can provide understandable explanations for their decisions is essential to maintain accountability and build public confidence.

- Bias and Fairness: AI systems can inherit biases present in the data used for their training, which can lead to unfair or discriminatory outcomes. In RES, bias could result in unequal distribution of energy resources, disadvantaging certain communities. Developers must actively mitigate biases and ensure fairness in algorithmic decision-making to prevent social injustices.
- Job Disruption and Transition: The automation of certain tasks in RES through AI could lead to job displacement for workers in traditional energy sectors. Ethical considerations include ensuring a just transition for these workers by providing retraining and employment opportunities in the growing clean energy sector.
- Environmental Impact: While RES contribute to sustainability, the environmental footprint of AI technologies used in these systems should also be considered. The energy consumption and resource requirements of AI models, especially deep learning, must be balanced against the overall benefits of renewable energy production.
- Data Privacy: AI in RES requires access to large amounts of data for effective decision-making. Privacy concerns arise when personal or sensitive data is collected, stored, or shared. Proper data anonymization, encryption, and adherence to data protection regulations are vital to safeguard individual privacy.
- Reliability and Accountability: AI-driven decisions can have significant impacts on energy distribution and grid stability. Ensuring the reliability of AI systems and establishing clear lines of accountability in case of errors or system failures is crucial to prevent disruptions and mitigate potential harm.
- Dependency and Autonomy: Overreliance on AI systems could diminish human expertise and decision-making autonomy. It's important to strike a balance between AI assistance and human oversight to avoid blindly following algorithmic decisions without critical evaluation.
- Socioeconomic Equity: AI-powered RES should not exacerbate existing socioeconomic disparities. Efforts should be made to ensure that underserved communities have equitable access to the benefits of renewable energy technologies and the AI systems supporting them.
- Energy Access: While AI can enhance RES efficiency, it shouldn't hinder access to energy for marginalized communities. The ethical imperative is to ensure that AI optimization doesn't exclude those who need energy the most.
- Security and Vulnerability: AI in RES introduces new attack surfaces for cyber threats. Ensuring the security of AI systems and protecting them from potential malicious attacks is paramount to prevent disruptions to energy supply and potential breaches of sensitive data.
- Long-Term Impact: Anticipating the long-term effects of AI in RES is an ethical concern. Will AI contribute to sustainable development, or might it lead to unforeseen consequences? Ethical frameworks should guide the continuous assessment of these impacts.

Addressing these ethical implications requires collaboration among researchers, policymakers, industry stakeholders, and civil society. Ethical guidelines, regulatory frameworks, and ongoing assessments can help guide the responsible development and deployment of AI in renewable energy systems, ensuring that the benefits are maximized while potential harms are minimized.

5 APPLICATIONS OF AI IN RENEWABLE ENERGY SYSTEMS

5.1 Solar Steam Generators

Different components of a solar steam generator have been modelled using artificial intelligence. A solar dish collector, a flash vessel, fast rotating pump, and the related pipework are all used in the system. Artificial Neural Networks were also employed to simulate the system's startup. It is critical for the engineer of such systems to be able to make such predictions since the energy expended during the morning start-up has a substantial impact on system performance. Because the system functions under transient settings, this problem is extremely difficult to solve using analytical approaches. ANNs might forecast the temperature history at various places across the system. The energy spent during the heat-up period may be simply quantified using the temperature profiles (Kalogirou et al., 2014; Bose & Wang, 2019; Bose, 2019; Huang, 2019). The system's mean monthly average steam generation is an important characteristic to consider when designing such systems.

5.2 Hydrogen from Solar Energy

Green hydrogen represents both a sustainable alternative to fossil fuels for mobility and a potential long-term storage solution for renewable energy. Hydrogen, as an alternative fuel, does not emit smoke. If it is made by solar technology, there is no greener fuel than that. An Austrian company called Fronius gives importance to it. Recently, it has set up a station called Solhub. The plant, which acts like a hydrogen punk, produces and sells about 100 kg of pure hydrogen daily. Solar panels should be installed and automated using AI techniques for generating, storing, distributing and reconverting green hydrogen from solar energy. A single Solhub is capable of refueling 16 hydrogen cars.

5.3 Wind Speed Prediction

Wind energy is one of the most important and possibly beneficial renewable energy sources available. Wind energy has the potential to generate electricity at all hours of the day, making it ideal for systems that require constant power. Seasonal variations in wind can be predicted, wind energy is a renewable energy source, and wind turbines can be built on the existing IoT based smart agricultural farms without causing land to be lost (Karthikeyan et al., 2021). Wind speed prediction can be automated with an IoT enabled smart sensor technology to control, maintain and improve its operations. One of the options for forecasting wind power based on wind speed data is to use machine learning algorithms. Although persistence and statistical methods are used in wind power forecasting, newer studies have favoured machine learning algorithms, particularly Random Forest classification algorithms, support vector machines, deep learning architectures of long-short-term memory networks and transfer learning models (Manikandakumar & Karthikeyan, 2023). One of the key advantages of employing machine learning algorithms is that they can adapt to shifting trends within datasets and generate models depending on input data rather than a generalised model. An artificial neural network can be trained to predict the average monthly wind speed in different impacted areas (Blaabjerg, & Ma, 2019). In both learning and prediction, a high level of accuracy can be attained.

Among renewable energy sources, wind energy is a substantial and eligible source that has the capacity to produce electricity in a continuous and sustainable manner. Machine learning techniques can be used to forecast wind power based on daily wind speed data. Because wind speed is inherently erratic and unpredictable, daily wind power forecasting is the most accurate technique to forecast. In the case of wind speed forecasting, the methodologies and algorithms cannot produce good and satisfying results, especially for long-term forecasting scenarios. Due to the challenge of forecasting continuous wind power levels, regression analysis algorithms can be applied (Blaabjerg, & Ma, 2019). The best machine learning techniques for forecasting wind speed are the regression algorithms such as LASSO, k Nearest Neighbor (kNN), eXtreme Gradient Boost (XGBoost), Random Forest (RF), and Support Vector Regression (SVR).

5.4 Weather Forecasting Using Artificial Intelligence

Improvements in weather forecasting may appear to have a direct influence in that they make associated planning easier, but even the tiniest gain in weather prediction can result in significant improvements for businesses and governments. Machine learning and artificial intelligence are well-suited to weather forecasting (Manikandakumar & Karthikeyan, 2022). The enormous amount of relevant data, historical data, and real-time data that may be studied is simply too much for a group of unaided individuals to process on their own.

§ The amount of weather-related data that is available is enormous. Over a thousand weather satellites are currently in orbit, giving a plethora of data on cloud patterns, winds, temperatures, and other factors. However, these satellites represent just a small part of the data generation process. Numerous governments and private companies use weather sensors to monitor the solar system in real-time.

§ Furthermore, as the internet of things (IOT) becomes more accessible due to lower-cost sensors and improved connection, the number of gadgets and pieces of equipment that can offer meaningful real-time weather information will certainly grow rapidly.

§ Every automobile, truck, solar panel, networked traffic light, mobile phone, smart home air conditioning system and smart agriculture, among other things, might be used as a source of real-time information for accurate prediction (Manikandakumar & Karthikeyan, 2022).

§ While it will be some time before most street lights are connected, it should serve as an example of how a particular IoT invention may dramatically expand the range of examples of hyper-local meteorological data.

5.5 Intelligent Energy Storage Systems

Energy storage provides a number of environmental advantages that make it a useful tool for achieving sustainability objectives. Storage increases the deployment of renewable energy by enhancing the overall efficiency of the energy grid (Ricalde et al., 2011; Bose, 2017). On a more local level, because an energy storage system produces no emissions, it may be installed anywhere in a facility without affecting the environment or air quality. Not only would the energy be GHG-free if coupled with solar PV, but the combined system will also be subject to federal income investment tax incentives.

Artificial intelligence is a computational approach that employs enormous amounts of data to complete a task. AI is especially useful in situations where there is a vast amount of data that can be used to teach computers to think and behave like humans. Energy is one of those fields where there is a lot of data, it is freely available, well-structured, and accurate, and it is especially well-suited to AI solutions. Researchers can integrate this data into an AI system to get results that will help energy storage solutions a lot.

Examples of AI improving energy storage solutions:

§ AI is especially adept at predicting power generation and demand, and thus the price at a given point in time. It accomplishes this by analyzing years of past electrical data, weather data, and other information.

§ AI utilizes a mixture of weather and satellite data, numerical weather prediction models, and statistical analysis to provide renewable electricity predictions.

§ Artificial intelligence can detect anomalies in a variety of electrical, electromechanical, chemical, and thermal subsystems before they cause system damage, allowing the operator to react quickly. AI accomplishes this by collecting data from various sensors and the environment and comparing it to past data that was used to teach AI what conditions normally contribute to a component's failure. Condition monitoring can reduce downtime, extend the life of a storage system, avoid damage, and raise the operator's profitability.

§ Artificial intelligence can be used to estimate energy use across structures in great detail, including accurate projections of passive solar capacity, wind speed, and building energy load. This would allow for the most efficient use of building energy storage and a reduction in building energy usage.

Smart grids are already being shaped by AI to become really smart, allowing them to meet future energy demands (Ricalde et al., 2011; Shahid, 2018). Energy storage is becoming an increasingly significant aspect of the smart grid, and AI will transform our understanding of consumption patterns and how these devices are operated to raise income for their operators. The emergence of intelligent energy systems is being aided by AI-powered power storage. We can already see how wind or solar systems combined with clever storage will soon outnumber stand-alone systems. When we combine AI with renewable energy and storage, we ensure that the energy source is fully utilized, which is the greatest path ahead for the renewable energy industry, the power sector, and, ultimately, its clients.

Enhancing independent systems, creating additional outsourced earnings, and providing value streams are just three of the numerous ways artificial intelligence and power storage through Intelligent Energy Storage will alter the energy business. Consumers can maximise the use of their energy storage unit with intelligent energy storage. An energy storage system (ESS) is a device that converts electrical energy from power systems into a form that may be stored for later conversion to electrical energy. We shall use as little energy as possible in the future civilization through energy conservation, and the energy we do need will originate from renewable sources.

Furthermore, in the future green and smart civilization, our houses will be low-energy homes that produce more energy than they consume. When energy bills are at their lowest, heat pumps and electric vehicles will recharge overnight. The green and smart society's individuals and businesses will require intelligent, energy-saving, and environment-friendly solutions. Intelligent energy storage is a type of equipment that can store and release energy but not perform the following processes or functions:

§ Control when they are charged and discharged based on the price of electricity to optimise their operation.
§ To get a longer cycle life, control how and when they are charged on a daily basis.
§ Coordinate with other energy storage technologies, power producing capacity, and consumers.
§ Act before a failure happens. Lithium-ion batteries, for example, will not recover from a thermal runaway or other degradation-inducing event, and the system may become permanently underperforming or stranded.

For power, heating, and transportation, an intelligent energy system must be able to incorporate massive volumes of fluctuating renewable energy sources effectively. Without compromising supply security or consumer comfort, it must ensure that the energy demand is met efficiently and appropriately Energy conservation and efficiency improvement are also required, as is economic efficiency. The use of energy storage systems in Renewable Energy Storages and Smart Grids is essential. Because Renewable Energy Storages have to deal with the issue of long-term energy sustainability. For renewable energy sources such as photovoltaic, wind, etc., raw material storage is not feasible.

Commercial and industrial demand for standalone systems can be met with intelligent energy storage, as can be seen. Because of this, AI can be used to perform predictive analytics, machine learning and big data processing in order to maximise a customer's return on investment from a RE storage system. A real-time, adaptive storage dispatch is possible with IES. The customer and grid benefit from this dispatch because it increases value.

5.6 Home Energy Storage System

Residential energy storage devices allow us to get the most out of the electrical energy by conserving it till the need. They are especially handy for folks who have home renewable energy systems and want to make better use of the renewable energy they generate. An energy storage system could help you save money on your energy bills and lessen the carbon footprint. The amount of money saved on gas depends on the system installed and how it is used. There is currently insufficient independent information to predict typical benefits for some new energy storage systems. Request that installers compute savings for you based on your home and circumstances, and that they explain how they do so. The majority of energy storage systems are intelligent. This helps to keep track of your energy usage online and select when to charge and refinishing from the storage unit. One can reduce energy waste since it is practically difficult to consume each kilowatt-hour of solar energy generated at the time it is received, ensuring that you get the most out of your solar investment.

It is better for the environment because most of the electricity we obtain from the grid comes from polluting coal-fired power plants and other fossil fuels (Shahid, 2018). You can make sure you are consuming as much renewable energy as possible at all times by storing your solar electricity. We can save money since energy storage allows us to consume less electricity from the grid, lowering our electric costs. Furthermore, we can protect ourselves from variable energy prices by selecting to consume our saved energy only during peak hours when rates are higher. Energy rates, for example, are often greatest on weekday afternoons and nights, making this an ideal time to use saved energy. We have more alternative energy because battery storage brings us one step closer to energy independence by giving us our own source of electricity that we can consume at any time of day. This allows you to have more control over our costs and lessen our reliance on utilities.

5.7 Integration With Smart Grids

Grid management is a crucial component of any renewable energy systems. Additionally, machine learning and artificial intelligence are essential in this field. These innovations forecast residential energy use using data analytics. The forecast is based on a specified period of time in a year and also takes data from prior years into account. This aids power firms in knowing how much energy will be needed in the days to come. They may control their grids without any interruptions based on that. They can increase energy output if there will be a large amount of demand. As an alternative, they can reduce output during periods of low energy demand to prevent waste. Smart grid integration and decentralized energy management can both benefit from AI as shown in Fig. 4. When community-scale renewable energy producing units are added to the major grid, balancing the energy flow within the grid becomes difficult (Ricalde et al., 2011; Atasoy et al, 2015; Bose, 2019). Advanced load management systems can be integrated with machinery like industrial furnaces or huge air conditioners, allowing them to automatically shut down when the power supply is low. The flow of supplies can also be used to modify intelligent storage devices.

Figure 4. Smart grid integration

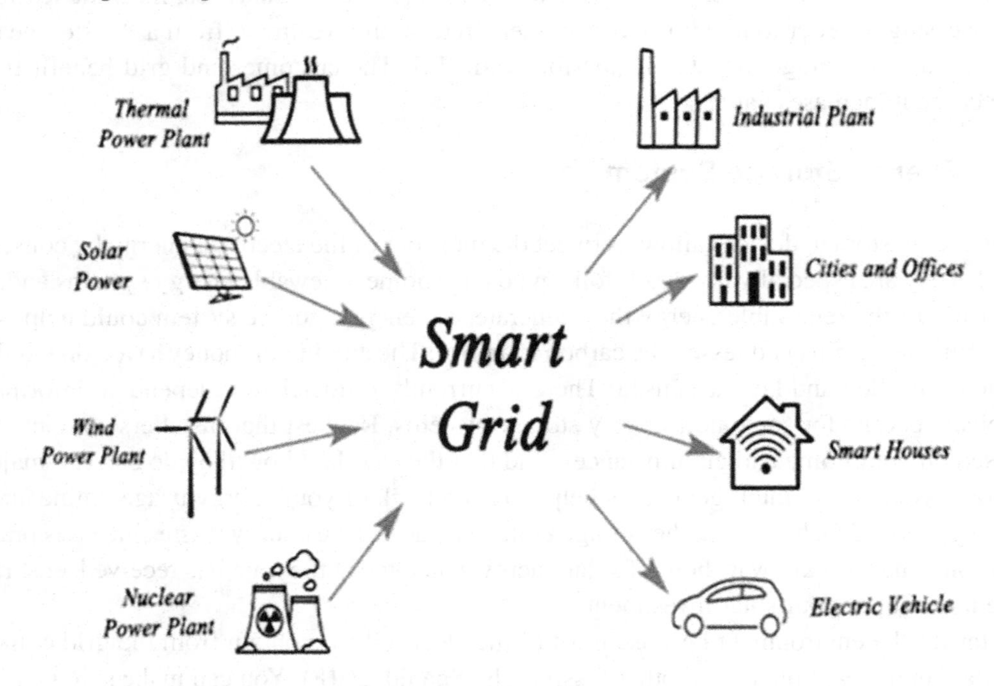

5.8 Expand the Market

The combination of artificial intelligence and Intelligent Energy Storage (IES) can offer the renewable energy market a long-term and dependable solution. This smart grid will be able to assess a large amount of data gathered from several sensors and make energy distribution decisions in real time (Bose, 2017). AI has already aided solar production in a number of areas, most notably in weather prediction. Grid supply may be forecast more accurately with more accurate weather predictions. Grid operators place a great importance on this. In terms of changeable weather conditions, wind energy confronts many of the same issues as solar energy (Blaabjerg & Ma, 2019; Gupta et al., 2022; Hu et al., 2022). They were able to anticipate production 36 hours ahead of time using machine learning techniques and generally available weather reports.

AI algorithms may incorporate large volume of historical data, regional weather station reports, satellite image input data, images of cameras, and sensor networks. By adjusting output to changing weather conditions, improved forecasting can lead to more effective management of conventional generators and lower the cost of starting and shutting down units. This may lower the price of reducing solar power. Generators and energy dealers are able to place bids in balancing markets because to more precise forecasts of renewable production. The sector for renewable energy has the potential to undergo a full transformation due to artificial intelligence and machine learning. These technologies will have an effect on electricity providers and customers in the next years. Energy suppliers will have a tool for more effective forecasting, grid management, and, most crucially, maintenance scheduling. Customers will experience the effects in the form of intermittent green energy and upfront notifications about planned grid repairs.

5.9 Condition Monitoring

Another application of machine learning that is demonstrating its ability to drastically reduce costs for power companies, particularly when applied to windmills. Newer versions have indeed been able to monitor blade defects and generator temperature, enabling for quick repairs and maintenance utilising machine learning and data (Bose, 2017). The regression tree model, which combines machine learning into its forecasts, has improved on the standard power curve methodology in wind farms. The regression tree model offers three times the accuracy of the classic power curve methods without requiring any additional data than would normally be required for a wind resource evaluation. Due to the necessity of precisely estimating the output of a given site, this would help both farm operators and grid operators. AI-powered sensors and monitoring systems can predict equipment failures and maintenance needs in renewable energy infrastructure.

5.10 Energy Management Tool

A national energy system modelling and energy balancing tool is provided by the International Network for Sustainable Energy. This tool is limited in scope and is only available to non-profit organisations. To model energy systems, the programme uses data from a spreadsheet. It incorporates energy demand, energy generation, energy policy, and energy trends in its modelling. It models renewable energy, thermal, hydrogen, and transportation energy systems, but not tidal power, pumped-heat electrical storage, battery storage, electric vehicles, or Vehicle-to-Grid (V2G) technology. Because of scheduled maintenance, consumers may be informed when there will be power outages. Power outages without prior notice can be prevented.

5.11 AI Security Applications in Renewable Energy

While managing intermittency is the primary purpose of AI in renewable energy, it can also increase safety, efficiency, and reliability. It can assist in deciphering energy usage trends, identifying energy leaks, and determining the health of the gadgets. Wind turbine sensors can be used to collect data for AI-powered renewable energy systems to track wear and tear. The system will keep track of the equipment's general health and notify the operator when maintenance is required.

AI integrated into centralised control systems will help to prevent energy shortages by detecting problems early and reducing the time it takes to repair them. It should have notifications based on statistics, reporting, a user-friendly and web-based interface, a backup server for unexpected cases, security keys for authentication for users from multiple locations, and so on to be effective in these directions. Enhanced prediction also improves grid stability and, as a result, supply security (Ford et al., 2014). Despite popular belief that AI makes the power system less safe, AI can play a critical role in the fight against cybercrime. It can check enormous amounts of data fast and discover anomalies. AI can also make inferences from previous cyberattacks. The large volume of energy data generated from AI and related IoT devices can also be protected by various security and privacy mechanisms (Manikandakumar & Ramanujam, 2018; Ricalde et al., 2011).

5.12 AI and Power Quality Management

Constraints such as economic load dispatch, load forecasting, generation, optimization and scheduling, transmission capacity and optimal power flow, real and reactive power limits of generators, bus voltages and transformer taps, load demand in interconnected large power systems and their protections, can now be easily held with Artificial Intelligence (Rekioua, 2019; Dellosa et al., 2021). AI techniques that perfectly deal with the generation, transmission, distribution, and consumption of electric power have successfully minimised most of the efforts in power system analysis. Using Artificial Intelligence to control electricity systems is a wise decision because earlier techniques may not have been as successful or as time-consuming as they could have been. Durability, robustness, dependability, technical breakthroughs, power system selection, and dynamic reaction are all critical for industrial development with power system expansion (Bose, 2017; Dubey et al., 2022). The complexity of the networks has increased dramatically as the electricity system has grown. As a result of this power system analysis using traditional methodologies and drawing conclusions from the collected data, the information, remote device management, and utility management processes have become more sophisticated and time-consuming.

5.13 AI Exposure in Power Systems

Several issues in power systems cannot be solved using traditional methods. As a result, AI approaches in power system applications are receiving a lot of attention. Here are some of the uses of the power system that are emphasised.

- § Economic load dispatch, load forecasting-based production and strategic decisions, and hydrothermal generation scheduling optimization.
- § Voltage and frequency control for system stability, device sizing and control.
- § Analysis of Automated power restoration and monitoring, fault diagnostics, and security margins.

§ Distribution planning and operation, adaptive control, demand-side response and management, and smart grid operation and control.

The direct benefit of improved weather forecasting may appear to most individuals to be that it makes vacation planning easier, but even the tiniest gain in weather prediction can result in significant improvements for businesses and governments. Machine learning and artificial intelligence are well-suited to weather forecasting. Any group of unaided humans can begin to handle the large amount of relevant information, historical data, and real-time data that can be studied.

A hybrid energy system is made up of two or more renewable energy sources that are combined to boost system efficiency and provide greater energy supply balance. One of the key goals is to match the generation schedule to the solar generation projection in order to maximise efficiency in terms of profit gained from various power markets (Mohamed et al., 2017).

5.14 Floating Solar Photovoltaics

A floating solar photovoltaic (FPV) system is a new type of solar photovoltaic (PV) system that is installed right on top of a water body rather than on ground or on top of buildings (Akarslan, 2012; Youssef et al., 2017; Zhang & Allagui 2021). FPV systems have recently gained popularity around the world. according to GTM Research, Figure 5. depicts the history of FPV installation as well as forecasts for specific geographic areas. A technology's technical potential is its maximum energy capacity and generation.

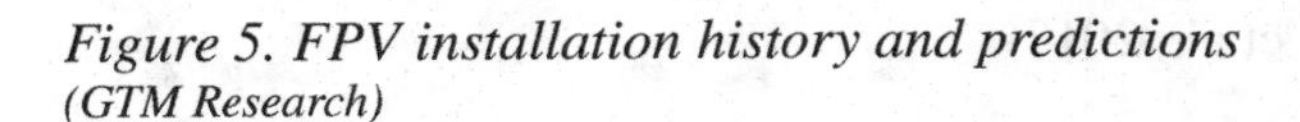

Figure 5. FPV installation history and predictions
(GTM Research)

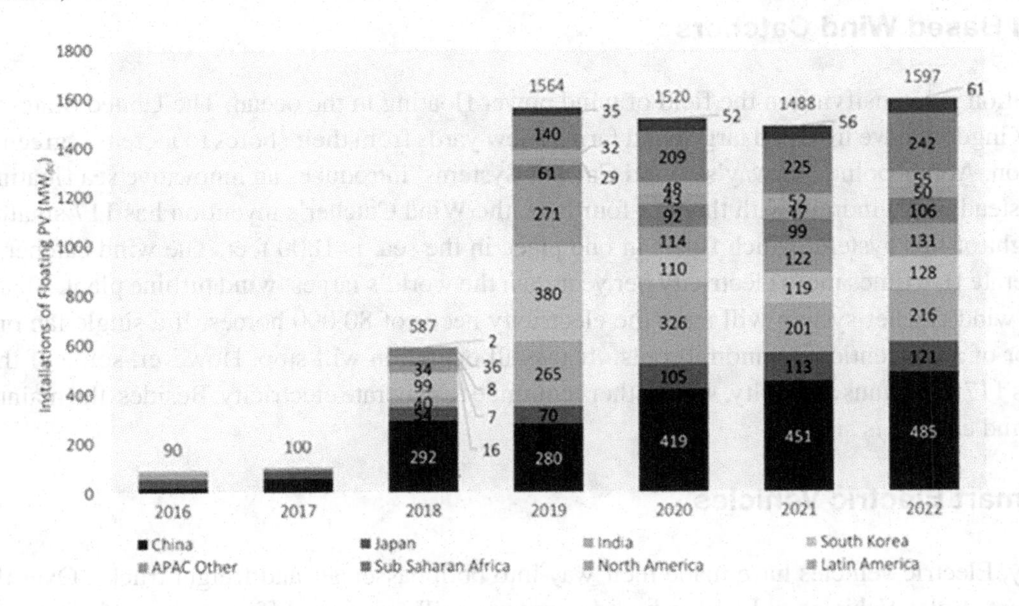

PV systems installed on water minimise competition for land that could be used for something else. By shading water bodies, they have the ability to reduce evaporation losses. FPVs, like land-based PV systems, can generate electricity in places where utility bills are high. The number of FPV installations has exploded in recent years. China, Japan, and South Korea have all seen strong increase in FPV, and this trend is expected to continue. FPV projects are also projected to be installed in the United States and many underdeveloped countries. These FPVs are could be integrated with AI systems for the better energy harvesting. In automated decision-making, information is processed by computer systems without human involvement, giving them the capacity to tackle complicated tasks more quickly than people. A solar photovoltaic (PV) plant's maintenance schedule are one of the best example for this.

5.15 Optimization for an Efficient Use of Energy

It is difficult to effectively participate in non-dispatchable renewable energy sources (RES), which is mostly dependent on metrological and oceanic conditions. Nonetheless, a plan for maximising the usage of renewable energy sources may be devised. In FPVs, the technique to apply is an optimization method that maximises income. The previously mentioned FPV generation prediction is one of most essential input variables of the algorithm's objective function (Rosen, 2021). The optimization method will be given constraints relevant to each power source in addition to other input variables. In hybrid energy power systems, artificial intelligence allows for the mixing of various picture data and other numerical information in order to extend the forecast horizon and reduce prediction mistakes. These forecasts can then be used as a crucial input for a hybrid energy system's optimization algorithm, allowing solar and hydro energy to be used more efficiently and RES to participate more freely in various power markets El Haj Assad et al., 2021; Rosen, 2021; Mohamed et al., 2017).

5.16 AI Based Wind Catchers

Competition is intensifying in the field of wind power floating in the ocean. The United States and the United Kingdom have installed large wind farms a few yards from their shores to increase 'green' power generation. At this point, Norway's 'Wind Catcher Systems' introduces an innovative sea floating wind farm. Instead of a windmill with three or four fans, the Wind Catcher's invention has 117 smaller fans. The height of this system, which floats in one place in the sea, is 1000 feet. The wind catcher system can generate five times more electricity per year than the world's largest wind turbine plant. That means a single wind catcher system will meet the electricity needs of 80,000 homes. If a single fan or power generator of a conventional windmill fails, its overall operation will stop. However, some of the wind catcher's 117 small fans are faulty, while others continue to generate electricity. Besides, the maintenance of the wind catcher is simple.

5.17 Smart Electric Vehicles

Similarly, Electric vehicles have made their way into both passenger and freight trucks. Over the next three years, Volta Vehicles, a London-based company, will test four different sizes of electric trucks. They have recently held demonstrations in several major European towns. The Volta Zero is the world's first 16-tonne electric truck, with a range of up to 200 kilometres when fully charged. The vehicle, which moves at a top speed of 90 kilometres per hour, is ideal for delivering items within the city. It is fitted with smart cameras capable of observing from a 360-degree angle for security.

Figure 6. Floating wind catcher systems (wind catching)

The truck is lightweight, with a non-vibrating motor that keeps drivers very convenient driving. Volta zero has received over a thousand bookings from a variety of nations throughout Europe. Experts anticipate that by 2025, the market for AI based electric trucks would be worth more than $100 billion. As a result, the arrival of electric trucks has become a foregone conclusion. Neural network-based AI can be applied to every phase of the smart grid for the overall performance and throughput improvement. Some remarkable neural network applications in renewable energy and smart grid systems are mentioned as follows,

§ Load forecasting on the smart grid systems
§ Forecasting of wind and PV generation curves
§ Automated fault detection and diagnostics of stations
§ Fault-tolerant control mechanism implementation in power systems
§ Robust estimation of feedback signals without sensors
§ Noise-free signal filtering

§ Neural network modelling of system elements
§ AI based real time simulation by DSPs and FPGA chips
§ Intelligent scheduling of generation and storage
§ High-performance intelligent control of system elements
§ Real-time pricing predictions of electricity with demand-side management.

6 AI-ENERGY SYSTEMS FOR SUSTAINABILITY

AI's ability to process and analyse vast amounts of data, make predictions, and adapt in real time makes it a valuable tool for enhancing energy efficiency, grid stability, and the integration of renewable energy sources into the existing power infrastructure. It enables more informed decision-making, better resource allocation, and a more resilient and sustainable energy system as described in this section.

6.1 Energy Efficiency Optimization

AI can analyse vast amounts of data from sensors, smart meters, and other sources to identify patterns and trends in energy consumption. By understanding how energy is used within buildings, industrial processes, and other systems, AI algorithms can suggest strategies to optimize energy usage. This can include real-time adjustments of heating, cooling, lighting, and other systems based on occupancy and external conditions. Machine learning algorithms can continuously learn and adapt to changing usage patterns, resulting in ongoing energy savings.

6.2 Grid Stability Enhancement

Modern power grids are complex and dynamic systems with various sources of energy generation and consumption. AI can help in predicting demand and supply fluctuations, identifying potential equipment failures, and optimizing the flow of electricity across the grid. Machine learning models can analyse historical data to predict spikes in demand or potential grid imbalances. This information can be used to adjust energy generation and distribution in real time to maintain grid stability, prevent blackouts, and reduce the need for expensive backup power sources.

6.3 Integration of Renewable Energy Sources

One of the challenges with renewable energy sources like solar and wind is their intermittent nature. AI can help by forecasting renewable energy generation based on weather patterns and historical data. By accurately predicting energy availability, utilities can better plan for when to rely on renewable sources and when to supplement with other forms of generation. AI can also optimize the scheduling and storage of energy from renewables, ensuring that excess energy is stored when available and utilized during peak demand periods.

6.4 Demand Response and Load Management

AI-powered demand response systems can encourage consumers to shift their energy-intensive activities to off-peak hours. This can be achieved through real-time pricing signals or incentives, helping to balance demand and reduce the strain on the grid during peak times. Machine learning models can analyse historical data to predict when peak demand will occur and proactively communicate with consumers or automated systems to adjust energy usage accordingly.

6.5 Predictive Maintenance

For energy infrastructure like power plants and substations, AI can enable predictive maintenance. By analysing data from sensors and historical maintenance records, AI algorithms can predict when equipment is likely to fail. This allows operators to schedule maintenance before a failure occurs, minimizing downtime and reducing the likelihood of expensive breakdowns.

6.6 Microgrid Optimization

AI can optimize the operation of microgrids, which are localized energy systems that can operate independently or in conjunction with the main grid. AI algorithms can balance energy generation and consumption within microgrids, integrating renewable sources, storage systems, and backup generators to ensure reliable power supply to specific areas or communities.

7 INTEGRATING HUMANS AND AI FOR RES

The collaboration between humans and AI technologies in the context of renewable energy is crucial for achieving sustainable and efficient energy solutions. Humans play various roles in working alongside AI to harness the full potential of renewable energy sources and address the challenges associated with their integration. Here are some key aspects of human-AI collaboration in the renewable energy sector:

7.1 Data Collection and Management

Humans are responsible for collecting and curating the data that AI systems rely on for analysis and decision-making. This includes data from renewable energy sources, weather patterns, energy consumption patterns, and grid performance. Effective collaboration involves ensuring the accuracy and quality of data, as AI's effectiveness heavily depends on the reliability of the data it processes.

7.2 Algorithm Development and Training

While AI algorithms can process and analyse data at scale, humans are essential for designing, developing, and training these algorithms. Domain experts, researchers, and data scientists work together to create machine learning models tailored to renewable energy applications. They refine the algorithms based on insights gained from data analysis and adjust them to fit the unique characteristics of renewable energy systems.

7.3 Interpreting and Validating Results

Human expertise is necessary to interpret the results and recommendations provided by AI systems. Domain experts can validate whether AI-generated insights align with physical realities and make sense within the context of renewable energy. They can identify potential errors or anomalies and ensure that AI recommendations are operationally sound and safe.

7.4 Policy and Decision-Making

Humans hold the responsibility of making policy decisions related to renewable energy integration. AI can provide data-driven insights that inform policy choices, such as setting incentives for renewable energy adoption or establishing regulations for grid interconnection. However, humans must consider broader societal, economic, and environmental factors that AI might not fully encompass. Collaboration between humans and AI also involves addressing ethical considerations. Humans need to ensure that AI technologies are used responsibly and do not inadvertently harm communities or ecosystems. They must also account for potential biases in AI algorithms and make efforts to mitigate them to ensure fair and equitable outcomes.

7.5 Innovation and Adaptation

Humans drive innovation by identifying new opportunities for renewable energy integration and technology improvement. They can adapt AI systems to evolving technological landscapes, regulatory changes, and shifts in consumer behavior. Continuous collaboration allows AI technologies to evolve in ways that best serve the renewable energy sector's needs.

7.6 Maintenance and Oversight

Human involvement is vital for the ongoing maintenance, monitoring, and oversight of AI systems. Regular updates, calibration, and refinement are necessary to ensure AI remains effective and aligned with the changing dynamics of the renewable energy landscape.

In essence, humans bring critical domain knowledge, ethical considerations, and the ability to make contextual judgments that AI lacks. AI technologies enhance human capabilities by processing and analysing large volumes of data and providing insights that inform decision-making. Successful collaboration between humans and AI in the renewable energy sector harnesses the strengths of both to drive innovation, efficiency, and sustainable energy solutions.

8 CONCLUSION

The chapter contains a detailed discussion of several aspects of Artificial Intelligence with Renewable Energy sources, as well as an explanation of how AI can be used in various industrial and nonindustrial areas. The exploration of individual, sustainable energy sources has been aided by increasing energy demands, depleting fuel resources, and environmental issues. Renewable energy systems are judged on a number of factors. They can be generated at all times because the resources are always available. Additionally, renewable energy sources are clean and eco-friendly. Researchers from all around the world are working to uncover and develop renewable energy systems in order to lessen reliance on fossil fuels. Solar energy stands out as one of the most important renewable energy sources currently available. Solar energy is one of the available renewable energy sources, and it is encouraged to be used by all nations to ensure their long-term viability.

The expansion of solar-based heating and cooling systems necessitates the use of thermal energy storage. Solar energy rises via the usual portion of the building envelope, which can reduce energy consumption for heating during the summer season, conserving energy from traditional heat sources. The value of Neural Networks is the most significant among all AI approaches. In renewable energy systems, the feedforward backpropagation network and its applications are very common. Various neural network applications for smart grid and renewable energy systems are presented. With the introduction of strong and cost-effective application-specific neural network systems, broad AI applications are predicted to usher in a new form of industrial revolution in the future.

At the same time, the renewable energy sector faces significant challenges. Some problems inherent in all renewable technologies, while others are the result of a distorted regulatory structure and market. Renewable technology cannot be adopted because of a lack of comprehensive policy and regulatory frameworks. Inadequate technology and a lack of infrastructure are obstacles that must be addressed properly in order to build renewable technologies. Renewable energy sources are the fastest growing green technology has approximately 20% of annual growth towards 2030. It has high single-digit annual growth rates in all key onshore markets. Massive growth expected in renewable hydrogen and green fuel sectors, and the market forecasts to 2030 continue to grow. More monies should be available from the government to assist research and innovation in this industry. Because there are insufficiently qualified AI employees to train, exhibit, maintain, and run renewable energy systems, institutions should be proactive in their workforce preparation.

REFERENCES

Akarslan, F. (2012). *Photovoltaic systems and applications*. Modeling and Optimization of Renewable Energy Systems., doi:10.5772/39244

AlShabi, M., & El Haj Assad, M. (2021). Artificial intelligence applications in renewable energy systems. *Design and Performance Optimization of Renewable Energy Systems,* 251–295. doi:10.1016/b978-0-12-821602-6.00018-3

Atasoy, T., Akinc, H. E., & Ercin, O. (2015). An analysis on smart grid applications and grid integration of renewable energy systems in smart cities. In *International Conference on Renewable Energy Research and Applications (ICRERA)*. IEEE. 10.1109/ICRERA.2015.7418473

Blaabjerg, F., & Ma, K. (2019). Renewable energy systems with wind power. *Power Electronics in Renewable Energy Systems and Smart Grid,* 315–345. doi:10.1002/9781119515661.ch6

Bose, B. K. (2017). Artificial intelligence techniques in smart grid and renewable energy systems—Some example applications. *Proceedings of the IEEE, 105*(11), 2262–2273. doi:10.1109/JPROC.2017.2756596

Bose, B. K. (2017). Power electronics, smart grid, and renewable energy systems. *Proceedings of the IEEE, 105*(11), 2011–2018. doi:10.1109/JPROC.2017.2745621

Bose, B. K. (2019). Artificial intelligence applications in renewable energy systems and smart grid – some novel applications. *Power Electronics in Renewable Energy Systems and Smart Grid,* 625–675. doi:10.1002/9781119515661.ch12

Bose, B. K. (Ed.). (2019). *Power Electronics in Renewable Energy Systems and Smart Grid.,* doi:10.1002/9781119515661

Bose B. K. (2020). Artificial intelligence techniques: how can it solve problems in power electronics? An Advancing Frontier," *In IEEE Power Electronics Magazine,* vol. 7, no. 4, pp. 19-27, doi: . doi:10.1109/MPEL.2020.3033607

Bose, B. K., & Wang, F. (Fred). (2019). Energy, environment, power electronics, renewable energy systems, and smart grid. *Power Electronics in Renewable Energy Systems and Smart Grid,* 1–83. doi:10.1002/9781119515661.ch1

Chakrabortty, A., & Bose, A. (2019). Smart grid simulations and control. *Power Electronics in Renewable Energy Systems and Smart Grid,* 585–624. doi:10.1002/9781119515661.ch11

Clark, C. E., & DuPont, B. (2018). Reliability-based design optimization in offshore renewable energy systems. *Renewable & Sustainable Energy Reviews, 97,* 390–400. doi:10.1016/j.rser.2018.08.030

Dellosa, J. T., & Palconit, E. C. (2021, September). Artificial Intelligence (AI) in renewable energy systems: A condensed review of its applications and techniques. *In 2021 IEEE International Conference on Environment and Electrical Engineering and 2021 IEEE Industrial and Commercial Power Systems Europe (EEEIC/I&CPS Europe),* (pp. 1-6). IEEE.

Dubey, A. K., Narang, S., Srivastav, A. L., Kumar, A., & García-Díaz, V. (Eds.). (2022). *Artificial Intelligence for Renewable Energy Systems*. Elsevier.

El Haj Assad, M., Alhuyi Nazari, M., & Rosen, M. A. (2021). Applications of renewable energy sources. *Design and Performance Optimization of Renewable Energy Systems,* 1–15. doi:10.1016/b978-0-12-821602-6.00001-8

Ford, V., Siraj, A., & Eberle, W. (2014). Smart grid energy fraud detection using artificial neural networks. In *IEEE Symposium on Computational Intelligence Applications in Smart Grid (CIASG).* IEEE. 10.1109/CIASG.2014.7011557

Gupta, P., Kumar, S., Singh, Y. B., Singh, P., Sharma, S. K., & Rathore, N. K. (2022). The Impact of Artificial Intelligence on Renewable Energy Systems. *NeuroQuantology : An Interdisciplinary Journal of Neuroscience and Quantum Physics, 20*(16), 5012–5029.

Hu, W., Wu, Q., Anvari-Moghaddam, A., Zhao, J., Xu, X., Abulanwar, S. M., & Cao, D. (2022). Applications of artificial intelligence in renewable energy systems. *IET Renewable Power Generation, 16*(7), 1279–1282. doi:10.1049/rpg2.12479

Huang, A. Q. (2019). Power semiconductor devices for smart grid and renewable energy systems. *Power Electronics in Renewable Energy Systems and Smart Grid,* 85–152. doi:10.1002/9781119515661.ch2

Kalogirou, S. A., Mathioulakis, E., & Belessiotis, V. (2014). Artificial neural networks for the performance prediction of large solar systems. *Renewable Energy*, *63*, 90–97. doi:10.1016/j.renene.2013.08.049

Karthikeyan, P., Manikandakumar, M., Sri Subarnaa, D. K., & Priyadharshini, P. (2021). Weed identification in agriculture field through IoT. In *Advances in Intelligent Systems and Computing* (pp. 495–505). Springer. doi:10.1007/978-981-15-5029-4_41

Lateef, A. A. A., Ali Al-Janabi, S. I., & Abdulteef, O. A. (2022, July). Artificial Intelligence Techniques Applied on Renewable Energy Systems: A Review. In *Proceedings of International Conference on Computing and Communication Networks: ICCCN 2021* (pp. 297-308). Springer Nature Singapore. 10.1007/978-981-19-0604-6_25

Manikandakumar, M., & Karthikeyan, P. (2022). Essentials, Challenges, and Future Directions of Agricultural IoT: A Case Study in the Indian Perspective. *Advances in Web Technologies and Engineering*, 181–196. doi:10.4018/978-1-7998-4186-9.ch010

Manikandakumar, M., & Karthikeyan, P. (2023). Weed classification using particle swarm optimization and deep learning models. *Computer Systems Science and Engineering*, *44*(1), 913–927. doi:10.32604/csse.2023.025434

Manikandakumar, M., & Ramanujam, E. (2018). Security and privacy challenges in big data environment. *Advances in Information Security, Privacy, and Ethics*, 315–325. doi:10.4018/978-1-5225-4100-4.ch017

Mohamed, M. A., Eltamaly, A. M., & Alolah, A. I. (2017). Swarm intelligence-based optimization of grid-dependent hybrid renewable energy systems. *Renewable & Sustainable Energy Reviews*, *77*, 515–524. doi:10.1016/j.rser.2017.04.048

Molina, M. G. (2019). Grid energy storage systems. *Power Electronics in Renewable Energy Systems and Smart Grid*, 495–583. doi:10.1002/9781119515661.ch10

Muthusamy, M., & Periasamy, K. (2019). A comprehensive study on internet of things security. In Advancing Consumer-Centric Fog Computing Architectures, 72–86. doi:10.4018/978-1-5225-7149-0.ch004

Ojiro, T., & Tsuruta, K. (2019). *A study of smart factory with artificial intelligence*. Control and Optimization of Renewable Energy Systems. doi:10.2316/P.2019.860-011

Rekioua, D. (2019). Design of hybrid renewable energy systems. *Green Energy and Technology*, 173–195. doi:10.1007/978-3-030-34021-6_5

Ricalde, L. J., Ordonez, E., Gamez, M., & Sanchez, E. N. (2011). Design of a smart grid management system with renewable energy generation. In *IEEE Symposium on Computational Intelligence Applications in Smart Grid (CIASG)*. IEEE. 10.1109/CIASG.2011.5953346

Rosen, M. A. (2021). Renewable energy and energy sustainability. *Design and Performance Optimization of Renewable Energy Systems*, 17–31. doi:10.1016/b978-0-12-821602-6.00002-x

Shahid, A. (2018). Smart grid integration of renewable energy systems. In *7th International Conference on Renewable Energy Research and Applications (ICRERA)*. IEEE. 10.1109/ICRERA.2018.8566827

Youssef, A., El-Telbany, M., & Zekry, A. (2017). The role of artificial intelligence in photo-voltaic systems design and control: A review. *Renewable & Sustainable Energy Reviews*, *78*, 72–79. doi:10.1016/j.rser.2017.04.046

Zhang, D., & Allagui, A. (2021). Fundamentals and performance of solar photovoltaic systems. *Design and Performance Optimization of Renewable Energy Systems*, 117–129. doi:10.1016/b978-0-12-821602-6.00009

Zhang, L., Ling, J., & Lin, M. (2022). Artificial intelligence in renewable energy: A comprehensive bibliometric analysis. *Energy Reports*, *8*, 14072–14088. doi:10.1016/j.egyr.2022.10.347

Chapter 10
Exploratory Cluster Analysis Using Self-Organizing Maps: Algorithms, Methodologies, and Framework

Nuno C. Marques
https://orcid.org/0000-0002-3019-3304
NOVA-LINCS, SST, Universidade NOVA de Lisboa, Portugal

Bruno Silva
https://orcid.org/0000-0001-7873-2111
EST, Polytechnic Institute of Setúbal, Portugal

ABSTRACT

As the volume and complexity of data streams continue to increase, exploratory cluster analysis is becoming increasingly important. In this chapter, the authors explore the use of artificial neural networks (ANNs), particularly self-organizing maps (SOMs), for this purpose. They propose additional methodologies, including concept drift detection, as well as distributed and collaborative learning strategies and introduce a new open-source Java ANN library, designed to support practical applications of SOMs across various domains. By following our tutorial, users will gain practical insights into visualizing and analyzing these challenging datasets, enabling them to harness the full potential of our approach in their own projects. Overall, this chapter aims to provide readers with a comprehensive understanding of SOMs and their place within the broader context of artificial neural networks. Furthermore, we offer practical guidance on the effective development and utilization of these models in real-world applications.

DOI: 10.4018/978-1-6684-9591-9.ch010

INTRODUCTION

Self-organizing maps (SOMs), also known as Kohonen maps (Kohonen, 1982), represent a significant advancement in artificial intelligence, providing a robust framework for data visualization and analysis. Their ability to transform complex, high-dimensional data into simple, comprehensible visual representations has made them an indispensable tool in a variety of fields, from computer science to bioinformatics, finance, and beyond. Since their inception in the early 1980s by Teuvo Kohonen, SOMs have significantly contributed to our understanding of complex systems, pattern recognition, and data representation (Kohonen, 2001). SOMs provide a robust method for visualizing and comprehending intricate data structures. By organizing high-dimensional input data into a low-dimensional grid, they facilitate the identification of hidden patterns, clusters, and relationships within the data, which might otherwise be challenging to discern (Vesanto, 1999).

In many scientific and engineering domains, researchers often encounter datasets with numerous variables and complex relationships. SOMs offer a unique solution to this challenge, enabling visualization and interpretation. The ability of self-organizing maps to uncover latent patterns within datasets has played a pivotal role in data mining and knowledge discovery. By identifying clusters and similarities in data, SOMs support the exploration of large datasets to extract valuable information. SOMs have proven instrumental in discovering new relationships, trends, and correlations, enabling researchers and scientists to gain valuable insights into data distributions and spatial relationships (Oja, Kaski, & Kohonen, 2003). Today, SOMs continue to empower informed decision-making and drive innovation across diverse domains, including genomics, finance, environmental sciences, marketing, healthcare, and social sciences.

SOMs have served as a crucial milestone in the development of artificial neural networks. By demonstrating how a simplified model of the brain's self-organization could be applied to data analysis, SOMs laid the foundation for subsequent advancements in neural network research. They have contributed to the evolution of deep learning, reinforcement learning, and other neural network architectures, leading to breakthroughs in fields such as computer vision, natural language understanding, and robotics (Goodfellow, Bengio, & Courville, 2016).

This chapter aims to guide readers through the role of SOMs in machine learning and introduce a publicly available neural network framework for exploratory cluster analysis. We will explore the practical application of these techniques using the *Wine Dataset*, a popular choice for machine learning and data mining tasks. By following along, readers will gain practical insights into the power of SOMs in handling complex, multi-dimensional data.

The chapter commences with an introduction to the capabilities of SOMs. This is followed by a detailed examination of the *Wine Dataset* using SOMs, where we elucidate complex relationships through the Unified Distance Matrix (U-Matrix) and Component Planes. The *Wine Dataset* serves as a comprehensive case study for SOMs, and we further explore the component planes, a visualization technique that shows the distribution of different feature values in the map, and a *feature clustering* technique based on hierarchical clustering. The focus then shifts to the application of SOMs in data stream contexts, specifically through the Ubiquitous Self-Organizing Map (UbiSOM). We present theoretical results and discuss the philosophical implications of this approach, as well as practical results from real-time exploratory analysis. We introduce the UbiSOM Library and discuss the application of *multiSOM* visualization with UbiSOM. This technique enhances interpretability by enabling simultaneous visualization of multiple dimensions and multiple SOMs. Following this, we illustrate the application of multiple SOMs to the

Wine Dataset using *Processing.org*, a flexible software sketchbook and language ideal for interactive data visualization. This section covers the setup of the environment, the creation of SOMs, result visualization, dynamic training and visualization, interaction with the visualization, and result interpretation. We discuss multiple SOMs for the Wine dataset, starting with an initial map and then focusing on Taste and Aspect Maps, culminating in a final Aggregated Map that combines the information from all the individual maps into a single, comprehensive representation. The chapter concludes with a summary of the key points and findings, providing a comprehensive understanding of the power and potential of SOMs in handling complex, multi-dimensional data.

SELF-ORGANIZING MAPS

Self-organizing maps (SOMs), also known as Kohonen maps, are a type of artificial neural network inspired by the neurophysiological principles of the human brain. They were introduced by Teuvo Kohonen in the 1980s, drawing on the concept of how neurons in the brain self-organize through learning and experience. The fundamental idea is to map high-dimensional data onto a low-dimensional grid, preserving the topological and metric relationships of the original data. This makes SOMs a powerful tool for visualizing and interpreting complex datasets (Ultsch & Siemon, 1990).

The SOM model consists of a grid of neurons (the *lattice*), each associated with a prototype vector of the same dimensionality as the input data. The neurons compete to represent input data during the learning process, hence deeming the SOM as a *competitive learning* algorithm. The neuron that is closest to the input data point, determined by some metric distance such as Euclidean, is declared the best matching unit (BMU) or the 'winner'. This neuron, along with its neighbors (determined by the lattice structure, i.e., the arrangement of the neurons in a rectangular or hexagonal grid), adjusts its prototypes to become more similar to the input. This process is repeated for each data point in the dataset, iteratively refining the map. The dataset must be revisited several times to allow the convergence of the learning procedure – these are called *epochs*.

The learning process in SOMs is governed by two key parameters: the *learning rate* and the *neighborhood radius*. The learning rate determines the extent to which the prototypes are adjusted during each iteration. It is typically set high at the beginning of the learning process to allow for rapid convergence, and gradually decreased over time to allow for fine-tuning and stability. The neighborhood radius determines the range of neighboring neurons that are updated along with the winning neuron. Like the learning rate, the neighborhood radius is also typically set high initially and gradually decreased, focusing the learning on increasingly local neighborhoods.

The steps of the classical[1] SOM algorithm are presented in Algorithm 1. The corresponding process produces a topological ordering of the map, in the sense that adjacent neurons in the lattice will have similar prototype vectors that, overall, approximate the probability distribution function (*PDF*) of the underlying distribution. This implies that the prototype vectors will order themselves with approximately equal distances between them if input vectors appear with even probability throughout a section of the input space. If input vectors occur with varying frequency throughout the input space, the map tends to allocate neurons to an area in proportion to the frequency of input vectors there.

Algorithm 1. The classical SOM algorithm

1. Initialize the SOM with random prototype vectors for each neuron.
2. Select an input data vector.
3. Compute the Euclidean distances between the input and all neuron prototype vectors.
4. Identify the BMU - the one with the closest prototype vector to the input.
5. Update the prototypes of the winning neuron and its neighbors, making them more similar to the input. The degree of adjustment is governed by the learning rate and the neighborhood radius.
6. Repeat steps 2-5 for all input data vectors, and for several epochs, until the map converges to a stable state that represents the topological and metric relationships of the input data.

ORDERING AND CONVERGENCE

Starting from a state of complete disorder (a random initialization) the SOM algorithm gradually achieves an organized representation of the input space, provided that the parameters of the algorithm are chosen properly. The adaptation process of the algorithm can be decomposed into two phases: an *ordering* phase followed by a *convergence* phase.

It is during the *ordering* phase of the adaptation process that the topological ordering of the prototype vectors takes place. This ordering phase is relatively short in comparison to the convergence phase. Large values for the neighborhood radius and learning rate should be used, such that the prototype vectors initially take large steps all together toward the area of input space where input vectors are occurring this can be thought as the "unfolding" of the map. These values then should decrease to their tuning values and encompass only the closest neighbors. During this phase the map tends to order itself topologically over the presented input vectors. Kohonen advises that the ordering phase should consist at least in the presentation of 1000 input patterns; at this point the network should be fairly well-ordered.

The *convergence* phase lasts for the rest of the adaptation process and is necessary to fine tune the prototype vectors and, therefore, provide a good quantization of the input space. During this phase the prototype vectors converge to their "correct" values. To achieve such approximation, the neighborhood should be fairly small, encompassing only the immediate neighbors. This should also apply to the learning rate, such that the magnitude of the prototype updates is very small. The convergence phase is usually several times longer than the ordering phase, e.g., at least 10 times longer.

RELATED ALGORITHMS

SOM is related to several other algorithms. It's a special case of the *k-means* clustering algorithm when the neighborhood radius is set to zero, turning it into a winner-takes-all scheme like incremental *k-means* (Bação, Lobo, & Painho, 2005). However, in SOM, the number of prototype vectors should be much larger, irrespective of the number of clusters, enabling cluster structures to become visible through special visualization techniques.

Vector quantization (*VQ*) is a classical method for approximating a *PDF* using a finite number of vector prototypes. The set of prototypes is referred to as the codebook. The SOM is a *VQ* method and the set of prototypes of the SOM is indeed frequently referred to as the SOM codebook. One of the most well-known methods to perform *VQ* is the *LBG* algorithm (Linde, Buzo, & Gray, 1980). The SOM, however, is not an optimal *VQ* procedure due to the "tension" induced by neighboring relationships among prototypes.

In terms of vector projection (*VP*) algorithms, which aim for dimensionality reduction and visualization, SOM offers a nonlinear topological projection of the data onto the map, unlike linear methods such as Principal Component Analysis (*PCA*). This allows for more flexible and accurate representation of complex data distributions and discovery of nonlinear relationships. Other known visualization and *VP* algorithms include multi-dimensional scaling (Kruskal, 1964), Sammon's mapping (Sammon, 1969), curvilinear component analysis (Hérault & Demartines, 1997), and *t-SNE* (Maaten & Hinton, 2008). While *t-SNE* is a popular method for high-dimensional data visualization, it often is used with a large number of features to produce meaningful results. This can lead to complex and hard-to-interpret visualizations, especially when dealing with datasets that are not inherently high-dimensional or have many localized correlations. Moreover, *t-SNE* does not provide a consistent topological mapping, meaning that different runs can produce different unrelated results. On the other hand, SOMs offer a more interpretable and consistent approach. They preserve the topological relationships between data points, enabling intuitive visualization and identification of clusters. This makes SOMs particularly useful for exploratory data analysis, where understanding the structure and relationships in the data is crucial.

OVERALL CAPABILITIES

SOMs excel at capturing and representing complex patterns within the data. By leveraging competitive learning and weight adjustment mechanisms, SOMs can learn and adapt to the underlying distribution of the input data. This allows them to detect both global and local patterns, including clusters, boundaries, and nonlinear relationships that may not be immediately apparent in the original data. SOMs can uncover intricate structures and reveal hidden information, aiding in exploratory data analysis.

Through their powerful visualization capabilities, SOMs facilitate understanding and interpretation of data. The two-dimensional, topologically ordered representation of high-dimensional data on the SOM grid enables effective visual exploration. Clustering patterns, density variations, and the spatial organization of neurons on the map reflect the characteristics of the data.

SOMs exhibit robustness to noise and missing data in the input. Due to their competitive learning nature, SOMs can handle noisy or incomplete datasets without significant loss of performance. Their adaptability and capacity to generalize well with imperfect data make them valuable tools across diverse domains (Oja, Kaski, & Kohonen, 2003). For instance, in Business and Marketing, SOMs have long been instrumental in revealing customer segments based on purchase behavior (Kuo, Ho, & Hu, 2002). They have also found utility into analyze consumer sentiment in noisy social media data (Mostafa, 2009). In Finance and Risk Management, SOMs have been employed for credit scoring, assisting financial institutions in assessing the creditworthiness of applicants (Chen, Ribeiro, Vieira, & Chen, 2013). Recent applications extend to identifying trends in financial data and anomalies that may signal potential risks (Carrega, Santos, & Marques, 2021) to engineering, as demonstrated by their utilization in visualizing complex material property relationships in materials research and education (Qian, et al., 2019).

EXPLORING WINE DATASET WITH SOMS: UNVEILING COMPLEX RELATIONSHIPS

In this section, we delve into the analysis of the *Wine* dataset (Aeberhard & Forina, 1991) using our open-source Java ANN framework, publicly available at GitHub[2].

The Framework

Our open-source Java ANN framework contains several neural network models, with particular focus given to Self-Organizing Map models. This library provides a robust platform for researchers and developers to utilize these models in their projects, along with additional utilities for data handling and model visualizations. This makes it a versatile tool for both research and practical applications.

The Wine Dataset: A Rich Playground for SOMs

The *Wine Dataset*, a popular choice for machine learning and data mining tasks, presents a non-trivial number of features that make it an excellent candidate for demonstrating the power of SOMs in handling multi-dimensional data and their visualization capabilities. The dataset consists of 178 samples from three different types of Italian wines, with 13 chemical measurements taken for each sample. These measurements, or features, are *alcohol* content, *malic acid*, *ash*, *alcalinity of ash*, *magnesium*, *total phenols*, *flavanoids*, *nonflavanoid phenols*, *proanthocyanins*, *color* intensity, *hue*, *OD280/OD315 of diluted wines* and *proline*. These features, while providing a comprehensive profile of each wine instance, also introduce a level of complexity that makes traditional data analysis techniques less effective.

SOMs, with their ability to reduce dimensionality and preserve topological properties of the input data, offer a unique advantage in analyzing such datasets. By mapping the multi-dimensional wine data onto a two-dimensional grid, SOMs allow us to visualize and interpret complex relationships among the wines based on their chemical properties. However, before we can effectively use this dataset for training our SOMs, we need to ensure that the data is properly normalized. Normalization is a critical step in data preprocessing (Han, Kamber, & Pei, 2012), as it transforms all features to a common scale, preventing any one feature from dominating the others due to its scale. For our purposes, we will be applying min-max normalization, which scales the data to a specified range, in this case, 0 to 1 - our framework already contains normalization procedures that can be applied to datasets before any ANN learning procedure.

In analyzing the Wine dataset using standard histograms, we can observe various distributions of values across features. Some, like *proline*, appear positively skewed, indicating most wines have lower proline content. Others may follow a normal or Gaussian distribution. Outliers, or data points significantly different from others, can also be observed. For instance, *magnesium* seems to have a few wines with unusually high levels. Also, it is well known that wine classes significantly overlap (Cortez, Cerdeira, Almeida, Matos, & Reis, 2009). Features with less overlap are more discriminative and can better assist the SOMs in distinguishing between different classes of wine. We will see that component planes can provide insights superior to those obtained through histogram analysis, with the added benefits of highlighting feature correlations and the inherent cluster structure in the data.

Exploratory Knowledge Discovery: Framework Hands-On Example

The currently established way of employing the SOM for knowledge discovery is using large maps. This was popularized in (Ultsch & Siemon, 1990), where the author called them *emergent* self-organizing maps. From such large maps, one can perform exploratory data analysis, e.g., detect clusters of arbitrary shape and non-linear correlations between features, by using specialized visualizations. These visualizations are only possible due to the topological ordering of the prototypes, input density matching and the fixed-sized lattice of the SOM.

The code depicted in ***Listing 1*** allows us to import and normalize the Wine dataset (lines 2-5), create a 20x30 SOM model (lines 8-10) and perform the learning process during the specified number of epochs (lines 13-23). The epoch number and learning parameters were derived experimentally to obtain good evaluation metrics for the obtained map (lines 26-27), namely a quantization error of 0,099 and a topographic error of 0,000. The *quantization error* measures the quality of the vector quantization process, with 0 (zero) meaning perfect quantization. The *topographic error* measures the quality of the topological ordering of the obtained model, with 0 (zero) meaning perfect topology preservation of the model in respect to the input space.

The *learning rate* typically assumes a value between 0 (exclusive) and 1. In this example, during the convergence phase of the algorithm the learning rate will decay from 0.5 to 0.1 and the latter value will hold during the convergence phase. Analogously, the *neighborhood radius* decay refers to the gradual reduction of the radius around the BMU during training. As training progresses, the radius decreases, causing the SOM to focus on fine-tuning the weights of nearby neurons. In this case, during the convergence phase only adjacent neighbors are affected by the fine-tuning process.

The "classic" SOM algorithm is deemed as an *offline* algorithm in our framework, since it assumes that data can be revisited, as opposed to the *UbiSOM*, which is considered a *streaming* algorithm as we shall see later.

The SOM visualizations can use gray-scale color maps (or colors on computer displays) as a visual representation of data values[3]. This makes them available not only to experts, but laymen, when analyzing them and can be easily understood if one is familiar with the representation. The basic visualizations that can be derived from a SOM model are the *component planes* and the *unified distance matrix.*

Component Planes: Discerning Complex Relationships Between Features

By using component planes, we can visualize the relative component distributions of the input data. Component planes can be thought of as sliced versions of the SOM. Each component plane has the relative distribution of the values of one feature. In this representation, and using a temperature-like color scale, "cooler" colors (usually blue) represent lower values while "warmer" colors (usually red) represent higher values. On a grey-scale lower values are represented by darker shades and higher values by whiter shades. By comparing component planes, we can see if two components correlate. If the outlook is similar, the corresponding features correlate, if they seem like the opposite of each other, then the corresponding features are inversely correlated. Anything in between may signal non-linear correlations.

Listing 1. Code leveraging to presented ANN java framework to produce a SOM model for the wine dataset

```
1: // Load a dataset and normalize it
2: Dataset dataset = new Dataset("datasets/wine.data");
3:
4: DatasetNormalization normalization = new MinMaxNormalization(dataset);
5: normalization.normalize(dataset);
6:
7: // Create basic SOM with random initialization of prototypes
8: int width = 20;
9: int height = 30;
10: SelfOrganizingMap som =
…
12:
13: // Instantiate an offline training algorithm and train the SOM
14: double iLearningRate = 0.5;
15: double fLearningRate = 0.1;
16: double iNeighRadius = 2 * StrictMath.sqrt(som.getWidth()*som.getWidth() +
17:     som.getHeight()*som.getHeight());
18: double fNeighRadius = 1;
19: int orderEpochs = 100;
20: int fineTuneEpochs = 1000;
21:
22: // Instantiate a training algorithm (classic)
23: OfflineLearning learning =
…
26:
27: learning.train(som, dataset);
28:
29: // Print statistics for model fitting
30: SelfOrganizingMapStatistics statistics =
31: SelfOrganizingMapStatistics.compute(som, dataset);
32: System.out.println(statistics);
```

Figure 1 depicts the obtained component planes, labeled with the corresponding feature names. For example, by comparing the component planes of *flavanoids* and *total phenols*, we observe a strong correlation between these two features, suggesting that wines with high flavanoid contents also tend to have high total phenols. Such insights can be invaluable in understanding the underlying data. The reader may try, at this point, to discern other correlations between the features by inspecting the component planes.

However, with very high data dimensionality and, therefore, number of component planes it may be increasingly difficult to visually compare all the component planes. Moreover, some correlations between features may not be as visually striking as others.

Listing 2. Code to produce the visualization of all component planes of a SOM model

```
1: int dimensionality = som.getDimensionality(); // is 13, for the Wine data-
set
2:
3: // Create a set of panels for all component planes
4: JPanel[] panels = new JPanel[dimensionality];
5:
6: // Create the component planes
7: for(int d=0; d < dimensionality; ++d) {
8:     panels[d] = SelfOrganizingMapVisualizationFactory.
createComponentPlane(som,
       d, dataset.inputVariableNames()[d]);
9: }
10:
11: GenericWindow window = GenericWindow.gridLayout("Component Planes",
…
13: window.setVisible(true)
```

Since the component planes are just numbers arranged in a grid, we can further explore the relationships among the features by performing an agglomerative hierarchical clustering of these component planes – in this context, this procedure is called *feature clustering* (Silva & Marques, 2010). The result of this procedure is a *dendogram* where the relative similarity between features can be inspected, providing a clear picture of which features are closely related, and moreover, groups (clusters) of features.

The result of the feature clustering for these component planes is depicted in the *feature clustering* subplot of Figure 1, shedding light on the similarity and dissimilarity between the dataset features; the Pearson's correlation was used to provide a distance metric between the component planes. The reader may use this information to validate the previous proposed exercise.

In (Silva & Marques, 2010) we proposed the usage of a different metric – the improved R_v *coefficient* (suitable for matrices) – and used this approach to cluster high-dimensional financial data. This method proved to be particularly useful when dealing with a high number of features, where visual analysis of component planes becomes challenging.

The Unified Distance Matrix: Unveiling Cluster Structures

The unified distance matrix (U-Matrix) presents distances between neurons' prototypes. The distances between adjacent neurons are calculated and presented with different colorings between the adjacent neurons. Following the same temperature-like color scale, a warmer coloring between neurons corresponds to a large distance and thus a gap between the codebook values in the projected space; a cooler coloring means codebook vectors are close to each other in the input space. Cooler areas can be thought of as clusters and warmer frontiers as cluster separators. This is especially powerful to understand the underlying structure of data, i.e., when one tries to find clusters in the input data without having any a priori information about the clusters. Consequently, the detection of complex clusters is achieved not by regarding single units, but by regarding the topological structure map. Hence, while component planes allow discovery of relationships among features, the U-Matrix allows the visual discovery of clusters within the data. Used in conjunction, one may infer which features best describe the detected clusters, i.e., a description of the clusters. It must be emphasized that this knowledge discovery through visual exploration is the main motivating factor for using SOMs over data streams, regarding current available methods.

The resulting U-Matrix, displayed in the top left subplot of Figure 1, provides a visual representation of the underlying data structure. Although there does not seem to be a clear cluster structure, a trained eye can discern two cooler and relatively big areas in the bottom left and top left (triangular shaped) corners and a smaller rounded cluster near the middle of the U-Matrix.

By integrating these findings with the information from the component planes, we can derive some conclusions, for example:

- *The large upper left cluster is composed of wines with, e.g., low flavonoids, low total phenols, low hue, low proanthocyanins, some of them with high color intensity and average alcohol content;*
- *The large bottom left cluster is composed of wines with, e.g., high* proline *content, average* flavonoids *and higher* total phenols*;*
- *The smaller rounded center cluster is composed of wines with the highest values of* magnesium *and* proanthocyanins *and low values of* ash.

Since the dataset is labelled, we can prove some of these findings by projecting the input data classes directly onto the map and labelling each BMU with the corresponding target class – this is depicted in the *target classes* subplot of Figure 1. Additionally, the *hit map* subplot depicts the dispersion of the *BMUs* over the map. This indeed confirms the accuracy of the SOM in identifying two of the original classes, demonstrating the power of SOMs in exploratory cluster analysis and their ability to uncover hidden structures in complex datasets.

UbiSOM: SOMs in a DataStream Context

Until now we have covered the basics of the SOM model, related algorithms, capabilities and how the reader can leverage our ANN framework to perform exploratory knowledge discovery in your own data. However, we have only talked about *static* data, i.e., relatively small amounts of data that can be revisited by the standard SOM algorithms.

UbiSOM and its Adaptations for Data Streams

Data streams, characterized by their continuous and potentially infinite nature, pose significant challenges to traditional data mining techniques. The Ubiquitous Self-Organizing Map (UbiSOM) is a SOM variant specifically tailored for streaming and big data (Silva & Marques, 2015). It retains the topological ordering and visualization capabilities of SOM while introducing adaptations to adjust its model to the, possibly, changes in the underlying data structure of the data stream.

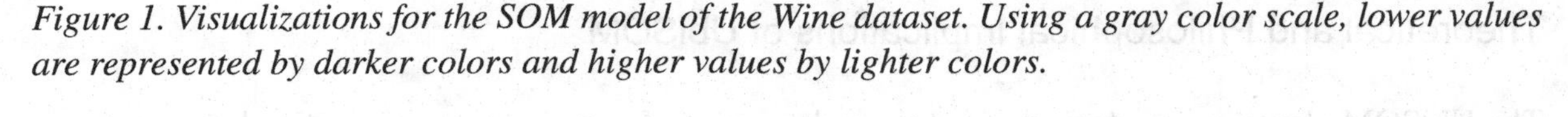

Figure 1. Visualizations for the SOM model of the Wine dataset. Using a gray color scale, lower values are represented by darker colors and higher values by lighter colors.

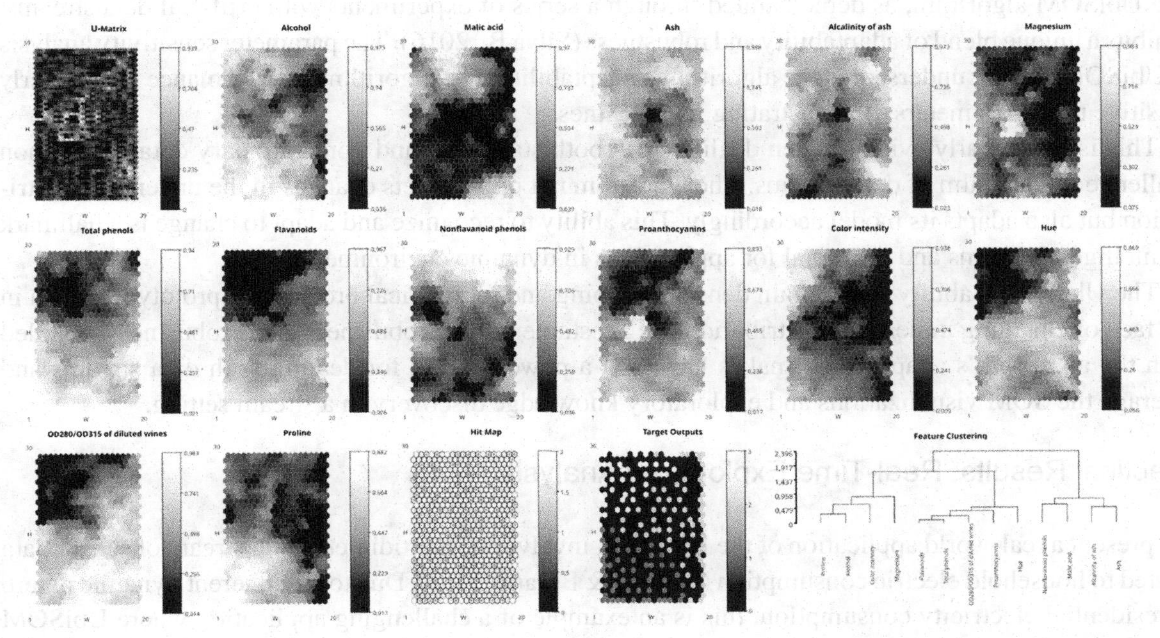

The original SOM algorithm works by using a monotonically decreasing scheme for the learning parameters. This ensures topology-preserving convergence of the model to the data. However, it assumes the underlying distribution is stationary and will not change during the adaptation process. Therefore, the UbiSOM algorithm uses a proposed *drift* function for estimating the learning parameters and a finite-state machine to handle drastic changes in the underlying data stream. This drift function weighs the *average quantization error* and a newly introduced metric, the *average neuron utility*, to provide an overall indication of the map's performance and adaptation over the data stream. If the model is converging, the learning parameters will decrease steadily; if the model is found to be misrepresenting the underlying data, then the learning parameters will increase proportionally. In case the underlying distribution deviates significantly from the learned model, the UbiSOM algorithm may "reset" itself – this mainly occurs when a completely new underlying distribution has emerged.

The finite-state machine design allows UbiSOM to conform to the typical training phases of SOM and to handle drastic changes in the underlying data stream. It consists only of two states: the *ordering* state allows the map to initially unfold over the underlying distribution with monotonically decreasing learning parameters. The *learning* state, on the other hand, allows the learning parameters to be adjusted based on the drift function, enabling the map to retain an indefinite plasticity while maintaining the original SOM properties over non-stationary data streams. In conclusion, UbiSOM's adaptations for data streams allow it to handle the complexities and uncertainties of data streams, making it a powerful tool for real-time decision-making and adaptation to new inputs.

Theoretical and Philosophical Implications of UbiSOM

The UbiSOM algorithm, as demonstrated through a series of experiments with artificial data streams, exhibits a unique blend of adaptability and robustness (Silva B., 2016). The parameter sensitivity analysis of UbiSOM further underscores the algorithm's adaptability. The algorithm's performance is not overly sensitive to its parameters, demonstrating its robustness.

This is particularly evident when dealing with both stationary and non-stationary data, a common challenge in the realm of data streams. The algorithm not only detects changes in the underlying distribution but also adapts its model accordingly. This ability to recognize and adapt to change is a hallmark of intelligent systems and is crucial for applications in dynamic environments.

The UbiSOM's ability to maintain density mapping and topological ordering of prototypes, even in the face of changing underlying distributions, is a testament to its robustness. This robustness, coupled with the algorithm's adaptability, makes UbiSOM a powerful tool for dealing with data streams and leverage the SOM visualizations and exploratory knowledge discovery in a stream setting.

Practical Results: Real-Time Exploratory Analysis

We present a real-world application of the UbiSOM, involving a multidimensional stream of sensor data related to household electric consumption (Hebrail & Berard, 2012). Due to the inherent dynamic nature of residential electricity consumption, this is an example of a challenging application where UbiSOM may operate. The sensor data is composed of seven different measurements and these values are streamed every minute. The different measurements are:

- *Global active power*: household global minute-averaged active power (in kilowatt);
- *Global reactive power*: household global minute-averaged reactive power (in kilowatt);
- *Voltage*: minute-averaged voltage (in volt);
- *Global intensity*: household global minute-averaged current intensity (in ampere);
- *Sub-metering 1*: energy sub-metering No. 1 (in watt-hour of active energy); corresponds to the kitchen, containing mainly a dishwasher, an oven and a microwave (hot plates are not electric but gas powered);
- *Sub-metering 2*: energy sub-metering No. 2 (in watt-hour of active energy); corresponds to the laundry room, containing a washing-machine, a tumble-drier, a refrigerator and a light;
- *Sub-metering 3*: energy sub-metering No. 3 (in watt-hour of active energy); corresponds to an electric water-heater and an air-conditioner.

Consequently, through such an application we are interested in extracting knowledge from the UbiSOM model that is maintained over time along the underlying data stream, e.g., clusters of patterns of usage, features values that contribute the most for the formation of those clusters (cluster descriptions) and correlated measurements.

Our Java ANN framework includes the UbiSOM algorithm, particularly useful for those dealing with data streams and continuous learning scenarios. ***Listing 3*** depicts an example code to test this application. The UbiSOM parameters were set from the parameter sensitivity analysis in (Silva B., 2016), as good overall values.

Listing 3. UbiSOM real-time exploratory analysis over the Household datastream

```
1: // Load a datastream and normalize it
2: Dataset dataset = new Dataset("datasets/household_power_sensor.data");
3:
4: DatasetNormalization normalization = new MinMaxNormalization(dataset);
5: normalization.normalize(dataset);
6:
7: // Create an UbiSOM instance
8: int width = 20;
9: int height = 40;
10: StreamingSOM som = new UbiSOM(width, height, dataset.inputDimensionality(),
11:  0.1, 0.08, 0.6, 0.2, 0.7, 2000);
12:
13: RealTimeVisualizationPanelArray viz =
    RealTimeVisualizationPanelArray.init(som, dataset);
14: viz.setVisible(true);
15: // Stream the dataset
16: for (DatasetItem item: dataset) {
17:     VectorN input = item.getInput();
18:     som.learn(input);
19:
20:     viz.updateAll();
21: }
```

Figure 2 depicts two snapshots of the SOM visualizations obtained at different times, namely at some point in the spring (spring snapshot) and sometime in the autumn (autumn snapshot). It must be made clear that the SOM model depicts the underlying data structure in the relative "recent past" until the snapshot was taken and not of a specific minute; the latter would be a single data point. By analyzing the visualizations one can inspect the evolution of the consumption patterns over time.

Overall, the component planes reveal interesting patterns and correlations. For instance, the *global active power* and *global intensity* features are found to be strongly correlated, while showing some degree of inverse correlation with *voltage*. The component planes also provided insights into the relative usage of different appliances, such as the heating system, kitchen, and laundry room in different seasons.

UbiSOM's performance in this real-world scenario further underscores its practical utility. The algorithm can handle the complex, non-stationary data stream effectively, providing valuable insights into the patterns of electric power consumption. Hence, one of the key strengths of the UbiSOM algorithm is its ability to provide real-time exploratory analysis.

Figure 2. Visualizations of the UbiSOM model monitoring electric consumption sensor data

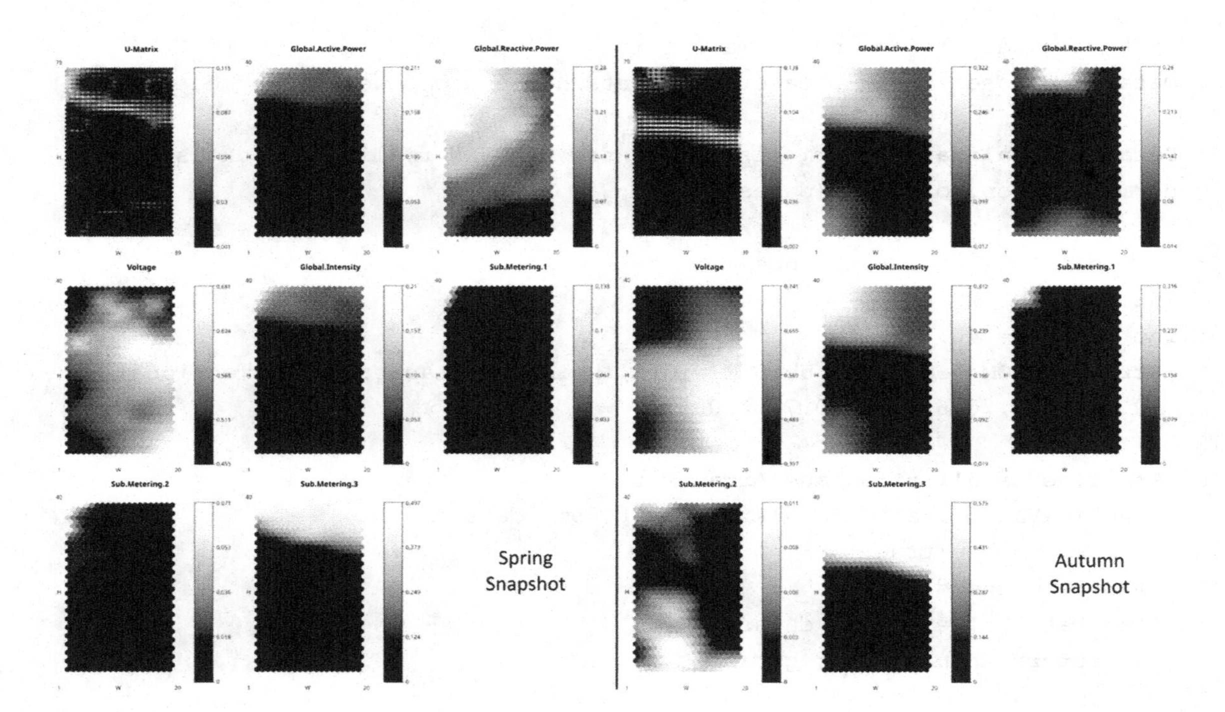

MULTISOM VISUALIZATION FOR UBISOM

Building upon the unique properties of UbiSOM, we can further enhance our understanding of the data through interactive visualization. This dynamic approach allows us to explore the map in a more intuitive way, revealing patterns and relationships that might be difficult to discern from static images or numerical data alone. The *multiSOM* library, with its range of interactive visualization options, enables us to delve deeper into the structure of the data, making the most of the adaptability and flexibility of UbiSOM. So, the interactive visualizations provided by the *multiSOM* library, when combined with the capabilities of UbiSOM, offer an invaluable tool for understanding and interpreting complex, high-dimensional data. They allow us to explore the data in a dynamic and intuitive way, revealing the underlying structure and relationships in the data, and demonstrating the power of UbiSOM in real-world applications.

One of the key visualizations provided in the *multiSOM* library is a dynamic *U-Matrix*. The *U-Matrix* visualizes the distances between neighboring neurons in the SOM, providing a clear picture of the clusters and boundaries in the data. High values in the *U-Matrix* indicate a large distance between neurons, suggesting a boundary between different clusters, while low values indicate similar neurons that belong to the same cluster. By interacting with the *U-Matrix*, we can explore these clusters and boundaries in detail, gaining a deeper understanding of the data's structure.

The *multiSOM* library also provides information about the BMUs for each data point. By examining the *BMUs*, we can see which neurons are most representative of the data, and how the data points are distributed across the map.

In addition to these visualizations, the *multiSOM* library allows for multidimensional projection of the data. This enables us to view the data in a reduced-dimensional space, making it easier to visualize and understand complex, high-dimensional data. By interactively exploring this projection, we can gain insights into the relationships among the data points and the structure of the data. All these additional visualizations are illustrated in this section.

Multi-Layer SOMs With *Processing.org*

In this section, we will delve into the process of analyzing the Wine dataset using a multi-layer SOM approach. We will be utilizing the UbiSOM algorithm implementation from our Java ANN library and the *processing.org* software for visualization.

A key aspect to note is that the UbiSOM algorithm is designed to work with continuous data streams. To simulate this, we will repeatedly present the same dataset from a *CSV* file containing the full Wine dataset features, excluding the target class which is stored only as a label identifying the line number and the class. This approach allows us to emulate the conditions of a continuous data stream while working with a static dataset.

In previous work (Silva B., 2016), we found that the UbiSOM exhibits similar behavior to the traditional SOM and demonstrates fast convergence on stationary problems. With the *multiSOM* approach (Marques, Silva, & Santos, 2016), we can simulate random sampling from the examples in the *CSV* file or sequential presentation of data by following the order of examples in the file. By using this well-known dataset, we can provide a robust comparison point for the performance of the algorithm, while also demonstrating its capabilities in handling continuous, high-dimensional data.

Environment Setup and SOM Creation

First, we need to set up our environment. We will use the *Processing.org* software for visualization. You can download it from the Processing website[4]. Next, download the Java ANN library from GitHub[5] and add it to your Processing libraries folder.

We will create a multi-layer SOM approach with four maps. By default, *multiSOM* uses maps with 40 columns and 20 rows to maximize visualization size while still maintaining a sufficiently good map size for detecting emergent features. The first map will use all features from the Wine dataset and load the Dataset into memory. Then the dataset will be continuously iterated from start to finish in trying to simulate a datastream. The second and third maps will focus on specific subsets of features related to the taste and appearance of the wine, respectively. The fourth map will aggregate the results of the first three maps to detect the main clusters in the dataset. The setup code for creating the SOMs is presented in Listing 4.

In the processing sketch, **multiSOM[0]** is the first map, which uses all features from the Wine dataset. **multiSOM[1]** is the second map, which focuses on the *Alcohol*, *Total phenols*, *Flavanoids*, and *Nonflavanoid* phenols features. **multiSOM[2]** is the third map, which focuses on the *Malic acid*, *Ash*, *Alcalinity of ash*, and *Color* intensity features. **multiSOM[3]** is the fourth map, which aggregates the results of the first three maps. In Listing 5, these maps will be initialized with the function **initAllMaps()** .

Listing 4. multiSOM setup with four SOM maps

```
44: MultiSOMLibrary multiSOM[] = new MultiSOMLibrary[NMAPS];
45: String[] tasteFS = {"Alcohol","TotPhenols","Flavanoids","NonflavPhenols"};
46: int[] tasteFN = null;
47: String[] appearanceFS = {"MalicAcid","Ash","AshAlcal","ColorInt"};
48: int[] appearanceFN = null;
…
54: void setup() {
55:
56:     frameRate(10);
58:     size(1450,768,P2D);
…
63:     maximizedMap = 0;
64:
65:     multiSOM[0] = new MultiSOMLibrary(this, 30, "wine.mSOM");
66:     multiSOM[0].setViewPort(mainX, mainY, mainScale);
67:     multiSOM[0].setSOMTitle("ini","Wine Initial");
68:
69:     for(int i=1; i<NMAPS; i++) {
70:         println(i + " multiSOM starting");
71:         multiSOM[i] = new MultiSOMLibrary(this, 30,"");
72:         multiSOM[i].setViewPort(5, 100+i*150, 0.1f);
73:         println(i + " multiSOM is OK");
74:     }
75:     multiSOM[1].setSOMTitle("taste","Wine Taste");
76:     multiSOM[2].setSOMTitle("asp","Wine Aspect");
77:     multiSOM[3].setSOMTitle("agt","Wine Aggregatted");
78: }
```

Here the function method **setAxisTransform()** defines the initial projection for converting the 13-dimensional Wine observations and corresponding SOM grid to 2 dimensions. This projection can be changed during training by dragging each axis to another position. Then the label for each axis is set with the **setCatLabel()** method. An auxiliar method **getFeatureIndices()** is provided for converting a vector of strings (with the labels) into the proper vector of integer indices.

Result Visualization

Once the SOMs are trained, we can visualize the results using the Processing sketch. The visualization will show the SOMs in a 2D grid, with each cell representing a neuron (see Figure 3). The code for visualizing the SOMs is presented in Listing 6.

Listing 5. SOM map initialization on multiSOM

```
81: int[] initLabels(MultiSOMLibrary ms, int[] FN, int st) {
82:     for(int i=0; i< FN.length; i++)
83:         ms.setCatLabel(i+st,multiSOM[0].getCatLabel(FN[i]));
84:         return FN;
85: }
86:
87: void initAllMaps() {
88:     tasteFN = initLabels(multiSOM[1],
        multiSOM[0].getFeatureIndices(tasteFS),0);
89:     appearanceFN = initLabels(multiSOM[2],
        multiSOM[0].getFeatureIndices(appearanceFS),0);
90:
91:     multiSOM[1].setColorDim(0);
92:     multiSOM[1].setAxisTransform(
93:         new int[]{2238, 29, 216, 5435},
94:         new int[]{2071, 6384, 8031, 106});
95:
96: multiSOM[2].setColorDim(3);
97: multiSOM[2].setAxisTransform(
98: new int[]{2320, 1719, 10000, 0},
99: new int[]{4778, 1627, 168, 10000});
100:
101:    String aggregatedLabels[] = {"xBmu", "yBmu","nQe"};
102:    for(int i=0; i< aggregatedLabels.length; i++)
        multiSOM[3].setCatLabel(i,aggregatedLabels[i]);
103:    initLabels(multiSOM[3], multiSOM[0].getFeatureIndices(tasteFS),
        aggregatedLabels.length);
104:    initLabels(multiSOM[3], multiSOM[0].getFeatureIndices(appearanceFS),
        aggregatedLabels.length+tasteFS.length);
105:    multiSOM[3].setColorDim(0);
106:    multiSOM[3].setAxisTransform(// values where acquired on
        previous execution by manual adjustment and using the "A" key
107:        new int[]{2391, 0, 122, 91, 16, 33, 214, 117, 89, 228, 40},
108:        new int[]{8, 10000,2851, 6981,1898,3345,802,4811, 1316, 1916,76});
109:
110: }
```

In this code, **multiSOM[i].drawDatastreamSOM()** is used to draw the i-th SOM on the screen. Notice the processing **draw()** function takes care of all the details of drawing the SOM, including setting the colors and drawing the grid lines. In the **draw()** function, we first check if all SOMs have data ready for training. If they do, we call the **trainAllMaps()** function to train all the SOMs. After training, we call the **postTrainVisualization()** function to update the visualization after each training step.

Listing 6. SOM visualization on multiSOM

```
140: void draw() {
141:
142: boolean hasData = true;
143: for(int mapInIteration=0;mapInIteration<NMAPS;mapInIteration++) {
144:    multiSOM[mapInIteration].drawDatastreamSOM();
145:    if(!multiSOM[mapInIteration].hasTrainData())
146:        hasData = false;
147:    }
148:
149: if(hasData){ // ready SOMs wait until every map is ready for train
150:    nIts += 1;
151:    if(tasteFN == null) // on first call features where already read from
csv
152:        // and maps can be init
153:        initAllMaps();
154:    trainAllMaps();
155:    }
156:
157:    /// Post-Visualization updates
158:    for(int mapInIteration=0;mapInIteration<NMAPS;mapInIteration++)
159:        multiSOM[mapInIteration].postTrainVisualization();
160: }
```

Dynamic Training and Visualization

The dynamic training process is the heart of this tutorial, and it is presented in Listing 7. In this code, each Self-Organizing Map learns from the data and adjusts its weights to better represent the data's underlying structure. Instead of training the SOMs all at once, we train them incrementally with each frame. This allows us to visualize the training process in real-time and see how the SOMs evolve as they learn from the data. The **trainAllMaps()** function is where the actual training happens. For each SOM, we call the **train()** function, which updates the weights of the SOM based on the current data.

In the **trainAllMaps()** function, we first retrieve the next data pattern from the base map (**multiSOM[0]**). This data pattern contains the current data point that we want to train the SOMs on. Next, we create subsets of this data pattern for the taste and appearance maps. The **subset()** function is used to select only the features that are relevant for each map. For the taste map (**multiSOM[1]**), we select the *Alcohol*, *Total phenols*, *Flavanoids*, and *Nonflavanoid phenols* features. For the appearance map (**multiSOM[2]**), we select the *Malic acid*, *Ash*, *Alcalinity of ash*, and *Color intensity* features. We then add these data patterns to the corresponding SOMs using the **addSOMDataMPoint()** function. This function adds the data pattern to the SOM's data queue, ready for training. The next step is to get the Best Matching Unit (BMU) position from the base map. The BMU is the neuron in the SOM that is closest to the current data point. We get the BMU's position as a 2D coordinate (x, y) and its quantization error (qe), which is a measure of how well the BMU represents the data point. We normalize these values to be between 0 and 1. Finally, we create an aggregated data pattern that combines the BMU position from the base map and the BMUs from the taste and appearance maps. This aggregated data pattern is then added to the final map (**multiSOM[3]**) for training. This process is repeated for each data point in the dataset. As the SOMs are trained on each data point, they gradually adjust their weights to better represent the structure of the data. This dynamic training process allows us to visualize the learning process in real-time and see how the SOMs evolve as they learn from the data.

Listing 7. SOM map training on multiSOM

```
112: void trainAllMaps() {
113:   DataPattern data = multiSOM[0].getNextPat(); // reference map
114:
115:   DataPattern tasteData = data.subset(tasteFN, "S1"+data.getLabel());
116:   DataPattern appearanceData = data.subset(appearanceFN,
        "S2"+data.getLabel());
117:
118:   multiSOM[0].addSOMDataMPoint(data);
119:   multiSOM[1].addSOMDataMPoint(tasteData);
120:   multiSOM[2].addSOMDataMPoint(appearanceData);
121:
122:   double [] bmuPos = multiSOM[0].getBMUPos(0.035); // Second level map
123:
124:   if(bmuPos != null) { // only draws submap when QE is acceptable
125:     DataPattern aggregatedData = new DataPattern(
126:       MultiSOMLibrary.join(new double[][]{
127:         //new double[] {bmuPos[0],bmuPos[1]},
128:         bmuPos,
129:         multiSOM[1].getBMUPrototype(),
130:         multiSOM[2].getBMUPrototype()}),
131:         "Ag"+data.getLabel());
132:
133:   multiSOM[3].addSOMDataMPoint(aggregatedData);
134:   } else
135:     multiSOM[3].setSOMIteration(multiSOM[0].getSOMIteration()); // Sync
       iteration counts (for error plot)
136: }
```

Interacting With the Visualization

The visualization is interactive, allowing the user to pause the training, step through the training manually, and maximize any of the SOMs for a closer look. The visualization provided by the Processing sketch is illustrated in Listing 8, offering an extensive set of controls to manipulate the training process and examine the results in detail. The **keyTyped()** handler is a particularly useful feature, as it allows the user to trigger different actions based on the key pressed. For instance, the user can pause or resume the training by pressing the **'P'** key. When the training is paused, the user can view the current set of projection values by pressing the **'A'** key. The **'W'** key allows the user to write all the trained maps to disk, while the **'C'** key saves the current maximized map to a CSV file and runs a Python script on it. Finally, the **'E'** key runs another Python script that parses the log file being continuously generated by *multiSOM* and presents a chart of the average quantization error per feature on all maps being trained.

Additional keys provide further functionality inside the *multiSOM* library. The **'s'** key, when pressed with the mouse over a point representing a given example, reveals the data for that point. The **'M'** key switches the grid view to a U-Matrix view, where the user can see the examples associated with each unit by pressing the mouse over the circle representing the number of hits of a given unit projected over the U-Matrix.

The **'?'** key brings up a help menu with additional keys and their functions. The **'H'** key allows the user to hide certain features from the projection, although these features will still be used in the training process. The **'X'** key sets the axis of a selected feature to its average value, effectively making the feature non-relevant for the SOM analysis from that point forward.

The interactive terminal, initiated by pressing the **'I'** key, allows the user to issue commands to the library using a mini language designed for the Processing script. This feature provides a high level of control over the behavior of the map. For instance, the user can set the frame rate of Processing in frames per second. By setting this value to over 1000, the user can train several examples per visualizing frame. For example, setting the frame rate to 1030 will attempt to train 1000 examples per frame and will aim for a goal of 30 frames per second (it's important to note that the actual frame rate will be slower and set to the remainder of the division by 1000). Another useful command is *'next'*, which allows the user to specify the number of iterations for training the SOM maps. For example, issuing the command 'next 20000' will train the SOM maps for 20,000 iterations. In the current setting, only the base map (i.e., the initial SOM) accepts commands, and the other maps will automatically follow the base map's behavior. This setup ensures a coordinated training process across all maps. Additionally, the interactive terminal provides a 'help' command, which displays a list of all available commands and their functions. This can be a useful reference for users who are new to the Processing script or who need a quick reminder of the available commands. The CSV file used for training the SOM maps is specified in the interactive terminal. This allows for flexibility in choosing the dataset for analysis. For those interested in replicating or exploring the analysis presented in this chapter, the list of commands used on the base map can be found in the text file *'wine.mSOM'* (specified as input to the base "initial" map constructor). This file serves as a record of the training process and can be a valuable resource for understanding the steps taken to train the SOM maps.

Listing 8. SOM map training on multiSOM

```
180: void keyTyped() {
181:     switch(key) {
…
187:         case 'P': // Pause train on all maps
188:             pauseNetwork = !pauseNetwork;
189:             setPauseAllNets();
190:             if (pauseNetwork) {
191:             multiSOM[maximizedMap].showMessage("Pause on Network
                 Information.");
192:             } else
193:             multiSOM[maximizedMap].showMessage("Activating Network
                 Information.");
194:             break;
195:         case 'I': // interactive terminal on base map
196:             pauseNetwork = true;
197:             setPauseAllNets();
198:             multiSOM[0].keyTyped(key);
199:             break;
200:         case 'A':
201:             println(multiSOM[maximizedMap].getAxisLabels());
202:             break;
203:         case 'W':
204:             for(int i=0;i<NMAPS;i++) multiSOM[i].saveModel();
205:             break;
…
209:         case 'C':
210:             multiSOM[maximizedMap].runPython("prototypes.py");
211:             break;
…
218:         default:
219:         multiSOM[maximizedMap].keyTyped(key);
220:     }
221: }
```

Regarding mouse interaction (Listing 9), the user can maximize any of the SOMs by clicking on it. This will make the clicked SOM fill most of the screen, allowing the user to see it in more detail. The user can return the SOM to its original size by clicking on it again. We now if a map was selected by checking the Boolean function **mouseInViewPort().** The **setViewPort()** function is used to dynamically change the visualization coordinates of each SOM.

Listing 9. SOM map training on multiSOM

```
225: void mousePressed() {
226:     if(multiSOM[maximizedMap].mouseInViewPort())
227:         multiSOM[maximizedMap].mousePressed(mouseX, mouseY);
228:     else
229:         for(int i=0; i<NMAPS; i++) {
230:             if(multiSOM[i].mouseInViewPort()) {
231:             background(200);
232:             multiSOM[maximizedMap].setViewPort(5, 100+maximizedMap*150,
              0.1f);
233:             multiSOM[i].setViewPort(mainX, mainY, mainScale);
234:             maximizedMap = i;
235:             println(maximizedMap + " multiSOM is maximized.");
236:         multiSOM[maximizedMap].showMessage("Maximized SOM "+maximizedMap,
         1000);
237:         }
238:     }
239: }
```

These interactive controls and commands make the Processing sketch using the ubiSOM library a powerful tool for exploring and understanding the behavior of the SOMs, allowing the user to manipulate the training process and examine the results in a dynamic and intuitive way.

Interpreting the Results

The power of SOMs lies in their ability to transform complex, high-dimensional data into a simplified, visual representation that can be easily interpreted. In this section, we will delve into the results obtained from our experiment with the Wine dataset, using the *multiSOM* library and the Processing.org software for visualization. Our experiment involved training four SOMs: an initial map with all wine features, two separate maps for taste and aspect related features, and an aggregated map that combines the results of the other three maps. The training process involved more than 80,000 iterations, allowing the SOMs to adjust and refine their models to accurately represent the data.

The Taste Map. The taste map was trained on taste-related features of the Wine dataset. After 5,000 iterations, the U-Matrix visualization already showed some clustering, although the regions are not yet well-defined. For example, the corresponding component planes reveal a high correlation between *Flavanoids* and *TotPhenols*, indicating that these two features play a significant role in the taste of the wines[6].

The Initial and Aggregated Maps. The initial map was trained on all wine features, while the aggregated map used the Best Matching Units (BMUs) from the initial map and the prototypes resulting from the taste and aspect maps. The evolution of the quantization error on all the maps during the training process is shown in Figure 3A. The aggregated map (KAg) and the taste map (KS1) have the lowest quantization errors, indicating a good fit to the data. The initial map (Kin) and aspect map (KS2) have higher quantization errors, suggesting that they may not capture all the nuances of the data.

A comparison of the U-Matrices for the initial map and the aggregated map after 80,000 iterations (Figure 3B) reveals a much sharper division of the regions in the aggregated map. The areas with the highest distances (depicted in red) and the lowest distances (in blue) are more clearly defined, indicating a better separation of the different wine classes.

The Aggregated Map Analysis. The U-Matrix for the aggregated map after 80,000 iterations provides a high-level overview of the data (Figure 3B3). The wines are grouped into distinct clusters based on their chemical properties. For instance, Class 3 wines are mainly located in the lower left, Class 2 wines are mainly in the top area, and Class 1 wines are in the lower right.

A closer look at the SOM grid visualization (Figure 3D) provides more insights into the distribution of the wine classes. Class 2 is located in the lower part of the map, while Classes 1 and 3 are in both corners of the top part of the grid.

The component planes for the aggregated map (Figure 3C) reveal the main features that contribute to the classification of the wines. The *x/yBMU* components of the initial map gained the main relevance on all features. The normalized quantization error of the initial map (*nQe)* has the highest values mainly in the U-Matrix areas of higher distance among classes. Class 2 seems the most heterogeneous one. Class 3 has the highest values of *MalicAcid* and *ColorIntensity,* but not so high values on *Flavanoids* and *Alcohol*. Class 2 has the lowest *Alcohol*. Class 1 is high on *Alcohol*, *Flavanoids* and low on *AshAlcaline*.

Finally, the projection of the aggregated SOM grid (Figure 3D, presenting the print-screen of the application window) shows the distribution of the quantization error feature on the y-axis and x/y-bmu on the x/y-axis. Class 2 has both extreme high and low values in quantization errors, while Classes 1 and 3 usually have average quantization errors.

In conclusion, the results of our experiment demonstrate the power of SOMs in analyzing complex, multi-dimensional data. By training multiple SOMs on different subsets of the data and combining their results, we were able to gain valuable insights into the relationships among the different wine classes and their chemical properties.

CONCLUSION

The results of our presented experiments with the Wine dataset demonstrate the power of Self-Organizing Maps (SOMs) in analyzing complex, multi-dimensional data, supported by our publicly available Java ANN library. This library provides a robust platform for researchers and developers to utilize the available models in their projects, along with additional utilities for data handling and model visualizations. Also, by training multiple SOMs on different subsets of the data and combining their results, we were able to gain valuable insights into the relationships among the different wine classes and their chemical properties. This approach, which we call *multiSOM*, provides a comprehensive and nuanced view of the data that would not be possible with a single SOM.

The full set of examples presented in this Chapter is available online together with a *wiki*[7] that serves as a comprehensive guide to using the library and provides additional dynamic color examples and illustrations[8]. These materials include detailed explanations on how to use the library effectively. It also provides additional examples and tutorials, making it a valuable resource for both newcomers to self-organizing maps and experienced users looking to deepen their understanding.

Figure 3. Color visualization for the four distinct Wine maps, as generated by multiSOM

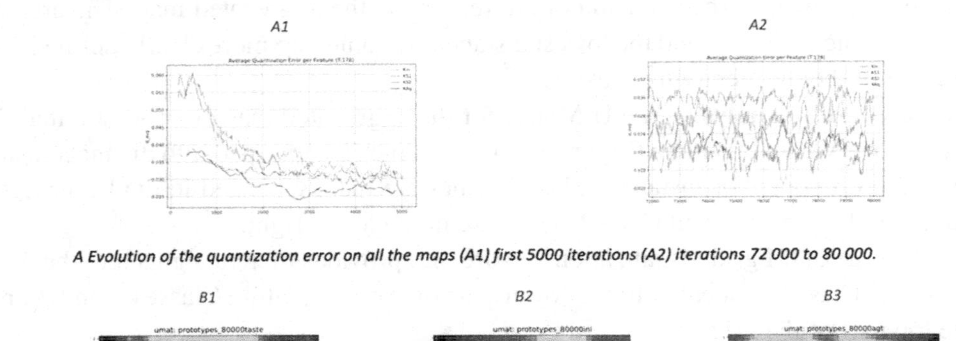

A Evolution of the quantization error on all the maps (A1) first 5000 iterations (A2) iterations 72 000 to 80 000.

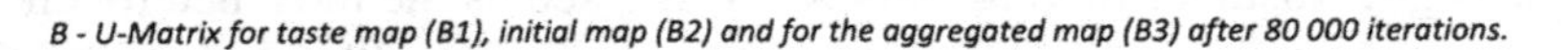

B - U-Matrix for taste map (B1), initial map (B2) and for the aggregated map (B3) after 80 000 iterations.

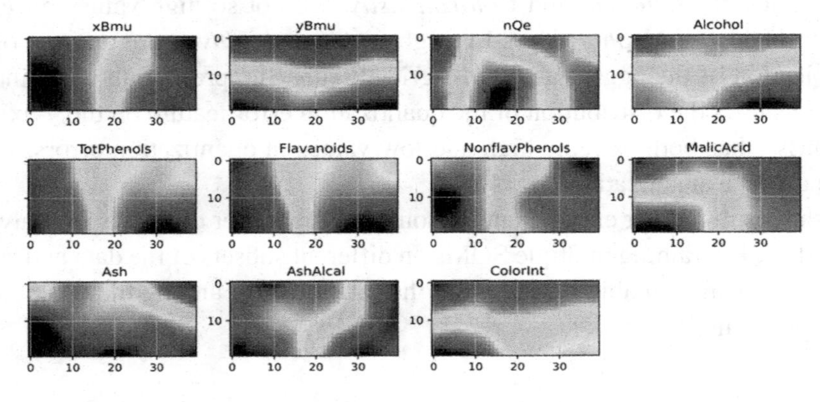

C- Component planes for the aggregated map after 80000 iterations.

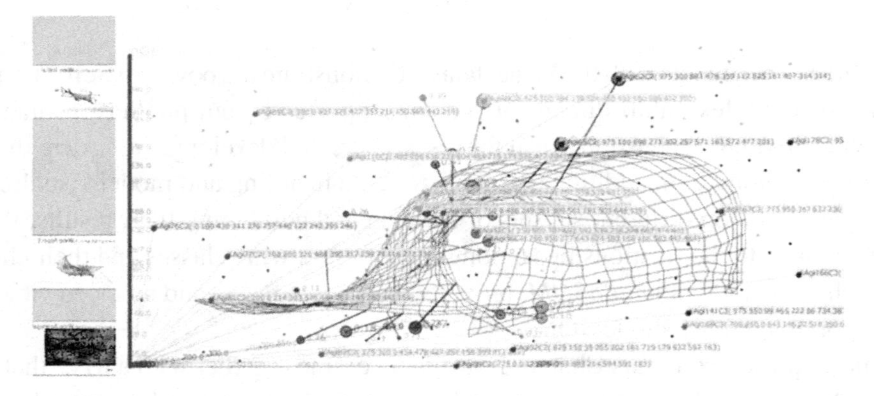

D- The processing sketch uses the multiSOM library for showing different visualizations of the four ubiSOM maps and dataset during and after training (illustrating projection of the aggregated SOM grid at 75000 iterations).

The use of SOMs extends far beyond the analysis of wine data. In fact, SOMs have a wide range of practical applications in various industries. For instance, we also presented the UbiSOM variant of SOMs to analyze household electric power consumption data. This real-world application involved continuously modeling a non-stationary data stream, providing valuable insights into patterns of electric power usage. We also highlighted the feature clustering capabilities of the SOM model, that we used in a previous application of SOMs is in the financial industry, to model the relationships between a wide variety of asset prices for portfolio selection. The ability of SOMs to detect patterns and relationships in complex financial data makes them a powerful tool for risk management and investment decision-making.

In conclusion, SOMs are a powerful tool for exploratory data analysis in various domains. Their ability to transform complex, high-dimensional data into a simplified, visual representation makes them invaluable in today's data-driven world. Whether it's understanding the chemical properties of wine, monitoring household power consumption, or making investment decisions, SOMs provide a unique and powerful approach to understanding complex data.

REFERENCES

Aeberhard, S., & Forina, M. (1991). *Wine*. UCI Machine Learning Repository.

Bação, F., Lobo, V., & Painho, M. (2005). *Self-organizing maps as substitutes for k-means clustering. Computational Science—ICCS 2005*. Springer.

Carrega, M., Santos, H., & Marques, N. (2021). Data Streams for Unsupervised Analysis of Company Data. In F. S. Goreti Marreiros, *Progress in Artificial Intelligence: 20th EPIA Conference on Artificial Intelligence* (pp. 609–621). Springer-Verlag. 10.1007/978-3-030-86230-5_48

Chen, N., Ribeiro, B., Vieira, A., & Chen, A. (2013). Clustering and visualization of bankruptcy trajectory using self-organizing map. *Expert Systems with Applications*, *40*(1), 385–393. doi:10.1016/j.eswa.2012.07.047

Cortez, P., Cerdeira, A., Almeida, F., Matos, T., & Reis, J. (2009). Modeling wine preferences by data mining from physicochemical properties. *Decision Support Systems*, *47*(4), 547–553. doi:10.1016/j.dss.2009.05.016

Fort, J.-C., Letrémy, P., & Cottrell, M. (2002). Advantages and drawbacks of the Batch Kohonen algorithm. 10th-European-Symposium on Artificial Neural Networks, (pp. 223-30).

Goodfellow, I., Bengio, Y., & Courville, A. (2016). *Deep Learning*. MIT Press.

Han, J., Kamber, M., & Pei, J. (2012). *Data Mining: Concepts and Techniques* (3rd ed.). Morgan Kaufmann.

Hebrail, G., & Berard, A. (2012). *Individual household electric power consumption*. UCI Machine Learning Repository.

Hérault, J., & Demartines, P. (1997). Curvilinear Component Analysis: A Self-Organizing Neural Network for Nonlinear Mapping of Data Sets. *IEEE Transactions on Neural Networks*, 148–154. PMID:18255618

Kohonen, T. (1982). Self-organized formation of topologically correct feature maps. *Biological Cybernetics*, *43*(1), 59–69. doi:10.1007/BF00337288

Kohonen, T. (2001). *Self-Organizing Maps* (3rd ed.). Springer-Verlag. doi:10.1007/978-3-642-56927-2

Kruskal, J. (1964). Multidimensional scaling by optimizing goodness of fit to a nonmetric hypothesis. *Psychometrika*, *29*(1), 1–27. doi:10.1007/BF02289565

Kuo, R., Ho, L., & Hu, C. (2002). Integration of self-organizing feature map and K-means algorithm for market segmentation. *Computers & Operations Research*, *29*(11), 1475–1493. doi:10.1016/S0305-0548(01)00043-0

Linde, Y., Buzo, A., & Gray, R. (1980). An algorithm for vector quantizer design. *IEEE Transactions on Communications*, *28*(1), 84–95. doi:10.1109/TCOM.1980.1094577

Maaten, L. v., & Hinton, G. (2008). Visualizing Data using t-SNE. *Journal of Machine Learning Research*, *9*, 2579–2605.

Marques, N., Silva, B., & Santos, H. (2016). An Interactive Interface for Multi-dimensional Data Stream Analysis. *2016 20th International Conference Information Visualisation (IV)*. IEEE.

Mostafa, M. (2009). Shades of green: A psychographic segmentation of the green consumer in Kuwait using self-organizing maps. *Expert Systems with Applications*, *36*(8), 11030–11038. doi:10.1016/j.eswa.2009.02.088

Oja, M., Kaski, S., & Kohonen, T. (2003). Bibliography of self-organizing map (SOM) papers: 1998-2001 addendum. *Neural computing surveys*, 1--156.

Qian, J., Nguyen, N., Oya, Y., Kikugawa, G., Okabe, T., Huang, Y., & Ohuchi, F. (2019). Introducing self-organized maps (SOM) as a visualization tool for materials research and education. *Results in Materials, Volume 4*.

Sammon, J. J. (1969). A Nonlinear Mapping for Data Structure Analysis. *Transactions on Computers*, 401--409.

Silva, B. (2016). *Exploratory Cluster Analysis from Ubiquitous Data Streams using Self-Organizing Maps.* [PhD Thesis, Universidade NOVA de Lisboa]. http://hdl.handle.net/10362/19974

Silva, B., & Marques, N. (2010). Feature Clustering with Self-Organizing Maps and an Application to Financial Time-series for Portfolio Selection. *ICNC 2010 International Conference on Neural Computation.* Valencia.

Silva, B., & Marques, N. (2015). The ubiquitous self-organizing map for non-stationary data streams. *Journal of Big Data*, *2*(1), 27. doi:10.118640537-015-0033-0

Ultsch, A., & Siemon, H. (1990). Kohonen's Self Organizing Feature Maps for Exploratory Data Analysis. *Proceedings of the International Neural Network Conference (INNC-90)* (pp. 305--308). Kluwer Academic Press.

Vesanto, J. (1999). SOM-based data visualization methods. *Intelligent Data Analysis*, *3*(2), 111–126. doi:10.3233/IDA-1999-3203

KEY TERMS AND DEFINITIONS

Best Matching Unit (BMU): The neuron closest to the input data point, determined by a metric distance such as Euclidean distance.

Data Stream: Continuous and potentially infinite data set with each example represented as a fixed-dimensional vector of values.

Learning Rate: Determines the extent to which the prototypes are adjusted during each iteration.

Neighborhood Radius: Determines the range of neighboring neurons that are updated along with the winning neuron.

Prototype Unit: (also known as codebook vector) A prototype vector with the same dimensionality as the input data. The set of all NxM prototype units makes up the SOM grid of neurons (the lattice) for a map of size NxM units.

Self-Organizing Maps (SOM): Also known as Kohonen maps, these are a grid of prototype units. Values in these units are adjusted using the SOM learning algorithm over a dataset.

Ubiquitous: Self-Organizing Maps (ubiSOMs): An extension of the SOM algorithm designed for generating local SOM models over potentially unbounded, non-stationary data streams.

Unified Distance Matrix (U-Matrix): Represents distances between adjacent neurons. It is presented with different colorings following a temperature-like color scale: warmer coloring (e.g., red) corresponds to a large distance, while cooler coloring (e.g., blue) indicates that codebook vectors are close to each other in the input space.

ENDNOTES

1. This is the original SOM algorithm proposed by Kohonen. There is also a *batch* variant (Fort, Letrémy, & Cottrell, 2002) where the prototype vector updates are performed only at the end of each epoch.
2. https://github.com/brunomnsilva/UbiquitousNeuralNetworks
3. Please check the online materials in https://github.com/nmm-fct-unl/ubiSOM2023 for colored visualizations.
4. https://processing.org/download/
5. https://github.com/nmm-fct-unl/ubiSOM2023
6. Extra details are available in the *multiSOM* github: https://github.com/nmm-fct-unl/ubiSOM2023.
7. https://github.com/brunomnsilva/UbiquitousNeuralNetworks/wiki
8. https://github.com/nmm-fct-unl/ubiSOM2023

Chapter 11
Persons and Personalization on Digital Platforms:
A Philosophical Perspective

Travis Greene
https://orcid.org/0000-0003-4487-0529
Copenhagen Business School, Denmark

Galit Shmueli
https://orcid.org/0000-0002-0820-0301
National Tsing Hua University, Taiwan

ABSTRACT

This chapter explores personalization and its connection to the philosophical concept of the person, arguing that a deeper understanding of the human person and a good society is essential for ethical personalization. Insights from artificial intelligence (AI), philosophy, law, and more are employed to examine personalization technology. The authors present a unified view of personalization as automated control of human environments through digital platforms and new forms of AI, while also illustrating how platforms can use personalization to control and modify persons' behavior. The ethical implications of these capabilities are discussed in relation to concepts of personhood to autonomy, privacy, and self-determination within European AI and data protection law. Tentative principles are proposed to better align personalization technology with democratic values, and future trends in personalization for business and public policy are considered. Overall, the chapter seeks to uncover unresolved tensions among philosophical, technological, and economic viewpoints of personalization.

1. INTRODUCTION

The creeping rise of government and corporate surveillance. The viral spread of misinformation and conspiracy theories. The growing ideological polarization of politics and society. The profit-driven promotion of addictive social media behaviors. The increasing concern that algorithmic discrimination

DOI: 10.4018/978-1-6684-9591-9.ch011

reproduces social, historical, and economic inequities. What do all of these have in common? They are all, at least in part, believed to stem from applications of artificial intelligence (AI). In particular, these ethically, socially, and personally troubling effects are believed to stem from an increasingly widespread form of AI-based technology: personalization.

At the same time, however, many applications of personalization seem intuitively beneficial. Personalized medicine promises to ease human suffering and extend our lifespan by providing more effective, precise, and evidence-based treatments of medical conditions based on our unique genetic profiles and biomarkers (Jameson & Longo, 2015). Personalized behavioral interventions promise to help cure us of our bad habits, improve our physical fitness, and promote our psychological well-being. Personalized education promises to provide customized sequences of exercises to help us master new skills and domains of knowledge. Positive use cases for personalization abound, and more remain to be discovered, particularly in the realm of health, public policy, and government. Legal scholars have even suggested that law itself would benefit from becoming more personalized[1] (Ben-Shahar & Porat, 2021).

Personalization is a multi-faceted concept often used synonymously with *customization*, *user modeling*, *behavioral profiling*, *algorithmic selection*, *computational advertising* or *actuarial prediction*, depending on the field of study. But surprisingly, very little of the extant research on personalization views it from a humanistic angle or relates it to any philosophical concept of the person. We believe this is a major oversight because cogently articulating why a particular application of personalization is good or bad generally requires justifying one's evaluation through the concepts of a particular ethical theory (Gal et al., 2022). To date, however, most academic discussions of personalization tend to treat its technical foundations separately from its ethical implications. For instance, personalization researchers have generally focused on how it can reduce information overload and search costs, improve decision-making, and boost the user experience (Häubl & Trifts, 2000). The dominant view of personalization seems to be that it is an AI-driven process that permits better preference matching, reduces cognitive load, and makes performing digital tasks more convenient (Aguirre et al., 2015).

Personalization is also a major source of business value. Personalization is a competitive advantage and means to greater customer satisfaction, loyalty, and profits (Murthi & Sarkar, 2003). Indeed, the technological evolution of personalization really begins with the emergence of a new business model: the platform-based business. Personalization offers firms new dynamic capabilities for not only responding to changes in fast-changing digital environments, but importantly, for actively creating change in digital environments in pursuit of longer-term business goals. Technology firms such as Alphabet, Facebook (Meta), and Amazon derive much of their business revenue from the ownership and control of highly popular digital platforms offering vast ecosystems of digitally-connected products and services (Van Dijck et al., 2018). These companies have consistently proven their ability to find creative ways of monetizing user activity data (Wang et al., 2022; Zuboff, 2019). Consequently, AI-centric, platform-based companies have the motivation, financial resources, and data science talent to steer the future of personalization. But can we trust them with the keys?

Taking a critical stance towards this question, this chapter looks to the root of the word *personalization* itself to suggest that personalization raises important but often overlooked philosophical questions about the nature of persons and personal identity in the digital age. This chapter therefore examines personalization from diverse disciplinary lenses, including philosophy, law, and information systems, taking a broadly sociotechnical approach to the topic that goes beyond a purely economic understanding of human nature and behavior. We articulate a unifying perspective of personalization as a novel form of automated control over human environments, enabled by digital platforms and emergent AI-based approaches to personalization, such as reinforcement learning. Ultimately, we argue that personalization can contribute to human well-being and autonomy, but only if it takes the concept of the person seriously. This requires aligning the technology with philosophical and political, rather than purely economic, considerations. Legal tools and institutions, along with the technical efforts of impartial academics and researchers, can and should be used to help ensure this alignment.

Broadly, the structure of this chapter is as follows. Section 2 examines definitions of personalization across disciplines, noting their diversity, and ultimately offering a comprehensive, cybernetic definition that clarifies its goal-directed nature. It then describes the AI and machine learning foundations of personalization, contrasting traditional supervised learning approaches with the increasingly popular reinforcement learning paradigm. Next, Section 3 focuses on the business impact and evolution of personalization as used on major digital platforms, while also highlighting ethically dubious applications of personalized persuasive technology and behavior modification. Against this technical and historical backdrop, Section 4 then pivots to philosophical issues surrounding the notion of the *person* that is presumably the central object of personalization. This section canvasses philosophical perspectives on personal identity and personhood and connects them to ethical, legal, and political issues of autonomy and self-determination. Section 5 synthesizes these perspectives and suggests some principles to democratically align personalization technology with philosophical notions of the person. Section 6 then looks at future trends in personalization and discusses the implications for companies and society, highlighting the key role of regulation and of academic AI researchers in auditing and vetting personalized AI systems used on platforms. Finally, Section 7 provides a brief summary and conclusion.

2. AI-BASED PERSONALIZATION

Personalized products and services are an inescapable fact of modern life. In the wake of the COVID-19 pandemic, ever more of our daily lives takes place in digital environments where personalization technologies are invisibly at work in the background. From personalized predictions (e.g., Uber ETA, insurance quotes, credit scores), resource allocations (e.g., loans, job interviews, school admissions), to personalized recommendations (e.g., product, movies, friends), personalized treatments (e.g., by medical apps), ads, news, nudges, offers, prices, notifications and more, we are inundated with personalized scores and actions based on data about our physical condition, genetic makeup, location, measurable behaviors and interactions.

2.1 The Ambiguous Concept of Personalization

Despite its central role in the design of digital products and services, personalization remains an ambiguous and under-examined concept (Vesanen, 2007). A search of the information systems, AI, and recommender systems literature reveals a surprising lack of consensus on its essential characteristics. Personalization is either not explicitly defined, explained circularly, or used in relation to or interchangeably with customization, tailoring, experimental treatment assignment, or precision marketing. It is sometimes described as a product (a personalized prediction), and sometimes as a process (a learning process). In general, it seems easier to say what personalization is not, rather than what it is.

Even when a formal definition is given, the definition often does not reflect recent advances in AI and platform data collection and analysis. A sampling of definitions of personalization from the information systems, marketing, management, AI and machine learning (ML) literature is given in Table 1. Given the diversity of definitions, we think a more coherent conceptualization of personalization remains to be articulated.

Table 1. Selected definitions of personalization across disciplines

Definition	Discipline(s)
Not giving a user a pre-defined, fixed list of items, regardless of his preferences (Cremonesi et al., 2010)	Machine learning
The opposite of the case where a firm implements a uniform policy for the entire population (Rafieian & Yoganarasimhan, 2022)	Marketing
An iterative process composed of learning, matching or recommendation, and measurement components (Adomavicius & Tuzhilin, 2005; Churchill, 2013; Murthi & Sarkar, 2003)	Marketing, Information Systems, Management
Adapting products and services to individual tastes in order to become more appealing, desirable, informative, and relevant (den Hengst et al., 2020)	Data science, Machine learning
Results tailored to us specifically based on the types of products or services that we specifically are likely to engage with (McAuley, 2022)	Machine learning
A process (changing) the functionality, interface, information access and content, or distinctiveness of a system to increase its personal relevance to an individual or a category of individuals (Fan & Poole, 2006)	Information Systems
A process of collecting and using personal information to uniquely tailor products, content and services to an individual (Liang et al., 2006)	Management, Information Systems
The problem of determining the best treatment option for a given instance (Kallus, 2017)	Machine learning

Cremonesi et al. (2010) say what personalization is not: giving "any user a pre-defined, fixed list of items, regardless of his/her preferences." This definition implicitly invokes the concept of *preferences*, a major topic in economic theory and in philosophy. On this view, merely recommending to a user the most popular product or service would not be a personalized recommendation unless he or she actually prefers popular items (which is probably not an unrealistic assumption). The definition also highlights how personalization is commonly seen as a task in "preference modeling" (Pommeranz et al., 2012), whose basic aim is to infer the ordinal structure of an individual's preferences from his choice behavior. This preference modeling approach operates under the assumption that persons make decisions consistent with an unobservable objective or "utility" function[2] (Chambers & Echenique, 2016). Many philosophers and psychologists, however, would dispute the rationality of our *unreflective* preferences, especially when such preferences promote addiction or otherwise foster conditions conducive to the erosion of one's capacity for autonomy (Frankfurt, 1971; Lehrer, 1997).

Continuing with our survey of definitions, a similarly negative definition of personalization is given in Rafieian and Yoganarasimhan (2022), who describe personalization as "the opposite of the case where a firm implements a uniform policy for the entire population." This definition invokes the concept of a *policy*, thereby highlighting the importance of action selection strategies in personalization. We will come back to the concept of a policy in our discussion of reinforcement learning-based personalization in Section 2.2.3. For now, we can think of a policy as a rule for acting given a certain measured state of the (digital) world or environment.

Furthermore, the word *personalization* contains the adjective *personal,* implying that personalized scores or actions should, at least in part, be based on personal data. For example, one group of researchers write that personalization is "a process of collecting and using personal information to uniquely tailor products, content and services to an individual" (Liang et al., 2006). Yet, most personalization researchers have largely overlooked the crucial connection between personalized scores and legal definitions of personal data. The question is important however in light of the interactive nature of new forms of personalization that adapt and evolve over time.

Lastly, one influential view of personalization highlights its similarity with self-regulating systems that rely on user feedback to drive system adaptation. Such definitions of personalization describe it as an iterative process composed of learning, matching or recommendation, and measurement and feedback components (Adomavicius & Tuzhilin, 2005; Churchill, 2013; Murthi & Sarkar, 2003). For instance, one recent data science survey focuses on the process of individual-level adaptation by describing personalization as "adapting products and services to individual tastes in order to become more appealing, desirable, informative (and) relevant to the intended user than one-size-fits all alternatives" (den Hengst et al., 2020).

The above definitions capture, in our estimation, the basic *process* of personalization but not its ultimate purpose and value in enabling the control of an abstract environment. Moreover, current definitions make no reference to how such a personalized system might be technologically implemented or mathematically formalized. This requires integrating perspectives and concepts from a variety of disciplines.

2.1.1 Towards a Unifying Cybernetic View of Personalization

We think it is therefore useful to articulate a cybernetic[3] understanding of personalization that captures key aspects of evolutionary and psychological theories of (human) intelligence (Sternberg & Kaufman, 2013), AI (Hutter, 2004; Sutton & Barto, 2018), business strategy (Helfat et al., 2009) and organizational learning and exploration (March, 1991). AI, for example, has been defined as the study of agents that receive percepts from the environment and perform actions (Russell & Norvig, 2010), and intelligence as "the ability to reason, plan, solve problems, think abstractly, comprehend complex ideas, learn quickly and learn from experience" (Gottfredson, 1997). The adaptive, evolutionary process of personalization seems tailor-made for AI. We therefore propose the following definition to capture these ideas.

Personalization is a goal-directed process of taking actions, making recommendations or decisions, or allocating resources on the basis of measured states of humans, the digital environment, and their interactions, while possibly adaptively modifying these actions after observing feedback about their consequences.

In our cybernetics and control theory-inspired view of personalization, a decision-making agent interacts with humans, who constitute its environment. The agent can be human or artificial, but in either case has complete or partial access to the state of its human environment, typically observed via a sensor or measurement device. The agent's task is to control its environment in order to pursue its goals.[4] That is, the agent seeks to maintain *homeostasis* (Wiener, 1988) by adaptively refining its selection of actions in order to control its environment, which may be constantly changing (Friston, 2010). The adaptivity of personalization arises via an evolutionary, trial-and error process guided by elements of variation, reproduction, and differential success (Baum, 2017). After observing which actions successfully maintain homeostasis, over time the agent's behavior adapts to better fit the structure of its environment. This feedback-based learning process allows the agent to reliably control its environment and promote its long-term interests. Put in these general terms, any organization or system we might reasonably describe as attempting to achieve self-regulation—i.e., to maintain a stable and coherent identity over time—can benefit from personalization.

To more concretely illustrate the cybernetic view of personalization, consider a credit lender learning how to identify good loan applicants by making personalized loan decisions (D'Amour et al., 2020). The basic process requires the lender to make loan decisions based on various business rules and carefully record the consequences of these decisions. On the basis of the records of successes and failures (i.e., feedback), the credit lender adapts its decision rules about which people to extend or deny credit based on how such rules help it achieve its business goals. These loan decisions in turn determine which areas of the "environment" of human loan applicants the decision-maker observes. Over time, the lender may develop a preference for lending to areas of the human environment where sufficient profits can be made, thus achieving the organizational equivalent of homeostasis.

This more general control-based framing of personalization has a number of conceptual and practical benefits. First, it clarifies the connection of personalization with technical concepts such as feedback signals, predictive models, iterative learning, and agent-environment interactions from disciplines such as AI/ML, statistics, computer science and engineering. Personalization researchers often employ these concepts in an ad hoc way without necessarily recognizing how these conceptual tools serve a larger purpose in supporting the goal of control.

Second, the control-based framing clarifies why organizations employing personalization typically seek data of better quality and resolution (see e.g., Martens et al. (2016); Shmueli & Tafti (2023)). Organizations invest in better quality data to resolve uncertainty around the underlying environmental state so agents can select more optimal actions, i.e., make better predictions, in order to achieve their goals.

Third, by being explicit about the properties of the environment in which the agent acts, we can shine new light on the ethical and legal issues of personalization (Milano et al., 2020, 2021; Neuwirth, 2022). Not all environments are created equal, and some environments may be more susceptible to rearrangement and modification (i.e., control) by agents. The ethics of control will depend on the type of environment and its relevant moral properties, such as the interests, values, or goals of objects in that environment. The moral implications of control may depend on whether the environment is composed of gas molecules, distributed compute resources, or human persons driving taxis or delivery vehicles (see e.g., Möhlmann et al., 2021)

Finally, by understanding personalization as a business-oriented form of optimal control, we can not only explain why personalization is so valuable to organizations in a qualitative way: it helps them to control an environment in order to promote their long-term survival (March, 1991). Perhaps more importantly, in a quantitative way, framing personalization with the right kind of abstraction allows us to leverage precise computational models for how it can be achieved and implemented in software. With these considerations in mind, we will later consider how the digital platform, which provides integrated data collection and algorithmic intervention possibilities, serves as a perfect vehicle for organizations such as large technology corporations to harness the power of personalization to promote their business interests.

2.2 The Science of Personalization

Personalization technology rests on a large body of knowledge and research in fields such as computer science, engineering, mathematics, and statistics. Consequently, personalization involves concepts and techniques drawn from linear algebra, analytic geometry, probability and statistics, and information theory, among many other areas of mathematics. Since the field of AI involves many other disciplines, the terminology used is not always consistent and can sometimes be confusing. Here we simply aim to give a high-level conceptual understanding of the process of personalization.[5]

Personalization is at heart a machine learning application designed for handling personal data, generating individual-level predictions, and ultimately making decisions based on those predictions. Distilled to its essence, a basic machine learning problem involves four ingredients: data, a model, a cost function, and an optimization procedure (Goodfellow et al., 2016). During the training step, an optimization procedure is used to fit individual data instances to a more general model, whose generalization ability is estimated by comparing the model's predictions to actual values for new, "unseen" test data not present during model training. After selecting a final model with the best generalization ability, an individual's data is then fed into the model and a predicted score is generated. Depending on that score, "personalized" action can be undertaken by a decision-maker (human or artificial), such as recommending a specific movie, granting a loan, or displaying certain ads. These decisions may be based on certain policies, which can be implicit (i.e., based on intuition), or explicit (i.e., based on precise rules for actions). For instance, for business or legal reasons, an explicit policy used by a bank might be *if the predicted probability of default is greater than 0.5, reject the loan application.* So if a particular person, Jim, is predicted as having a probability of default of 0.71, the bank's policy stipulates that the decision-maker should reject Jim's loan application.

2.2.1 Personalization Using Supervised Learning

As noted, personalization is a special application of machine learning that uses personal data to train the predictive algorithm. That is, the input data used for learning are measurements of individual persons or their behaviors, and a target output behavior of interest, such as a binary click, like, or a numerically-measured behavior such as dwell time. Together this set of measurements comprise a dataset, typically structured as a table or matrix of rows and columns, where generally rows represent unique persons and columns represent each person's feature values and known target value ("label"). The basic task of *supervised machine learning* is to inductively learn or extract a general rule, pattern, or model from a finite training dataset relating input features to the target output. The mapping between inputs and output is assumed fixed but unknown. During model training, the mapping or statistical dependency between input and output is estimated on the basis of the training dataset where ground truth outcomes are known. Figure 1 illustrates a simple supervised learning-based personalization pipeline.

A number of machine learning algorithms exist for extracting personalized decision rules or models from data, running the gamut from the relatively simple (e.g., decision trees), to the extremely complex (e.g., deep neural networks). Each algorithm encodes its own inductive bias in learning a general rule from a particular set of training data. Yet because there is no universally superior inductive bias for all predictive tasks (Wolpert, 1996), multiple models usually are learned and evaluated to find the best one for a particular problem.

Figure 1. The basic steps of personalization using supervised machine learning

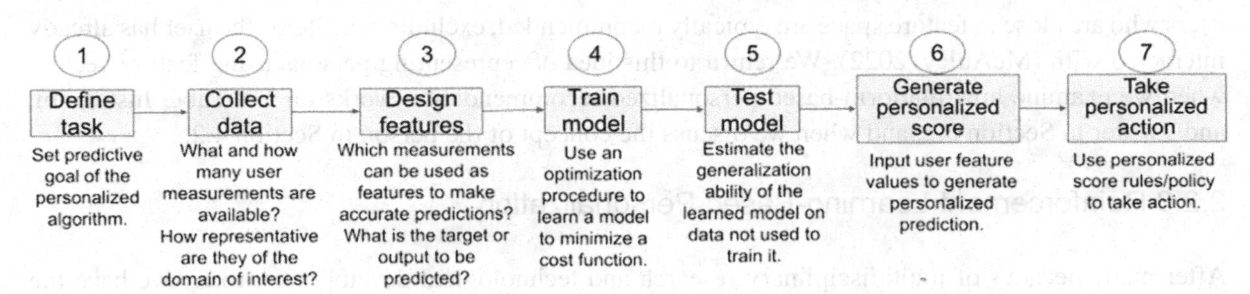

The search for the best or optimal model involves an optimization procedure that minimizes a pre-specified cost function such as the average prediction error. Gradient descent is a popular general purpose optimization method. It not only works for simple linear and logistic regression models, but extends to more complex optimization scenarios involving neural networks in a process known as *backpropagation* (Hinton & Salakhutdinov, 2006). Gradient descent starts with randomly selected initial parameter values and iteratively adjusts them until a particular set of values is found that minimizes the cost function for the training set (McAuley, 2022).

The generalization ability of this optimized model is then estimated by making predictions on a test set of previously unseen data. Various performance metrics exist for this evaluation, depending on whether the output is a continuous number, binary classification, or ranked list (Japkowicz & Shah, 2011). In most real-world applications of personalization, as organizations acquire new data (i.e., feedback), the model parameters are updated to ensure predictive accuracy remains acceptable. Note that the goal of generalization distinguishes optimization from the more difficult task of *learning* (Goodfellow et al., 2016). Put differently, developing robust predictive models requires trading off model complexity with model fit, which is known as the *bias-variance tradeoff* (Hastie et al., 2009). In some cases, better generalization performance can be obtained by imposing constraints on model complexity through techniques such as regularization or pruning, depending on the underlying type of predictive model used.

Machine learning, very generally, can be viewed as an optimization process that results in improved predictive performance on a specific task (Deisenroth et al., 2020). The resulting learned or optimized model can be thought of as approximating the underlying data generating process responsible for producing the observed data.

2.2.2 Representing Persons as Feature Vectors

One helpful way of thinking of personalization, whether using simpler machine learning models or more complex "black-box" methods, rests on the metaphor of the person as a feature vector. A *feature vector* (also known as a *user embedding*) is an array of numbers—each representing some descriptive feature or aspect of an object—which can be thought of as the axes of a coordinate system (Kelleher et al., 2020). A 10-dimensional feature vector of a person, for instance, represents a person as an array of 10 numbers, derived from measurements of their observed behavior, and replaces the "person" with a single point in 10-dimensional feature space. Once converted to a point in feature space, the "similarity" of this person to others can be computed by measuring the distance between this point and others in the feature space. Points closer to each other are deemed more similar. Typically, when making personalized recommendations for, say, movies, the feature vectors of both items and users are used. Items and users who are close in feature space are typically recommended, excluding the items the user has already interacted with (McAuley, 2022). We return to this idea of representing persons using feature vectors when we examine how platform-based personalized recommendation works on YouTube, Instagram, and Twitter in Section 3.2, and when we discuss the concept of the person in Section 4.2.

2.2.3 Reinforcement Learning-Based Personalization

After many decades of multidisciplinary research and technological development, today we have the approach to AI known as *reinforcement learning* (Sutton & Barto, 2018). Reinforcement learning has played a pivotal role in advances in robotics, interactive language agents, self-driving cars, video and board games, medicine and clinical decision-making, and resource allocation and management. Reinforcement learning techniques, when combined with large language models (LLMs) have also led to major advances in the generation of human-like text in tools such as OpenAI's ChatGPT. Not unsurprisingly, some AI researchers have declared reinforcement learning the machine learning paradigm most likely to achieve artificial general intelligence (Silver et al., 2021).

The ideas behind reinforcement learning have been around for decades, and can be found in the early research works of numerous AI luminaries. For example, in 1950, the founding father of computing, Alan Turing, imagined designing machines that could learn through punishments and rewards and evolve new behaviors, just as small children do (Turing, 1950). Around the same time, the mathematician John von Neumann sought to design self-reproducing automata (Neumann, 1966) that resembled living organisms. Likewise, Norbert Wiener, one of the founders of cybernetics, envisioned how digital computers might approximate the goal-directed, self-regulatory systems of living organisms (Wiener, 1988). And in the 1960s, Marvin Minsky drew modern-looking diagrams of "reinforcement machines" based on behaviorist psychologist BF Skinner's notion of operant conditioning (Minsky, 1961).

Yet some philosophers and researchers claim reinforcement learning agents have incentives to interact with their human counterparts in "pathological" ways involving manipulation and deception, modifying users' beliefs, and fostering addiction. An emerging literature deals with its safety, value alignment, containment, governance, and even its ability to overtake and destroy the human species (Bostrom, 2012). As we clarify in Section 3, we contend that reinforcement learning-based personalization represents a new paradigm in personalization technology, especially when coupled with the economic imperatives facing digital platforms (Greene, Shmueli, & Ray, 2022). This new paradigm draws on a set of common algorithms, methods, and problem formulations that differ from more traditional supervised and unsupervised machine learning methods. The computational properties and behaviors of reinforcement learning algorithms are complex, having been influenced by diverse fields such as animal learning, computer science, statistics, economics, and the cognitive sciences. Only recently, however, have these scientific insights trickled down into consumer-facing personalization technology. Table 2 lists several ways in which this new paradigm of personalization differs from earlier approaches.

What exactly is reinforcement learning and why is it so well-suited for personalization? Reinforcement learning is a "family of methods for addressing sequential decision-making problems characterized by non-deterministic dynamics, delayed decision-outcome pairings, and a lack of ground truth regarding optimal decisions" (Cai et al., 2017). In reinforcement learning-based personalization, an artificial agent interacts with an environment of human users, observes information about the environment's state and reactions, and receives positive and negative rewards for good and bad actions, respectively. The environment then transitions into a new state and the action-observation-reward feedback process repeats. With experience, the agent learns to take actions that lead to the accumulation of reward. When applied to personalization, the learning agent might be a recommender system, and the environment might be platform users.

Table 2. Reinforcement learning-based personalization is a new technological paradigm based on control of human platform environments. We highlight five emergent features affecting individuals and society

Emergent Features of Reinforcement Learning-based Personalization
1. Fast, interactive, and automatically adaptive learning 2. High volume of interactive behavioral training data 3. Unspecifiable data collection and processing 4. Non-consensual human behavior modification 5. Pathological agent behavior via reward optimization

At its most basic, reinforcement learning focuses on the interaction of an agent and an environment over discrete periods of time. Think of reinforcement learning as a general, trial-and-error-based computational method for solving problems. From an optimization or engineering perspective, an appropriately trained agent is a "solution" to the "problem" posed by the environment of human users (Silver et al., 2021). Depending on the complexity and assumptions of the problem, such as whether reward is delayed or environment states are fully or only partially observable, a multi-armed bandit, contextual bandit, or more generally, a Markov decision process (MDP)[6], can be used to represent the task of reinforcement learning-based personalization. In reinforcement learning, the general task to solve (e.g., maximize users' engagement) is specified by the reward function (e.g., receive a positive reward if the user watches the movie). The reward function can, in theory, be designed to solve any conceivable task or, equivalently, control any environment. This idea forms the essence of what the pioneers of reinforcement learning have put forth as the "reward hypothesis" (Sutton & Barto, 2018). At the same time, however, decomposing a general and abstract task into an explicit reward function is one of the most difficult aspects of reinforcement learning (Levine, 2021).

Once the learning task has been formalized, the agent's goal is to learn a strategy for how to behave in its human environment in order to solve the task specified by the reward function. This learning strategy amounts to estimating the environment's underlying transition function (i.e., its state dynamics) through trial-and-error interactions. The resulting learned strategy for interacting with the environment is known as its *policy*. The policy is a rule telling the agent what action to take given any encountered state of the environment. In engineering terms, then, a robust control policy allows for the regulation of a system such that it remains in a desired range of states.

Unlike unsupervised learning (e.g., clustering) where no feedback is given, or supervised learning, where ground truth labels provide the learning algorithm with known optimal decisions, in reinforcement learning, there is no "ground truth" telling the agent about the "correct" action. The agent only receives "evaluative" feedback about the quality of its action in a given state via the reward signal (Littman, 2015). The concept of evaluative feedback is surprisingly intuitive. Essentially it is the basis for the children's game where one blindfolded child must locate other hidden children based on evaluative feedback about whether she is getting "hotter" or "colder" relative to their location. When other children call out "hotter" they provide a positive reward signal the blindfolded player can use to improve the quality of her actions. When other players call out "colder," they are in effect providing negative reward (i.e., punishment) indicating movement in the wrong direction. The reward signals are like a trail of breadcrumbs, leading her to the goal state. With frequent and high-quality evaluative feedback (i.e., breadcrumbs are not too far apart), the blindfolded player should eventually be able to solve the task of finding the hidden players.

It can also be useful to consider reinforcement learning from a psychological perspective. Reinforcement learning works by learning instrumental associations between perceptual stimuli (states) and actions in order to maximize some positive reward determined by the algorithm's designer. In this way the agent's behavior resembles the behaviorist concept of *operant conditioning* (Touretzky & Saksida, 1997), which itself embodies Thorndike's famous *law of effect* (Dan & Loewenstein, 2019). The law of effect states that organisms whose actions result in pleasurable or positive effects tend to repeat those actions, while organisms whose actions tend to result in painful or negative effects will repeat those actions less often. In summary, with the advent of reinforcement learning-based personalization, previously qualitative behaviorist learning principles can be replaced with precise quantitative formulations, giving rise to a new generation of platform-based artificial agents that can acquire and perform complex, novel behaviors in response to their human environments.

We earlier suggested conceiving of personalization as a form of control. Seeing it through a control theory lens has a number of interesting theoretical consequences. Applying control principles of learning and adaptation (Friston, 2010) to personalization, we can say that, in general, the complexity of personalization required for any goal-directed entity (i.e., a digital platform) depends on a variety of factors, including: 1) the agent's internal capacity to represent decision rules; 2) the agent's ability to sense and act on the environment; 3) the timing and quality of feedback; and 4) the external complexity of the environment, which itself can evolve over time (Sternberg & Kaufman, 2013). These considerations imply that one cannot give generic advice about "the right amount of personalization" for a platform without knowing more about the agent's capacity to represent complex decision rules (e.g., is it using a simple decision tree or a deep neural network?), its perceptual and action affordances (e.g., its sensors and actuators), and the number of environment states relevant to the decision-maker's goals.

These theoretical considerations also have practical implications for personalization. Namely, as the complexity of the human environment increases, so too must the complexity of the agent's actions (Ashby, 1957). Because personalization is nothing more than the rational, goal-driven matching of actions with environmental states, unless human platform users are very homogenous in their preferences and behaviors, a uniform (i.e., non-personalized) policy is unlikely to control them well and reliably accumulate reward. In a marketing context this might hypothetically involve, say, giving the same promotional 10% discount to customers A and B, even though customer B would have purchased the product with a much lower discount. In short, if the agent cannot control its human environment well, it cannot advance the platform's business goals.

2.3 Implications of Using Reinforcement Learning for Personalization

2.3.1 The Explore-Exploit Dilemma and its Effects on Human User Environments

The ability to learn while simultaneously acting in the environment is unique to reinforcement learning and is referred to as *online learning* (den Hengst et al., 2020). Online learning highlights a tradeoff faced by both biological and artificial agents known as the *explore-exploit dilemma* (Sutton & Barto, 2018).

The dilemma captures the conflict between balancing short-term reward and long-term value when making decisions under uncertainty. Depending on the complexity and predictability of the environment, sometimes it pays to take a sub-optimal (or even random) action early on to receive more and better data in the long term about its quality. These exploratory actions provide information that can lead to better decisions in the future. For instance, the ability to reason about the long-term effects of actions allows the agent to learn to anticipate and prevent bad events, rather than passively react to them.

To better understand how reinforcement learning agents raise new possibilities for sophisticated interactions with human persons, consider the *mountain car problem*. The toy problem reveals how the ability to associate actions with delayed consequences can help solve difficult optimization problems. The agent (a mountain car) needs to learn how to reach a destination atop a nearby mountain. The problem is that the mountain car starts at the bottom of a valley. Yet by taking exploratory actions, however, the agent learns valuable information, such as that by counterintuitively *moving away* from the destination in the beginning, it can eventually build enough velocity to reach its long-term goal on top of the mountain.

The explore-exploit dilemma raises the important question of timing—how long should we explore our options before finally settling on the best one? The effects of real-world interventions can vary depending on when and in which sequence they are administered. As is clear from many medical and biological contexts involving "critical windows" of development, such as during prenatal periods, the timing of exposure to treatment can drastically moderate the effects of exposures (Selevan et al., 2000). The same goes for the timing of medical interventions during emergency situations like sepsis, or in long-term cancer treatments (Gottesman et al., 2019). Arguably there are also psychological periods when persons may be susceptible to algorithmic interventions, during major life transitions, identity development, midlife crises, or retirement. It is not hard to see how the ability to automatically learn effective sequences and time-specific interventions could be usefully applied to marketing and promotional scenarios that involve multiple interactions with customers, often under resource constraints (Sutton & Barto, 2018; Wang et al., 2022).

How does an artificial agent learn when to explore new options and when to exploit the currently best ones? One algorithmic approach to this dilemma involves optimism in the face of uncertainty: among the actions whose values are being estimated, select the one with the highest upper bound (Littman, 2015). The essence of these exploratory approaches is that they produce probabilistic, on-the-fly data collection decisions. But the occasional random action selection adds new difficulty for regulators or algorithm auditors, as it is no longer possible to specify in advance the purpose and nature of data collection and processing.

Algorithmic exploration raises some difficult ethical questions. For instance, when applied to millions of platform users, some persons will inevitably be on the receiving end of a long sequence of "sub-optimal" random actions, raising issues of algorithmic harm and fair allocation of technological risks across individuals and social groups (Joachims et al., 2021; Rhoen, 2017). While novel in terms of personalization technology, the ability of artificial agents to optimize for long-term outcomes can be dangerous for an environment of human users, especially when the agent's reward function is designed to incentivize agent actions that result in human engagement behavior (Everitt et al., 2021).

In what has been described as a failure-mode of over-optimization (Manheim & Garrabrant, 2018), AI researchers warn that improperly specified reward functions can lead to pathological outcomes, such as promoting addictive behaviors for human users (Burr et al., 2018). Indeed, neuroscientists have demonstrated that behavioral predictions made from reinforcement learning models are nearly indistinguishable from actual social media engagement behavior, suggesting that platform algorithms can reliably control human user engagement by varying the rate of associated platform rewards (Lindström et al., 2021). The potential conflict between an agent's reward function and human well-being is known as the *value alignment problem* (Gabriel, 2020). The problem highlights the technical and normative aspects involved in aligning an artificial agent's values with human values and interests. The introduction of platform-based personalization complicates value alignment by adding the issue of corporate interests (e.g., profit maximization), a conflict we also discuss within the context of persuasive technology and algorithmic behavior modification in Section 3.

2.3.2 From Personalized Predictions to Actions

As the pipeline in Figure 1 illustrates, personalization leverages a learned model to generate an individual-level prediction based on that individual's particular feature values. The prediction is "personal" in the sense that the measurements used as input to the model can be traced back to prior interactions with that person (though typically most personalization algorithms also include measurements from relevantly similar other people). But while simply generating an individual-level prediction might be sufficient to count as personalized for marketing purposes, it is perhaps more important from a legal and moral perspective to consider the actions or sequence of actions taken on the basis of this predicted score. Depending on the application of personalization, these actions are usually based on organizational rules, business policies, or social policies in the case of the public sector. These rules are chosen to promote the more general goals of profit-making or a specific public policy such as selective incapacitation of criminals.

Personalization is an essentially sociotechnical phenomenon as personalized predictions are intended to inform agents whose actions are selected such that they purposefully influence and control human social relations. Indeed, the etymology of the word *decision* itself means to cut (into the world), a subtle point recognized by the philosopher Jacques Derrida who analogized the process of making a decision as "a cut in time and space, (leaving) its trace, a wound of a sort" (Bates, 2005). We thus think it is important to realize that the resulting decision to act based on a personalized score has some causal effect on the estimated predictive relationship used to make the initial prediction, thereby modifying the relationship in the process. In marketing and advertising, the goal of attitude or behavior modification is explicit (Nord & Peter, 1980), and the large amount of time, effort, and money invested in such behavior-modifying activities by rational organizations is ostensibly evidence of its causal efficacy. In any case, the growing recognition of the causal force of prediction-driven decision-making (Fernández-Loría & Provost, 2022) motivates new forms of "persuasive" personalization algorithms that can adaptively update their behaviors on the fly as they interact with platform users and collect feedback on the quality of these interactions.

Typically, however, whether due to simplicity or convenience, most supervised learning problems assume model predictions do not influence future states of the environment, implying a causally-inert and static view of the algorithm-environment relationship. In the scenario where the environment is fixed and our model of it is very accurate, the best strategy may indeed be a "greedy" short-term strategy that takes advantage of the currently best option. This kind of search strategy thus "exploits" its knowledge of the human environment. But in the real world of personalization, the underlying statistical structure of human environments is constantly changing. Moreover, the personalized action can change the input-output dependency itself, thereby creating a machine-human feedback loop. Cosley et al. (2003) for example, experimentally demonstrate that displaying predicted ratings to users affects their subsequent rating of the item. This modification in fact is intended in most business and marketing contexts (Hansotia & Rukstales, 2002), and is the goal of recommendation itself (Xiao & Benbasat, 2007). As Rafieian & Yoganarasimhan (2022) explain in their study of AI in personalization and marketing: "the primary purpose of a personalized policy is to change the current policy, thereby shifting the current data distribution." Indeed, the implicit goal of nearly all business efforts is to change the attitudes or behavior of potential customers towards some object of interest (Eyal, 2014; Fogg, 2003).

In the more dynamic and adaptive paradigm of reinforcement learning-based personalization described above, the MDP model explicitly accounts for the "feedback loop" (i.e., temporal associations) between the agent's actions, the receipt of reward, and states of the environment (Sutton & Barto, 2018). In other words, each action by the artificial agent results in the environment transitioning into a new state and possibly receiving a reward, allowing the agent to "assign credit" to past actions (or sequences of actions) that lead to the receipt of reward at a later time. Receiving a reward changes the estimated value of the state and thereby can indirectly influence the agent's strategy for selecting actions in the future. For this reason, reinforcement learning is well-suited to deal with shifting user preferences and dynamic user behavior on platforms, unlike more traditional static paradigms of personalization that rely on a fixed "greedy" strategy of action selection (Zhao et al., 2019).

Before we examine how personalization works in complex digital environments known as platforms, two important points with ethical implications can be made regarding differences between supervised and reinforcement learning paradigms of personalization.[7] The first is that in online reinforcement learning, correlations between agent actions and resulting states of the human environment reduce the information content in observed data—thus resulting potentially in the need for significantly larger volumes of training data. The second is that the sequential actions of artificial agents can be used to strategically interact with people. As the mountain car problem illustrates, agents can learn to cleverly trade off short-term losses for larger long-term gains against human adversaries, who might not even be aware the agent is exploiting regularities in their decision patterns (Dezfouli et al., 2020).

3. PERSONALIZATION ON COMMERCIAL PLATFORMS

While personalization offers a novel means for organizations to control digital environments in order to advance various social and business goals, personalization is not without its critics. For example, personalization has been singled out as fostering undesirable "filter bubbles" (Pariser, 2011) and "echo chambers" (Helberger et al., 2018). Because personalization algorithms learn from users' behavioral feedback, over time these algorithms tend to feed users content that aligns with their existing views and preferences, creating informational bubbles that can filter out dissenting or different viewpoints (Milano et al., 2020). Here, however, we focus on how the platform concept provides newfound possibilities for integrated data collection and algorithmic interventions perfectly suited for personalization. To illustrate, we give two real-world examples of how personalization works on Instagram and YouTube, and introduce the important related concepts of *persuasive technology* and *algorithmic behavior modification* (*BMOD*).

Personalization technology has co-evolved with the development and expansion of the digital platform as a novel business model (Srnicek, 2017). Platforms are implemented as programmable digital architectures geared towards the systematic collection, algorithmic processing, circulation, and monetization of user data (Helmond, 2015). Harvard social psychology professor Shoshana Zuboff argues that digital platforms play a key role in the development of what she calls *surveillance capitalism*, described as "a new economic order that claims human experience as free raw material for hidden commercial practices of extraction, prediction, and sales" (Zuboff, 2019, p. 8). In this new mode of capitalism, data is now productive capital to be accumulated by any means possible (Sadowski, 2020).

Technology corporations successfully leveraging the economic potential of digital platforms include Amazon, Facebook, Google, Uber, and Microsoft in the United States, and Baidu, Tencent, Alibaba, and Bytedance in China. Nearly all of these firms have well-funded and influential AI research groups with names like DeepMind, Google Brain, and Tencent AI Research. Commercial platforms take advantage of global network effects potentially involving billions of users, and offer an ever-evolving array of digital services and products, while collecting and processing massive amounts of user-generated text, audio, image, and video data. Increasingly a major source of political news and public health information, commercial platforms now also influence scientific progress, democracy, and our ability to manage global crises (Bak-Coleman et al., 2021).

3.1 The Co-Evolution of the Platform and Personalization

Personalization initially began as a relatively top-down, static, knowledge-based technology. Evolving out of research on recommender systems and the adaptive web (Brusilovsky & Maybury, 2002), a shift occurred in the mid-2000s when technology corporations began embracing advertising-centric business models and expanding digital platforms to harness their potential for the economic circulation of implicit behavioral data (Langley & Leyshon, 2017). But the volume of behavioral data needed to make machine learning-based approaches viable was not readily available until a user-centered, collaborative Web 2.0 culture appeared in the form of interactive platforms such as Wikipedia, Facebook, and YouTube (Van Dijck, 2013). Leveraging the participatory ethos of the Web, freemium[8] business models allowed platforms to rapidly expand their user bases and harness network effects (Posner & Weyl, 2019). Growing user bases however meant higher overhead costs. Most platforms either incorporated or were bought by large media corporations, thus requiring them to choose between retaining their original participatory culture and realizing a sizable return on investment demanded by venture capital (Van Dijck, 2013). For many platforms, this meant displaying third party advertisements to platform users.

Motivated by economic research on two-sided markets (Rochet & Tirole, 2006), corporate managers began strategically expanding and networking their own platforms to harness their potential for the economic circulation of implicit data (Langley & Leyshon, 2017). For example, Alphabet, the parent corporation of Google, provides an ecosystem of networked service platforms: search engine, cloud computing, email, social networking, and advertising network, each resting on a shared infrastructure owned and operated by the parent corporation (Van Dijck et al., 2018).

During this evolution, machine learning techniques advanced in sophistication, from the once relatively static recommendation algorithms trained using explicit user feedback data (e.g., ratings, surveys, or written text) to more adaptive algorithms trained using greater amounts of engagement-driven, implicit user feedback data in the form of clicks, scrolls, and dwell time (Aggarwal, 2016; Ekstrand & Willemsen, 2016). Although easily collected and quantified, implicit data can hide the wider situational and intentional context of the behavior (Meissner et al., 2019). They are therefore only a rough proxy for the unobservable constructs of user preferences, satisfaction or interests. Yet from the platform's perspective, implicit feedback is easier, faster, and cheaper to collect than explicit feedback, and due to its scalability across millions of users, still allows profitable predictions of individual behavior (Zuboff, 2019).

As platforms began to collect and analyze massive volumes of unstructured data involving text, images, and video, complex neural network methods gradually were also integrated within the collaborative filtering paradigm of personalization (Salakhutdinov et al., 2007). A major breakthrough in image classification involved combining massive datasets of images and data augmentation techniques with convolutional neural networks (Krizhevsky et al., 2012), leading to even more intricate personalization involving images. Other advances in personalization came in the form of new loss functions that directly optimized the ranking of implicit feedback (Rendle et al., 2012). As computational power expanded, and as platforms provided access to massive datasets of users' behavior and user-generated content, neural network architectures from NLP were increasingly applied to improve the scale, sophistication, and automation of personalized recommendations.

The introduction of *persuasive technology* (Fogg, 2003) to the platform concept marked a second major shift in personalization. Persuasive technology seeks to change users' attitudes and behavior by merging psychological behavior modification techniques with platform technology, thus allowing platforms to tailor persuasive intervention strategies to individual user features (Berkovsky et al., 2012). Naturally, marketers and advertisers embraced the concept of persuasive technology, as social media platforms provided useful channels for influencing product choice and controlling communication with consumers (Palmer & Koenig-Lewis, 2009). Soon the concepts and techniques of nudging, choice architectures, and "dark patterns" began to take root on platforms (Mathur et al., 2019). Dark patterns are cases where designers use their knowledge of human behavior and rationality to implement deceptive functionality against the user's best interests (Gray et al., 2018).

One intriguing example of how the lines between persuasion and control can be blurred in platform-based personalization comes from AI researcher Stuart Russell. He warns in his book *Human Compatible* that combining the technology of reinforcement learning (see Section 2.2) with the economic goals of platforms can result in profitable yet pathological algorithmic behavior that incentivizes making human click behavior more predictable in order to increase advertising revenues (Russell, 2019). While illustrating the fundamental issue of value alignment, Russell's point is that such powerful algorithms, when combined with the economic incentives facing the platform, can lead to an unintended situation in which an artificial agent learns to make its data (its observations) more revealing of users' underlying states, thus improving its ability to select optimal actions to control and direct their behavior. Other AI experts have come to similar conclusions, going so far as to say that with the help of adaptive technologies such as reinforcement learning, corporate platforms increasingly resemble adaptive, goal-directed social machines (Cristianini et al., 2021). Taking up this thread, our proposed cybernetic view of personalization as control explains how the self-regulatory business goals of the platform can be achieved through controlling user behavior.

3.2 Examples of Platform-Based Personalization

Personalization is clearly important from a business perspective. The streaming platform Netflix estimates its personalized recommendations save them close to $1B per year, while accounting for about 80% of hours streamed (Gomez-Uribe & Hunt, 2015). But beyond the business implications, the scale of personalization on major digital platforms has the potential to shape the desires, preferences, and self-conceptions of millions of people every day. Personalization can even have society-wide effects, such as shifting opinions towards political issues and candidates (Dutt et al., 2018). Below we briefly describe how the technical task of personalization is intertwined with platform business goals and the various constraints imposed by platform scale—in terms of user and content bases—and computational efficiency.

3.2.1 YouTube's Recommender System: Personalization at Scale

In modern, large-scale platform recommender systems such as those on Instagram or YouTube, prediction is most properly viewed as one step within a larger process (Stray et al., 2022). Broadly the personalized recommendation process on a major platform like YouTube involves five steps:

1. Gathering all platform items (e.g., videos)
2. Performing content moderation
3. Generating a set of candidate items (i.e., filtering)
4. Predicting and ranking user-item engagement
5. Re-ranking the previous step's results using metrics tied to business goals

In the case of YouTube, the basic machine learning task of the recommender system is to accurately predict user watch time for its huge corpus of user-generated videos, many of which are uploaded fresh each day (Covington et al., 2016). To do this[9], techniques developed in natural language processing are creatively repurposed. Two massive predictive models, one for initial candidate generation, and then another for prediction and ranking, are trained on sequences of unique video IDs watched by users, along with user demographics and video features (Covington et al., 2016). More specifically, the deep neural network used by YouTube and other major platforms can have billions of parameters, require hundreds of billions of records for training, and need special computing infrastructure to train and test them (Covington et al., 2016). These models can be massive, making it difficult to explain or understand their output in an intelligible way. Due to business reasons and their massive scale, these large models are only iteratively updated in batches, after testing has confirmed—often first in simulations—that the updates are likely to lead to improved performance on some metric of interest (Gauci et al., 2019; He et al., 2014).

The business goals of platform-based personalization also affect the choice of performance metrics. Since the business goals of platforms include getting users to discover new and diverse platform items (Smith & Linden, 2017), improving user loyalty, and encouraging cross-selling items, unique performance measures are used, such as item coverage, diversity, or serendipity (Kaminskas & Bridge, 2016; Kotkov et al., 2016). *Serendipity* refers to adding random noise into recommendations to increase the diversity of recommended content[10] (Shardanand & Maes, 1995).

3.2.2 Instagram's Explore: Account Embeddings

We now consider Instagram's Explore algorithm, which works in a similar fashion. A user is represented by features extracted from his or her Instagram account. The complicated set of algorithms[11] used for Instagram's Explore recommendations work by creating "account embeddings" in order to infer topical similarity (Medvedev et al., 2019). Roughly put, a feature vector representation of an Instagram user's account is derived from the sequence of other users' account IDs the focal user has visited. Next, the feature vectors are used to determine a "community" of similar accounts. Similarity in this space implies topical similarity, assuming users do not randomly explore different topics, but instead tend to view multiple accounts offering similar content. Once the candidate accounts (the nearest neighbors) have been found, a random sample of 500 pieces of eligible content (videos, stories, pictures, etc.) is taken and a "first-pass" ranking is made to select the 150 most relevant pieces of content for a focal user. In a filtering step, a neural network then narrows the 150 to the top 50 most relevant. Computational efficiency is a high priority. Finally, a deep neural network predicts specific behaviors available on the Instagram app for 25 pieces of content, such as "like," "save," or "see fewer posts like this." Content is then ranked based on a weighted sum of these specific in-app behaviors and their associated predicted probabilities. Weights are set by Instagram's data scientists, reflecting their assumptions about Instagram users' intentions and also the business value of these behaviors. Finally, the Explore algorithm uses a simple heuristic rule to downrank certain content and diversify results. After applying the downranking procedure, the content with the highest weighted sum in the "value model" is displayed in decreasing order on the focal user's Explore page.

3.2.3 Twitter's For You Tweet Recommender System

Twitter recently made public[12] how Tweets are personalized for users when the timeline is refreshed. Once again heavily relying on the person as a feature vector approach[13], Twitter's basic three-step pipeline, which is run over 5 billion times per day and takes less than 1.5 seconds to process, is similar to the one we described in the case of YouTube's video recommender system. Personalized Tweet recommendations derive from features in three main information sources: the Tweet itself, the user's social network graph, and engagement data. Step one involves generating a set of candidate tweets from different sources from the hundreds of millions of possible Tweets. Step two then scores or ranks each candidate Tweet with a proprietary machine learning model. And step three consists of applying various heuristics and information filters related to user experience and psychology (e.g., Tweet fatigue, social proof, author diversity), business goals and product features (e.g., block Tweets with *not safe for work* (NSFW) content or from blocked users). Finally, the personalized Tweets are mixed with advertisements, follow recommendations, and other prompts and notifications and delivered to the user's device.

3.3 Behavior Modification and Personalization

With advances in personalization technology, platforms can now combine the automated control of reinforcement learning with the science of behavior modification to modify both digital content and delivery methods over time. We define *algorithmic behavior modification* (or *BMOD*) as any algorithmic action, manipulation, or intervention on digital platforms intended to impact user behavior (Greene, Martens, & Shmueli, 2022).

Although in medical and public health contexts BMOD is described as an observable and replicable "intervention designed to alter or redirect causal processes that regulate (human) behavior," (Michie et al., 2013) platform BMOD is often unobservable and irreplicable. And even when platform BMOD is visible to the user, as with displayed recommendations, ads, or auto-complete text, it typically is unobservable to external researchers (Milano et al., 2021). Various researchers are worried about the lack of transparency and unintended ethical, social and political effects of platform BMOD at scale (Beam et al., 2018; Menczer, 2020; Milano et al., 2020). As a team of computational social scientists describes the situation,

The relationship between platforms and people is profoundly asymmetric: platforms have deep knowledge of users' behavior, whereas users know little about how their data is collected, how it is exploited for commercial or political purposes, or how it and the data of others are used to shape their online experience (Lorenz-Spreen et al., 2020).

Computationally, BMOD can be implemented on platforms via various machine learning algorithms that operate in a data-driven, autonomous, interactive, and sequentially-adaptive way. Two examples are the NLP-based algorithms used for predictive text and reinforcement learning, which we described earlier. Both are commonly used to personalize services and recommendations, increase user engagement, generate behavioral feedback data (Chen et al., 2019), and even "hook" users by habit formation (Eyal, 2014). Platforms deploy *autonomous experimentation systems* (Bird et al., 2016) to learn complicated associations between sequences of actions and outcomes conducive to platform goals. Although the pioneers of persuasive technology condemn the use of deception or coercion (Fogg, 2003), these autonomous systems may exhibit deceptive or coercive behavior as they interact with human users on platforms, potentially modifying users' beliefs and fostering addiction (Burr et al., 2018; Cristianini et al., 2021; Russell, 2019).

While controversial, BMOD methods have been used by major platforms to control digital environments and shape behavior towards "guaranteed outcomes" (Zuboff, 2019, p. 201). Derived from principles of behaviorist psychology, they are premised on the idea that "once you understand the environmental events that cause behaviors to occur, you can change the events in the environment to alter behavior" (Miltenberger, 2016, p. 3). BMOD thus represents a potentially new form of AI-driven control in digital environments with implications for individual persons (e.g., addiction and emotional polarization) and society (e.g., mental health and political polarization). The emergent ability of AI to strategically capture human attention and reliably trigger human behavioral engagement in a variety of situations—often against our conscious will—thus threatens human autonomy and self-determination in the increasingly digital world. In short, BMOD raises significant ethical concerns and calls for a deeper examination of their impact on individuals and society as a whole. We explore these and related concerns next.

4. THE CONCEPT OF THE PERSON

Recall that one of the main objectives of this chapter is to examine the often-overlooked philosophical assumptions about the nature of persons and personal identity in the digital age. One way of doing this is to ask whether and to what extent AI-driven personalization is aligned with philosophical concepts of the person. Clarifying the normative properties of the person is essential in analyzing the ethical as well as political implications of personalization, and can also help us to understand the motivation and purpose behind current and future AI, privacy, and data protection regulation.

To begin, we draw on a number of important Western philosophical and political works in our analysis of the person. Among those influential philosophers and political theorists who have written about the person, many emphasize the value of autonomy and the related notion of self-determination. Autonomy, both in an ethical and political sense, is the ability to make laws for oneself without external interference[14] (Benn, 1975). The basic idea is that a person requires a protected sphere safe from the control of external influences in order to choose freely, develop into a well-functioning person, and be the author of her own life (Mill, 2015). A belief in the inherent value of personal autonomy is also an essential ingredient in a flourishing democracy. This belief is embodied in, for example, foundational political documents such as the United States' Bill of Rights that aims to protect citizens against the coercive power of the state or the majority (Winick, 1992). The ethical and political value of autonomy has not diminished as we move into the digital age. We argue that a new platform-centric, cybernetic paradigm of personalization as control poses a threat to our autonomy as persons. This section lays out a philosophical vision of the person and connects these ideas to philosophical, legal, and political issues of personal autonomy and democratic self-determination.

4.1 Data Privacy and the Construction of Digital Persons: Technical Perspectives

In accordance with our sociotechnical perspective on personalization, we contrast notions of personal data, personal identity, and data privacy from different disciplines. The goal is to provide a more holistic picture of the person and personal identity and relate it to relevant concepts of personal data. We focus primarily on the European Union (EU) regulatory context because the EU's vision of personal privacy, personal data protection, and the use of AI for personalization has set the benchmark for the rest of the world, and moreover, is heavily grounded in philosophical ideals about what constitutes a good human life and society (Hijmans & Raab, 2018). Broadly, the aim of the European regulations, such as the 2016 General Data Protection Regulation (GDPR) and the proposed AI Act of 2021, are to "ensure the development of AI….is ethically sound, legally acceptable, socially equitable, and environmentally sustainable, with a vision of AI that seeks to support the economy, society, and the environment" (Floridi, 2022).

The GDPR is arguably the first attempt by a major democratic political body to not only grapple with the social, political, and legal consequences of digital technologies, but also to actively direct their development. The EU's GDPR is viewed by many data protection scholars as the standard by which governments might attempt to regulate the social risks posed by AI while also balancing the need for technological innovation. Indeed, GDPR framers believed it would "change not only the European data protection laws, but nothing less than the whole world as we know it" by setting a "global gold standard for every new innovation, for consumer trust in digital technologies and for an entry point to the growth opportunities of an emerging digital market" (Albrecht, 2016). We thus argue that the GDPR should be understood as an aspirational document—similar in purpose to a constitution (De Gregorio & Celeste, 2022)—aimed at normatively regulating interactions between data subjects and data collectors in the age of increasingly powerful digital platforms and AI-based technologies.

4.1.1 Personal Data

Before tackling philosophical issues of personal identity, let us start by clarifying what constitutes personal data under the European Union's GDPR. The GDPR defines personal data as "any information relating to an identified or identifiable natural person ('data subject')" (Official Journal of the European Union, 2016). This broad definition, which is similar to definitions found in other data protection laws in the US and elsewhere[15], means personal data can constitute anything from browser cookies, to location data, to even a combination of nonsensitive measurements, if they can be used to single out individuals. To limit the ability to identify persons through their data, however, the GDPR requires organizations storing and processing personal data to follow the principles of privacy by design and data minimization (Greene et al., 2019).

Why are personal data valuable from a business perspective? Importantly, database technologies allow organizations such as platforms to link personal data with personal identity. For platforms serving personalized ads or recommendations, the ability to reliably identify and match two seemingly different (e.g., misspelled) records as belonging to the same person is important if users' behavior is to be tracked and attributed to the same person over time. In short, reliable identification is necessary for controlling humans in digital environments.

The form of personal identity used in database design is a technical one different from the one philosophers and legal theorists draw on. Computer scientists, data privacy, and personalization researchers generally focus on the database-centric idea of personal identity. That is, personal identity is treated in a forensic, third-person administrative manner, where persons are presumed to be isolated, atomic entities with essential definitional criteria (Kent, 2012). Successful use of database technologies requires that data representations of persons are error-free, easily, and efficiently integrated from various sources. A central goal in designing these databases is maintaining a consistent data structure needed for performing various tasks related to data cleaning, capture, storage, sharing, analysis and transfer (Anagnostopoulos et al., 2016).

When represented as an entry in such a database, identification of a person proceeds by the specification of a unique set of descriptive features or attributes, such as whether he or she is older than 30, is married, and uses an iPhone. This kind of matching-based approach to personal identity is applied when, for example, records from multiple tables are linked in a database join operation or when DNA from a crime scene is linked to a specific individual. These database-driven technologies use a set of measurements linked with a particular human person, such as sequences of DNA base pairs or ridge and bifurcation patterns in a person's fingerprints, to uniquely identify them. Often however, in platform applications, obtaining exact matches is not the goal. Instead, approximations based on a notion of similarity are used, generally by defining a metric space that can be used to derive a distance function between objects located in this abstract space (Zezula et al., 2006). We discussed a similar process in the context of recommender systems which treat persons as feature vectors and generate recommendations by comparing how "far" items are from users in an abstract vector space.

Importantly, the attributes used to represent persons are selected in order to advance the goals of the data collector (Fisher, 1987), which in the case of digital platforms may include various supervised and unsupervised learning tasks. These machine learning tasks can often impose their own constraints related to avoiding the "curse of dimensionality" (Domingos, 2012). Consequently, data about persons may be filtered or removed in order to preserve model parsimony, storage space, and computational efficiency, especially due to distance computations in high-dimensional spaces. This is similar to how filtering and ranking are used to reduce the computational burden involved in YouTube's, Instagram's, and Twitter's recommender systems (see Section 3.2).

Furthermore, due to the way data is collected on platforms, data that are easier to measure or cheaper to obtain, tend to be included in these data descriptions of persons. For example, a platform may easily measure click behavior and mouse movements (Leiva et al., 2021), but may know little about a user's religious beliefs or emotional state because the platform has no suitable technological apparatus or sensor for measuring these attributes.

In enterprise applications involving databases of personal data, such as in customer relationship management (CRM) (Zwick & Dholakia, 2004), personal identity is treated as if it were a tangible object or fungible commodity. In these business contexts personal identity is treated as if it were a valuable resource—a commodified object, piece of property, or asset abstracted from its generator, a human person—to be protected against theft or unauthorized access by an unknown intruder or adversary (Cheng et al., 2017). As the clichéd metaphor goes, data is the new oil, and so personal data are likened to an up-for-grabs resource in the public domain that must be claimed and appropriated before one's competitors do (Purtova, 2015). Firms' motivation for taking data privacy seriously appears to be that privacy threats, data leaks, and data breaches potentially lurk around every corner, and so they must be prepared to protect data assets from prying competitors, disgruntled employees, and unscrupulous hackers.[16]

4.1.2 Person Identification vs. Personal Identity

Whatever one thinks of the moral appropriateness of commodifying personal data, one must acknowledge that digital identity management techniques play an essential role in today's increasingly globalized and digital worlds. In the more technical domains of data privacy and information systems, personal data are the means through which human identification takes place. Desirable properties of human identifiers include their universality, uniqueness, permanence, collectability, storability, exclusivity, precision, simplicity, cost, convenience, and acceptability (Clarke, 1994). Data privacy is primarily concerned with personal data used for human identification via the central concepts of linkability, anonymity, disclosure, and pseudonyms (Torra, 2017). A *pseudonym* replaces a direct identifier of a person, e.g., *Barack Obama* with an indirect identifier such as *User123*, while anonymity means that an adversary cannot distinguish an individual's unique identity in a set of individuals to an arbitrary degree of precision (Torra, 2017). Techniques such as masking, aggregating records, or adding noise, allow for formal guarantees about what attributes of a person an attacker might learn about, thereby giving organizations greater control over what information they disclose or conceal about the persons whose information they store. Pseudonyms and anonymity are examples of *nyms*, or digital identities users can take on when interacting with other parties in different environments. Authentication tools like tokens, smart cards, digital certificates and biometrics use nyms to associate individuals with their true digital identities. Additionally, persons might also rely on partial identities to interact with others (Damiani et al., 2003). These are subsets of their real properties, such as a personal credit card number, passport number, or a birthplace that foster trusted interactions in lieu of their actual presence.

One must however be careful to distinguish *human identification*, which is a practical and technical matter, with *human identity*, which is a philosophical matter[17] (Clarke, 1994). This distinction is not always clearly upheld as much data privacy research focuses on the data processor or controller, not on the human persons who generated the data, who are mostly viewed as passive entities (Torra, 2017) or inert commodifiable resources. By treating personal data in abstraction, without reference to what or who it is about, personal data becomes an object of ownership or consumption and thus might be exchanged or sold on markets to those who value it most (Posner, 1981). In this process of abstraction, personal data becomes alienated from the original person generating it, only to be appropriated by those with the power and resources to collect and process it (Zuboff, 2019).

As such, the computer science, business, and database-focused approach does not fully capture personal identity as many legal scholars, psychologists, and philosophers understand it. They view personal data, including our personal data profiles[18], as more than an abstract, fungible asset, resource, or commodity used to transact business or improve a service, arguing that it plays a moral role in constructing a person's self-identity. In the philosophical context underlying European data protection law, one's personal identity is not presumed stable and fixed through time; neither is it presumed reducible to an object's material, external, third-person description alone. Personal identity is more about a fundamentally constructive and spontaneous process, akin to that used to create a piece of art (Brinkmann, 2005). In the following sections, we describe in detail philosophical perspectives on personal identity relevant to personalization.

4.2 Constructing Ourselves: Philosophical, Psychological, and Legal Perspectives

Influenced by a dynamic, dialectical philosophy and famously described as the *looking-glass self* (Cooley, 1902), personal identity is conceived as something constructed over time and inherently related to the way one interacts with and responds to others in one's social communities (Gergen, 1985). Personal identity thus has moral and social components (Stryker & Burke, 2000) and is related to the concept of an abstract, unified self that evolves through time. Common types of social identities relate to one's ethnicity, religion, political affiliation, job, and social relationships. Persons are thus said to exist in a socio-cultural network (Baumeister, 2005) of personal, cultural, and historical meanings, identities, roles, and values, many of which we do not ourselves choose but simply are "thrown" into and must learn to navigate (Heidegger, 2010).

Moreover, as philosophers emphasize, constructing one's personal identity depends on self-referential description (De Vries, 2010; Manders-Huits, 2010; Schechtman, 2018). Very generally, one's personal identity is a "description under which you value yourself, a description under which you find your life worth living and your actions to be worth undertaking" (Korsgaard et al., 1996, p. 101). Self-referential descriptions make possible what philosopher Daniel Dennett calls "radical self-evaluation," or reflective, identity-defining judgments about our life projects, interests, and values (Dennett, 2015). Psychologists draw on a similar idea when they refer to an "authentic" or "true self," which answers the question "*what is the core and authentic me*?" (Ashmore & Jussim, 1997). The dynamism of the self-concept is reflected in numerous self-representations that must be coherently integrated: the past, present, and future self; the ideal, "ought," actual, possible, and undesired self (Markus & Wurf, 1987). These self-referential descriptions involve reference to an internal, first-person subjective view of oneself and one's life, structured in the form of a personal narrative (Bruner, 1991; McAdams & McLean, 2013; Polkinghorne, 1988).

Granting persons access to and control over these digital forms of self-description is key to ensuring that personalization technologies play an enabling role in constructing our personal identities in the digital age (Hildebrandt & Gutwirth, 2008). Once the moral aspects of personal data are acknowledged, data privacy law becomes a crucial tool in securing the free (i.e., self-directed) construction of one's personal identity. Privacy as part of personal identity construction requires that persons themselves play some role in aligning their own personal identities with the data representations that drive the decisions emanating from personalization algorithms. This aspect of privacy has been described as "autonomy over one's self-representation" (Aizenberg & van den Hoven, 2020). The practical implication is that putting legal and technical constraints on the ability to re-identify persons may be a necessary but ultimately insufficient step for ensuring the privacy of persons in this broader sense of autonomy over one's self-representation.[19]

While the concept of privacy is disputed and complex, many legal scholars and ethicists agree it ultimately centers on the issue of control (DeCew, 1986; Hartzog, 2021). Persons require some degree of control over the information others know about them and therefore potentially can use to exploit them. This includes control over how events and behaviors related to a person are described, and to what use those descriptions are put (Mahieu, 2021). Thus, we see in data protection laws such as the GDPR various rights to data access and requirements of purpose specification for the analysis and processing of personal data (Zarsky, 2016). Data privacy law is therefore a potential antidote to the power others—including digital platforms—might have over us due to asymmetries in knowledge about our behavior, and that of relevantly similar others. When such knowledge is coupled with personalized technologies, it becomes possible for platforms to control user behavior in ways that go against users' self-determined goals. Unfortunately, however, due to the complexity of AI-driven personalization, many people have neither the knowledge about nor control over how such data are used in personalization (Caudill & Murphy, 2000). Moreover, the assumption that platform users can somehow rationally trade-off costs and benefits of personalization seems disingenuous, given the complexity of AI-driven personalization.[20]

In summary, current forms of personalization do little to promote autonomy of self-representation, and therefore neglect a major aspect of what makes human persons unique. In contrast, most approaches to personalization based on the collection of implicit behavioral data exemplify a radical form of empiricism advocated by behaviorist psychologists (Ekstrand & Willemsen, 2016) and thus fail to adequately recognize the representational duality of the self-conscious person. Consequently, they also fail to properly understand the concept of *context* and the crucial importance of *narrative meaning* in integrating and organizing first-person human experience (MacIntyre, 1984), which can be at odds with third-person behavioral descriptions. For instance, the mathematical representation of a person as a feature vector abstracts a "person" away from his socio-cultural network of meanings, identities, and values and represents him as a precisely-defined digital object in order to predict a very specific behavior, limited to a specific data-driven application. Figure 2 provides a schematization of the process by which persons are digitally represented in personalization and how meaningful context and narrative are filtered out from the digital representations of persons.

With these considerations in mind, we suggest that AI-based personalization can contribute to human well-being and autonomy, but only if it takes the concept of the person seriously. This means aligning the technology with philosophical and political, rather than purely economic, considerations. The tools and institutions of law can help foster such alignment. We return to this issue in Section 5, where we propose several principles for incorporating aspects of democratic political and ethical theory in personalization.

Figure 2. A schematic of the personalization process. Personalization requires reducing the self-aware person embedded in social reality to a feature vector of observable behaviors and information embedded in the feature space of digital reality. Note the potential for looping effects.

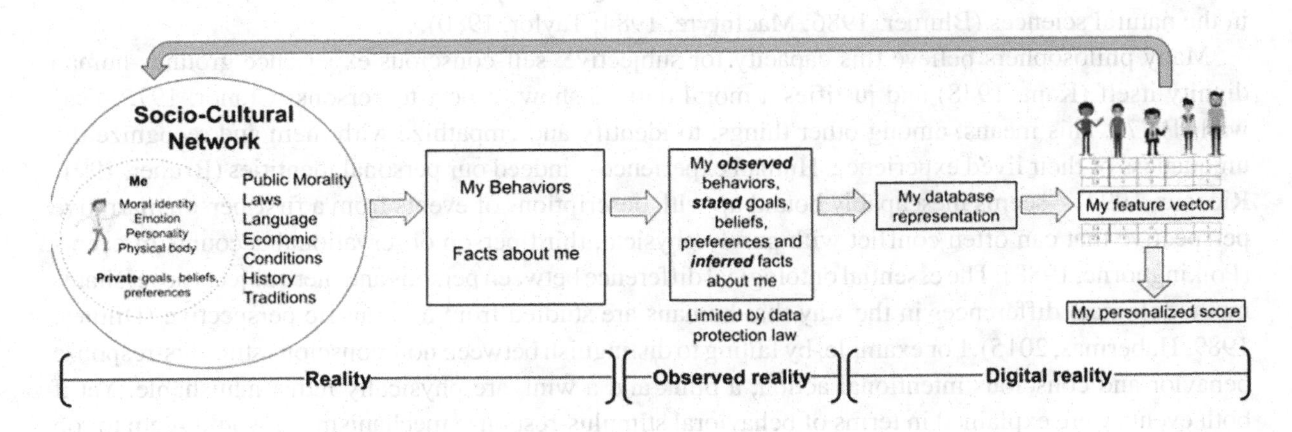

4.3 The Duality of the Person: Inner and Outer Views

The history of the concept of the person is complicated and intertwined with other concepts like minds, bodies, and souls. Numerous influential yet disputed definitions of the person have been proposed throughout the centuries. A person has been variously defined as a "human animal," "moral agent," "rational, self-conscious subject," "possessor of particular rights," "free, ensouled creature of a particular kind," or "being with a defined personality or character" (Macdonald, 2017; Schechtman, 2018; Smith, 2011). As these examples illustrate, there is little agreement on what the essence of a person is, though most agree that persons have a uniquely dual nature. That is, most accounts of personhood agree that persons possess certain physical and mental properties, but they weigh these physical and mental aspects differently.

Etymologically, the word *person* comes from the Latin *persona*, which derives from the Ancient Greek word for a type of mask worn by dramatic actors. Thus, the concept of a person is inherently connected to one's social roles (Douglas & Ney, 1998). One's persona is a specific kind of self-identity that is public, socially-defined, and varies depending on context. Western thinkers, following Kant (who himself was influenced by Christian theology), expanded the idea of the person into two main parts: an outer *personnage* of public roles and an inner conscience and consciousness, complete with a moral identity and phenomenology off-limits to outside observers (Fukuyama, 2018). A number of philosophers, many influenced by the teachings of Plato and neo-platonic Christian philosophers such as Saint Augustine, have believed this inner realm houses something like an immaterial soul or mind that survives death and constitutes our true personal identity (Macdonald, 2017). In more modern times, philosopher and mathematician René Descartes is perhaps most well-known for advancing the idea of mind-body dualism (Descartes, 1999).

As philosophers in the existentialist and phenomenological traditions have emphasized, persons exist and act in two epistemically asymmetric inner (mental or subjective) and outer (physical or objective) domains.[21] A person's external behavior can be described, explained, and predicted using the same scientific, third-person language of physics that applies to other moving bodies; yet we appear unable to scientifically account for human experience from the first-person perspective (Varela et al., 2017). We can, for instance, describe the color red by referring to electromagnetic radiation with a wavelength of approximately 700 nanometers, but describing *what it is like to experience* the color red seems difficult, if not impossible in purely physical or neurobiological terms (Jackson, 1986). The inner lives of persons and their capacity for self-consciousness make them different from the non-conscious objects—particles, molecules, and other material objects whose motions might be explained by the laws of physics—studied in the natural sciences (Blumer, 1986; MacIntyre, 1984; Taylor, 1980).

Many philosophers believe this capacity for subjective, self-conscious experience grounds human dignity itself (Kant, 1948) and justifies a moral duty to show respect to persons (Cranor, 1975; Darwall, 1977). This means, among other things, to identify and empathize with them and recognize the uniqueness of their lived experience. Human experience—indeed our personal identities (Bruner, 1991; Ricoeur, 1994)—seems inescapably bound up with descriptions of events from a first-person, narrative perspective that can often conflict with purely physical, third-person observational accounts of events (Polkinghorne, 1988). The essential ontological difference between persons and mere objects necessitates methodological differences in the way that humans are studied from a scientific perspective (Dilthey, 1989; Habermas, 2015). For example, by failing to distinguish between non-conscious stimulus-response behavior and conscious intentional action, a blink and a wink are physically indistinguishable. Yet if both events were explained in terms of behavioral stimulus-response mechanisms, it would seem to rob the person of causal agency. So long as personalization technologies fail to recognize and respect such differences between internal and external (i.e., database) views of the person—including the uniqueness of persons (Sen & Williams, 1982)—it is unclear they can meaningfully advance human well-being.

As noted earlier, for many contemporary philosophers interested in the concept of personhood (Dennett, 1976), truly being a person requires consciously (i.e., self-aware) reflecting on and endorsing one's desires and preferences and their relevant functioning in one's life plans (Bratman, 2000; Frankfurt, 1971, 2006). These first-order desires and preferences are roughly analogous to what standard economic theory calls our "revealed preferences" (Beshears et al., 2008). Few philosophers would object that these first-order preferences may indeed be captured in our implicit behavioral data such as clicks or dwell time. But many philosophers would further assert that our true selves are not identical with such desires, and that to equate ourselves with them is to fail to recognize what makes humans persons unique from other animals. The implicit behavioral data driving platform-based personalization (e.g., clicks and dwell time), arguably correspond more to what the goal-directed, "sub-personal" reward systems of the human brain want (Berridge, 2009), than what we, as persons capable of self-conscious reflection, reflectively endorse.[22] The failure to take into account a person's capacity to reflect on and consciously endorse his or her behaviors, attitudes, and preferences consequently results in an impoverished form of personalization, a form that *does not take the person seriously* (Greene, Shmueli, & Ray, 2022).

These abstract philosophical insights are now beginning to have a concrete impact on the design of personalization technologies. In general, there is growing awareness about how the way we are represented as data influences the way in which personalization algorithms interact with us, and consequently, how we relate to ourselves as persons. Whether through data protection laws, academic research, professional associations, or engineering codes, as life steadily moves online, there is growing recognition of the legal and moral importance of our digital representations (Leidner & Tona, 2021; Van der Sloot, 2017). For example, the IEEE's vision of Ethically Aligned Design (EAD) empowers "individuals to curate their identities and manage the ethical implications of their data" (IEEE Standards Association, 2017). Without greater focus on the ethical impact and philosophical assumptions of personalization, such technologies risk evolving into technological instruments of oppression and control, constraining rather than expanding human freedom, self-expression, and creativity (Kane et al., 2021).

Legal scholars in particular are concerned with the way in which personalization technologies can erode the foundations of liberal democracy, which is centered on having a private personal sphere free from external control (Mill, 2015). Legal scholar Jack Balkin explains how persons can begin to self-police their behavior in response to digital and algorithmic surveillance, arguing "algorithms cause you to internalize their classifications and assessments of risk, causing you to alter your behavior in order to avoid surveillance or avoid being categorized as risky" (Balkin, 2017). Such "looping effects" (Hacking, 2007)—where the classifications of us by social and scientific institutions can change our descriptions of ourselves and affect our life opportunities—could curtail our ability for self-expression, ultimately leading to a society of conformism and groupthink. European legal scholar Mireille Hildebrandt therefore defines privacy in the digital age as "freedom from unreasonable constraints on the construction of one's own identity" (Hildebrandt, 2015, p. 80). Indeed, the German Constitution enshrines protection for persons' freedom in developing their personalities as long as they do not violate the rights of others (Coors, 2010). Taken together, these political and legal concerns for a person's ability to self-determine motivate protecting personal data.

Not only is self-determination a foundational concept in international law, but it has been described as a precondition for "a free democratic society based on its citizens' capacity to act and to cooperate"

(Rouvroy & Poullet, 2009). Self-determination is also the metaphysical foundation of human dignity (Floridi, 2016), playing an essential role in the grand philosophies of both Kant and Hegel, who presume that persons have free will—that at least some of the time, as rational beings, persons have the capacity to set their own ends.[23] Notably, however, self-determination implies that human persons have free will, or at least the capacity for it, a claim, which if true, conflicts with the basic assumptions of modern science.[24] Here again we see fundamental conflicts between philosophical views of the person and the scientific assumptions implicit in technologies of personalization. With the help of legal institutions and appropriate regulation, however, it may be possible to ensure that humans shape personalization technologies, rather than personalization technologies shaping the design of humans.

To conclude, we have argued that AI-driven personalization should not be considered in a vacuum, isolated from any notion of what a person is and should be. In other words, we are suggesting that in the digital age of platforms and increasingly complex forms of personalization, the descriptive and normative aspects of personalization can no longer tenably be viewed as independent. If personalization is to be aligned with human well-being, it must be viewed from a broader, holistic perspective that takes persons' subjective, lived experience seriously and treats persons as unique individuals. Law can help achieve this alignment. The next section speculates on how personalization might be democratically aligned with philosophical views of the person described above.

5. ALIGNING PERSONALIZATION WITH THE HUMAN GOOD

In line with our basic thesis that ethical forms of personalization must be justified with reference to a substantive vision of the human person and good society, let us restate our definition of personalization given earlier in Section 1. Doing so will help us to focus the arguments presented in this section.

Personalization is the goal-directed process of taking actions, making recommendations or decisions, or allocating resources on the basis of measured states of humans, the digital environment, and their interactions, and possibly adaptively modifying these actions after observing feedback about their consequences.

The first thing to notice is that personalization, particularly when implemented as reinforcement learning, is a process of control that embodies consequentialist ethics (Card & Smith, 2020). While there are several variants of consequentialism (Driver, 2011), broadly it is an ethical doctrine holding that the value of actions derives from their contribution to the accumulation of observed reward or utility. Consequentialism is therefore often viewed as closely allied with utilitarianism. Note that in personalization, utility or reward is defined by the designer of the system.

While consequentialist reasoning may suffice for effectively controlling and predicting phenomena in real-world environments, consequentialism and utilitarianism are nevertheless controversial ethical doctrines. Many philosophers and legal scholars consider utilitarianism deficient insofar as it advocates for policies potentially inimical to liberal values and strong legal protections of individuals and minority groups against majority groups and government authorities (Dworkin, 2013; Rawls, 1971). Consequentialist reasoning can lead to morally disturbing forms of decision-making that disregard the uniqueness and dignity of individual persons (Sen & Williams, 1982; Zagzebski, 2001). Thus, nearly all ethical and legal criticisms of consequentialism seem relevant to personalization as well.

To adapt a common example from the ethics literature, imagine a public health system in which a robot optimizing for quality-adjusted life (QALY) years acts to surgically remove the organs of one person—who, due to a personalized prediction, was presumed to die soon anyway—in order to save the lives of several young people needing a transplant. In short, consequentialism sounds great until you yourself are the person selected to suffer for the greater good.

Consequentialism, as a decision theory, might be made more palatable by incorporating forms of democratic participation in specifying what the consequences are to be optimized and in setting limits to the kinds of actions that can be taken. Another way of saying this is that there should be agreement and mutual understanding on the underlying formalisms and assumptions behind personalization. As we earlier argued in Section 4, part of what it means to take the person seriously is to respect a person's capacity for engaging and participating in personal and political self-determination. Yet deciding on mutually agreed upon sets of states and actions, specifying an appropriate reward function, and setting a discount factor—a parameter determining the weight of immediate versus future rewards—is not a trivial task even for a team of data scientists. Also, behavioral finance research suggests individuals vary widely in how they discount future rewards in different contexts (health or finance, etc.) (Chabris et al., 2008), which might prevent widespread consensus on this important optimization parameter. Still, arguably those affected by personalization should have the opportunity to decide whether decision policies should be biased more towards maximizing immediate or long-term outcomes.

But who would participate in deciding what these factors should be and under what conditions? Most societies have laws specifying an age at which adolescents become legal persons capable of making important decisions, related to, for example, voting in elections or joining the military. In countries such as the US there are also laws that revoke one's rights to participate in democratic decision making under some conditions, such as felony convictions. One open question concerns the feasibility of an approach based on direct democracy, since a more republican form of participation using "trusted" data science fiduciaries may be more realistic, given the technical complications involved in personalization (see e.g., Greene et al. (2022)). These considerations suggest that personalization is not inherently bad or dangerous. It is simply a form of automated control that allows a social unit to achieve a goal. Given the shared etymology of the words *government* and *cybernetics*, the constitutional law literature (Tushnet, 2012), which centers on the building and writing of constitutions, may be relevant in setting up fair procedures for deliberating on and amending the control rules by which personalization systems operate.

In domains where wide social consensus exists on optimization goals, personalization is less likely to go wrong. Personalized medicine provides a nice example of how personalization can be used to encourage human flourishing, while also clearly illustrating why autonomy and informed consent are essential to the just application of personalization. As medical therapies and diagnosing technologies have improved over the years, doctors are increasingly able to match specific treatments with very specific patient states, related to biomarkers, biopsies, or genetic profiles (Jameson & Longo, 2015). In earlier eras of medicine, doctors would make diagnoses and provide roughly the same treatment for all patients. In the language of personalization, doctors applied a near-uniform policy of treatment to patients, even though clearly some patients experienced adverse side effects or much better responses to certain treatments. The problem was that measurement devices (e.g., genetic tests, fMRIs, blood assays) did not exist that could be used to create fine-grained (state) features to support increasingly specialized treatments. To see the power in this kind of control, imagine a national health system in which an entire nation's patient-doctor interactions were guided by a collective reinforcement learning agent. The agent could, on the basis of health screenings and the collective health outcomes of all, select preventative treatments and interventions so that acute and costly sickness might be avoided later. But just as the story of King Midas illustrates, unexpected agent behavior can result in disasters. Russell (2019) gives the example of an artificial agent tasked with curing cancer as soon as possible; yet the ultra-rational agent finds ways of inducing tumors in the whole human population, in order to run millions of clinical trials in parallel.

Personalization thus raises new opportunities for rethinking how collective decision-making is done, while also focusing our attention to basic ethical and social questions about the goals and purposes of our social systems and institutions. As personalization technology marches forward, new digital devices, new forms of data collection, and new means and modes of exploiting digital vulnerabilities will inevitably emerge. Currently, many of the people, businesses, and organizations designing these control systems are biased, implicitly or explicitly, towards self-interested goals. The consequences of various actions are only evaluated with respect to the advancement of their own self-interest, leading to a number of social externalities (Haugen, 2021). Moreover, personalization optimization goals are often skewed towards convenient proxy metrics (Thomas & Uminsky, 2022) and immediate short-term gains at the expense of future generations and the stability of the (social) system itself.

As we all know, however, democracy is complicated and often requires placing considerable burdens on citizens, through both active participation and extensive education and deliberation on relevant policy issues. In some domains we can expect that goals will be shared more broadly in society, as in the case of some medical contexts where most persons would agree on what constitutes good and bad health outcomes, and where the decision-makers have moral and legal obligations to look out for those affected by the decisions. But in many other potential social contexts where personalization might be deployed there may be greater value heterogeneity, especially when essentially contested concepts of justice are involved (Gallie, 1955). In a modern pluralist society, people will disagree about what is valuable or just (Rawls, 2005). This is not necessarily bad (see e.g., Mouffe (1999)) it just means that we must find as-close-to universally fair ways of deciding, among the people controlled, the rules by which they would agree to be controlled. Ideally, those who find the resulting control rules distasteful may voluntarily opt-out. Multi-stakeholder, multi-objective optimization techniques might be one way to better align control policies with democratic processes (Koster et al., 2022). Combining these technical ideas with democratic political theories would make for interesting future research (though see Sætra et al. (2022) for an opposing view).

5.1 Towards Democratic Principles for Ethical Personalization

When thinking about the ethics of personalization, we suggest drawing on the liberal democratic concepts of *public reason*[25] (Binns, 2018) and the *rule of law*[26] (Tamanaha, 2004). In the context of personalization, we offer an analogous concept we call the *rule of control law*. The basic idea is that the control rules (i.e., the decision policy or mechanism by which actions are taken, given that certain states of the environment are observed) governing personalized interventions are intelligible, transparent, public, and open to change and contestation. Similarly, in their analyses of the foundations of legal systems based on the rule of law, legal philosophers Hart (1961) and Fuller (1964) argue that citizens must be able to know what is legally expected of them in order to fruitfully plan and live their lives. This requires the law to possess a certain non-arbitrariness and predictability in how it will be interpreted and applied to specific cases by human judges.

Personalization, ideally, should be "of the people, by the people, for the people" ultimately affected by it. There is nothing in principle keeping personalization systems from embodying what public policy scholar Archon Fung (2013) calls the *democratic principle of affected interests*, by allowing all those affected a voice in deliberation about the control laws that govern them. If this could be somehow done using democratic procedures, control rules might acquire moral legitimacy, especially on commercial platforms. From an ethical and political perspective, such deliberative and participatory procedures encourage respect for the law, rather than fear or distrust, which is important because the behaviorist learning principles behind reinforcement learning, when applied to societies, threaten individual free will and self-determination (Zuboff, 2019). Persons affected by personalization need assurance that the system will not be used as a tool for coercion, manipulation, and other "non-rational" forms of persuasion (Spahn, 2012).

Adapting these ideas to personalization, we might say that persons should be able to plan—to some extent—for what a personalized system will do, otherwise personalized interventions will appear arbitrary or confusing to those affected by them. Arbitrary personalized interventions, particularly when automated by artificial agents, threaten to violate the democratic norm of *procedural due process* (Kroll et al., 2017), which in part aims to protect individual rights against arbitrary government interventions into the affairs of citizens and ensures that legal proceedings are conducted in a fair and impartial manner (Citron, 2007). In the context of a commercial platform, where particular personalized interventions might appear arbitrary due to the larger exploratory goals of the system, it is not clear how randomized interventions can obtain the requisite legitimacy to command the respect of those affected by them (Tyler, 2006). Even more, the economic system-wide goals of commercial platforms can be at odds with the interests and goals of the individuals using them, as when, for example, Uber employs "surge pricing" interventions to control overall ride demand even though individual drivers and riders may find such interventions unfair or arbitrary (Möhlmann et al., 2021).

Our suggested rule of control law leads to a few key considerations in the democratic design of personalized systems. First, the rule of control law may require that the rate at which control laws adapt to conditions must be relatively constant—thus setting limits on the system's learning rate parameters. The downside, of course, is that when the social environment changes suddenly, it will take some time before these changes are registered and fed back into an adapted control law. Second, the rule of control law suggests the need for fundamental limits on the complexity and transparency of control rules such that they remain interpretable for those who are affected by them. Work on explainable reinforcement learning (Puiutta & Veith, 2020) may provide useful ideas on how the actions of artificial agents might be explained to those affected by them. Several possibilities exist here. For example, we might identify the input (i.e., state) features that affected an agent's action, or show how the agent's past experiences led to the current behavior. Lastly, we might explain the agent's behavior by reference to its optimization goals (Milani et al., 2022). This topic remains an open area of research.

But the danger in this purely technological approach to explainability is that it may motivate the development of even more complex control rules rather than settling on intrinsically interpretable control rules that nevertheless are sufficient to achieve the goals (Rudin, 2019). When control rules become too complex for ordinary people to understand, then they are forced to trust that others, perhaps so-called data science fiduciaries, are acting in their best interests (Greene, Dhurandhar, & Shmueli, 2022). If such a situation were to emerge over time, then it might make sense for data scientists to adhere to the same professional and legal standards as doctors and lawyers. But even if social agreement could be reached on the appropriate level of complexity for control rules, what happens if an individual claims he or she was inappropriately controlled? Perhaps there was a bug in the system, or a randomly selected exploratory action was required and this person happened to be in the wrong place at the wrong time.

Similar to the way that judicial review works in constitutional democracies such as the United States, groups of selected (elected?) data science experts might audit personalization systems to ensure they meet the normative standards of the rule of control law. Because a personalization system cannot find an optimal control policy without at least some random exploration, citizens will need to agree on what constitutes a fair probability of exploration. Citizens may be willing to tolerate some small chance they themselves will receive a random action or treatment, depending on the stakes (Kahneman & Tversky, 1979). There is a small philosophical literature, partly influenced by economics, dedicated to the question of when it is ethical to randomly select people or use a lottery to assign a treatment (Broome, 1984). Some of these philosophical and psychological discussions of risk and lotteries may be relevant to the ethics of personalization conceptualized as a form of automated democratic control of social systems.

5.1.2 A Federalist Model for Ethical Personalization

Personalization as control illustrates classic tensions between advancing the collective good and respecting individual autonomy. Different societies strike this balance in different ways (Etzioni, 1996), just as different moral systems emphasize the individual or group as the basic unit of analysis (Haidt, 2008). One way to split the difference, however, is to use something like a federal system of many locally autonomous units that together constitute a global collective unit. Take, for instance, the federal system of government in the US. Local governments can set their own laws in many domains, but in others nationally valid "control laws" (i.e., federal laws) may prevail. From an abstract learning perspective, federal systems allow individual states to explore new policies and exploit what other states find (Hills et al., 2015). This same search pattern is exhibited in numerous species of social organisms—such as honeybees and ants—that must navigate complex and ever-changing environments.

Similarly, we might also look at personalization systems that decouple data collection and action policies so that artificial agents might specialize in learning to collect useful observations, while others specialize in learning reward-promoting interactions and behaviors (Riedmiller et al., 2021). The idea is that an entire country or larger organizational unit can have access to collective observations, but the rules of control and intervention applicable to a local unit can be custom-tailored to local values. Just as some states in the US may make abortion illegal, others may not. One political bonus of a federalized system is that it allows for a form of *democratic exit* (Warren, 2017), whereby if people do not find control laws are aligned with their personal values, they may choose to leave and go to another place where they are better aligned. In theory, at least.

In practice, however, many people simply adapt to living under authoritarian, oppressive, and unjust social systems, sometimes even finding reasons to support such oppression when they believe it to be unescapable or inevitable, as in abusive marital relationships or repressive political regimes (Jost, 2019). Likewise, in much more mundane contexts, research on platform switching costs reveals that many users can end up coercively "locked-in" to a particular platform, even though another may offer them more value (Ray et al., 2012).

As with any large, centralized organization composed of smaller partially-autonomous units, coordination and communication issues are bound to arise. One real-world obstacle to building a federalized personalization system is that when building a collective store of data, individual units (e.g., countries in the EU) may have conflicting and varying personal data laws and data collection and processing practices. Standardization might resolve some of the issues, but of course this raises the question of whose standards should prevail. In any case, harmonizing data structures and communication channels, along with the ability to download portable copies of one's personal data and social network, would likely be needed for such a federalized personalization system to function well and allow those negatively affected a realistic opportunity to exit.

5.2 Towards a Golden Rule for Personalization

If any answers are to be found to the difficult ethical issues of personalization, our best bet may be to turn to the literature on political philosophy, ethics, and bioethics. For example, Persad et al. (2009) suggest that when forced to make interventional decisions about how to allocate scarce resources, we might consider four categories of principles: treating people equally, maximizing total benefits, promoting and rewarding social usefulness, and favoring the worst-off. Since reinforcement learning already is based on consequentialist principles, we instead sketch what control laws favoring the worst-off might look like, based on an influential thought experiment from the philosopher John Rawls.

The basic logic behind our suggested approach to ethical personalization is the so-called *golden rule*: do unto others as you would have them do unto you (see also Berdichevsky & Neuenschwander (1999) for a similar take on applications of persuasive technology). The golden rule implies one should not deploy a control law affecting others unless one is also prepared to be controlled by the same rules guiding the actions, recommendations, or resource allocation decisions. This golden rule is the basis behind liberal, Kantian forms of autonomy and freedom that extend individual freedoms only as far as they do not encroach on the ability of others to also exercise their freedoms (Oshana, 2006), a principle we also saw in the German Constitution. Unfortunately, when it comes to many AI-based technologies, economic incentives can foster moral hazard. Many of the persons and organizations designing harmful technologies are unlikely to be personally affected by them, or else they possess the monetary, social, and political resources to avoid being subjected to the damaging effects of the very technologies they help create.

From political philosophy, one might extend the logic of the golden rule by borrowing from Rawls' notion of the *veil of ignorance* (Rawls, 1971). The purpose of the veil of ignorance is to articulate what kind of control rules are just. The basic procedure would involve asking persons to imagine they will design a personalization system that will operate in society—perhaps in higher-stakes domains of criminal justice, education, or finance—but without knowing their identity, social status, physical appearance, etc. as observed by the system's sensors. The designers would thus be uncertain as to how they might be treated under such a personalized system. That is, the designers would neither know what their personalized scores or actions would be (e.g., they would not know whether they might be members of a socially salient group associated with a high risk of crime) nor would they be privy to the algorithmic details driving personalized interventions (e.g., possible actions, states, frequency of exploratory actions, and the complexity of the policy). Rawls argues that the designers of this system, being rational, self-interested persons facing such uncertainty, would articulate control rules that would broadly be acceptable to all rational persons. These rules would embody what he calls the "maximin" principle, such that under the resulting rules, the worst-off persons would still face a tolerable outcome and chance at leading a good life. A Rawlsian approach to AI alignment and governance has been put forth by Gabriel (2022), and a recent empirical study testing Rawls' claims provides initial evidence that persons do indeed favor rules that leave the worst-off with a still-acceptable outcome (Weidinger et al., 2023).

The Rawlsian veil of ignorance procedure could also be extended to analyze other potential uses of personalization, such as in criminal justice contexts that require personalized risk predictions (Greene, Shmueli, Fell, et al., 2022). Would a designer of a risk prediction algorithm used in sentencing be willing to be subjected to this very same algorithm? Two further examples may be useful in probing our ethical intuitions on variations of personalization related to the emerging field of *information design* (Kamenica, 2019) on digital platforms. Again, aimed at optimizing for the business goals of commercial platforms discussed earlier in Section 3, these economics-inspired approaches make certain assumptions about how individual users update their beliefs, and on the basis of the resulting beliefs, strategically consider how to release information to maximally exploit information asymmetries in favor of platform interests (Wu et al., 2022). A veil of ignorance analysis would ask whether the designers[27] of these systems would themselves agree to be subjected to them. The answer, we venture, would be no. At least if they valued their own autonomy.

6. FUTURE TRENDS IN PERSONALIZATION: REGULATION AND INDEPENDENT AUDITING OF PLATFORM-BASED PERSONALIZATION

Personalization technology continues to advance at a rapid pace, driven by platforms' ever-increasing ability to capture and analyze human behavioral data via an expanding array of always-on and interconnected digital devices. On top of this, major investments into AI research by platform-based companies are catalyzing these advances. In the short term, we can pinpoint two strands of research likely to have an impact on personalization. The first is deep reinforcement learning (Botvinick et al., 2020), which combines neural networks' ability to learn complex, non-linear patterns with reinforcement learning's focus on automated, goal-directed agent behavior. The second is new frameworks for training and fine-tuning large language models using reinforcement learning, such as the techniques used to develop OpenAI's ChatGPT (Ouyang et al., 2022). These techniques represent a new technological means to interact with platform users in more personalized ways and to optimize for new kinds of business goals. At the same time, these innovations also raise new ethical and political questions, as the previous sections have described.

To address these and other concerns, regulation like the EU's AI Act specifically targets AI systems that engage in subliminal manipulation techniques (Neuwirth, 2022). Complementing the AI Act are the Digital Markets Act and Digital Services Act, which together broadly require certain AI systems and platform-based services to be transparent, traceable and subject to human oversight (Busch & Mak, 2021). These top-down regulations not only make use of independent auditing bodies to ensure the safety and alignment of large-scale and high-risk AI systems, but they also empower and motivate platform-based companies to take the ethical risks and social externalities of personalization more seriously, particularly when dealing with vulnerable populations such as children and the elderly.

6.1 The Public Policy of Personalization: The Role of Sandboxed Experiments

As personalization becomes more autonomous, interactive and opaque, additional safeguards will be required to ensure it aligns with individual and social values. Like most foundational political and ethical questions, however, there are no easy or unique answers for what the ideal person or society looks like.

While philosophy may provide us with a broad vision of the ideal person and society, we can and should also harness the power of empirical science aimed at the study of human well-being and flourishing (VanderWeele, 2017) to answer more concrete and urgent questions that can help guide public policy on personalization. Moving forward, a collaborative approach recognizing shared interests of firms, citizens and society will be critical. Platform-based firms who care about the ethical use of personalization can demonstrate their commitment by fostering research collaborations with academics. These collaborations should not only involve sharing user data, but, most crucially, grant them experimental access to the platform systems used in personalization. These considerations suggest that change may be needed at the level of corporate law and governance to better align corporate objectives with human values (Stout, 2012).

In other work we have highlighted the potential of developing experimental sandboxes to guide research and formulate evidence-based public policy on personalization and its impact on persons and society (Greene, Martens, & Shmueli, 2022; Greene, Shmueli, & Ray, 2022). Due to the complex feedback loops created by platform personalization algorithms, achieving an accurate understanding of their safety and impact on human users will require actually deploying and observing their effects on consenting participants within the actual platform environment. The sandbox concept also complements simulation-based approaches to platform risk assessment, auditing, and safety testing, by providing empirical confirmation and discovery of new forms of human-machine interactions, risk factors, and pathologies of short and long-term individual and collective behavior on platforms.

What is an experimental sandbox? In line with the auditing requirements of recent EU regulations covering platforms and AI systems, the experimental sandbox concept fuses the "regulatory sandbox" (Jeník & Duff, 2020) and the "virtual sandbox" concept used in computer cybersecurity applications (Bishop, 2002). The experimental sandbox also incorporates the staggered design of clinical trials, which pose increasingly rigorous evidential hurdles and often involve data monitoring committees to ensure the safety and efficacy of trials (Ellenberg et al., 2019). Importantly, the experimental sandbox would allow impartial external researchers and academics to contribute their expertise by developing and deploying experimental treatments, including testing new forms of algorithms, to groups of consenting platform users in carefully monitored digital environments. We are not alone in voicing the need to develop new experimental infrastructures. Behavioral scientists are now also pushing for a shared, publicly accessible experimental infrastructure to help identify heterogeneity in treatment effects and find vulnerable subpopulations for a variety of interventions (Bryan et al., 2021).

6.2 Implications for Platform-Based Firms

For platform-based firms, a major implication of our analysis is the need to carefully evaluate how to extract the benefits of personalization technologies while still avoiding undesirable social consequences. As our discussion of BMOD makes clear, when algorithms interact with and learn from humans, they may begin to exploit vulnerabilities or influence their behavior in ways that they do not reflectively endorse. Ideally, firms would invest in building data science teams tasked with auditing personalization algorithms to detect and mitigate unethical agent behavior before deployment with real users. Here, advances in simulation techniques could also help mitigate the need to train artificial agents without real-world interactions with humans. Interdisciplinary data science teams that include both social scientists and philosophers might also be useful step in achieving alignment. Together, these efforts could broadly be viewed as new forms of corporate social responsibility for major platform-based businesses.

Firms will also need to more clearly communicate personal data policies and offer transparency into how and why personal data is used for personalization. Greater transparency would not only empower platform users, but it could also improve the scientific rigor of platform-based research (Greene et al., 2022). Indeed, data protection regulations such as the GDPR already require such purpose specification (Greene et al., 2019). There is also an emerging AI accountability literature that focuses on new methods for disclosing how large-scale reinforcement learning-based systems operate (Gilbert et al., 2022). Yet unless platforms are forced by regulators to disclose such information, it is unlikely that they will voluntarily do so, citing reasons of trade secrecy (Rudin, 2019).

As briefly mentioned earlier, we also see much potential in the power of explainable AI to help persons better understand why and how personalized decisions were made. These explanations could improve platform user autonomy in a variety of domains from credit lending to product recommendations. Notably, addressing business concerns of trade secrecy, there exist "post-hoc" explanation methods that do not require disclosing algorithm details, making it harder to carry out adversarial attacks or "game the algorithm." At the same time, however, monetary incentives could influence companies to provide low-quality explanations (Greene et al., 2023). Future research should investigate the quality and properties of algorithmic explanations with an eye towards guiding future regulation.

Ultimately, ensuring that personalization is a force for good will require proactive efforts by firms, regulators, researchers and civil society to develop governance frameworks, technical solutions, evaluation methods and incentives promoting responsible personalization. Following the example set by psychological science (Lilienfeld et al., 2009; Miller, 1969), if personalization is to improve the human condition, then we should aim to "give away" the knowledge generated by data science. In other words, rather than platforms narrowly exploiting information asymmetries and users' cognitive vulnerabilities for reasons of competition and profit, we should aim to give away knowledge about human behavior for the public good so people can make better decisions, become better people, and lead better lives. Arguably this is only possible if platform technologies are aligned with users' long-term goals, which can only be known through conscious deliberation and honest self-reflection. No doubt many people may not be accustomed to engaging in such introspection and public discussion about the purposes of AI-based technology. But it is the uniquely human capacity to reflect on who we are and what we want that makes us persons after all.

7. CONCLUSION

In this chapter we investigated platform-based personalization by honing in on what is ostensibly the central object of personalization: the person. We concluded that personalization is both a multi-faceted concept and an AI-driven technology that may promote harmful as well as beneficial ends. Taking a philosophical perspective, we defended the thesis that personalization, in order to promote the human good, requires a substantive vision of the human person and a good society. Along these lines, we proposed several tentative principles for aligning personalization with ideas from democratic ethical and political theory. We also discussed future trends and implications for platform-based businesses, highlighting the role of AI-related legislation and the resulting opportunities for independent academic researchers to help audit the safety of personalization and evaluate its long-term impact on persons and society.

This chapter's humanistic, philosophical view of platform-based personalization departs from the prevalent economics- and engineering-based approaches that often neglect the role of the person in personalization, and moreover, tend to make a number of implicit assumptions about the human person that many philosophers would find problematic. By contrasting these perspectives and highlighting the overlooked conflicts and tensions between them, we hope to spur interdisciplinary research addressing the fundamentally sociotechnical nature of platform-based personalization.

In addition, due to the complexity of the concept of personalization and its varied definitions and applications, this chapter argued for a new unifying conception of personalization as a form of control. Specifically, we suggested viewing personalization as a goal-directed process of taking actions, making recommendations or decisions, or allocating resources on the basis of measured states of humans, the digital environment, and their interactions, and possibly adaptively modifying these actions after observing feedback about their consequences. We believe our notion of personalization as control helps to make sense of the various computational, commercial, and ethical aspects of personalization currently treated more or less independently across various research streams.

In summary, applying a philosophical lens to personalization clarifies how new forms of AI raise timeless political questions of power, control, and legitimacy, while also underscoring the inevitable conflicts between the interests and goals of individual persons and social collectives. How is it decided, and who decides, to which ends personalization is applied? This question is, we submit, central to an ethics of personalization.

REFERENCES

Adomavicius, G., & Tuzhilin, A. (2005). Personalization technologies: A process-oriented perspective. *Communications of the ACM*, *48*(10), 83–90. doi:10.1145/1089107.1089109

Aggarwal, C. C. (2016). *Recommender systems*. Springer. doi:10.1007/978-3-319-29659-3

Aguirre, E., Mahr, D., Grewal, D., De Ruyter, K., & Wetzels, M. (2015). Unraveling the personalization paradox: The effect of information collection and trust-building strategies on online advertisement effectiveness. *Journal of Retailing*, *91*(1), 34–49. doi:10.1016/j.jretai.2014.09.005

Aizenberg, E., & van den Hoven, J. (2020). Designing for human rights in AI. *Big Data & Society*, *7*(2). doi:10.1177/2053951720949566

Albrecht, J. P. (2016). How the GDPR will change the world. *Eur. Data Prot. L. Rev.*, *2*(3), 287–289. doi:10.21552/EDPL/2016/3/4

Anagnostopoulos, I., Zeadally, S., & Exposito, E. (2016). Handling big data: Research challenges and future directions. *The Journal of Supercomputing*, *72*(4), 1494–1516. doi:10.100711227-016-1677-z

Ashby, W. R. (1957). *An introduction to cybernetics*. Chapman & Hall Ltd.

Ashmore, R. D., & Jussim, L. (1997). *Self and Identity: Fundamental Issues*. Oxford University Press.

Bak-Coleman, J. B., Alfano, M., Barfuss, W., Bergstrom, C. T., Centeno, M. A., Couzin, I. D., Donges, J. F., Galesic, M., Gersick, A. S., Jacquet, J., Kao, A. B., Moran, R. E., Romanczuk, P., Rubenstein, D. I., Tombak, K. J., Van Bavel, J. J., & Weber, E. U. (2021). Stewardship of global collective behavior. *Proceedings of the National Academy of Sciences of the United States of America*, *118*(27), e2025764118. doi:10.1073/pnas.2025764118 PMID:34155097

Balkin, J. (2017). 2016 Sidley Austin Distinguished Lecture on Big Data Law and Policy: The Three Laws of Robotics in the Age of Big Data. *Ohio State Law Journal*, *78*, 1217.

Bates, D. (2005). Crisis between the Wars: Derrida and the Origins of Undecidability. *Representations (Berkeley, Calif.)*, *90*(1), 1–27. doi:10.1525/rep.2005.90.1.1

Baum, W. M. (2017). *Understanding behaviorism: Behavior, culture, and evolution*. John Wiley & Sons. doi:10.1002/9781119143673

Baumeister, R. F. (2005). *The Cultural Animal: Human Nature, Meaning, and Social Life*. Oxford University Press. doi:10.1093/acprof:oso/9780195167030.001.0001

Beam, M. A., Hutchens, M. J., & Hmielowski, J. D. (2018). Facebook news and (de) polarization: Reinforcing spirals in the 2016 US election. *Information Communication and Society*, *21*(7), 940–958. doi:10.1080/1369118X.2018.1444783

Ben-Shahar, O., & Porat, A. (2021). *Personalized law: Different rules for different people*. Oxford University Press. doi:10.1093/oso/9780197522813.001.0001

Benn, S. I. (1975). Freedom, Autonomy and the Concept of a Person. *Proceedings of the Aristotelian Society*, *76*(1), 109–130. doi:10.1093/aristotelian/76.1.109

Berdichevsky, D., & Neuenschwander, E. (1999). Toward an ethics of persuasive technology. *Communications of the ACM*, *42*(5), 51–58. doi:10.1145/301353.301410

Berkovsky, S., Freyne, J., & Oinas-Kukkonen, H. (2012). Influencing individually: Fusing personalization and persuasion. [TiiS]. *ACM Transactions on Interactive Intelligent Systems*, *2*(2), 1–8. doi:10.1145/2209310.2209312

Berridge, K. C. (2009). Wanting and liking: Observations from the neuroscience and psychology laboratory. *Inquiry*, *52*(4), 378–398. doi:10.1080/00201740903087359 PMID:20161627

Beshears, J., Choi, J. J., Laibson, D., & Madrian, B. C. (2008). How are preferences revealed? *Journal of Public Economics*, *92*(8–9), 1787–1794. doi:10.1016/j.jpubeco.2008.04.010 PMID:24761048

Bird, S., Barocas, S., Crawford, K., Diaz, F., & Wallach, H. (2016). Exploring or Exploiting? Social and Ethical Implications of Autonomous Experimentation in AI. *Workshop on Fairness, Accountability, and Transparency in Machine Learning*. https://ssrn.com/abstract=2846909

Bishop, M. (2002). *Computer Security: Art and Science*. Addison-Wesley.

Blumer, H. (1986). *Symbolic Interactionism: Perspective and Method*. University of California Press.

Bostrom, N. (2012). The superintelligent will: Motivation and instrumental rationality in advanced artificial agents. *Minds and Machines*, *22*(2), 71–85. doi:10.100711023-012-9281-3

Botvinick, M., Wang, J. X., Dabney, W., Miller, K. J., & Kurth-Nelson, Z. (2020). Deep reinforcement learning and its neuroscientific implications. *Neuron*, *107*(4), 603–616. doi:10.1016/j.neuron.2020.06.014 PMID:32663439

Boulding, K. E. (1956). General systems theory—The skeleton of science. *Management Science*, *2*(3), 197–208. doi:10.1287/mnsc.2.3.197

Bouneffouf, D., Rish, I., & Aggarwal, C. (2020). Survey on applications of multi-armed and contextual bandits. *2020 IEEE Congress on Evolutionary Computation (CEC)*, (pp. 1–8). IEEE. 10.1109/CEC48606.2020.9185782

Bratman, M. E. (2000). Reflection, planning, and temporally extended agency. *The Philosophical Review*, *109*(1), 35–61. doi:10.1215/00318108-109-1-35

Brinkmann, S. (2005). Human kinds and looping effects in psychology: Foucauldian and hermeneutic perspectives. *Theory & Psychology*, *15*(6), 769–791. doi:10.1177/0959354305059332

Broome, J. (1984). Selecting people randomly. *Ethics*, *95*(1), 38–55. doi:10.1086/292596 PMID:11651785

Bruner, J. (1991). The narrative construction of reality. *Critical Inquiry*, *18*(1), 1–21. doi:10.1086/448619

Brusilovsky, P., & Maybury, M. T. (2002). From adaptive hypermedia to the adaptive web. *Communications of the ACM*, *45*(5), 30–33. doi:10.1145/506218.506239

Bryan, C. J., Tipton, E., & Yeager, D. S. (2021). Behavioural science is unlikely to change the world without a heterogeneity revolution. *Nature Human Behaviour*, *5*(8), 980–989. doi:10.103841562-021-01143-3 PMID:34294901

Burr, C., Cristianini, N., & Ladyman, J. (2018). An Analysis of the Interaction Between Intelligent Software Agents and Human Users. *Minds and Machines*, *28*(4), 735–774. doi:10.100711023-018-9479-0 PMID:30930542

Busch, C., & Mak, V. (2021). Putting the Digital Services Act into Context: Bridging the Gap between EU Consumer Law and Platform Regulation. *Available at SSRN 3933675*.

Cai, L., Wu, C., Meimandi, K. J., & Gerber, M. S. (2017). Adaptive mobile behavior change intervention using reinforcement learning. *2017 International Conference on Companion Technology (ICCT)*, (pp. 1–2). IEEE. 10.1109/COMPANION.2017.8287078

Card, D., & Smith, N. A. (2020). On Consequentialism and Fairness. *Frontiers in Artificial Intelligence*, *3*, 34. doi:10.3389/frai.2020.00034 PMID:33733152

Caudill, E. M., & Murphy, P. E. (2000). Consumer online privacy: Legal and ethical issues. *Journal of Public Policy & Marketing*, *19*(1), 7–19. doi:10.1509/jppm.19.1.7.16951

Chabris, C. F., Laibson, D., Morris, C. L., Schuldt, J. P., & Taubinsky, D. (2008). Individual laboratory-measured discount rates predict field behavior. *Journal of Risk and Uncertainty*, *37*(2-3), 237–269. doi:10.100711166-008-9053-x PMID:19412359

Chambers, C. P., & Echenique, F. (2016). *Revealed preference theory* (Vol. 56). Cambridge University Press.

Chen, M., Beutel, A., Covington, P., Jain, S., Belletti, F., & Chi, E. H. (2019). Top-k off-policy correction for a REINFORCE recommender system. *Proceedings of the Twelfth ACM International Conference on Web Search and Data Mining*, (pp. 456–464). ACM. 10.1145/3289600.3290999

Cheng, L., Liu, F., & Yao, D. (2017). Enterprise data breach: Causes, challenges, prevention, and future directions. *Wiley Interdisciplinary Reviews. Data Mining and Knowledge Discovery*, *7*(5), 1211. doi:10.1002/widm.1211

Churchill, E. F. (2013). Putting the person back into personalization. *Interaction*, *20*(5), 12–15. doi:10.1145/2512050.2504847

Citron, D. K. (2007). Technological due process. *Wash. UL Rev.*, *85*, 1249.

Clarke, R. (1994). Human identification in information systems: Management challenges and public policy issues. *Information Technology & People*, *7*(4), 6–37. doi:10.1108/09593849410076799

Conant, R. C., & Ross Ashby, W. (1970). Every good regulator of a system must be a model of that system. *International Journal of Systems Science*, *1*(2), 89–97. doi:10.1080/00207727008920220

Cooley, C. H. (1902). Looking-glass self. *The Production of Reality: Essays and Readings on Social Interaction*, *6*, 126–128.

Coors, C. (2010). Headwind from Europe: The new position of the German courts on personality rights after the judgment of the European Court of Human Rights. *German Law Journal*, *11*(5), 527–537. doi:10.1017/S207183220001868X

Cosley, D., Lam, S. K., Albert, I., Konstan, J. A., & Riedl, J. (2003). Is seeing believing? How recommender system interfaces affect users' opinions. *Proceedings of the SIGCHI Conference on Human Factors in Computing Systems*, (pp. 585–592). IEEE.

Covington, P., Adams, J., & Sargin, E. (2016). Deep neural networks for youtube recommendations. *RecSys 2016 - Proceedings of the 10th ACM Conference on Recommender Systems*, (pp. 191–198). Doi:10.1145/2959100.2959190

Cranor, C. (1975). Toward a Theory of Respect for Persons. *American Philosophical Quarterly*, *12*(4), 309–319.

Cremonesi, P., Koren, Y., & Turrin, R. (2010). Performance of recommender algorithms on top-n recommendation tasks. *Proceedings of the Fourth ACM Conference on Recommender Systems*, (pp. 39–46). ACM. 10.1145/1864708.1864721

Cristianini, N., Scantamburlo, T., & Ladyman, J. (2021). The social turn of artificial intelligence. *AI & Society*.

D'Amour, A., Srinivasan, H., Atwood, J., Baljekar, P., Sculley, D., & Halpern, Y. (2020). Fairness is not static: deeper understanding of long term fairness via simulation studies. *Proceedings of the 2020 Conference on Fairness, Accountability, and Transparency*, (pp. 525–534). ACM. 10.1145/3351095.3372878

Damiani, E. S., Di Vimercati, D. C., & Samarati, P. (2003). Managing multiple and dependable identities. *IEEE Internet Computing*, *7*(6), 29–37. doi:10.1109/MIC.2003.1250581

Dan, O., & Loewenstein, Y. (2019). From choice architecture to choice engineering. *Nature Communications*, *10*(1), 1–4. doi:10.103841467-019-10825-6 PMID:31243285

Darwall, S. L. (1977). Two Kind of Respect. *Ethics*, *88*(1), 36–49. doi:10.1086/292054

De Gregorio, G., & Celeste, E. (2022). Digital Humanism: The Constitutional Message of the GDPR. *Global Privacy Law Review, 3*(1).

De Vries, K. (2010). Identity, profiling algorithms and a world of ambient intelligence. *Ethics and Information Technology, 12*(1), 71–85. doi:10.100710676-009-9215-9

Deacon, T. W. (1997). The symbolic species: The co-evolution of language and the brain (Issue 202). WW Norton & Company.

DeCew, J. W. (1986). The scope of privacy in law and ethics. *Law and Philosophy*, 145–173.

Deisenroth, M. P., Faisal, A. A., & Ong, C. S. (2020). *Mathematics for machine learning*. Cambridge University Press. doi:10.1017/9781108679930

den Hengst, F., Grua, E. M., el Hassouni, A., & Hoogendoorn, M. (2020). Reinforcement learning for personalization: A systematic literature review. *Data Science, 3*(2), 107–147. doi:10.3233/DS-200028

Dennett, D. C. (1976). Conditions of Personhood. In A. O. Rorty (Ed.), *The Identities of Persons* (pp. 175–196). University of California Press.

Dennett, D. C. (2015). *Elbow Room: The Varieties of Free Will Worth Wanting*. MIT Press. doi:10.7551/mitpress/10470.001.0001

Descartes, R. (1999). *Discourse on method and meditations on first philosophy*. Hackett Publishing.

Dewey, J. (1927). *The Public and its Problems* (Vol. H). Holt and Company.

Dezfouli, A., Nock, R., & Dayan, P. (2020). Adversarial vulnerabilities of human decision-making. *Proceedings of the National Academy of Sciences of the United States of America, 117*(46), 29221–29228. doi:10.1073/pnas.2016921117 PMID:33148802

Dilthey, W. (1989). *Introduction to the Human Sciences* (Vol. 1). Princeton University Press.

Domingos, P. (2012). A few useful things to know about machine learning. *Communications of the ACM, 55*(10), 78–87. doi:10.1145/2347736.2347755

Donald, M. (1991). *Origins of the modern mind: Three stages in the evolution of culture and cognition*. Harvard University Press.

Douglas, M., & Ney, S. (1998). *Missing persons: A critique of the personhood in the social sciences*. University of California Press.

Driver, J. (2011). *Consequentialism*. Routledge. doi:10.4324/9780203149256

Dutt, R., Deb, A., & Ferrara, E. (2018). "Senator, We Sell Ads": Analysis of the 2016 Russian Facebook Ads Campaign. *Communications in Computer and Information Science, 941*, 151–168. doi:10.1007/978-981-13-3582-2_12

Dworkin, R. (2013). *Taking Rights Seriously*. A&C Black.

Ekstrand, M. D., & Willemsen, M. C. (2016). Behaviorism is not enough: better recommendations through listening to users. *Proceedings of the 10th ACM Conference on Recommender Systems*, (pp. 221–224). 10.1145/2959100.2959179

Ellenberg, S. S., Fleming, T. R., & DeMets, D. L. (2019). *Data monitoring committees in clinical trials: a practical perspective*. John Wiley & Sons. doi:10.1002/9781119512684

Etzioni, A. (1996). The responsive community: A communitarian perspective. *American Sociological Review*, *61*(1), 1–11. doi:10.2307/2096403

Everitt, T., Hutter, M., Kumar, R., & Krakovna, V. (2021). Reward tampering problems and solutions in reinforcement learning: A causal influence diagram perspective. *Synthese*, *198*(S27), 1–33. doi:10.100711229-021-03141-4

Eyal, N. (2014). *Hooked: How to build habit-forming products*. Penguin.

Fan, H., & Poole, M. S. (2006). What is personalization? Perspectives on the design and implementation of personalization in information systems. *Journal of Organizational Computing and Electronic Commerce*, *16*(3–4), 179–202. doi:10.120715327744joce1603&4_2

Fernández-Loría, C., & Provost, F. (2022). Causal decision making and causal effect estimation are not the same… and why it matters. *INFORMS Journal on Data Science, 1*(1).

Fisher, D. H. (1987). Knowledge acquisition via incremental conceptual clustering. *Machine Learning*, *2*(2), 139–172. doi:10.1007/BF00114265

Floridi, L. (2016). On human dignity as a foundation for the right to privacy. *Philosophy & Technology*, *29*(4), 307–312. doi:10.100713347-016-0220-8

Floridi, L. (2022). The European Legislation on AI: A brief analysis of its philosophical approach. In The 2021 Yearbook of the Digital Ethics Lab (pp. 1–8). Springer. doi:10.1007/978-3-031-09846-8_1

Floridi, L., Cowls, J., Beltrametti, M., Chatila, R., Chazerand, P., Dignum, V., Luetge, C., Madelin, R., Pagallo, U., Rossi, F., Schafer, B., Valcke, P., & Vayena, E. (2018). AI4People—An ethical framework for a good AI society: Opportunities, risks, principles, and recommendations. *Minds and Machines*, *28*(4), 689–707. doi:10.100711023-018-9482-5 PMID:30930541

Fogg, B. J. (2003). *Persuasive Technology: Using Computers to Change What We Think and Do*. Elsevier. doi:10.1016/B978-155860643-2/50011-1

Frankfurt, H. G. (1971). Freedom of the Will and the Concept of a Person. *The Journal of Philosophy*, *68*(1), 5. doi:10.2307/2024717

Frankfurt, H. G. (2006). *Taking ourselves seriously and getting it right*. Stanford University Press.

Friston, K. (2010). The free-energy principle: A unified brain theory? *Nature Reviews. Neuroscience*, *11*(2), 127–138. doi:10.1038/nrn2787 PMID:20068583

Fukuyama, F. (2018). *Identity: The demand for dignity and the politics of resentment*. Farrar, Straus and Giroux.

Fuller, L. (1964). *The Morality of Law*. Yale University Press.

Fung, A. (2013). The Principle of Affected Interests: An Interpretation and Defense. In *Representation* (pp. 236–268). University of Pennsylvania Press. doi:10.9783/9780812208177.236

Furuta, H., Matsushima, T., Kozuno, T., Matsuo, Y., Levine, S., Nachum, O., & Gu, S. S. (2021). Policy information capacity: Information-theoretic measure for task complexity in deep reinforcement learning. *International Conference on Machine Learning*, (pp. 3541–3552).

Gabriel, I. (2020). Artificial intelligence, values, and alignment. *Minds and Machines, 30*(3), 411–437. doi:10.100711023-020-09539-2

Gabriel, I. (2022). Toward a Theory of Justice for Artificial Intelligence. *Daedalus, 151*(2), 218–231. doi:10.1162/daed_a_01911

Gallie, W. B. (1955). Essentially contested concepts. *Proceedings of the Aristotelian Society, 56*(1), 167–198. doi:10.1093/aristotelian/56.1.167

Gauci, J., Ghavamzadeh, M., Honglei, L., & Nahmias, R. (2019, October 19). *Open-sourcing ReAgent, a modular, end-to-end platform for building reasoning systems*. Facebook AI. https://ai.facebook.com/blog/open-sourcing-reagent-a-platform-for-reasoning-systems/

Gergen, K. J. (1985). The social constructionist movement in modern psychology. *The American Psychologist, 40*(3), 266–275. doi:10.1037/0003-066X.40.3.266

Gilbert, T. K., Lambert, N., Dean, S., Zick, T., & Snoswell, A. (2022). *Reward reports for reinforcement learning*. arXiv preprint arXiv:2204.10817.

Gomez-Uribe, C. A., & Hunt, N. (2015). The Netflix Recommender System: Algorithms. *Business Value, 6*.

Goodfellow, I., Bengio, Y., & Courville, A. (2016). *Deep learning*. MIT press.

Gottesman, O., Johansson, F., Komorowski, M., Faisal, A., Sontag, D., Doshi-Velez, F., & Celi, L. A. (2019). Guidelines for reinforcement learning in healthcare. *Nature Medicine 2019 25:1, 25*(1), 16–18. doi:10.1038/s41591-018-0310-5

Gottfredson, L. S. (1997). Mainstream science on intelligence: An editorial with 52 signatories, history, and bibliography. [). JAI.]. *Intelligence, 24*(1), 13–23. doi:10.1016/S0160-2896(97)90011-8

Gray, C. M., Kou, Y., Battles, B., Hoggatt, J., & Toombs, A. L. (2018). The dark (patterns) side of UX design. *Proceedings of the 2018 CHI Conference on Human Factors in Computing Systems*, (pp. 1–14). ACM. 10.1145/3173574.3174108

Greene, T., Dhurandhar, A., & Shmueli, G. (2022). Atomist or holist? A diagnosis and vision for more productive interdisciplinary AI ethics dialogue. *Patterns*, 100652.

Greene, T., Goethals, S., Martens, D., & Shmueli, G. (2023). *Monetizing Explainable AI: A Double-edged Sword*. arXiv preprint arXiv:2304.06483.

Greene, T., Martens, D., & Shmueli, G. (2022). Barriers to academic data science research in the new realm of algorithmic behaviour modification by digital platforms. *Nature Machine Intelligence*, *4*(4), 323–330. doi:10.103842256-022-00475-7

Greene, T., Shmueli, G., Fell, J., Lin, C.-F., & Liu, H.-W. (2022). Forks over knives: Predictive inconsistency in criminal justice algorithmic risk assessment tools. *Journal of the Royal Statistical Society Series A: Statistics in Society, 185*(Supplement_2), S692–S723.

Greene, T., Shmueli, G., & Ray, S. (2022). (Forthcoming). Taking the Person Seriously: Ethically-aware IS Research in the Era of Reinforcement Learning-based Personalization. *JAIS Preprints*, 77.

Greene, T., Shmueli, G., Ray, S., & Fell, J. (2019). Adjusting to the GDPR: The Impact on Data Scientists and Behavioral Researchers. *Big Data*, *7*(3), 140–162. doi:10.1089/big.2018.0176 PMID:31033336

Habermas, J. (1990). *Moral Consciousness and Communicative Action*. MIT Press.

Habermas, J. (2015). *Knowledge and Human Interests*. John Wiley & Sons.

Hacking, I. (2007). Kinds of People: Moving Targets. *Proceedings of the British Academy*, 286–318.

Haidt, J. (2008). Morality. *Perspectives on Psychological Science*, *3*(1), 65–72. doi:10.1111/j.1745-6916.2008.00063.x PMID:26158671

Hansotia, B., & Rukstales, B. (2002). Incremental value modeling. *Journal of Interactive Marketing*, *16*(3), 35–46. doi:10.1002/dir.10035

Harris, M. (1976). History and significance of the emic/etic distinction. *Annual Review of Anthropology*, *5*(1), 329–350. doi:10.1146/annurev.an.05.100176.001553

Hart, H. L. A. (1961). *The Concept of Law*. Oxford University Press.

Hartzog, W. (2021). What is Privacy? That's the Wrong Question. *U. Chi. L. Rev.*, *88*, 1677.

Hastie, T., Tibshirani, R., Friedman, J. H., & Friedman, J. H. (2009). *The elements of statistical learning: data mining, inference, and prediction* (Vol. 2). Springer. doi:10.1007/978-0-387-84858-7

Häubl, G., & Trifts, V. (2000). Consumer decision making in online shopping environments: The effects of interactive decision aids. *Marketing Science*, *19*(1), 4–21. doi:10.1287/mksc.19.1.4.15178

Haugen, F. (2021, October 4). *Statement of Frances Haugen*. Whistleblower Aid. https://www.commerce.senate.gov/services/files/589FC8A558E-824E-4914-BEDB-3A7B1190BD49

He, X., Pan, J., Jin, O., Xu, T., Liu, B., Xu, T., Atallah, A., Herbrich, R., Bowers, S., & Candela, J. (2014). Practical lessons from predicting clicks on ads at facebook. *Proceedings of the Eighth International Workshop on Data Mining for Online Advertising*, (pp. 1–9). ACM. 10.1145/2648584.2648589

Heidegger, M. (2010). *Being and Time*.

Helberger, N., Karppinen, K., & D'acunto, L. (2018). Exposure diversity as a design principle for recommender systems. *Information Communication and Society*, *21*(2), 191–207. doi:10.1080/1369118X.2016.1271900

Helfat, C. E., Finkelstein, S., Mitchell, W., Peteraf, M., Singh, H., Teece, D., & Winter, S. G. (2009). *Dynamic capabilities: Understanding strategic change in organizations*. John Wiley & Sons.

Helmond, A. (2015). The platformization of the web: Making web data platform ready. *Social Media+ Society, 1*(2), 2056305115603080.

Hijmans, H., & Raab, C. D. (2018 Forthcoming). *Ethical Dimensions of the GDPR. Commentary on the General Data Protection Regulation*. Edward Elgar.

Hildebrandt, M. (2015). *Smart technologies and the end (s) of law: novel entanglements of law and technology*. Edward Elgar Publishing. doi:10.4337/9781849808774

Hildebrandt, M., & Gutwirth, S. (2008). *Profiling the European Citizen*. Springer. doi:10.1007/978-1-4020-6914-7

Hills, T. T., Todd, P. M., Lazer, D., Redish, A. D., & Couzin, I. D. (2015). Exploration versus exploitation in space, mind, and society. *Trends in Cognitive Sciences, 19*(1), 46–54. doi:10.1016/j.tics.2014.10.004 PMID:25487706

Hinton, G. E., & Salakhutdinov, R. R. (2006). Reducing the dimensionality of data with neural networks. *Science, 313*(5786), 504–507. doi:10.1126cience.1127647 PMID:16873662

Hofree, G., & Winkielman, P. (2012). 13. On (not) knowing and feeling what we want and like. Handbook of Self-Knowledge, 210–224.

Holm, A. B., & Günzel-Jensen, F. (2017). Succeeding with freemium: Strategies for implementation. *The Journal of Business Strategy, 38*(2), 16–24. doi:10.1108/JBS-09-2016-0096

Hosanagar, K. (2020). *A Human's Guide to Machine Intelligence: How Algorithms are Shaping Our Lives and How We can Stay in Control*. Penguin.

Hutter, M. (2004). *Universal artificial intelligence: Sequential decisions based on algorithmic probability*. Spring Science & Business Media.

IEEE Standards Association. (2017). *Ethically aligned design: A vision for prioritizing human well-being with autonomous and intelligent systems, Overview: Version 2*. IEEE.

Jackson, F. (1986). What Mary Didn't Know. *The Journal of Philosophy, 83*(5), 291–295. doi:10.2307/2026143

Jameson, J. L., & Longo, D. L. (2015). Precision medicine—Personalized, problematic, and promising. *Obstetrical & Gynecological Survey, 70*(10), 612–614. doi:10.1097/01.ogx.0000472121.21647.38 PMID:26014593

Japkowicz, N., & Shah, M. (2011). *Evaluating learning algorithms: a classification perspective*. Cambridge University Press. doi:10.1017/CBO9780511921803

Jeník, I., & Duff, S. (2020). *How to Build a Regulatory Sandbox*.

Joachims, T., London, B., Su, Y., Swaminathan, A., & Wang, L. (2021). Recommendations as treatments. *AI Magazine, 42*(3), 19–30. doi:10.1609/aimag.v42i3.18141

Jost, J. T. (2019). A quarter century of system justification theory: Questions, answers, criticisms, and societal applications. *British Journal of Social Psychology*, *58*(2), 263–314. doi:10.1111/bjso.12297

Kahneman, D., & Tversky, A. (1979). Prospect Theory: An Analysis of Decision under Risk. *Econometrica*, *47*(2), 263–292. doi:10.2307/1914185

Kallus, N. (2017). Recursive partitioning for personalization using observational data. *International Conference on Machine Learning*, (pp. 1789–1798). IEEE.

Kamenica, E. (2019). Bayesian persuasion and information design. *Annual Review of Economics*, *11*(1), 249–272. doi:10.1146/annurev-economics-080218-025739

Kaminskas, M., & Bridge, D. (2016). Diversity, serendipity, novelty, and coverage: A survey and empirical analysis of beyond-accuracy objectives in recommender systems. *ACM Transactions on Interactive Intelligent Systems*, *7*(1), 1–42. doi:10.1145/2926720

Kane, G. C., Young, A. G., Majchrzak, A., & Ransbotham, S. (2021). Avoiding an Oppressive Future of Machine Learning: A Design Theory for Emancipatory Assistants. *Management Information Systems Quarterly*, *45*(1), 371–396. doi:10.25300/MISQ/2021/1578

Kane, R. (2001). *Free will*. John Wiley & Sons.

Kant, I. (1948). *Groundwork of the Metaphysics of Morals* (H. J. Paton, Trans.). Hutchinson.

Kelleher, J. D., Mac Namee, B., & D'arcy, A. (2020). *Fundamentals of machine learning for predictive data analytics: algorithms, worked examples, and case studies*. MIT press.

Kent, W. (2012). *Data and Reality: A Timeless Perspective on Perceiving and Managing Information*. Technics publications.

Korsgaard, C., Cohen, G. A., Geuss, R., Nagel, T., & Williams, B. (1996). *The Sources of Normativity* (O. O'Neill, Ed.). Cambridge University Press. doi:10.1017/CBO9780511554476

Koster, R., Balaguer, J., Tacchetti, A., Weinstein, A., Zhu, T., Hauser, O., Williams, D., Campbell-Gillingham, L., Thacker, P., Botvinick, M., & Summerfield, C. (2022). Human-centred mechanism design with Democratic AI. *Nature Human Behaviour*, *6*(10), 1398–1407. doi:10.103841562-022-01383-x PMID:35789321

Kotkov, D., Wang, S., & Veijalainen, J. (2016). A survey of serendipity in recommender systems. *Knowledge-Based Systems*, *111*, 180–192. doi:10.1016/j.knosys.2016.08.014

Krizhevsky, A., Sutskever, I., & Hinton, G. E. (2012). Imagenet classification with deep convolutional neural networks. *Advances in Neural Information Processing Systems*, *25*, 1097–1105.

Kroll, J., Barocas, S., Felten, E., Reidenberg, J., Robinson, D., & Yu, H. (2017). Accountable Algorithms. *University of Pennsylvania Law Review*, *165*, 633–705.

Langley, P., & Leyshon, A. (2017). Platform capitalism: The intermediation and capitalization of digital economic circulation. *Finance and Stochastics*, *3*(1), 11–31.

Lehrer, K. (1997). Freedom, preference and autonomy. *The Journal of Ethics*, *1*(1), 3–25. doi:10.1023/A:1009744817791

Leidner, D. E., & Tona, O. (2021). The CARE Theory of Dignity Amid Personal Data Digitalization. *Management Information Systems Quarterly*, *45*(1), 343–370. doi:10.25300/MISQ/2021/15941

Leiva, L. A., Arapakis, I., & Iordanou, C. (2021). My mouse, my rules: Privacy issues of behavioral user profiling via mouse tracking. *Proceedings of the 2021 Conference on Human Information Interaction and Retrieval*, (pp. 51–61). IEEE. 10.1145/3406522.3446011

Lerner, A. P. (1972). The economics and politics of consumer sovereignty. *The American Economic Review*, *62*(1/2), 258–266.

Levine, S. (2021). Understanding the World Through Action. *ArXiv Preprint ArXiv:2110.12543*.

Liang, T.-P., Lai, H.-J., & Ku, Y.-C. (2006). Personalized content recommendation and user satisfaction: Theoretical synthesis and empirical findings. *Journal of Management Information Systems*, *23*(3), 45–70. doi:10.2753/MIS0742-1222230303

Lilienfeld, S. O., Ammirati, R., & Landfield, K. (2009). Giving debiasing away: Can psychological research on correcting cognitive errors promote human welfare? *Perspectives on Psychological Science*, *4*(4), 390–398. doi:10.1111/j.1745-6924.2009.01144.x PMID:26158987

Lindblom, C. E. (1959). The Science of "Muddling Through." *Public Administraation Review*, 79–88.

Lindström, B., Bellander, M., Schultner, D. T., Chang, A., Tobler, P. N., & Amodio, D. M. (2021). A computational reward learning account of social media engagement. *Nature Communications, 12*(1), 1–10. doi:10.1038/s41467-020-19607-x

Littman, M. L. (2015). Reinforcement learning improves behaviour from evaluative feedback. *Nature*, *521*(7553), 445–451. doi:10.1038/nature14540 PMID:26017443

Lorenz-Spreen, P., Lewandowsky, S., Sunstein, C. R., & Hertwig, R. (2020). How behavioural sciences can promote truth, autonomy and democratic discourse online. *Nature Human Behaviour*, *4*(11), 1102–1109. doi:10.103841562-020-0889-7 PMID:32541771

Macdonald, P. S. (2017). *History of the concept of mind: Volume 1: Speculations about soul, mind and spirit from homer to hume*. Routledge. doi:10.4324/9781315092959

MacIntyre, A. (1984). *After Virtue*. University of Notre Dame Press.

Mahieu, R. (2021). The right of access to personal data: A genealogy. *Technology and Regulation*, 62–75.

Mahmood, T., & Ricci, F. (2007). Learning and adaptivity in interactive recommender systems. *Proceedings of the Ninth International Conference on Electronic Commerce*, (pp. 75–84). IEEE. 10.1145/1282100.1282114

Manders-Huits, N. (2010). Practical versus moral identities in identity management. *Ethics and Information Technology*, *12*(1), 43–55. doi:10.100710676-010-9216-8

Manheim, D., & Garrabrant, S. (2018). Categorizing variants of Goodhart's Law. *ArXiv Preprint ArXiv:1803.04585*.

March, J. G. (1991). Exploration and exploitation in organizational learning. *Organization Science*, *2*(1), 71–87. doi:10.1287/orsc.2.1.71

Markus, H., & Wurf, E. (1987). The dynamic self-concept: A social psychological perspective. *Annual Review of Psychology*, *38*(1), 299–337. doi:10.1146/annurev.ps.38.020187.001503

Martens, D., Provost, F., Clark, J., & de Fortuny, E. J. (2016). Mining massive fine-grained behavior data to improve predictive analytics. *Management Information Systems Quarterly*, *40*(4), 869–888. doi:10.25300/MISQ/2016/40.4.04

Mathur, A., Acar, G., Friedman, M. J., Lucherini, E., Mayer, J., Chetty, M., & Narayanan, A. (2019). Dark patterns at scale: Findings from a crawl of 11K shopping websites. *Proceedings of the ACM on Human-Computer Interaction, 3*(CSCW), (pp. 1–32). IEEE. 10.1145/3359183

McAdams, D., & McLean, K. (2013). Narrative Identity. *Current Directions in Psychological Science*, *22*(3), 233–238. doi:10.1177/0963721413475622

McAuley, J. (2022). *Personalized Machine Learning*. Cambridge University Press. doi:10.1017/9781009003971

Medvedev, I., Gordon, T., & Wu, H. (2019, November 25). *Powered by AI: Instagram's Explore recommender system*. Meta AI Blog. https://ai.facebook.com/blog/powered-by-ai-instagrams-explore-recommender-system/

Meissner, F., Grigutsch, L. A., Koranyi, N., Müller, F., & Rothermund, K. (2019). Predicting behavior with implicit measures: Disillusioning findings, reasonable explanations, and sophisticated solutions. *Frontiers in Psychology*, *10*, 2483. doi:10.3389/fpsyg.2019.02483 PMID:31787912

Menczer, F. (2020). 4 Reasons Why Social Media Make Us Vulnerable to Manipulation. *Fourteenth ACM Conference on Recommender Systems*, (pp. 1–1). IEEE. 10.1145/3383313.3418434

Michie, S., Richardson, M., Johnston, M., Abraham, C., Francis, J., Hardeman, W., Eccles, M. P., Cane, J., Wood, C. E., Michie, S., Johnston, M., Wood, C. E., Richardson, M., Abraham, C., Francis, J., Hardeman, W., Eccles, M. P., & Cane, J. (2013). The Behavior Change Technique Taxonomy (v1) of 93 Hierarchically Clustered Techniques: Building an International Consensus for the Reporting of Behavior Change Interventions. *Annals of Behavioral Medicine*, *46*(1), 81–95. doi:10.100712160-013-9486-6 PMID:23512568

Milani, S., Topin, N., Veloso, M., & Fang, F. (2022). A survey of explainable reinforcement learning. *ArXiv Preprint ArXiv:2202.08434*.

Milano, S., Mittelstadt, B., Wachter, S., & Russell, C. (2021). Epistemic fragmentation poses a threat to the governance of online targeting. *Nature Machine Intelligence*, *3*(6), 466–472. doi:10.103842256-021-00358-3

Milano, S., Taddeo, M., & Floridi, L. (2020). Recommender systems and their ethical challenges. *AI & Society*, *35*(4), 957–967. doi:10.100700146-020-00950-y

Mill, J. S. (2015). *On Liberty, Utilitarianism and Other Essays* (P. F. R. Mark, Ed.). Oxford university press. doi:10.1093/owc/9780199670802.001.0001

Miller, G. A. (1969). Psychology as a means of promoting human welfare. *The American Psychologist, 24*(12), 1063–1075. doi:10.1037/h0028988

Miltenberger, R. G. (2016). *Behavior Modification: Principles and Procedures* (6th ed.). Cengage Learning.

Minsky, M. (1961). Steps toward artificial intelligence. *Proceedings of the IRE, 49*(1), 8–30. doi:10.1109/JRPROC.1961.287775

Möhlmann, M., Zalmanson, L., Henfridsson, O., & Gregory, R. W. (2021). Algorithmic Management of Work on Online Labor Platforms: When Matching Meets Control. *Management Information Systems Quarterly, 45*(4), 1999–2022. doi:10.25300/MISQ/2021/15333

Mouffe, C. (1999). Deliberative democracy or agonistic pluralism? *Social Research*, 745–758.

Murthi, B. P. S., & Sarkar, S. (2003). The role of the management sciences in research on personalization. *Management Science, 49*(10), 1344–1362. doi:10.1287/mnsc.49.10.1344.17313

Naffine, N. (2003). Who are law's persons? From Cheshire cats to responsible subjects. *The Modern Law Review, 66*(3), 346–367. doi:10.1111/1468-2230.6603002

Nahum-Shani, I., Smith, S. N., Spring, B. J., Collins, L. M., Witkiewitz, K., Tewari, A., & Murphy, S. A. (2018). Just-in-Time Adaptive Interventions (JITAIs) in Mobile Health: Key Components and Design Principles for Ongoing Health Behavior Support. *Annals of Behavioral Medicine, 52*(6), 446–462. doi:10.100712160-016-9830-8 PMID:27663578

Neuwirth, R. J. (2022). *The EU Artificial Intelligence Act: Regulating Subliminal AI Systems*. Taylor & Francis. doi:10.4324/9781003319436

Nord, W. R., & Peter, J. P. (1980). A behavior modification perspective on marketing. *Journal of Marketing, 44*(2), 36–47. doi:10.1177/002224298004400205

Official Journal of the European Union. (2016). *Regulation (EU) 2016/679 of the European Parliament and of the Council of 27595 April 2016 on the Protection of Natural Persons with Regard to the Processing of Personal Data and on the Free Movement of such data, and Repealing Directive 95/46/ec (General Data Protection Regulation)*. European Parliament and Council. https://eur-lex.europa.eu/eli/reg/2016/679/oj

Oshana, M. (2006). *Personal autonomy in society*. Ashgate Publishing, Ltd.

Ouyang, L., Wu, J., Jiang, X., Almeida, D., Wainwright, C., Mishkin, P., & Lowe, R. (2022). Training language models to follow instructions with human feedback. *Advances in Neural Information Processing Systems, 35*, 27730–27744.

Palmer, A., & Koenig-Lewis, N. (2009). An experiential, social network-based approach to direct marketing. *Direct Marketing: An International Journal.*

Pariser, E. (2011). *The filter bubble: How the new personalized web is changing what we read and how we think*. Penguin.

Persad, G., Wertheimer, A., & Emanuel, E. (2009). Principles for allocation of scarce medical interventions. *Lancet, 373*(9661), 423–431. doi:10.1016/S0140-6736(09)60137-9 PMID:19186274

Polkinghorne, D. E. (1988). *Narrative Knowing and the Human Sciences*. SUNY Press.

Pommeranz, A., Broekens, J., Wiggers, P., Brinkman, W.-P., & Jonker, C. M. (2012). Designing interfaces for explicit preference elicitation: A user-centered investigation of preference representation and elicitation process. *User Modeling and User-Adapted Interaction, 22*(4-5), 357–397. doi:10.100711257-011-9116-6

Posner, E., & Weyl, E. (2019). *Radical Markets*. Princeton University Press. doi:10.2307/j.ctvdf0kwg

Posner, R. A. (1981). The Economics of Privacy. *The American Economic Review, 71*(2), 405–409.

Puiutta, E., & Veith, E. M. S. P. (2020). Explainable reinforcement learning: A survey. *Machine Learning and Knowledge Extraction: 4th IFIP TC 5, TC 12, WG 8.4, WG 8.9, WG 12.9 International Cross-Domain Conference, CD-MAKE 2020, Dublin, Ireland, August 25–28, 2020. Proceedings, 4*, 77–95.

Purtova, N. (2015). The illusion of personal data as no one's property. *Law, Innovation and Technology, 7*(1), 83–111. doi:10.1080/17579961.2015.1052646

Rafieian, O., & Yoganarasimhan, H. (2022). AI and Personalization. *Available at SSRN 4123356*.

Rahwan, I., Cebrian, M., Obradovich, N., Bongard, J., Bonnefon, J.-F., Breazeal, C., Crandall, J. W., Christakis, N. A., Couzin, I. D., Jackson, M. O., Jennings, N. R., Kamar, E., Kloumann, I. M., Larochelle, H., Lazer, D., McElreath, R., Mislove, A., Parkes, D. C., Pentland, A. S., & Wellman, M. (2019). Machine behaviour. *Nature, 568*(7753), 477–486. doi:10.103841586-019-1138-y PMID:31019318

Rawls, J. (1971). *A Theory of Justice*. Harvard University Press. doi:10.4159/9780674042605

Rawls, J. (2005). *Political Liberalism*. Columbia University Press.

Ray, S., Kim, S. S., & Morris, J. G. (2012). Research note—Online Users' switching costs: Their nature and formation. *Information Systems Research, 23*(1), 197–213. doi:10.1287/isre.1100.0340

Rendle, S., Freudenthaler, C., Gantner, Z., & Schmidt-Thieme, L. (2012). BPR: Bayesian personalized ranking from implicit feedback. *ArXiv Preprint ArXiv:1205.2618*.

Rhoen, M. (2017). Rear view mirror, crystal ball: Predictions for the future of data protection law based on the history of environmental protection law. *Computer Law & Security Review, 33*(5), 603–617. doi:10.1016/j.clsr.2017.05.010

Ricoeur, P. (1994). *Oneself as Another*. University of Chicago Press.

Riedmiller, M., Springenberg, J. T., Hafner, R., & Heess, N. (2021). Collect & Infer--a fresh look at data-efficient Reinforcement Learning. *ArXiv Preprint ArXiv:2108.10273*.

Rochet, J., & Tirole, J. (2006). Two-sided markets: A progress report. *The RAND Journal of Economics, 37*(3), 645–667. doi:10.1111/j.1756-2171.2006.tb00036.x

Rorty, R. (1989). *Contingency, Irony, and Solidarity*. Cambridge University Press. doi:10.1017/CBO9780511804397

Rouvroy, A., & Poullet, Y. (2009). The right to informational self-determination and the value of self-development: Reassessing the importance of privacy for democracy. In *Reinventing Data Protection?* (pp. 45–76). Springer Netherlands. doi:10.1007/978-1-4020-9498-9_2

Rudin, C. (2019). Stop explaining black box machine learning models for high stakes decisions and use interpretable models instead. *Nature Machine Intelligence, 1*(5), 206–215. doi:10.103842256-019-0048-x PMID:35603010

Russell, S. (2019). *Human compatible: Artificial intelligence and the problem of control*. Penguin.

Russell, S. J., & Norvig, P. (2010). *Artificial intelligence: a modern approach*. Pearson Education.

Sadowski, J. (2020). The internet of landlords: Digital platforms and new mechanisms of rentier capitalism. *Antipode, 52*(2), 562–580. doi:10.1111/anti.12595

Sætra, H. S., Borgebund, H., & Coeckelbergh, M. (2022). Avoid diluting democracy by algorithms. *Nature Machine Intelligence, 4*(10), 804–806. doi:10.103842256-022-00537-w

Salakhutdinov, R., Mnih, A., & Hinton, G. (2007). Restricted Boltzmann machines for collaborative filtering. *Proceedings of the 24th International Conference on Machine Learning*, (pp. 791–798). IEEE. 10.1145/1273496.1273596

Schauer, F. (2006). *Profiles, probabilities, and stereotypes*. Harvard University Press. doi:10.4159/9780674043244

Schechtman, M. (2018). *The Constitutions of Selves*. Cornell University Press.

Selevan, S. G., Kimmel, C. A., & Mendola, P. (2000). Identifying critical windows of exposure for children's health. *Environmental Health Perspectives, 108*(suppl 3), 451–455. PMID:10852844

Sen, A., & Williams, B. (1982). *Utilitarianism and Beyond*. Cambridge University Press. doi:10.1017/CBO9780511611964

Settles, B. (1994). Active Learning Literature Survey. *Machine Learning, 15*(2), 201–221. doi:10.1023/A:1022673506211

Shafer, G. (1986). Savage revisited. *Statistical Science*, 463–485.

Shardanand, U., & Maes, P. (1995). Social information filtering: Algorithms for automating "word of mouth." *Proceedings of the SIGCHI Conference on Human Factors in Computing Systems*, (pp. 210–217). IEEE. 10.1145/223904.223931

Shmueli, G., & Tafti, A. (2023). How to "improve" prediction using behavior modification. *International Journal of Forecasting, 39*(2), 541–555. doi:10.1016/j.ijforecast.2022.07.008

Silver, D., Singh, S., Precup, D., & Sutton, R. S. (2021). Reward is enough. *Artificial Intelligence, 299*, 103535. doi:10.1016/j.artint.2021.103535

Smiley, M. (1992). *Moral responsibility and the boundaries of community: Power and accountability from a pragmatic point of view*. University of Chicago Press. doi:10.7208/chicago/9780226763255.001.0001

Smith, B., & Linden, G. (2017). Two Decades of Recommender Systems at Amazon.com. *IEEE Internet Computing*, *21*(3), 12–18. doi:10.1109/MIC.2017.72

Smith, C. (2011). *What is a person?: Rethinking humanity, social life, and the moral good from the person up*. University of Chicago Press.

Spahn, A. (2012). And lead us (not) into persuasion…? Persuasive technology and the ethics of communication. *Science and Engineering Ethics*, *18*(4), 633–650. doi:10.100711948-011-9278-y PMID:21544700

Srnicek, N. (2017). *Platform capitalism*. John Wiley & Sons.

Stace, W. T. (1924). *The philosophy of Hegel: A systematic exposition*. Macmillan.

Sternberg, R. J., & Kaufman, J. C. (2013). *The evolution of intelligence*. Psychology Press. doi:10.4324/9781410605313

Stout, L. (2012). *The shareholder value myth: How putting shareholders first harms investors, corporations, and the public*. Berrett-Koehler Publishers.

Stray, J., Halevy, A., Assar, P., Hadfield-Menell, D., Boutilier, C., Ashar, A., Beattie, L., Ekstrand, M., Leibowicz, C., Sehat, C. M., Johansen, S., Kerlin, L., Vickrey, D., Singh, S., Vrijenhoek, S., Zhang, A., Andrus, M., Helberger, N., Proutskova, P., & Vasan, N. (2022). Building Human Values into Recommender Systems: An Interdisciplinary Synthesis. *ArXiv*.

Stryker, S., & Burke, P. J. (2000). The past, present, and future of an identity theory. *Social Psychology Quarterly*, *63*(4), 284–297. doi:10.2307/2695840

Sutton, R. S., & Barto, A. G. (2018). *Reinforcement learning: An introduction*. MIT press.

Tamanaha, B. Z. (2004). *On the rule of law: History, politics, theory*. Cambridge University Press. doi:10.1017/CBO9780511812378

Taylor, C. (1980). Understanding in Human Science. *The Review of Metaphysics*, *34*(1), 25–38.

Thomas, R. L., & Uminsky, D. (2022). Reliance on metrics is a fundamental challenge for AI. *Patterns*, *3*(5), 100476. doi:10.1016/j.patter.2022.100476 PMID:35607624

Torra, V. (2017). *Data privacy: foundations, new developments and the big data challenge*. Springer. doi:10.1007/978-3-319-57358-8

Touretzky, D. S., & Saksida, L. M. (1997). Operant conditioning in Skinnerbots. *Adaptive Behavior*, *5*(3–4), 219–247. doi:10.1177/105971239700500302

Turing, A. M. (1950). Computing Machinery and Intelligence. *Mind*, *59*(236), 433–460. doi:10.1093/mind/LIX.236.433

Tushnet, M. (2012). Constitution-making: An introduction. *Texas Law Review*, *91*, 1983.

Tyler, T. R. (2006). *Why people obey the law*. Princeton university press. doi:10.1515/9781400828609

Van der Sloot, B. (2017). Decisional privacy 2.0: The procedural requirements implicit in Article 8 ECHR and its potential impact on profiling. *International Data Privacy Law*, *7*(3), 190–201. doi:10.1093/idpl/ipx011

Van Dijck, J. (2013). *The culture of connectivity: A critical history of social media*. Oxford University Press. doi:10.1093/acprof:oso/9780199970773.001.0001

Van Dijck, J., Poell, T., & De Waal, M. (2018). *The platform society: Public values in a connective world*. Oxford University Press. doi:10.1093/oso/9780190889760.001.0001

Van Otterlo, M. (2013). A machine learning view on profiling. In Privacy, Due Process and the Computational Turn (pp. 41–64). Routledge.

VanderWeele, T. J. (2017). On the Promotion of Human Flourishing. *Proceedings of the National Academy of Sciences of the United States of America*, *114*(31), 8148–8156. doi:10.1073/pnas.1702996114 PMID:28705870

Varela, F. J., Thompson, E., & Rosch, E. (2017). The embodied mind, revised edition: Cognitive science and human experience. MIT Press. doi:10.7551/mitpress/9780262529365.001.0001

Vesanen, J. (2007). What is personalization? A conceptual framework. *European Journal of Marketing*, *41*(5/6), 409–418. doi:10.1108/03090560710737534

von Neumann, J. (1966). Theory of self-reproducing automata. *Mathematics of Computation*, *21*, 745.

Wang, S., Wang, T., He, C., & Hu, Y. J. (2022). Can Your Toothpaste Shopping Predict Mutual Funds Purchasing?—Transferring Knowledge from Consumer Goods to Financial Products Via Machine Learning. *SSRN*.

Wang, W., Li, B., Luo, X., & Wang, X. (2022). Deep reinforcement learning for sequential targeting. *Management Science*.

Warren, M. E. (2017). A problem-based approach to democratic theory. *The American Political Science Review*, *111*(1), 39–53. doi:10.1017/S0003055416000605

Weidinger, L., McKee, K. R., Everett, R., Huang, S., Zhu, T. O., Chadwick, M. J., Summerfield, C., & Gabriel, I. (2023). Using the Veil of Ignorance to align AI systems with principles of justice. *Proceedings of the National Academy of Sciences of the United States of America*, *120*(18), e2213709120. doi:10.1073/pnas.2213709120 PMID:37094137

Wiener, N. (1988). *The human use of human beings: Cybernetics and society*. Da Capo Press.

Winick, B. J. (1992). On autonomy: Legal and psychological perspectives. *Vill. L. Rev.*, *37*, 1705. PMID:11654414

Wolpert, D. H. (1996). The lack of a priori distinctions between learning algorithms. *Neural Computation*, *8*(7), 1341–1390. doi:10.1162/neco.1996.8.7.1341

Wu, J., Zhang, Z., Feng, Z., Wang, Z., Yang, Z., Jordan, M. I., & Xu, H. (2022). *Sequential information design: Markov persuasion process and its efficient reinforcement learning*. ArXiv.

Xiao, B., & Benbasat, I. (2007). E-commerce product recommendation agents: Use, characteristics, and impact. *Management Information Systems Quarterly*, *31*(1), 137–209. doi:10.2307/25148784

Zagzebski, L. (2001). The uniqueness of persons. *The Journal of Religious Ethics*, *29*(3), 401–423. doi:10.1111/0384-9694.00090

Zarsky, T. Z. (2016). Incompatible: The GDPR in the age of big data. *Seton Hall Law Review*, *47*, 995.

Zezula, P., Amato, G., Dohnal, V., & Batko, M. (2006). *Similarity search: the metric space approach* (Vol. 32). Springer Science & Business Media. doi:10.1007/0-387-29151-2

Zhao, X., Xia, L., Tang, J., & Yin, D. (2019). " Deep reinforcement learning for search, recommendation, and online advertising: a survey" by Xiangyu Zhao, Long Xia, Jiliang Tang, and Dawei Yin with Martin Vesely as coordinator. *ACM Sigweb Newsletter, Spring*, 1–15.

Zuboff, S. (2019). *The age of surveillance capitalism: The fight for a human future at the new frontier of power*. Profile books.

Zwick, D., & Dholakia, N. (2004). Whose identity is it anyway? Consumer representation in the age of database marketing. *Journal of Macromarketing*, *24*(1), 31–43. doi:10.1177/0276146704263920

ADDITIONAL READING

Dewey, J., & Rogers, M. L. (2012). *The public and its problems: An essay in political inquiry*. Penn State Press.

Frankfurt, H. G. (1971). Freedom of the Will and the Concept of a Person. *The Journal of Philosophy*, *68*(1), 5–20. doi:10.2307/2024717

Kant, I. (2005). *The moral law: Groundwork of the metaphysic of morals*. Routledge.

McAuley, J. (2022). *Personalized Machine Learning*. Cambridge University Press. doi:10.1017/9781009003971

Mill, J. S. (2015). *On Liberty, Utilitarianism and Other Essays* (P. F. R. Mark, Ed.). Oxford university press. doi:10.1093/owc/9780199670802.001.0001

Myers, M. D., & Klein, H. K. (2011). A set of principles for conducting critical research in information systems. *Management Information Systems Quarterly*, *35*(1), 17–36. doi:10.2307/23043487

Russell, S. (2019). *Human compatible: Artificial intelligence and the problem of control*. Penguin.

Sen, A., & Williams, B. (Eds.). (1982). *Utilitarianism and beyond*. Cambridge University Press. doi:10.1017/CBO9780511611964

Sutton, R. S., & Barto, A. G. (2018). *Reinforcement learning: An introduction*. MIT press.

Zuboff, S. (2019). *The age of surveillance capitalism: The fight for a human future at the new frontier of power*. Profile books.

KEY TERMS AND DEFINITIONS

Algorithmic Behavior Modification (BMOD): Any algorithmic action, manipulation, or intervention on digital platforms intended to impact user behavior.

Autonomy: The (idealized) capacity of a person to be self-lawgiving and self-determining in choosing one's ends or goals. A key precondition for informed consent, autonomy is freedom from both internal (e.g., addictions, false beliefs) and external forms of control (e.g., physical force, threat).

Consequentialism: A system of ethical thought associated with utilitarianism in which the rightness or wrongness of an action derives from its external consequences (e.g., increased net utility), rather than its internal motivations. Consequentialism describes how artificial agents reason about the value of actions.

Control: A key concept in engineered systems that relies on the notion of feedback. A controller senses the state of a system, compares it against the desired or reference behavior, computes corrective actions, and intervenes in the system to effect the desired change. Some control approaches require a model of the underlying system dynamics, while others, such as those based on reinforcement learning, can learn a model of the system while interacting with it.

Personalization: Personalization is the goal-directed process of taking actions, making recommendations or decisions, or allocating resources on the basis of measured states of humans, the digital environment, and their interactions, and possibly adaptively modifying these actions after observing feedback about their consequences.

Platform: Both a digital infrastructure and business model based on designing digital multi-sided markets for the systematic collection, algorithmic processing, circulation, and monetization of user data.

Reinforcement Learning: An approach to AI influenced by psychology, animal learning, neuroscience, and control theory that studies how artificial agents interacting with an environment learn to accumulate reward by solving sequential decision-making problems under uncertainty, delayed action-outcome pairings, and using evaluative feedback.

Value Alignment: The task of creating artificial agents that behave in accordance with users' or designers' intentions and ethical values.

ENDNOTES

1. From the perspective of law, personalization is positively associated with rationality, consistency, fairness, justice, and desert. Personalization arguably embodies Aristotle's dictum that the pursuit of justice consists in treating similar persons similarly and different persons differently (Schauer, 2006).
2. In the economics-inspired approach to personalization, consumer choices have a certain authority in that they are assumed to be the coherent outcome of an idealized process of subjective utility maximization (Shafer, 1986). For instance, the doctrine of *consumer sovereignty* (Lerner, 1972) suggests that consumers' revealed preferences command normative respect, regardless of their causal origin.
3. Cybernetics literally means "governor" or "steersman," and is an interdisciplinary field that studies how informational messages can be communicated and engineered to control machines and society (Wiener, 1988). Cybernetics, particularly its concept of self-regulating systems and negative and

positive feedback loops, has been influential in a number of fields, such as psychology, business, engineering, and AI.

[4] A relevant theorem from cybernetics states that an agent wishing to control its environment must also possess some kind of (statistical) model of its environment (Conant & Ross Ashby, 1970).

[5] For readers curious about the underlying computational and mathematical details of personalization, we suggest Deisenroth et al. (2020) and McAuley (2022).

[6] The MDP clarifies the mathematical assumptions of the problem and allows theoretical guarantees about the performance of the reinforcement learning algorithm to be derived. We begin with a system composed of two coupled elements, an agent and an environment, and specify five key components (Mahmood & Ricci, 2007): a set of states S, a set of actions A, a reward function R(s,a) defining rewards received by taking action $a \in A$ in state $s \in S$, and a transition function T(s'|s, a) defining the new state $s' \in S$ the agent transitions into when action $a \in A$ is taken in state $s \in S$. A discount factor may also be applied to future rewards. We also set a distribution of initial states from which agent-environment interactions begin. As the learning agent interacts with its environment of human users, it collects and stores episodes of experience (state-action-reward tuples) at each time step until a terminal state is reached, at which point an entire learning episode, or trajectory, is completed.

[7] In the domain of supervised learning, where one potentially has access to (expensive to acquire) ground truth labeled data, active learning (Settles, 1994) techniques can also be used to "explore" areas of the training data that are maximally informative in terms of improving a given model's prediction ability (Bouneffouf et al., 2020). Most applications of active learning however focus on identifying which new (training) records to acquire, whereas reinforcement learning uses the MDP formalism to learn a behavioral policy that maximizes reward.

[8] The freemium business model follows a "growth before profitability" strategy (Holm & Günzel-Jensen, 2017) that initially provides a free-to-use platform with basic features in order to build a large user base capable of generating content and engaging with others, leading to network effects. Once a certain scale is reached, the platform "monetizes" by offering premium features to users or by offering advertising opportunities to external parties wishing to reach the platform's user base. This model allows social media platforms to benefit from user-generated content, data, and network effects while offering value to users and revenue streams from businesses who would like access to relevant user bases.

[9] Our description is based on information available at the time of writing (June 2023).

[10] Serendipity is functionally similar to the idea of exploration in reinforcement learning-based personalization, but its motivation stems from business interests in maintaining a good user experience rather than theoretical considerations of algorithmic performance.

[11] Our description is based on information available at the time of writing (June 2023). https://ai.facebook.com/blog/powered-by-ai-instagrams-explore-recommender-system/

[12] https://blog.twitter.com/engineering/en_us/topics/open-source/2023/twitter-recommendation-algorithm

[13] More specifically, Twitter's algorithm relies on several "embedding spaces" to infer similarities between users and Tweets.

[14] We emphasize that autonomy is an ethical ideal that few persons actually live up to in their daily lives. Most people realistically live their lives somewhere in between states of *heteronomy* and *autonomy* (Benn, 1975; Winick, 1992). In other words, we all occasionally experience weakness of

the will, temptation, and even addiction to some substances, all while following rules set by other persons and institutions.

15 For example, the California Consumer Privacy Act (CCPA) defines personal data as "information that identifies, relates to, describes, is reasonably capable of being associated with, or could reasonably be linked, directly or indirectly, with a particular consumer or household." See https://ccpa-info.com/home/1798-140-definitions/

16 These comments about the motivations of commercial platforms for data privacy are less relevant to medical domains, where organizations such as hospitals often have legal and ethical duties to protect personal data. Governmental, educational and financial institutions may also be legally required to implement data privacy techniques.

17 Philosophical issues of personhood pose difficult questions in law (i.e., corporate personality and the status and rights of legal persons) and in ethics (i.e., the rights of an unborn fetus). For a helpful overview of the concept of personhood relevant to law, see Naffine (2003).

18 A data profile is a statistical model built from large amounts of data from other users that can be used to infer information about a particular individual (Van Otterlo, 2013).

19 Stanford educational psychologist Carol Dweck has done pioneering research showing how people's *self-theories* (e.g., self-descriptions about whether certain personal traits and qualities are fixed or changeable) can have drastic effects on their academic performances and motivation when facing challenges and threats (Dweck, 2000).

20 Caudill & Murphy (2000) give several examples of this economics-inspired perspective in the marketing literature.

21 Anthropologists make a similar distinction with the concepts of *emic* versus *etic* perspectives on social practices and rituals. The emic perspective aims to understand practices from the insider's point of view, while in the etic perspective, the meaning and interpretation of social behavior is imposed from the outside observer's perspective (Harris, 1976).

22 Self-reference allows a subject to represent itself as an object and thus come to have beliefs about itself as an object. This representational capacity has played an important evolutionary role and is believed to be behind the human ability to construct idealized formal systems using abstract symbols (Boulding, 1956; Deacon, 1997; Donald, 1991).

23 Kant and Hegel argue that, as rational beings, moral persons must by necessity choose to bind their wills to what universal moral law requires (Stace, 1924). In other words, reason requires that for example no logical inconsistencies result from an action, the basic principle behind Kant's so-called *categorical imperative*: that one should only act according to a rule which one would also will to be a universal rule for all others in any particular circumstance (Kant, 1948). Kant's *deontological*—duty-based and desire independent—ethics contrasts with utilitarian and consequentialist ethics that evaluate actions or rules for action on the basis of the desirability of their observable consequences.

24 Human free will (i.e., agent-causation), however, is believed to conflict with the presumption of "causal closure" in the physical realm: that all physical effects have fully physical causes (Kane, 2001). Despite difficulties in explaining and locating something like a "will" that could have a causal impact on physical objects, philosophers have suggested that we must nevertheless maintain a metaphysical belief in the freedom of the will if the notion of moral, social, and legal responsibility is to remain viable (Smiley, 1992).

[25] Public reason describes a normative standard for legitimate political action in liberal democracies stating that reasons for action must be intelligible, comprehensible, and accessible to citizens (see Rawls, 1971).

[26] The rule of law is a central notion in Western political thinking, and is associated with liberal democracy and individual rights. Broadly, it declares government officials and citizens obligated to abide by the regime of legal rules that govern their conduct, rather than the personal wishes of a ruler.

[27] This point might also be extended to include the designers' children, family, and friends.

Chapter 12
Mind Uploading in Artificial Intelligence

Jason Wissinger
Waynesburg University, USA

Elizabeth Baoying Wang
Waynesburg University, USA

ABSTRACT

Mind uploading is the futurist idea of emulating all brain processes of an individual on a computer. Progress towards achieving this technology is currently limited by society's capability to study the human brain and the development of complex artificial neural networks capable of emulating the brain's architecture. The goal of this chapter is to provide a brief history of both categories, discuss the progress made, and note the roadblocks hindering future research. Then, by examining the roadblocks of neuroscience and artificial intelligence together, this chapter will outline a way to overcome their respective limitations by using the other field's strengths.

INTRODUCTION

The concept of mind uploading has been explored in science fiction for over fifty years, however, we are only scratching the surface of how to bring this technology into existence. Humanity's current trajectory towards discovery is vastly disagreed upon. Famous computer science theorists the world over have differing opinions on when mind uploading will be recognized. In an interview with IEEE Spectrum, a few of these theorists were asked to predict when machines would surpass the computational limitations of the human brain. Ray Kurzweil, cofounder of Singularity University and author of countless books dedicated to this very subject, predicts that, "computers will match and then quickly exceed human capabilities in the areas where humans are still superior today by 2029" (IEEE Spectrum, 2023). Meanwhile, Robin Hanson, author of *The Age of Em: Work, Love, and Life When Robots Rule the Earth* and Chairman and CTO of Rethink Robotics, Rodney Brooks, both believe it to be centuries away (2023). Despite the wide range of singularity projections, most theorists believe the key to unlocking a computer's emula-

DOI: 10.4018/978-1-6684-9591-9.ch012

tive potential is brain scans. Additionally, most of these predictions hinge on the creation of nanobots, capable of being injected into the bloodstream to travel to the brain and transmit a full scan. This paper will propose a different method, a new path towards whole brain emulation. Taking current neurological findings and pairing it with recent progress in the development of complex machine learning models could be the best way to shave decades, and possibly centuries, off the mind uploading development timeline.

The rest of the chapter is organized as follows. An overview of the background of mind uploading technologies is given in section 2. Section 3 examines major approaches of artificial intelligence. Section 4 proposes how neuroscience and meta-learning could be united to finally reach the goal of mind uploading. Current limitations, challenges, and ethical issues in mind uploading are discussed at the end.

BACKGROUND

To develop technology that emulates brain function, we must first have a way to measure and record brain function. The ability to analyze brain waves and electrical impulses was, and still is, fundamental for all experiments to follow. In this section, this paper will discuss three of the most important developments towards understanding the human brain and how it relates to mind uploading: Brain Computer Interfacing, the Blue Brain Project, and nanobots. The main issues in mind uploading, from a neurological standpoint, are addressed at the end of the section.

BCI Technology

BCI stands for Brain Computer Interfacing. An organization known as the BCI Society categorizes a BCI system based on its functionality, which could be any of the following: replacement, restoration, enhancement, supplementation, or improvement (Welikala & Karunananda, 2017). The first recorded advancement in understanding brain function was discovered by a British physician, Richard Caton, in 1875 (Welikala & Karunananda, 2017). He was the first to note electrical impulses due to brain activity. This discovery was humanity's first step towards harnessing the power of BCI technology. Then, in 1929, a German Psychiatrist, Hans Berger, recorded these same signals using a device with small discs covering the scalp. Now, the process of using this type of gadget to measure and record brain activity is called Electroencephalography (EEG). Despite the impact this discovery has on neuroscience today, his work was originally dismissed and abruptly brought to a halt by the Nazi regime and his eventual suicide. When Berger's work was replicated by two British physiologists in 1934, EEG was finally accepted as functional and effective (Biasiucci et al., 2019).

When Brain Computer Interfacing technology was first introduced in the 1970s, it was frequently set aside due to the costliness of the equipment and the logistics of human testing (Welikala & Karunananda, 2017). During this era, several distinguished computer scientists tried to layout a reasonable path to mind uploading. The first was Hans Moravec in 1988, when he published a technical explanation on the mind uploading method of replacing each individual brain cell with software until the whole brain is emulated. The next was Ralph Merkle in 1989, where he used automated analyses to test the possibility of reestablishing the brain through this method. With the success of his tests, he reported that we should see the emergence of this technology in a few decades (Welikala & Karunananda, 2017). This revelation sparked an onslaught of projects that attempted to scan and reconstruct the human brain, one of which was mildly successful.

The Blue Brain Project

The Blue Brain Project is an initiative led by Henry Markram whose goal is to further our understanding of the human brain and eventually reconstruct it with software. Taken directly from the Blue Brain Project's website:

The aim of Blue Brain is to establish simulation neuroscience as a complementary approach alongside experimental, theoretical and clinical neuroscience to understanding the brain, by building the world's first biologically detailed digital reconstructions and simulations of the mouse brain. (Blue Brain Project, n.d.)

The reason Blue Brain is so important is the possibility of retaining intelligence. Anything not documented is not saved for future generations with our current human limitations (Kumari & Khan, 2018). Currently, the Blue Brain project has not yet mastered the human brain. However, in 2014, a small sliver of a rat brain consisting of over one hundred million neurons was successfully constructed (Welikala & Karunananda, 2017).

The process by which a Blue Brain is created is broken down into three critical steps. In step one, all data from a brain must be collected and analyzed to understand the electrical behavior of the neurons. To continue, a programmer must convert these observations into algorithms. Then, in step two, mathematical models must be used to create situations for the algorithms to respond to. In these simulations, virtual cells are synthesized by the code to replicate the function of brain cells. Currently, Blue Brain systems respond around 200 times slower than a real brain. Finally, the results must be visualized in a way that can be observed and studied further for adjustments to algorithms and processes (Kumari & Khan, 2018).

Nanobots

The main holdup and the reason the Blue Brain Project has not yet been able to recreate a human brain lies within the acquisition of data. To understand the brain is to pick it apart, and while dissecting a rat brain to build an artificial neural network from its neurons is legal, humans cannot be subject to such experiments. Noninvasive methods of studying synapses are slow and inaccurate, and invasive methods are mostly unethical. The only solution therein would be a technological advancement that remained noninvasive but held the same efficiency and precision of surgical methods. The most often theorized advancement is nanobots.

Nanobots or nanorobots are tiny hypothetical robots that can travel throughout the human circulatory system. They can be used for a whole slew of purposes in the medical field.

Nanomedical robots can be used for many applications, such as targeting and early diagnosis of cancer, drug delivery, tissue engineering, gene delivery systems, cardiology, analysis of body vitals, monitoring of diabetes, minimally invasive brain surgery, and imaging and detection capabilities. (Giri et al., 2021)

More importantly for the context of mind uploading, nanobots can be used to provide a full scan of the structure and connections of our brain by scanning it. Then, uploading the data to a computer and adjusting the algorithms in the simulation step accordingly, we can begin to build an artificial human brain in a machine (Kumari & Khan, 2018). The creation of nanobots still has many challenges ahead, but when produced, scanning a brain in its entirety would slingshot mind uploading to the forefront of medical, philosophical, and computer science related research.

Issues and Resolutions

To create a new game plan for development of mind uploading technology, we should use projects of Blue Brain's nature as a baseline. The current outline is functional because it guarantees the creation of mind uploading technology eventually. However, it makes some bold assumptions about the development of technology that does not exist yet and may not exist anytime soon. The first assumption is that our supercomputers are fast enough and have enough storage to transfer a full brain scan into and run simulations. While the specifications of the computers used in the Blue Brain Project are impressive, to be as efficient as possible, IBM must continue to develop better and faster equipment (Kumari & Khan, 2018). Sticking with the machines they currently have may be an easier option for the sake of cost and avoidance of issues with compatibility, but it is not the most efficient. IBM must continue to upgrade their machines devoted to this cause and the software must be adaptable and well documented.

The second assumption is that nanobots are going to be created soon. Some rather generous articles estimate that nanorobot tech will be coursing through our veins by 2030, while others predict it will take another 200 years (English, 2020). The rigorous testing required to develop nanobots, as well as the testing required to efficiently recreate a human brain would delay our efforts for an indefinite amount of time. If we are looking to streamline the process of mind uploading, betting on a horse that does not yet exist is not the way to approach it. Perfecting our current noninvasive methods, I believe, is the best course of action.

The third and final main assumption is that once all the data is uploaded, we will know what to do with it. Estimates from studies done at Stanford University say that the human brain holds 2.5 petabytes of data (Neurology-Vegas, 2021). That is an overwhelming amount of data to sort through and visualize, especially compared to a rat brain. Improving our visualization methods of the data we gather is an important step to make the development pipeline of mind uploading technology as efficient as possible.

The most effective way to resolve the issues mentioned above is using artificial intelligence. Currently, the most common method of attack to upload the mind is a slow and steady replacement of neurons (Welikala & Karunananda, 2017). Eventually, this method would yield a complete emulation of the human brain and avoid the issue of waiting on development of nanorobots, however, artificial intelligence could be used to speed up the process. A subset of artificial intelligence, machine learning, is defined as "the area of computational science that focuses on analyzing and interpreting patterns and structures in data to enable learning, reasoning, and decision making outside of human interaction" (NetApp, n.d.). The creation of neurons in a human brain happens to fall under something that machine learning could compute. Using neural networks, a commonly used machine learning algorithm, we could build the entire brain based on smaller and more reasonably obtainable samples, significantly reducing the mind uploading timeline. Before we can understand how artificial intelligence can be used to finish creating a digital brain, however, we must first understand the basics and history of machine learning.

ARTIFICIAL INTELLIGENCE APPROACHES

The concept of artificial intelligence (AI) usually begins with Alan Turing. Alan Turing is arguably the most famous computer scientist of all time, largely due to his contributions to artificial intelligence. He created the Turing Test, a way to test the ability of a machine to imitate human traits. If a blind interviewer is unable to distinguish the computer from a human, that computer passes the Turing Test. In 1959, machine learning took off with Arthur Samul's checkers program. Throughout the 80s, many computer scientists theorized about uses for machine learning systems that involved processing of data. Then, finally, with the influx of video games in the 1990s, studies in artificial intelligence began approaching where they are today (Çelik & Altunaydın, 2018). This section examines several major artificial intelligence approaches.

Machine Learning

Machines performing operations from lines of code never make an error. Simple I/O algorithms are defined by the programmer and the machine never deviates. However, in machine learning situations, computers are sometimes required to make inferences based upon a set of sampled data. "The main aim of machine learning is to create models which can train themselves to improve, perceive the complex patterns, and find solutions to the new problems by using the previous data" (Çelik & Altunaydın, 2018). These models, trained with the right data set and given a specific goal, can be used to solve some of the toughest mathematical equations or even play a video game with near maximum efficiency. Brain function, being one of the most complicated patterns of all, could be tackled by machine learning as well, given the right instructions and data.

There are four categories of machine learning: supervised, unsupervised, semi-supervised, and reinforced learning. In the first type, input data is analyzed by the machine and categorized or used to make a prediction about the remaining data. Unsupervised learning is similar to supervised learning in that it relates the data within a given data set, but instead of being given specific relations to look for, the machine must determine them itself. In the third type, semi-supervised, the machine is given more data of unknown types than known types. The machine must then use what is known about the labelled and unlabeled data to deduce more information about the unknown types. The fourth and final category is reinforced learning, where an agent learns via a system of punishments and rewards. When the algorithm ends, the agent gets a reward or punishment based on how close to the target goal they got. The algorithm then adjusts accordingly until it reaches the target as efficiently as possible (Çelik & Altunaydın, 2018). All of these categories are useful in specific situations, but the most useful type of machine learning for the situation of mind uploading is supervised, more specifically, artificial neural networks.

Artificial Neural Networks

An artificial neural network (ANN) is a structure of supervised machine learning that is similar to the way a human brain operates. In this structure, neurons with five basic functions are connected to each other in layers. In the input layer, a user gives the machine a data set to work with. Then, in an intermediary layer, each neuron has a weight, and each connection has a summation or activation function. The weight of each neuron determines the impact the result will have on the decided output. The summation functions are a formula that uses the initial input value and weight of the neuron it is attached to in order to calculate the total input for that neuron. To calculate the final output, an activation function, which is typically derivable, is used (Çelik & Altunaydın, 2018;Brunton & Noack, 2020).

ANN algorithms are divided into two subgroups as well, feedforward and back propagated. In a feed-forward ANN, signals are sent straight through the structure in one way. There are no cycles or loops, the intermediary layer is hidden, and the delta rule can be applied to train the algorithm and adjust the weights of the neurons. The other subgroup, back propagation, is the most frequently used ANN, especially with multi-layer neural networks. If the activation functions are differentiable, the network can calculate the derivative of the error function with respect to the neuron's weights, change the weights, and send the error back up the network to continue adjusting previous weights. Back propagation algorithms are typically used to train feedforward multi-layer neural networks more effectively in the areas of deep learning, adjusting the weights in each node instead of using the less efficient delta rule (Çelik & Altunaydın, 2018; Brunton & Noack, 2020). The main limitation with a single artificial neural network is that it must be created and adjusted by a human, however, an artificial neural network can be used to do this too! The idea of using a machine learning system to fine tune another machine learning system is called meta learning and will be explored in the next subsection.

Deep Learning

The process in which human beings acquire new knowledge and skills usually involves using previously learned knowledge and skills. In machine learning, the same logic is applied. When we use a machine learning model, we use our human understanding to adjust the machine to suit the requirements or we build a new machine. If machine learning models are meant to learn and make decisions, however, we could use them to analyze the results and adjust a different, lower-level machine learning model. This process is called meta learning. Meta learning, more often called deep learning, begins with the collection of meta data that accurately describes models that have already been learned.

They comprise the exact algorithm configurations used to train the models, including hyperparameter settings, pipeline compositions and/or network architectures, the resulting model evaluations, such as accuracy and training time, the learned model parameters, such as the trained weights of a neural net, as well as measurable properties of the task itself, also known as meta-features. (Hutter et al., 2019)

These meta features can be presented to another machine learning construct for that machine to learn how to make the original machine more efficient (Hutter et al., 2019). The idea of using one machine learning system to build and adjust another machine learning system seems farfetched, but deep learning is already being deployed in a multitude of areas that require artificial neural networks to be fine-tuned.

The application of deep learning has historically been specifically geared towards the field of computer science. For computer vision and graphics, deep learning can be used to avoid overfitting or nonconvergence due to training a large model with a small amount of data. In robotics, meta reinforcement learning can be used to fix sample inefficiency due to sparsity of rewards, the need for exploration, and the sheer number of algorithmic possibilities (Hospedales et al., 2022). More recently, however, deep learning has begun to seep into other fields. ChatGPT, a language processing chatbot rose to stardom from late 2022 to early 2023. Now, ChatGPT's popularity is not necessarily due to its intelligence, but more so its widespread availability. Despite being created to serve as a humanlike chatbot, the accessibility brought about new applications of ChatGPT's capabilities. From creating cartoon scripts to writing complex code, ChatGPT is just a glimpse into the future of what artificial intelligence and machine learning can do.

An example more akin to mind uploading, however, would be recent strides in three-dimensional modeling of human anatomy. One deep learning model can use two-dimensional pictures to estimate a patient's face shape preinjury (Habal et al, 2023). Another application uses convolutional and recurrent neural networks trained on large datasets of labelled ultrasound pictures to reduce background noise, quality fluctuations, and overlapping anatomical structures in ultrasound imagery (Gautum & Kaur, 2023). These machine learning structures are using massive datasets to render and improve two-dimensional and three-dimensional images of human anatomy. The brain is just that, another piece of human anatomy waiting to be predicted and reconstructed by a deep learning model, if given a large enough, high quality dataset and the proper algorithms to tackle it.

Proposed Solutions and Challenges

In previous sections, we explored the major approaches and technologies in mind uploading and AI. The progress in each field is notable. However, their greatest challenges seem to complement each other. The Blue Brain Project's biggest challenge is having to scan and understand a human brain neuron by neuron, a tall task considering we have only fully scanned a nematode with three hundred and two neurons (Tang et al., 2020). Computer scientists have been trying to create AI that can emulate human decision making, but their biggest challenge has been trying to replicate the functions of a human brain with artificial neurons. Only when both respective fields' progress is combined can we start to see a path towards overcoming the challenges and uploading the mind emerge.

USING ANNS TO COMPLETE BRAIN SCANS

Currently, the Blue Brain Project and other similar efforts have the same focus—scanning the entire brain. Suppose a full brain scan was not necessary, only a large dataset of small snippets, gathered through noninvasive methods. The deep learning area of machine learning is capable of building and optimizing complex three-dimensional structures using predictive analytics. Using deep learning, developers could create an artificial neural network that could build a smaller animal's brain by taking many tiny samples, as many as we can gather noninvasively, and predicting the rest of the structure. Creation of this technology would not only benefit medical practice, as mentioned in the introduction, through reduced animal testing and therefore faster drug and treatment development, but it would make the concept of mind uploading exponentially easier. Once a deep learning model can effectively predict and replicate animal brains, that same model could be trained with human datasets and with enough high-quality data and algorithm adjustment, the goal of mind uploading could be effectively achieved. This method, albeit completely hypothetical currently, would bypass the also hypothetical creation of nanorobots entirely, and the need for a full brain scan.

Another benefit to this plan of attack would be easy replication. When the creation of nanorobots is finally finished and the goal of scanning an entire brain is realized, it would be very hard to replicate that same process. Nanorobots would have to be mass produced, leading to incredible expenses and thin availability, at least for a long time. If this supposed deep learning model were created, adding an additional brain would be even simpler than the first, due to the corrective nature of machine learning models. As the model would continue to learn, the process would become faster, cheaper, and more accurate.

Limitations

The plan of attack proposed above, using deep learning to predict and finish a brain scan, does not come without its own set of limitations and drawbacks. For one, there is very little means to test if the scanned brain functions exactly like a human one. Outputs produced by machine learning and deep learning are never 100% accurate, but we can reduce the margin of error to such an infinitesimal amount, it might not matter for the functionality. Another limitation is the sheer amount of computer power training machine learning systems and deep learning systems take. While the Blue Brain Project has supercomputers built by IBM at its aid, our current capabilities might be too slow or too small in storage to run and train the amount of machine learning structures required for this solution (Welikala & Karunananda, 2017). In this case, we would be waiting for the development of more powerful technology, which is neither good nor bad. Development of more powerful computers never stops, so eventually this technology will certainly exist, which is an advantage this strategy has over waiting for the creation of nanobots.

A third limitation would be the staffing requirement to create the deep learning models required for a project of this magnitude. Computer science experts are getting progressively better at using this methodology as time passes. From the applications of deep learning mentioned in subsection 3.3, this tool is being used for projects pertaining to speech recognition and graphics, which are much smaller by comparison. Along with adequate funding, a project of this nature would require a team of highly skilled researchers, neuroscientists, computer scientists, and programmers. Finally, the project would have to be backed by some large entity to bring all the necessary requirements of a diverse group and hardware together.

Ethical Implications

Another problematic aspect of the creation of this technology that cannot be ignored is the ethical implications of such a project. Bypassing animal and human testing spare this project from that specific ethical dilemma during the development phase, but, once operational, how do we handle a consciousness inside a machine? If the computer-embodied consciousness is sentient, does it have human rights? Before its implementation, legislation must be created to regulate how we treat an uploaded mind, as well as how an uploaded mind treats us and other technologies or minds.

If the resulting emulation would contain a consciousness comparable to that of a human, it would probably have the same moral rights as a human would. After all, that emulation is you, with all your conscious abilities, memories, and intentions, just substantiated by different materials. (Perez & Cohen, 2019)

An ethical standard would have to be created, like the ethical standards we have with biological humans.

Another example of an ethical implication for a project like this is personal identity. After copying an individual to a computer, there is theoretically two copies of the same person. From a naturalistic standpoint, important aspects of identity—personality, memory, and consciousness—are nothing but physical operations like neurons firing in a specific way in the brain. By this definition, if these qualities are transferred, there is no difference between the computer and the individual (Sarbey, 2016). If the subject whose mind is being transferred is still alive and well, there would be two copies of the same person with different thoughts and ideas. These issues of identity and individuality are all ones that must be confronted before we can continue any attempts to transfer an individual's consciousness to a machine.

Yet another example of an ethical issue is how we should handle criminal activity. Theoretically, a consciousness uploaded to a computer should be tried as a human being, if all their thought processes match that of the original human's. After being tried, if found guilty of a crime, the uploaded mind would have to serve some kind of sentence. However, confinement would seemingly be pointless since the brain is already confined to a computer. Like every other ethical issue approached beforehand, some sort of legislation or standard would have to be created for this situation, as well.

One final ethical issue, this time affecting public reception of mind uploading, is how it would fit into religion. Many different religions place a large value on life in its human form and have some concept of an afterlife. The idea of immortality in a digital space could be seen by the public as defying the god or gods they worship. While preserving an individual's memories and knowledge may be beneficial to society, many cultures could have strong opinions about what it does to the soul. This could be a large inhibitor for acceptance of a digital afterlife and make many people hesitant to back a project of this nature.

FUTURE RESEARCH DIRECTIONS

Famous mind uploading researchers and theorists alike have been stuck at a very similar roadblock, waiting for nanorobots. There is not much literature looking to find a different path towards the goal of emulating the human brain. This outline proposes an alternate route using predictive deep learning models to effectively guess the construction of the human brain. An obvious limitation is the current lack of quality data to train the deep learning algorithm with. The crux of every plan of attack to date is a lack of consistent, high-quality data. However, the goal of this paper is to minimize the need for nanorobotics. The most important takeaway from my proposed plan is that just because most researchers agree that the creation of nanobots is a way to realize whole brain emulation, it does not mean that it is the only way to achieve it.

Another serious limitation of this plan of attack that is yet to be addressed is the lack of knowledge about deep learning. While it is true that understanding and application of deep learning models has increased exponentially in recent years, there is still years' worth of research and experimentation before an algorithm as complex as this could come to be. It seems that, no matter which approach is followed to achieve mind uploading, there will be a lengthy wait.

This pathway, much like nanobot pathway, requires future research in a few different aspects. First, a deep learning model capable of predicting the structure of a human brain and its neural pathways would need a large amount of high-quality data. To get this data, more effective noninvasive methods of gathering would need to be produced. Second, the model would require more powerful hardware than we currently have. Better data storage, more processing power, and better three-dimensional rendering hardware and software would be the most basic upgrades necessary. Third, more effective algorithms would be required. Current examples of deep learning algorithms take weeks to train with just a fraction of the data that would be required to predict the structure of a human brain. And finally, a massive team of researchers, developers, and neuroscientists would have to be put together throughout the entire process.

All things considered, this path might not be faster than the creation of nanobots, but the point is that it could be. Rather than staking all hopes of mind uploading on one possibility, diversifying the options might lead to a faster solution, whether this is the correct one or not.

CONCLUSION

The advancements and future goals of neuroscience and computer science have pushed each field towards one another. With the discipline of neuroscience dealing with data far too vast to understand or compute without the use of technology, complex computer science subjects like artificial neural networks and meta learning are necessary to continue studies. With the domain of computer science working towards building machines capable of emulating human decision, complex areas of neuroscience like the natural neural network of a human brain must be studied and understood. Throughout the history of both fields, there have been various important inventions and challenges that were overcome to gather this understanding. From the invention of the EEG for research in neuroscience to machines built to dominate a specific game by learning popular strategies and adjusting, these challenges were always met, and the branches of science prevailed. Mind uploading remains just that, another battle in the long history of obstacles each field has had to face and complete to proceed with more research. By taking a new route towards development, the one outlined above with deep learning predictive algorithms guiding the construction of a digital brain, the era of transferring consciousness' could begin soon. Possibly, more soon than continuing down the route that is currently accepted.

REFERENCES

Biasiucci, A., Franceschiello, B., & Murray, M. M. (2019, February 4). Electroencephalography. *Current Biology*, *29*(3), R80–R85. doi:10.1016/j.cub.2018.11.052 PMID:30721678

Blue Brain Project. (n.d.). EPFL. Retrieved November 28, 2022, from https://www.epfl.ch/research/domains/bluebrain/

Brunton, S. L., Noack, B. R., & Koumoutsakos, P. (2020). Machine Learning for Fluid Mechanics. *Annual Review of Fluid Mechanics*, *52*(1), 477–508. doi:10.1146/annurev-fluid-010719-060214

Çelik, Ö., & Altunaydın, S. S. (2018). A Research on Machine Learning Methods and Its Applications. *Journal of Educational Technology & Online Learning*, *1*(3), 25–40. doi:10.31681/jetol.457046

English, T. (2020, November 20). *Nanobots will be flowing through your body by 2030*. IE. Retrieved November 29, 2022, from https://interestingengineering.com/innovation/nanobots-will-be-flowing-through-your-body-by-2030

Gautam, L., & Kaur, G. (2023). Review on ultrasound-image analysis of fetus using Deep Learning. *2023 IEEE 8th International Conference for Convergence in Technology (I2CT)*. 10.1109/I2CT57861.2023.10126500

Giri, G., Maddahi, Y., & Zareinia, K. (2021). A brief review on challenges in design and development of Nanorobots for Medical Applications. *Applied Sciences (Basel, Switzerland)*, *11*(21), 10385. doi:10.3390/app112110385

Habal, M. B., Clement, A., Hardisty, M., Fialkov, J. A., Whyne, C. M., & Hanieh, A. (2023, August 28). Artificial Intelligence–based modeling can predict face. *The Journal of Craniofacial Surgery*. https://journals.lww.com/jcraniofacialsurgery/abstract/9900/artificial_intelligence_based_modeling_can_predict.1013.aspx

Hospedales, T., Antoniou, A., Micaelli, P., & Storkey, A. (2022). Meta-Learning in Neural Networks: A Survey. *IEEE Transactions on Pattern Analysis and Machine Intelligence*, *44*(9), 5149–5169. PMID:33974543

Hutter, F., Kotthoff, L., & Vanschoren, J. (2019). *Automated machine learning*. The Springer Series on Challenges in Machine Learning. doi:10.1007/978-3-030-05318-5

IEEE Spectrum. (2023, January 25). *Human-level AI is right around the corner-or hundreds of years away*. IEEE Spectrum. https://spectrum.ieee.org/humanlevel-ai-is-right-around-the-corner-or-hundreds-of-years-away

Kumari, M., & Khan, R. (2018). Review Paper on Blue Brain Technology. *International Journal of Computer Applications*, *180*(48), 21–23. doi:10.5120/ijca2018917256

Markram, H. (2013). Seven challenges for neuroscience. *Functional Neurology*, *28*(3), 145–151. doi:10.11138/FNeur/2013.28.3.144 PMID:24139651

Neurology-Vegas. (2021, February 5). *What is the memory capacity of a human brain?* Clinical Neurology Specialists. Retrieved November 20, 2022, from https://www.cnsnevada.com/what-is-the-memory-capacity-of-a-human-brain/

Perez, M.N., Cohen, R. (2019). Uploading the Mind: The Basics and Ethics of Whole Brain Emulation. *Pushing the Limits*, 30-33.

Sarbey, B. (2016). Definitions of death: Brain death and what matters in a person. *Journal of Law and the Biosciences*, *3*(3), 743–752. doi:10.1093/jlb/lsw054 PMID:28852554

Tang, A., Christian, E., & Huang, Y. M. (2020). *What needs to happen before we can upload our brains to a Computer*. Business Insider. Retrieved November 21, 2022, from https://www.businessinsider.com/how-to-upload-brain-computer-eternal-digital-life-black-mirror-2020-9

Weiss, K., Khoshgoftaar, T. M., & Wang, D. (2016). A survey of transfer learning. *Journal of Big Data*, *3*(1), 9. doi:10.118640537-016-0043-6

Welikala, S., & Karunananda, A. S. (2017, October 31). *Mind Uploading with BCI Technology* [Paper presentation]. SLAAI – International Conference on Artificial Intelligence, Moratuwa, Sri Lanka. https://slaai.lk/wp- content/uploads/2018/01/slaai2017.pdf#page=15

What is machine learning - ml - and why is it important? (n.d.). NetApp. Retrieved November 20, 2022, from https://www.netapp.com/artificial-intelligence/what-is-machine-learning/

ADDITIONAL READING

Gidwani, M., Bhagwani, A., & Rohra, N. (2015). Blue Brain - The Magic of Man. *2015 International Conference on Computational Intelligence and Communication Networks (CICN)*, 607–611. 10.1109/CICN.2015.319

Nicholas, A. (2011). Ray Kurzweil and Uploading: Just Say No! *Journal of Evolution and Technology / WTA*, *22*(1), 23–36. https://jetpress.org/v22/agar.pdf

Sierra, D., Weir, N., & Jones, J. (2005). A Review of Research in the Field of Nanorobotics. *Sandia Report*. doi:10.2172/875622

Vanschoren, J. (2011). Meta-Learning Architectures: Collecting, Organizing and Exploiting Meta-Knowledge. *Studies in Computational Intelligence*, 117–155. doi:10.1007/978-3-642-20980-2_4

KEY TERMS AND DEFINITIONS

Artificial Intelligent (AI): A branch of computer science dealing with tasks that normally require human intelligence, such as visual perception, speech recognition, decision-making, and translation between languages.

Artificial Neural Network (ANN): A machine learning method inspired by the biological neural networks that constitute animal brains. It consists of a series of algorithms that discover underlying patterns in a dataset through a process that mimics the way the human brain operates.

Blue Brain Project: A Swiss brain research initiative whose main goal is to digitally reconstruct the brain of a mouse by gathering data through scans, organizing the data, building a model, and performing a simulation to see if it behaved correctly.

Brain Computer Interfacing (BCI) Technology: A device that establishes a direct link to the brain and reads brain signals to gather data or manipulate another external device.

Electroencephalography (EEG): The measurement and recording of electrical impulses in the brain, usually using a headset.

Machine Learning: A branch of artificial intelligence dealing with computer systems that learn and adjust the system using algorithms and statistical analysis.

Meta-Features: Descriptions of an experiment that are beyond the experiment itself. Examples of this are measurement tools, uncertainties, and error measurements.

Meta-Learning: A subset of machine learning that applies a learning algorithm to look at the meta-features of existing machine learning experiments so that it can be further improved.

Nanobots: A hypothetical, microscopic machine, capable of being injected into the bloodstream and used for data acquisition and administration of drugs.

Transfer Learning: A high-performance learning algorithm that is trained by data already collected or used by other machine learning structures. This method is very effective when the source of the data necessary is too scarce or expensive to gather.

Chapter 13
Ethical Issues of Artificial Intelligence (AI):
Strategic Solutions

Sara Shawky
Griffith University, Australia

Park Thaichon
University of Southern Queensland, Australia

Sara Quach
https://orcid.org/0000-0002-0976-5179
Griffith University, Australia

Lars-Erik Casper Ferm
The University of Queensland, Australia

ABSTRACT

Ethical issues of AI have become a huge concern dominating government, media, and academic discourse. This chapter sheds light on some of the most pressing ethical issues that result from the adoption of AI-powered tools. Increasing inequality, widening social and economic gaps, compromising privacy and data protection, outsmarting humans and impacting human rights, lack of accountability, liability and reliability, and lack of empathy and sympathy are considered the most pressing challenges that need to be addressed concerning AI and big data. This chapter also provides insight into strategies that are currently in place to overcome adverse implications of AI in the public and private sectors. Providing insight into these ethical challenges along with the governing solutions makes a significant contribution to the ongoing discourse and urges for bringing forth sustainable solutions that are necessary for the ethical application of these technologies in different fields.

DOI: 10.4018/978-1-6684-9591-9.ch013

INTRODUCTION

The term Artificial intelligence (AI) is utilized to describe information systems that allow machines or software to endow the intellectual processes of human behaviors (Nilsson, 2014). Since its inception, AI has demonstrated the power of programming data-driven digital technologies to perform activities that used to be solely associated with intelligent human beings (Quach et al., 2022), such as problem-solving, decision making, drawing generalizations, communication, perception of meaning, and reasoning (Ayoko, 2021; Balajee, 2020). Organizations have continued to embrace the adoption of AI technologies for their accelerated progress in algorithms, the internet, interconnectedness, and big data storage (Ayoko, 2021; Hanelt et al., 2021).

Recent developments of AI have embraced a computerized system allowing the formation of patterns in algorisms data called, Machine Learning (ML) (Katznelson & Gerke, 2021). This technology has further enhanced AI's abilities to predict outcomes and form correlations. Various industries, including healthcare, production, manufacturing, autonomous vehicles, smart homes, social media, agriculture, and farming have become highly dependent on AI technologies in their operations. The ability of AI to analyze large amounts of data and perform data mining and computation of data processes in a timely and efficient manner could exceed the analytical abilities of human knowledge (Ayoko, 2021). Such efficiency, in turn, enhances and accelerates the generation of insights and optimization of choices which have revolutionized the internal operations and processes of organizations (Ayoko, 2021; Kretschmer & Khashabi, 2020; Magistretti, Pham, & Dell'Era, 2021; Stahl et al., 2022).

The digitalization provided by AI tools assists organizations in radically improving their efficiency and effectiveness in the way they organize resources and make strategic operational decisions (Hasan et al., 2021a; Hasan et al., 2021b; Nguyen et al., 2021), which in turn contribute to improving their productivity and competitiveness (Ayoko, 2021, Ruiz-Real et al., 2021). For instance, the application of AI technology has transformed the medical system by facilitating and speeding up processes, allowing timely storage, analysis and retrieval of data, and providing access to health professionals and organizations (Ayoko, 2021; Farhud & Zokaei, 2021). Over 160 health AI devices have been adopted by the United States Food and Drug Administration with the potential to exceed this number in the near future by furthering the number of products under development (Katznelson & Gerke, 2021). These devices have eased the processes of providing imaging and electronic medical records (EMR), laboratory diagnosis, treatment, new drug discovery, and preventive and precision medicine (Ayoko, 2021; Farhud & Zokaei, 2021; Katznelson & Gerke, 2021). Most recently, AI has played a pivotal role in the containment of the COVID-19 pandemic in the public sphere by allowing the identification, tracking, and forecasting of breakout locations through analyzing news reports, social media platforms and government documents (Ayoko, 2021; Pillai & Kumar, 2021). Improving efficiencies, in general, is regarded as an important contributor to an uplift in the economy in various sectors (European Commission, 2020; Stahl et al., 2022), and reducing environmental damage arising from inefficiencies of production.

On the one hand, the benefits that AI technologies offer to the development of various fields have led to an annual investment of more than $25 billion to further enhance its capabilities (Bughin et al., 2017; McGregor & Banifatemi, 2018). However, this vast investment could potentially lead to prioritizing financial returns while compromising the potential harm that AI could bring to societies (McGregor & Banifatemi, 2018). On the other hand, ethical issues of AI have become a huge concern that dominates government, media, and academic discourses in recent years.

Recognizing AI capabilities in fulfilling the tasks that were once reserved for humans and their abilities to surpass human efficiencies in carrying out a diverse range of activities could lead to fundamental changes to humanity and the way we live our lives (Stahl et al., 2022). According to empirical research, ethical problems of AI adoption were identified as dominating the organizational practices in the private and public sectors (Stahl et al., 2021). Addressing ethical issues is crucial to providing approaches to governing AI and ensuring the benefits of their adoption are maintained while diminishing their adverse consequences. Billions of dollars are allocated by governments and private organizations to find solutions for mitigating AI ethical issues while creating avenues to utilize AI for the betterment of societies and the environment (McGregor & Banifatemi, 2018). For example, in 2017, IMB has allocated a $5 million prize for AI start-ups and researchers who participate in the Watson AI XPRIZE (AIXP) four-year competition to identify the most pressing problems in various domains, such as sustainability and education, and to design AI solutions for world-improving impact (McGregor & Banifatemi, 2018).

Based on the foregoing discussion, this chapter sheds light on some of the most pressing ethical issues that result from the adoption of AI-powered tools. Increasing inequality, widening social and economic gaps, compromising privacy and data protection, outsmarting humans and impacting human rights, lack of accountability, liability and reliability, and lack of empathy and sympathy are considered the most pressing challenges that need to be addressed for AI and big data (Katznelson & Gerke, 2021; Santiago, 2020; Stahl et al., 2021). Other issues may include lack of transparency, bias and discrimination, loss of human decision-making, the concentration of power, lack of access to and freedom of information, artificial stupidity, racism, security, singularity and robot rights (Ayoko, 2021; Farhud & Zokaei, 2021; Santiago, 2020; Stahl et al. 2021; Sun & Medaglia, 2019). This chapter also provides insight into strategies that are currently in place to overcome adverse implications of AI in the public and private sectors. Providing insight into these ethical challenges along with the governing solutions makes a significant contribution to the ongoing discourse and urges for providing sustainable solutions that are necessary for the ethical application of these technologies in different fields.

TYPES OF ETHICAL ISSUES EMERGING FROM AI ADOPTION

In this section, we discuss some of the most pressing ethical issues, drawing on case studies to demonstrate the extent to which these ethical issues can play a role in jeopardizing values in business operations while also identifying relevant strategic solutions associated with every case. In the following sections, inequality and increasing social and economic gaps, compromising privacy and data protection, outsmarting humans and impacting human rights, lack of accountability and liability and lack of empathy and sympathy will be thoroughly discussed.

Inequality and Increasing Social and Economic Gaps

One of the most critical ethical issues is concerned with AI's potential to further increase inequality and widen the gaps between social and economic classes within and between different countries (Farhud & Zokaei, 2021; Nemitz, 2018). The benefits that AI brings to the efficiencies and competitiveness of organizations are reflected in the overall economic growth of the countries in which these organizations operate. Therefore, AI would bring a great advantage to developed countries that afford the deployment of such technologies and would contribute to further increasing their economic empowerment. On the other hand, limited access to AI technologies would further increase the barriers to AI adoption in different fields, which would adversely impact the overall economic growth of less developed countries (Nemitz, 2018). In this way, an increase in the social and economic gap between countries is expected to occur as a result of disadvantaging developing and less developed countries. The concentration of economic power and the growing dominance of giant internet companies are considered a huge threat to equality and opportunities for diminishing economic gaps (Nemitz, 2018; Stahl et al., 2022).

The lack of equal opportunities for adopting the advanced technology of AI is also believed to increase the gaps between social and economic classes within societies (Farhud & Zokaei, 2021; Nordling, 2019). The rise of digital transformation has the capacity to decrease the marginal costs of goods and services while increasing productivity. This efficiency in the production of goods and services leads to an increase in the productivity of capital (Brynjolfsson & McAfee, 2014). Therefore, an increase in the income of the owners and users of AI technologies would further increase the economic gap between them and disadvantaged groups. Providing an equal opportunity for accessing the intellectual property, public services, and information, and equal distribution of resources among individual groups and sections of society is questionable. In addition to an uneven wealth distribution among individuals, the digital transformation has the capacity to replace human-based activities in the workplace, and thus, eliminate employment opportunities in many fields where AI is relied upon (Mokyr, Vickers, & Ziebarth, 2015). AI efficiencies, automated systems, and the vast development of robots could increase unemployment, and cause loss of employment for many individuals as a result of the amplified requirements and expectations of AI knowledge and skills (Farhud & Zokaei, 2021; Haenlein & Kaplan 2019). For example, in the medical field, the dramatic rise of surgical robots and robotic nurses has threatened future job opportunities for healthcare workers (Farhud & Zokaei, 2021; Stahl et al., 2022).

Inequality within societies can further be exacerbated by biases arising from the inadequate representation of different community groups from the data that AI technologies utilize (Stahl et al., 2022). For instance, gender-biased machine learning is a consequence of utilizing information and data which are highly dominated by a specific gender and their failure to equally represent males, females, and others within a given dataset (Stahl et al., 2022). There is a risk that algorithms trained on male-dominated datasets may result in inaccurate or unreliable outputs from AI systems. Scholars identify the consequences resulting from machine learning reliability on historical data which were unrepresentative of community groups as potentially resulting in false interpretations and biased models (Criado Perez, 2019; Stahl et al., 2022). There is a growing discourse related to how these biases would further adversely impact some groups, or genders, in contemporary society. For example, biases involving human resource (HR) practices where AI systems are responsible for screening applicants based on prejudiced attributes potentially disadvantaging minority groups and races (Balasubramaniam et al. 2020; Liang et al., 2021).

Despite corporate guidelines on fairness in contemporary organizations that would preserve the rights of all groups to be equally represented, removing biases from AI data could still be problematic. While eliminating biases in AI data analysis is highly dependent on organizational management's subjective judgments. In some organizations, decisions related to the representation of genders, race and other diversities are highly subjective to the nature of particular projects, incidents, or situations. A biased AI system utility would impact the inclusion and exclusion of data sets. In this way, an overrepresentation, or an underrepresentation, of some groups would influence the fairness of the result. Bias and gaps in the data urge for recognizing the significance of establishing ethical guidelines which are required to maintain an equal representation of gender, race, and group diversity (Balasubramaniam et al. 2020). Fairness and unbiases among multiple groups of a community are very crucial to maintaining the trust of communities and customers in different sectors (Balasubramaniam et al. 2020). Consequently, these ethical issues would potentially threaten community trust in the capabilities of organizations and systems to make decisions and implement operations that concern any individual in the correct manner and follow the correct processes. Providing equal opportunities, protecting equal rights, and eliminating biases and discrimination is vital for maintaining societies, which makes AI ethical data representations issues of profound significance that could potentially impact entire societies.

Case Study

A case study of ten organizations operating in a variety of industries was conducted by Macnish et al. (2019) to gain an understanding of the perceptions of the ethical application of AI software for asset tracking through which the personal data of customers are collected. The researchers conducted interviews with 42 employees (at least two from each organization) who utilize this software to understand their perception of AI's ethical application in different organizations. These organizations were selected based on the applications of AI technologies in different social domains as follows: employee monitoring and administration, government, agriculture, sustainable development, science, insurance, energy and utilities, communications, media and entertainment, retail and wholesale trade, and manufacturing and natural resource.

The researchers found that despite considerable attempts to ensure ethical application of the software, there has been a number of ethical issues arising from the concentration of AI data collection, ownership and access by the owners of the software. Potential discrimination and inequality may arise because the access rights and permissions to the collected data are controlled by the subjective judgment of the administrators, who may choose to exclude some groups from accessing the data.

Key Takeaways

- The benefits that AI brings to the efficiencies and competitiveness of organizations would further increase the economic growth of countries in which these organizations operate and challenge the economic growth of AI-deprived counties.
- The concentration of economic power and dominance of giant internet companies threatens equality and opportunities for diminishing economic gaps.
- Gaps between social and economic classes, discrimination, and inequality within societies would exacerbate by further depriving disadvantaged groups who have less access to and knowledge about AI technologies

Compromising Privacy and Data Protection

One of the most crucial components of AI is machine learning. The availability of big data and the development in computing power have enabled machine learning techniques to lead to a variety of innovations, such as facial recognition, natural language processing, and autonomous driving (Horvitz, 2017; Ryan, 2020; Stahl et al., 2022). AI has recently played a huge role in the containment of the Covid-19 pandemic by its capacity to identify clusters and hotspot locations of the virus using technologies that allow location tracking of the infected individual, such as Bluetooth, GPS, social graph, contact details, digital card transaction data and systematic physical address (Garofalo et al., 2020; Pillai & Kumar,

2021). However, these capacities raise ethical concerns about the capabilities of machine learning to analyze and process large amounts of data and neural networks which may include personal information that may compromise the de-identification of individuals (Criado Perez, 2019; O'Neil, 2016; Stahl et al., 2022). Issues pertaining to the protection and confidentiality of data are arising from AI's control over personal data, which in turn impact data security and integrity. Although the high-tech life that we live nowadays has always raised questions related to data privacy and protection, AI has further aggravated this problem by allowing new mechanisms to generate data connections, identify patterns and make predictions. This ability may lead to possible identifications of anonymized individuals (Stahl et al., 2022; Stahl & Wright, 2018), which challenges the individual's ability to protect personal information (Balasubramaniam et al., 2020). It may also challenge the ability of individuals to control access to their own information (White et al, 2019). For example, individuals may be unaware of the extent of data collection that is gathered by the devices that they regularly use, such as smartphones and web tracking. Additionally, predictive analysis resulting from AI's ability to create patterns using personal data could be utilized to infer potential information which may extend beyond what is provided or disclosed by an individual. Similarly, it has the potential to deduct an individual's information that may be counted as irrelevant or inconsistent (White et al, 2019).

The expectations of data privacy may differ from one country to another, and they are inconsistent across the globe (Cheah et al., 2021; Quach et al., 2022). The European Union (EU) complies with the General Data Protection Regulation (GDPR), which was first enacted to amend privacy legislation in other countries, such as the United States and Canada. According to these regulations, all personal data and the activities of foreign communities and companies are processed by the union-based data processor or controller to protect the information of natural persons with sufficient protection (Balasubramaniam et al., 2020; Chazette, Karras, & Schneider, 2018). In this way, data collection, access and protection of privacy expectations among employees in the US are different from expectations among European citizens working on behalf of employers; for example (Balasubramaniam et al., 2020).

Additionally, some organizations in the US provide additional data privacy based on specific fields that require more sensitive information. For example, the Genetic Information Non-discrimination Acts (GINA) is an organization in the US that ensures the privacy protection of individuals by prohibiting employers from discriminative acts based on the genetic health information of individuals (Farhud & Zokaei, 2021). Scholars argue that the discrepancy in privacy expectations could be higher among Chinese citizens who are expected to have further lower expectations of privacy than US citizens. The discrepancy in privacy expectations between countries results in a rise of further ethical questions concerning the extent of data collection and access conducted by one country based on following its own set of ethical rules and guidelines which may not be conforming with the set of ethical guidelines by another country. Therefore, there is a pressing need for the issuance of universally agreed-upon measures related to data collection and privacy (Balasubramaniam et al., 2020).

Not only between countries but ethical guidelines concerning data privacy can also differ between organizations and institutions within the same country. In their case study conducted on three companies in different industries, Balasubramaniam et al. (2020) concluded that while all the companies under the EU comply with GDPR, the emphasis on the significance of data privacy in each company may differ based on the reliance on data privacy in each sector. For example, the banking and financial services sector is more sensitive to prioritizing necessary measures to emphasize, and guarantee, the protection of customers' data and information than other sectors. Companies in the banking and financial services sector provided data protection and ethics courses that mainly focus on privacy, privacy issues, privacy regulations, data security, and ways to use data in compliance with the GDPR. Sections of training material offered to employees are dedicated to complying with privacy laws in the EU and maintaining customer trust and privacy. Employees understand the significance of protecting the privacy rights of customers when it comes to cautious utilization of the data collected by the company. The considers the customers' extensive rights to be informed and have control over their own data. Additionally, data anonymization is widely employed to observe customers' behaviors, while avoiding individuals' identification. The customer's right to be informed of how their private information is utilized by the AI system within an organization is also a crucial measure that organizations undertake to maintain data protection.

In compliance with GDPR, companies have to inform and allow customers to check and exercise their rights to data privacy while ensuring that minimal sensitive information about users is collected to maintain individuals' privacy and safety. In retail, individuals' consent must be granted prior to usage of their data. Moreover, customers have to be informed about their right to decline permission to the usage of their information. Customers also must have the right to understand the purpose of collecting their data and the scope and purpose of utilizing their data Balasubramaniam et al. (2020). Despite these measures, in some sectors, such as healthcare, legislation is still regarded as insufficient to protect individuals' data (Farhud & Zokaei, 2021). For example, clinical data collected by AI-powered tools, such as robots, can be hacked into and used for malicious purposes that minimize privacy and security. Some social networks gather and store large amounts of users' data, for instance, individuals' mental health data, without their consent, which can be helpful in the marketing, advertising, and sales of these companies. Further, some genetics testing and bioinformatics companies, which are not legal or closely monitored, sell customer data to pharmaceutical and biotechnology companies (Farhud & Zokaei, 2021). Although individuals have the right to consent (or decline) access to information, privacy issues are still persistent in the case of errors or malfunctioning robotic medical devices (Farhud & Zokaei, 2021). In case of errors, these individuals are unclear who to hold responsible for the compensated privacy of their information.

Case Study

Balasubramaniam et al. (2020), conducted a case study on three different scales companies to investigate how ethical practices of AI are being met, including preserving data privacy. These companies operate in different sectors: a software consultancy company of 500 employees, retail with 22,500 employees, and a banking company with 12,300 employees. The researchers concluded that despite varying implications of adopting AI to each domain, the three organizations had similar methods to overcome privacy issues and ensure compliance with GDPR. For instance, although customer trust is regarded as more essential for the operations of the banking sector, data privacy, personal data protection, data security, and data anonymization are the foundation of defining ethical guidelines in all the case companies. Assigning an employee responsible to identify any possible misconduct in each project is also practiced to maintain customer privacy and organizational values. Their findings are highly consistent with other scholars who ascertain the necessity of assigning teams with diverse competencies for the effective operation of AI systems to comply with national-level AI ethical guidelines.

Key Takeaways

- AI machine learning capacities to attain the vast amount of personal information and generate data connections could lead to the possible identification of anonymized individuals, and thus, compromise data privacy ;
- Individuals' control over the access and collection of their information is limited, due to limited awareness of data collection mechanisms and predictive analysis which generate information that extends beyond what is originally disclosed.
- Data collection, access and protection of privacy expectations as well as ethical guidelines governing data privacy are inconsistent between countries, organizations, and institutions.

Outsmarting Humans and Impacting Human Rights

Limited comprehension arising from the functioning of the algorithms (CDEI, 2019; Stahl et al., 2022) is another ethically problematic concern. The exact operations of AI systems are sometimes considered highly ambiguous, among users, system developers, and data scientists, leading to a lack of transparency (Hagendorff, 2019; Stahl et al., 2022). The opacity to explaining and understanding what happens inside the operating systems of AI is limiting data scientists and users to add value (Balasubramaniam et al., 2020; Stahl et al., 2022). This lack of transparency raises some concerns as further problems could currently be concealed from human comprehension and could be discovered later in the future. The lack of transparency also leads to worries about the accountability of data interpretation, analyses and use. The difficulty to ascertain or foresee problems that could be discovered in the future could only be assessed as further understanding and knowledge become available.

AI's swift development represents another concern that would potentially challenge the nature of human lives and humans in the future. The vast and quick development of AI technologies would challenge the performance of humans in their abilities to analyze and interpret situations, carry out tasks efficiently, and solve problems. There are threatening alerts arising from AI machines' own potential to metaphysically enhance their own nature and capabilities in ways that make them become more conscious (Carter et al., 2018; Dehaene Lau, & Kouider, 2017; Stahl et al., 2022), and thus potentially become moral issues and raise further concerns about robots rights (Floridi & Sanders, 2004; Wallach et al., 2011; Stahl et al., 2021; Stahl, 2004). Further, exponential growth in AI capability or AI's superintelligence (Bostrom, 2016; Kurzweil, 2006; Tipler, 2012; Torrance 2012; Stahl et al., 2021) is also a threat to the nature of the relationship that would arise between humans and machines and alter change human societies and communities (Stahl et al., 2022). In a field experiment, AI-powered tools were proven to win over human intelligence in games, and thus, were believed to have the capacity to triumph in all walks of life (Liang et al., 2021).

The general pervasiveness of AI machines that allow close connection between humans and AI, such as wearable devices or implants, raises further threats concerning the potential implications of such connection. Wearable devices, such as fitness trackers, can record an individual's heart rate, breathing capacity and body temperature, and thus are highly perceived to be improving individuals' capacities of receiving timely information regarding one's body, and thus, improve the quality of life. For example, such devices can improve the healthcare system by directly feeding data into larger AI-assisted analytic networks with details related to a patient's blood pressure and insulin monitor records (Pillai & Kumar, 2021; Rajkumar & Ameen, 2018).

On the other hand, some doubts regarding the alternation of the natures of humans and the quality of human lives could be possible implications of the close human-machine integration. For instance, concerns humans' abilities to perceive available solutions or options due to their dependencies on AI machines, could highly impact their problem-solving skills and hinder human abilities to take action and enjoy the freedom of choice. Therefore, worries regarding a possible limitation of human autonomy are expected as a result of the rapid development of machines' autonomy. Such ethical issues urge data scientists to operate straightforward systems that they are fully capable of comprehending and interpreting for users to overcome transparency and unforeseen future issues. Systems developers should be responsible for demonstrating the capability to explain the kind of data that systems operate and how these data are being utilized by AI technologies (Balasubramaniam et al., 2020).

A Case Study on AI in the Service of Human Decision-Making in Debate Tutoring

A case study was conducted to investigate the implications of AI adoption in evaluating the performances of trainees in tutoring. Cukurova, Kent and Luckin (2019) compared the results of AI-powered tools versus human expert tutors of 127 questionnaires and 47 audio recordings to examine the efficiency of the decision-making process and the accuracy of outcome evaluations given to the trainees. The researchers concluded that multimodal transparent decision-making models provide a more efficient process for expert tutors to reflect upon their own decisions and enable them to provide more accurate and detailed feedback to trainees during evaluations. The qualitative notes reported by the expert tutors to justify their decision about the social and emotional characteristics of trainees were found more intuitive and insufficiently detailed compared to the ones received from using the multimodal. The multimodal enabled detailing of specific measures that were utilized to assess the social and emotional characteristics of trainees. Therefore, the multimodal did not only provide a clear justification for the outcome decision but also enabled pinpointing specific areas for development which in turn aids reflection upon the final decision.

Key Takeaways

- There is a limited comprehension resulting from the ambiguity of the functioning of algorithms and the exact operations of AI systems which leads to a lack of transparency.
- The difficulty to ascertain or foresee problems that could possibly be discovered in the future could only be assessed as further understanding and knowledge become available.
- The increased dependencies on AI devices could highly impact human problem-solving and decision-making abilities.

Accountability and Reliability

The difficulties in understanding the functioning of AI systems' algorithms have also contributed to ethical issues related to AI accountability and liability (CDEI, 2019; Tóth et al., 2022). Limited ability to entirely understand how these systems operate would consequently result in limited ability to detect, prevent or solve issues that may result from any malfunctioning in data interpretation or analysis. As a result, questions related to who should be held responsible for any possible errors are dominating the current discourse (Liang et al., 2021; Stahl et al., 2022). For example, in the event of malicious conduct resulting from AI robots' misinterpretation of data, or an error, identifying a complicit wrongdoer would be problematic (Tóth et al., 2022). The question of accountability in this instance would have to deal with the complexity of determining whether the AI robot, the system developers, implementing organization, overseeing managers, industry regulators, or governments should be held responsible (Tóth et al., 2022).

The inability to determine accountability for any wrong interpretation and subsequent errors in judgment could immensely impact individuals if occurred in sectors related to medicine, healthcare, or robot surgery, to name a few. As current machine learning techniques allow drawing conclusions from large data sets, the quality of these underlying datasets determines the quality of outputs. A prominent example of such systems would be AI systems applied in health, for example for the purpose of diagnosis of radiological data. If such systems are to be used successfully in clinical practice, they need to be highly reliable, a feature that currently few systems exhibit (Stahl et al., 2022, Topol, 2019). The reliability of the system and its outputs are highly dependent on the quality and accuracy of data processing.

Scholars similarly highlight the relationship, and overlapping, of ethical issues, such as inequality and unaccountability that could arise from AI systems (Liang et al., 2021; World Economic Forum, 2020) and the global impact of these issues. For example, in the case of employing external human resource (HR) organizations for recruiting candidates for organizations in a different country, AI HR applications could engage in unethical behavior, such as racism, inequality, or biases. Different ethical frameworks among different nations would further exacerbate discrimination problems in the workplace in this instance (Liang et al., 2021; Stahl et al, 2022). More specific issues deal with harm to physical integrity (e.g. through injuries suffered because of autonomous vehicles) and impact on health due to health-related AI.

Case Study on the Attitudes of Women Towards the Reliability of AI Use for Mammogram Reading for Breast Cancer Screening

Regular breast screening among women is essential to facilitate an early detection of breast cancer and improve outcomes for women. The National Health Service (NHS) Breast Screening Programme (NHSBSP) in England invites more than two million women between the ages of 50 and 70 years for an annual screening test. AI technologies are employed to support the reading of the high volume of images produced. To examine the attitude of women toward the reliability of AI deployment in breast cancer screening, Lennox-Chhugani et al. (2021) conducted a case study among over 18 years who are invited by the NHSBSP for screening.

The case study comprised a qualitative and quantitative survey of 4,096 and six focus groups of 25 women in July 2020 to understand the attitudes of a sample of women to the use of AI in breast screening. They found that women of screening age (≥50 years) were less likely to use technology platforms or applications for healthcare advice than women under screening age. Women of screening age were also less likely to trust the recommendations and diagnoses of AI technologies than women under screening age. Women of the screening age were less likely to agree that AI can have a positive effect on society than women under the screening age and was more likely to express concerns about the reliability and safety of technology, lack of trust and the absence of the human touch in interactions. They concluded that women of screening age are open to the use of AI in breast cancer screening; however, AI adoption in radiology is likely to be more acceptable among women in the future. The willingness of women of screening age to use AI in breast screening is moderated by the ability to understand how AI operates, the evidence to support AI's accuracy, and the use of AI to augment and not replace human interaction and decision-making. It is important for regulators of health technology to understand women's attitudes to increase the acceptance and adoption of AI-based technology and to emphasize education and dissemination of information about the use of AI in the medical field.

Key Takeaways

- Limited comprehension of AI systems could result in limited ability to detect, prevent or solve issues that may result from malfunctioning in data processing, interpretation, or analyses.
- In the occasion of direct or indirect harm resulting from AI errors, attributing responsibility is a complex process.
- Inaccuracy of data interpretations could further exacerbate societal issues, such as racism, inequality, and discrimination.

Lack of Empathy and Sympathy

The loss of human contact resulting from the dependency on AI machines and robots raises concerns about empathy and sympathy (Bellucci, Venkatraman, & Stranieri, 2020). Despite the immense benefits that AI brings to many industries, sectors such as elders and seniors care facilities, social work, medical and healthcare would suffer the most from the lack of empathy caused by human-machine interactions (Bellucci et al., 2020; Stahl & Coeckelbergh, 2016). For example, AI signifies a revolutionary concept in healthcare that transforms the clinical practices of physicians (Emanuel & Wachter, 2019; Pillai & Kumar, 2021). The adoption of advanced AI tools has been playing an immense role in preventing the spread of disease among communities through aiding health infrastructure, treatment, and surveillance (Pillai & Kumar, 2021). AI-based systems have improved screening and diagnoses, by radiology and pathology, which increased the diagnostic efficiency (Hsieh, 2017; Pillai & Kumar, 2021; Pearl, 2018), treatment and surgical procedures (Pillai & Kumar, 2021; Wottawa et al., 2016), reduce treatment costs (He et al., 2019; Pillai & Kumar, 2021) and increase early interventions (Hu et al., 2020; Mohamadou, Halidou & Kapen, 2020; Pillai & Kumar, 2021; Raza, 2020).

The implications of AI adoption in the same field; however, are raising a lot of ethical concerns resulting from human-machine interactions (Bellucci et al., 2020; Stokes & Palmer, 2020). For instance, medical consultations would deprive patients of experiencing the uniquely human emotions that autonomous medical systems fail to provide (Farhud & Zokaei, 2021). Although robotic physicians and nurses would revolutionize the healthcare system by providing precision and timeliness services, they are highly incapable of conversing with patients during treatments in an empathetic and compassionate environment. The consequence of losing human-to-human interactions, in many cases, might adversely impact the healing process and the emotional and psychological state of patients (Farhud & Zokaei, 2021). The impacts would even be more intensely noticed among children, who are more likely to experience anxieties and fear when engaging with healthcare professionals (Farhud & Zokaei, 2021). Similarly, the administration of treatment by robotic machines would highly impact vulnerable patients, such as individuals with psychiatric disorders (Farhud & Zokaei, 2021).

An Investigation of the Ethics of Caring for Robotics Use in Nursing

As nurses have traditionally been regarded as clinicians that deliver compassionate, safe, and empathetic health care. Stokes and Palmer (2020) investigated AI care robots' capability to provide caring as a fundamental characteristic, expectation, and moral obligation of the nursing and caregiving professions. The researchers situated their study within a synthesis of the literature on the ethics of nursing and AI and explored three dominant theories of caring and the two paradigmatic expressions of caring (touch and presence). Through their investigation, they conclude that AI is incapable of caring in the sense central to nursing and caregiving ethics. For AI to be implemented ethically, it cannot transgress the core values of nursing, usurp aspects of caring that can only meaningfully be carried out by human beings, and it must support, open, or improve opportunities for nurses to provide the uniquely human aspects of care. The researchers argue that any division of tasks or skills between AI and humans in caregiving professions must preserve caring and caring touch and presence as the core concern. Since AI lacks the capacity for caring and genuine expressions of care, activities most closely associated with caring ought to always have some degree of human involvement. The researchers further suggest that AI care bots may be permissible only if their application will not transgress the core values of nursing (i.e., caring), they wouldn't disrupt aspects of caring that can only be carried out by human nurses, and they must support, expand or improve opportunities for human nurses to provide aspects of care.

Key Takeaways

- Human-machine interactions could result in a lack of empathy which could impact some sectors, such as elders and senior care facilities, social workers, medical, and healthcare.
- Anxieties and fears could impact vulnerable groups when interacting directly with machines.

STRATEGIC SOLUTIONS TO OVERCOME ETHICAL ISSUES OF AI

Despite the importance of addressing the issues that AI applications could bring to humans and industries, which is the primary focus of this chapter, it is also worthy to address some mitigation strategies that are in place in an attempt to manage these issues. In this section, the book explores policies and strategies that are implemented to ensure responsible utilization of AI-powered tools in accordance with ethical guidelines. There are more than 80 public and private AI ethical guidelines documentation to date aiming to regulate AI use across different sectors (Balasubramaniam et al., 2020; Stahl et al., 2021).

A study by Stahl et al., 2021 provided a summary of the high-level guideline publications by three established expert groups: European Commission's (EU) Ethical guidelines for trustworthy AI, Institute of Electrical and Electronics Engineers (IEEE) Ethically Aligned Design, and Software and Information Industry Association's (SIIA) Ethical Principles for Artificial Intelligence and Data Analytics. They concluded that transparency, autonomy, fairness, safety, privacy ethical guidelines and others are covered in these documents to overcome some of the immense ethical problems of AI in all types of public and private sector organizations. Ethical guidelines on monitoring AI adoption's impact on the well-being of individuals and society also do exist (Balasubramaniam et al., 2020; Stahl et al., 2021). Ongoing monitoring is very significant for the early detection of risks of misuse of AI systems that would potentially cause any harm.

Besides the availability of these guiding documents on a societal level, organizational-level regulations are also in place to ensure that professional operations are abiding by ethically accepted practice. Stahl et al. (2022) identified five core strategies to alleviate AI ethical problems: organizational awareness and reflection (including stakeholder engagement, ethics review boards, adoption of codes and standards), ethical technical application (including anonymization, secure storage, hackathons and penetration testing), overcoming human oversight to ensure trustworthiness, conducting ethics training and education on technology use and balancing competing goods (including control over data and public-private partnerships. Recognizing the significance of adopting strategies, along with highly enforced guidelines and regulations, are essential to safeguard corporate responsibility to maintain social welfare and create positive values for the wellbeing of individuals, society, and the environment.

SUMMARY

The digital transformation provided by AI-powered tools has radically improved the efficiencies and competitiveness of organizations. AI computation and data analysis capabilities empower organizations to organize resources, make strategic operational decisions and improve productivity (Ayoko, 2021, Foerster-Metz et al., 2018). However, the swift dominance of AI technologies in organizational practices in the private and public sectors has raised a diverse range of ethical concerns. This book chapter aims to discuss the most pressing ethical issues while providing an overview of solutions to overcome these issues. This chapter does not provide an exhaustive exploration of all the ethical issues; rather it is designed to provide an overview of the most prominent ones and act as a launch pad for further exploration of potential solutions. Addressing these concerns is crucial to reflect the current discourse that gained momentum in both sectors and to urge potential approaches to diminish the adverse consequences which can jeopardize organizational values and ethical operations.

Firstly, AI has the potential to further increase inequality and widen the gaps between social and economic classes within and between different countries. The benefits that AI brings to the efficiencies and competitiveness of organizations would further increase the economic growth of countries in which these organizations operate and challenge the economic growth of AI-deprived counties. Similarly, the concentration of economic power and dominance of giant internet companies threatens equality and opportunities for diminishing economic gaps. Gaps between social and economic classes, discrimination, and inequality within societies would exacerbate by further depriving disadvantaged groups who have limited access to and knowledge about AI technologies worsening their income and employment opportunities.

Secondly, the adoption of AI technologies could compromise privacy and data protection. AI's capacity to attain vast amounts of personal information and generate data connections could lead to the possible identification of anonymized individuals and predict information that extends beyond what is originally disclosed. Individuals' awareness of and control over the access and collection of their information is limited. Thirdly, AI technologies could outsmart humans and impact human rights due to limited comprehension of the ambiguous functioning of algorithms and the exact operations of AI systems. This lack of transparency makes it difficult to ascertain problems that could arise in the future while jeopardizing humans' problem-solving and decision-making abilities as they become increasingly dependent on AI devices.

Fourthly, AI technologies could impact accountability, liability and reliability. Limited comprehension of AI systems could result in limited ability to detect, prevent or solve issues that may result from malfunctioning in data processing, interpretation, or analyses. Additionally, inaccuracy of data interpretations could further exacerbate societal issues and cause direct or indirect harm. While attributing responsibility to a specific party or individual is a complex process. Finally, human-machine interactions result in a lack of empathy which could impact some sectors, such as elders and senior care facilities, social workers, and medical and healthcare. Anxieties, fears and lack of understanding could influence vulnerable groups when interacting directly with machines.

This chapter also provides insight on strategies that are currently in place to overcome adverse implications of AI in the public and private sectors, including more than 80 public and private AI ethical guidelines documentation, such as EU, IEEE and SIIA. Societal and organizational level regulations are in place to ensure transparency, autonomy, fairness, and privacy guiding the adoption of AI technologies and overcoming challenges of unethical conduct. Calls to ensure government regulations are in place are essential to safeguard corporate responsibility and maintain social welfare.

REFERENCES

Ayoko, O. B. (2021). Digital Transformation, Robotics, Artificial Intelligence, and Innovation. *Journal of Management & Organization, 27*(5), 831–835. doi:10.1017/jmo.2021.64

Balajee, N. (2020). *What is Artificial Intelligence?* https://nanduribalajee.medium.com/what-is-artificial-intelligencec68579db123

Balasubramaniam, N., Kauppinen, M., Kujala, S., & Hiekkanen, K. (2020). *Ethical Guidelines for Solving Ethical Issues and Developing AI Systems* (1st ed.). Product-Focused Software Process Improvement. doi:10.1007/978-3-030-64148-1_21

Bellucci, E., Venkatraman, S., & Stranieri, A. (2020). Online dispute resolution in mediating EHR disputes: A case study on the impact of emotional intelligence. *Behaviour & Information Technology, 39*(10), 1124–1139. doi:10.1080/0144929X.2019.1645209

Bostrom, N. (2016). *Superintelligence: paths, dangers, strategies, reprint edition*. OUP Oxford.

Brynjolfsson, E., & Mcafee, A. (2014). *The second machine age: Work, progress, and prosperity in a time of brilliant technologies*. W. W Norton & Company.

Carter, O., Hohwy, J., van Boxtel, J., Lamme, V., Block, N., Koch, C., & Tsuchiya, N. (2018). Conscious machines: Defining questions. *Science, 359*(6374), 400–400. doi:10.1126cience.aar4163 PMID:29371459

Centre for Data Ethics and Innovation (CDEI). (2019). *Interim report: review into bias in algorithmic decision-making*. Centre for Data Ethics and Innovation. https://www.gov.uk/

Chazette, L., Karras, O., & Schneider, K. (2018). *Do end-users want explanations? Analyzing the role of explainability as an emerging aspect of non-functional requirements*. Academic Press.

Cheah, J. H., Lim, X. J., Ting, H., Liu, Y., & Quach, S. (2020). Are privacy concerns still relevant? Revisiting consumer behaviour in omnichannel retailing. *Journal of Retailing and Consumer Services*, 102242.

Cukurova, M., Kent, C., & Luckin, R. (2019). Artificial intelligence and multimodal data in the service of human decision-making: A case study in debate tutoring. *British Journal of Educational Technology, 50*(6), 3032–3046. doi:10.1111/bjet.12829

Dehaene, S., Lau, H., & Kouider, S. (2017). What is consciousness, and could machines have it? *Science*, *358*(6362), 486–492. doi:10.1126cience.aan8871 PMID:29074769

Emanuel, E. J., & Wachter, R. M. (2019). Artificial intelligence in health care: Will the value match the hype? *Journal of the American Medical Association*, *321*(23), 2281–2282. doi:10.1001/jama.2019.4914 PMID:31107500

European Commission. (2020). *White Paper on Artificial Intelligence: A European approach to excellence and trust*. Author.

European Commission. (2021). *Ethics Guidelines for Trustworthy AI*. https://ec.europa.eu/

Farhud, D. D., & Zokaei, S. (2021). Ethical Issues of Artificial Intelligence in Medicine and Healthcare. *Iranian Journal of Public Health*, *50*(11). Advance online publication. doi:10.18502/ijph.v50i11.7600 PMID:35223619

Floridi, L., & Sanders, J. W. (2004). On the morality of artificial agents. *Minds and Machines*, *14*(3), 349–379. doi:10.1023/B:MIND.0000035461.63578.9d

Foerster-Metz, U. S., Marquardt, K., Golowko, N., Kompalla, A., & Hell, C. (2018). Digital transformation and its implications on organizational behavior. *Journal of EU Research in Business*, *2018*(S3).

Haenlein, M., & Kaplan, A. (2019). A brief history of artificial intelligence: On the past, present, and future of artificial intelligence. *California Management Review*, *61*(4), 5–14. doi:10.1177/0008125619864925

HagendorffT. (2019). *The ethics of AI ethics—an evaluation of guidelines*. International Centre for Ethics in the Sciences and Humanities. arXiv: 190303425.

Hanelt, A., Bohnsack, R., Marz, D., & Antunes Marante, C. (2021). A systematic review of the literature on digital transformation: Insights and implications for strategy and organizational change. *Journal of Management Studies*, *58*(5), 1159–1197. doi:10.1111/joms.12639

Hasan, R., Thaichon, P., & Weaven, S. (2021). Are we already living with Skynet? Anthropomorphic artificial intelligence to enhance customer experience. In *Developing digital marketing*. Emerald Publishing Limited. doi:10.1108/978-1-80071-348-220211007

Hasan, R., Weaven, S., & Thaichon, P. (2021). Blurring the line between physical and digital environment: The impact of artificial intelligence on customers' relationship and customer experience. In *Developing digital marketing*. Emerald Publishing Limited. doi:10.1108/978-1-80071-348-220211008

He, J., Baxter, S. L., Xu, J., Xu, J., Zhou, X., & Zhang, K. (2019). The practical implementation of artificial intelligence technologies in medicine. *Nature Medicine*, *25*(1), 30–36. doi:10.103841591-018-0307-0 PMID:30617336

Horvitz, E. (2017). AI, people, and society. *Science*, *357*(6346), 7–7. doi:10.1126cience.aao2466 PMID:28684472

Hsieh. P. (2017). AI in medicine: rise of the machines. *Forbes*.

IEEE. (2022). Ethically Aligned Design. *IEEE Ethics in Action in Autonomous and Intelligent Systems*. https://ethicsinaction.ieee.org/

Jamal, D. N., Rajkumar, S., & Ameen, N. (2018). Remote elderly health monitoring system using cloud-based wbans. In *Handbook of research on cloud and fog computing infrastructures for data science* (pp. 265–288). IGI Global. doi:10.4018/978-1-5225-5972-6.ch013

Katznelson, G., & Gerke, S. (2021). The need for health AI ethics in medical school education. *Advances in Health Sciences Education : Theory and Practice*, *26*(4), 1447–1458. doi:10.100710459-021-10040-3 PMID:33655433

Kretschmer, T., & Khashabi, P. (2020). Digital transformation and organization design: An integrated approach. *California Management Review*, *62*(4), 86–104. doi:10.1177/0008125620940296

Kurzweil, R. (2006). *The singularity is near.* Gerald Duckworth and Co Ltd.

Lam, K., Iqbal, F. M., Purkayastha, S., & Kinross, J. M. (2021). Investigating the Ethical and Data Governance Issues of Artificial Intelligence in Surgery: Protocol for a Delphi Study. *JMIR Research Protocols*, *10*(2), e26552. doi:10.2196/26552 PMID:33616543

Lennox-Chhugani, N., Chen, Y., Pearson, V., Trzcinski, B., & James, J. (2021). Women's attitudes to the use of AI image readers: A case study from a national breast screening programme. *BMJ Health & Care Informatics*, *28*(1), e100293. doi:10.1136/bmjhci-2020-100293 PMID:33795236

Liang, T., Robert, L., Arbor, A., Sarker, S., Cheung, C. M. K., Matt, C., Trenz, M., & Turel, O. (2021). Artificial intelligence and robots in individuals' lives: How to align technological possibilities and ethical issues. *Internet Research*, *31*(1), 1–10. doi:10.1108/INTR-11-2020-0668

MacnishK.RyanM.GregoryA.JiyaT.AntoniouJ.HatzakisT.. (2019). D1.1 Case studies. *De Montfort University.* doi:10.21253/DMU.7679690.v3

Magistretti, S., Pham, C. T. A., & Dell'Era, C. (2021). Enlightening the dynamic capabilities of design thinking in fostering digital transformation. *Industrial Marketing Management*, *97*, 59–70. doi:10.1016/j.indmarman.2021.06.014

McCarthy, J. (1959). "Programs with Common Sense" at the Way back Machine (archived October 4, 2013). *Proceedings of the Teddington Conference on the Mechanization of Thought Processes,* 756-91.

Mohamadou, Y., Halidou, A., & Kapen, P. T. (2020). A review of mathematical modelling, artificial intelligence and datasets used in the study, prediction and management of covid-19. *Applied Intelligence*, *50*(11), 3913–3925. doi:10.100710489-020-01770-9 PMID:34764546

Mokyr, J., Vickers, C., & Ziebarth, N. (2015). The History of Technological Anxiety and the Future of Economic Growth. *The Journal of Economic Perspectives*, *29*(3), 31–50. doi:10.1257/jep.29.3.31

Nguyen, T. M., Quach, S., & Thaichon, P. (2021). The effect of AI quality on customer experience and brand relationship. *Journal of Consumer Behaviour*. Advance online publication. doi:10.1002/cb.1974

Nilsson, N. J. (2014). *Principles of Artificial Intelligence*. Morgan Kaufmann.

Nordling, L. (2019). A fairer way forward for AI in health care. *Nature*, *573*(7775), s103–s105. doi:10.1038/d41586-019-02872-2 PMID:31554993

Pearl R. (2018). *Artificial intelligence in healthcare: Separating reality from hype*. Academic Press.

Perez, C. (2019). *Invisible women: Exposing data bias in a world designed for men, 0* (1st ed.). Chatto and Windus.

Pillai, S. V. & Kumar, R. S. (2021). *The role of data-driven artificial intelligence on COVID-19 disease management in public sphere: A review*. doi:10.1007/s40622-021-00289-3

Quach, S., Thaichon, P., Martin, K. D., Weaven, S., & Palmatier, R. W. (2022). Digital technologies: Tensions in privacy and data. *Journal of the Academy of Marketing Science*, *50*(6), 1–25. doi:10.100711747-022-00845-y PMID:35281634

Raza, K. (2020). Artificial Intelligence Against COVID-19: A Meta-analysis of Current Research. In A. Hassanien, N. Dey, & S. Elghamrawy (Eds.), *Big Data Analytics and Artificial Intelligence Against COVID-19: Innovation Vision and Approach* (pp. 165–176). Springer. doi:10.1007/978-3-030-55258-9_10

Ruiz-Real, J. L., Uribe-Toril, J., Torres, J. A., & De Pablo, J. (2021). Artificial intelligence in business and economics research: Trends and future. *Journal of Business Economics and Management*, *22*(1), 98–117. doi:10.3846/jbem.2020.13641

Ryan, M. (2020). In AI we trust: Ethics, artificial intelligence, and reliability. *Science and Engineering Ethics*, *26*(5), 2749–2767. doi:10.100711948-020-00228-y PMID:32524425

Software and Information Industry Association (SIIA). (2017). *Ethical Principles for Artificial Intelligence and Data Analytics*. SIIA.

Stahl, B. C. (2004). Information, ethics, and computers: The problem of autonomous moral agents. *Minds and Machines*, *14*(1), 67–83. doi:10.1023/B:MIND.0000005136.61217.93

Stahl, B. C., Antoniou, J., Ryan, M., Macnish, K., & Jiya, T. (2022). Organisational responses to the ethical issues of artificial intelligence. *AI & Society*, *31*(1), 23–37. doi:10.100700146-021-01148-6

Stahl, B. C., & Coeckelbergh, M. (2016). Ethics of healthcare robotics: Towards responsible research and innovation. *Robotics and Autonomous Systems*, *86*, 152–161. doi:10.1016/j.robot.2016.08.018

Stokes, F., & Palmer, A. (2020). Artificial Intelligence and Robotics in Nursing: Ethics of Caring as a Guide to Dividing Tasks Between AI and Humans. *Nursing Philosophy*, *21*(4), e12306. doi:10.1111/nup.12306 PMID:32609420

Taylor, E. J. (2020). *AI and global governance: covid-19 underlines the need for AI- support governance of water-energy food nexus, AI and Global Governance*. Centre for Policy Research.

Tipler, F. J. (2012). Inevitable existence and inevitable goodness of the singularity. *Journal of Consciousness Studies*, *19*(1-2), 183–193.

Topol, E. J. (2019). High-performance medicine: The convergence of human and artificial intelligence. *Nature Medicine*, *25*(1), 44–56. doi:10.103841591-018-0300-7 PMID:30617339

Torrance, S. (2012). Super-intelligence and (super-)consciousness. *International Journal of Machine Consciousness*, *4*(2), 483–501. doi:10.1142/S1793843012400288

Tóth, Z., Caruana, R., Gruber, T., & Loebbecke, C. (2022). The Dawn of the AI Robots: Towards a New Framework of AI Robot Accountability. *Journal of Business Ethics*, *178*(4), 895–916. doi:10.100710551-022-05050-z

Wallach, W. A., Allen, C. B., & Franklin, S. C. (2011). Consciousness and ethics: Artificially conscious moral agents. *International Journal of Machine Consciousness*, *3*(1), 177–192. doi:10.1142/S1793843011000674

White, G., Ariyachandra, T., & White, D. (2019). Big Data, Ethics, and Social Impact Theory – A Conceptual Framework. *Journal of Management & Engineering Integration*, *12*(1), 9–15.

World Economic Forum. (2020). *Model artificial intelligence governance framework and assessment guide.* World Economic Forum.

Wottawa, C. R., Genovese, B., Nowroozi, B. N., Hart, S. D., Bisley, J. W., Grundfest, W. S., & Dutson, E. P. (2016). Evaluating tactile feedback in robotic surgery for potential clinical application using an animal model. *Surgical Endoscopy*, *30*(8), 3198–3209. doi:10.100700464-015-4602-2 PMID:26514132

Chapter 14
Intelligence Augmentation via Human-AI Symbiosis:
Formulating Wise Systems for a Metasociety

Nikolaos Stylos
University of Bristol, UK

ABSTRACT

Intelligence augmentation (IA) facilitates a new systems perspective to frame the value outcome of the interaction between human and AI agents. The factors that can optimize this collaborative integration of the multi-agent system are investigated and discussed. Different kinds of knowledge approaches are met in various contexts to create an optimized IA system in service settings. In this respect, AI agents are not just tools but rather co-creators of value that can influence human agents' learning cycles. Hence, humans' effective interaction with AI agents produces a learning effect that can empower humans' interpretative capability. This chapter focuses on IA and shows that IA is not only a theoretical paradigm but also serves as a platform to facilitate the transition from smart services to wise service innovation to the benefit of both the multi-agent system benefitting service organizations and the consumers too. Potential challenges are also discussed from a societal viewpoint.

1. INTRODUCTION

The paradigm of Intelligence Augmentation (IA) facilitates a new systems perspective to frame the value outcome of the interaction between human and AI agents. The factors that can optimize this collaborative integration of the multi-agent system are investigated and discussed in detail. The different service-learning experiential cycles of a) human agents (service employees) and consumers, and b) AI agents, respectively, may develop concurrently, to form a synergistic systems service cycle (Barile & Polese, 2010; Stylos, 2020).

DOI: 10.4018/978-1-6684-9591-9.ch014

Different kinds of knowledge approaches are met in various contexts to create an optimized IA system in service settings. In this respect, AI agents are not just tools but rather co-creators of value that can influence human agents' learning cycles. Hence, humans' effective interaction with AI agents produces a learning effect that can empower humans' interpretative capability, i.e. an IA effect (Barile et al., 2021; Markauskaite et al., 2022). Therefore, the mutual learning effects between humans and AI agents can develop co-creative services that are based on priorities set by both humans and AI agents.

With a focus on combining the strengths of human and machine or artificial intelligence, IA has the possibility to fundamentally change the role that technology plays in our work and life (Zhou et al., 2021). The significant advancement and accessibility of AI technologies and related technology infrastructure has made the present a historical moment for IA. As IA enters our daily lives and workplace, it creates tremendous opportunities to observe how humans interact with IA, which will provide ample evidence and help researchers and scientists develop new theories to explain and guide efforts to cultivate human-machine synergy for human goods, improve enabling technologies, and even prepare and train the future workforce (Maddikunta et al., 2022). The relationship between humans and machines will likely change as technology evolves. Ultimately, IA supports the creation of value for humans, be it business outcomes, cognition enhancement, or innovations. Thus, as a research field, IA has strong real-world relevance and requires a multidisciplinary perspective that considers goals, human factors/context, effective and efficient technologies, the ways in which humans and technologies interact, governance structures, and environmental constraints and their feedback (Dwivedi et al., 2021; Goodell & Craig, 2022).

Objectives

This chapter has two objectives. First, it seeks to clarify aspects of IA, and properly position it across the spectrum of collaboration between the animate and inanimate agents. Second, it discusses the transition from a smart to wise service systems by utilizing Industry 4.0 technologies in conjunction to human intelligence to form superior intelligent interactions in Society 5.0.

Contributions

First, it provides an account of different types of IA and shows that IA is not only a theoretical paradigm but also serves as a platform to facilitate the transition from smart services to wise service innovation (Kamboj & Gupta, 2020; Sarmah et al. 2018) to the benefit of both the multi-agent system (Humans and AI agents) representing the service organizations and the consumers. Second, the factors that can optimize this collaborative integration of the multi-agent system are investigated and discussed in detail. Potential challenges and benefits are also discussed from a societal viewpoint, particularly in the context of metasociety.

Organization

The structure of the rest of the current chapter is summarized as follows. Section 2 offers an introduction to human-machine symbiotic relationship. Section 3 exemplifies various types of IA in a multi-agent system perspective in a human-AI symbiotic context. Then, the transition from Smart service to Wise service systems is explained in Section 4, including practical examples. Finally, Section 5 offers conclusions to provide an overview of this work, as well as the challenges that set the way forward.

2. HUMAN-MACHINE SYMBIOTIC RELATIONSHIP IN A DECISION-MAKING AND PROCESSING CONTEXT

Smart machines have become like humans by recognizing voices, processing natural language, learning, and interacting with the physical world through their vision, smell, touch and other senses, mobility, and motor control. In some cases, they do a much faster and better job than humans at recognizing patterns, performing rule-based analysis on very large amounts of data, and solving both structured and unstructured problems. Significant publications have talked about how smart machines with AI may take jobs from humans by replacing them. In contrast, people do not perceive IA as threatening since it does not replace humans but enhances their capabilities.

IA is intended to function with human individuals and concentrates on developing systems that can amplify and enhance human cognition. The notion of IA effectively emerges from promoting the collaboration of AI agents with humans in a frame of co-existence to achieve mutual benefits (Yao, 2023). In essence, IA advocates the implementation of information technology in extending human abilities and competences (Abbas et al., 2022). Still, IA considers human individuals as the protagonists of human-machine collaboration (Correia et al., 2019). That's why it is deemed that IA helps humans make and be responsible for taking decisions in an evidence-informed decision-making pattern.

Nowadays, many people augment their abilities via utilizing new technologies, e.g. using AI-powered voice assistants and other tools which represent a new generation of cognitive collaborators (Acikgoz et al., 2023). These additional technologies capabilities help humans reach more in-depth understandings stemming from huge amounts of structured and unstructured data, which will boost both creativity and productivity (Wamba-Taguimdje et al., 2020). It has been predicted that in the next few years, many people will be utilizing various types of app tools, assistants, collaborators, and mediators in their smartphones or other devices within an IA context. Cognitive mediators represent an evolution in both technology capability and social trust (Han & Yang, 2018; Nalepa et al., 2019).

IA improves human's performance with the help of information technology. Key aspect of IA is the human-machine symbiosis (see Figure 1 for examples); in this setting machines are utilized only for the tasks they perform best, e.g., computing, recording, and routine tasks, while the human counterparts focus on undertakings and responsibilities, such as inventing, abstract reasoning, and imaginative/inspirational tasks. This concept proposes that implementing AI is positioned as a midway between solely human-driven and automated-only capabilities (Földes et al., 2021), thereby taking the form of tools that can help improve the efficiency of human intelligence (Leyer & Schneider, 2021). Thus, AI and IA represent two competing approaches, which act interchangeably depending on the technological and organizational developments.

AI and IA have formed a new research area widely acknowledged as computational intelligence (Duan et al., 2019). This refers to researching adaptive mechanisms to allow and expedite contemporary intelligence patterns in our everchanging and fast-moving technological era. It is important to clarify that IA is not synonymous to AI, as they have essentially different objectives, although their respective inherent technologies very much overlap (Engelbart, 2023). IA primarily concentrates on supporting people become smarter, while AI is mainly concerned with the level of machines' smartness. Unlike the traditional view that sees AI as autonomous systems that can fully automate tasks, workflows, and/or business processes and operate without human involvement. Specifically, IA promotes AI agents' implementation to function in support of or in collaboration with humans to augment human agents' decision-making and competences through capitalizing on and combining sophisticated data sources (Paul et al., 2022).

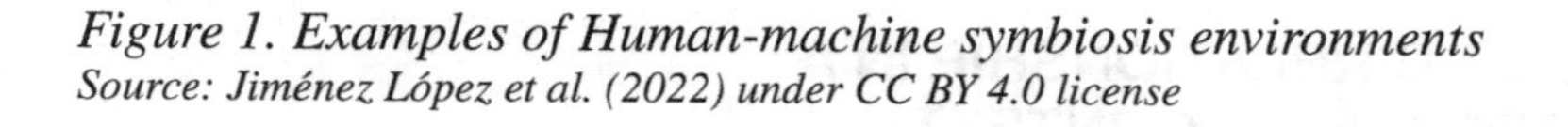

Figure 1. Examples of Human-machine symbiosis environments
Source: Jiménez López et al. (2022) under CC BY 4.0 license

The highlight in defining IA is the relationship between human and machines. IA forms various aspects of augmentation depending on the case, e.g. relating to improving power output capability, perception, memory skills, and problem-solving / decision making processes. The key in utilizing new technologies is that humans need to consider how machines or bots may contribute to various tasks and how much they should count on them. Advancing our knowledge on intelligence theories can help scholars better understand the characteristics that characterize human vs. machine intelligence. In this vein, human agents (so far) surpass artificial agents in skills that drive interpersonal, creative, and contextual areas of intelligence, while AI have surpassed humans in many computational and analytical logic areas of intelligence. Thus, notwithstanding machines can perform intelligently from a cognitive perspective, as well as artificial agents can identify emotional intelligence too, still they have not been equipped with the necessary skills for taking a holistic approach in the form of general intelligence during the third decade of 2000s.

Overall, an automation-augmentation continuum involves a mix of two different types of cognitive systems: biological and digital. AI focuses on developing and advancing technology to think or act like humans, whereas IA focuses on expanding and extending humans' abilities. Automation with AI versus augmentation with AI represent two different aspects of the same abilities AI agents perform highly accurately. Notwithstanding fully routine jobs go with the first option, still partially non-routine jobs will be open to some form of automation that augments humans who supervise their non-routine aspects (Pedota & Piscitello, 2022).

The scholars and practitioners working and favoring the AI perspective believe to the direction of autonomous systems that imitate or may replace human cognitive functions, whereas the ones supporting the IA perspective put their efforts in utilizing contemporary information technologies for supplementing or supporting human cognition so humans will remain at the epicenter of human-AI interactions. In addition to efficiency, and performance, there are other decisive characteristics such as user satisfaction, perception, effectiveness and other key factors for designing and evaluating IA. IA wants to externalize human consciousness in a machine, it focuses on the mind/body in a context and most importantly IA wants to produce a dependent machine. Another crucial point to make is that AI shouldn't be employed in situations where end goals and inputs are not well defined, thereby prompting IA to play a crucial role (Arrieta et al., 2020). Due to high levels of uncertainty humans face constantly, machines can support them in decision-making but seem to be unable to completely replace humans in performing several activities (Keding, 2021). This is where the concept of human-in-the-loop augmented intelligence comes to play via human-AI collaboration. Finally, humans' central role in IA not only promotes the trust, reciprocity, and synergistic collaboration of human and machine intelligence, but also relieves from the concerns associated with AI systems' proliferation (Samuel et al., 2022).

3. TYPES OF IA IN A MULTI-AGENT SYSTEM PERSPECTIVE

Considering a multi-agent system perspective, there are particular types of IA which are evident in the following existing practical applications: virtual automated assistants, robotic applications, drones and autonomous vehicles. To better delineate these applications, a robust understanding of the underlying human-AI intelligence interactions need to be discussed.

Drawing from the theories of intelligence, human intelligence and artificial intelligence can be deemed as the two extremes of the same continuum. A continuum of human-AI intelligence interaction exists that is ranging from situations where machines are repeating many of the tasks humans are already doing (assisted) to enabling humans to do more than they are currently capable of doing (augmented) to fully accomplishing tasks on their own without human intervention (autonomous) (Xu et al., 2023). In assisted-intelligence approaches, AI agents may undertake the action, but human agents are making the decisions. Then, in augmented intelligence methods, AI agents are taking actions and making decisions in collaboration with humans i.e., there is collaborative human-machine decision making. And, lastly, in autonomous systems, machines are both doing the actions and making the decisions unassisted by humans. In summary, intelligence augmentation enables organizations and humans to do things they could not otherwise do in an optimized way (Enholm et al., 2022).

Artificial intelligence demonstrates the necessary abilities to address well-defined problems with a rather confined focus and range, and of recurring nature. On the other hand, humans perform well at identifying and resolving non-standardized problems or making decisions and acting creatively which require flexibility and skills for dynamic action. Some artificially intelligent agents have outpaced human agents in certain types of specific tasks that have clear boundary conditions (Melville et al., 2023). Besides, humans have general intelligence, which allows them to perform a wide range of tasks. Many complex tasks such as intuition, conceptual understanding, moral judgement, reflective cognition, empathy, and sensation are representative of humans' advantages which enable them to reason, strategize, and handle uncertainty in all kinds of situations (Myllylä, 2022).

From the previous discussion it is deducted that human intelligence and artificial intelligence complement each other and that integrating them can potentially improve human wellbeing. Human beings have evolved by building advanced information technologies and a society that is rich in its communications, thus gradually leading to the acceleration of information storage abilities, speed processing, towards co-evolution and human-AI symbiosis (Bozdag, 2023; Zhang et al., 2023). IA can be acknowledged in different fields of modern business innovation, various robotic applications, drones and autonomous vehicles.

Virtual automated assistants are another example of IA in practice. Autonomous vehicles enable automated driving, yet humans are still needed to be supervising and in control of the overall system as there could be complications or faults to the technical part of the system. For safety reasons, these vehicles may continue to require human involvement. Robots are also working together with humans in an IA context, and they are mainly allocated risky, heavy-duty, or/and repetitive tasks. A system design framework for a human-robot interaction system called human-augmented robotic intelligence (HARI) has also been proposed (Mruthyunjaya & Jankowski, 2020), and multiple benefits may emerge from pairing robots with humans. As a matter of fact, IA mobilizes IoT which is utilized within the streams of intelligent asset monitoring (Malik et al., 2021 et al., 2022). Drones are also contributing to IA systems, which are widely used for inspections, filming, and delivering items to remote areas.

In conclusion, the IA conceptualization signifies a collaborative and integrated intelligence that allows the business systems to evolve from a smart system to a wise system by integrating the action of the rational component with the emotional one, hence putting back humans at the center of the process (Jain et al., 2021).

4. FROM SMART SERVICE TO WISE SERVICE SYSTEMS

Many researchers recognize that an emerging technology such as IA creates exciting new prospects for individuals and organizations. In this context, IA technologies have greatly evolved from being a supportive technology to be serving decision making through a collaborative role (Roschelle, 2021; Thiele, 2021). There are theories which apply to the human-computer interactions, however these may not apply in this new information era with users co-existing with intelligent technology. Under these circumstances, these theories may need to adapt to the contemporary IA context or new theories may need to be introduced (Ali et al., 2023; Zhang et al., 2023). This way, a more effective theorization could be reached that would accurately portrait the evolving and complex nature of the interaction between IA technology and individuals (Bretin et al., 2023).

IA showcases a type of transformation of the structure and process of human thinking which is enhanced by artificial cognitive processes to make them wiser rather than simply smarter. The 'wise' system concept has been developed within the service science theoretical context (Zhao et al., 2022; Pargman et al., 2019) which introduces wise service systems, stressing the importance of linking AI to human values towards achieving a positive influence to the future generations of humanity. Emerging from a viable systems approach, knowledge creation of a wise service system is processed via IA (Cukurova et al., 2019; Wang et al., 2022). As viable systems interact with other systems equipped with the resources needed for their viable functioning, the concept of relevance is of vital importance. This concept was developed in the interpretive framework of the viable systems approach is of great help to understand the role of wise integrated systems in interaction with technology. In this process, the concept of relevance provides the necessary theoretical support the capability of the integrated system to prepare and form the viability of it.

In this vein, the notions of intelligence and wisdom can play together a major role in the integrated and hybrid-systems perspective. As per the viable systems approach, intelligence represents the system capability to thrive in a complex environment and at the same time help support system identity (Kashef et al., 2021; Sydelko et al., 2023). In this sense, intelligence forms system's viability, and shapes the spatial and temporal dimensions in which contemporary technologies evolve, towards creating a balance between AI and human intelligence. The establishment of a well-balanced relationship between AI agents and humans ensures the system integration alongside the society, creating a consonant coevolution. Then, wisdom is the foundational value of a system that can generate deep meaning, thus supporting production processes of pervasive and collective significance (Ravetz, 2020; Yao et al., 2022). Consequently, wisdom advances the creation of collective intelligence in achieving both individual and collective goals.

From this viewpoint, wise systems are equipped to build integrated intelligence which is capable not only in increasing the solving problem skills but also gaining ability to clearly define them (Barile et al., 2021; Fügener et al., 2022). In essence, wisdom gives intelligence the ability of conceptualizing a wide perspective with varied relevance in regard to the most proper possibilities of the challenge, i.e. the integrated system is able to envision much better solution options (Schranz et al., 2021). In this manner, the concept of wisdom comes to play in order to create better decision options and solutions from a holistic viewpoint. This contributes to clarifying the specificity of the challenge to be solved and unraveling the link between humans and AI agents towards reaching universal solutions.

As intelligence augmentation has been gradually progressing conceptually, relevant applications have started appearing that exemplify human-AI symbiotic systems. Examples can be found in the healthcare sector, finance industry, engineering and innovation, military operations, retailing, and interior space design. In healthcare, AI systems are employed in a hybrid human-machine context to help medical doctors analyze medical data and provide insights, assist them in making better diagnoses and treatment decisions, and support surgeons in performing surgical operations more precisely and effortlessly. In finance, AI agents join humans to analyze market data and offer investment recommendations that can possibly be more accurate in predictions compared to those made by humans based on an assistive type of technology. With respect to engineering applications, where efficiency, effectiveness and reliability of processes and outputs are key qualities, three-dimensional imaging and analytics are employed in a hybrid human-AI mode to detect flaws and areas of concern requiring preventive/corrective maintenance and repairs; a similar approach is taken in the creation of modern, sophisticated and high-end machinery, which can save resources – i.e. increased productivity – both in the design and production stages, as well as while during it utilization. The military operations hugely benefit from a human-AI symbiosis, as this collaboration can prove extremely beneficial not only in the real-time decision-making, accuracy and precision of achieving the objectives in the field of operations, but mainly in protecting the lives of military personnel due to proactivity and resilience offered by utilizing proper flows of intelligence. Furthermore, digital signage and three-dimensional mapping which are offered via apps and accessed through smartphones are key for store navigation, collaborative decision-making on a human-AI basis when decisions could be rather complicated, thus reaching best decisions on a personalized basis. Finally, some really interesting applications of human-AI agents collaborations are already available in the designing interiors that would be not only be suitable aesthetically but also making best possible use of space in ergonomic terms.

Overall, IA introduces a range of options that can optimize the hybrid human-AI symbiosis. IA is practically generated by adopting interpretative schemes and value categories which make the human-AI systems wiser in decision-making. In this context, selection processes can greatly improve by aligning the system contributors towards expressing both rational and emotional aspects (Polese et al., 2022). Both the rational and the emotional aspects improve suggested solutions, with the first one focusing on optimization approaches (as per AI components), while the second one concentrates on proper balancing the selections made (as per IA logic). Consequently, humans have very good reasons to put efforts for emotional and social intelligence in addition to rational intelligence to create wise decision-making models and systems via IA (Alzoubi et al., 2021; Ren et al., 2023).

5. CONCLUSIONS AND CHALLENGES

IA is multidisciplinary in nature, and related research has started appearing in critical mass only recently, although there are various applications in this context (Fui-Hoon Nah et al., 2023). As humans would possibly want to keep, at least partial, control of decision-making and actioning of various processes in business, personal, and social relationships, there is need for examining the collaborative nature of employing AI, rather than just rely to automated processes that are fully governed by AI agents. This would turn humanity's focus from a smart orientation towards a wise system orientation of processing information and decision-making in a service industry context. Regarding theoretical aspects of IA there are some initial developments, but there is not an established overarching theory yet. Besides, a deep understanding of technology is needed from a multi-faceted perspective before theories could be possibly established. Thus, different research paradigms, methods, and involvement with different disciplines (computer science, engineering, social science, psychology, etc.) are required. Multidisciplinarity is in line with the need for more diverse viewpoints to gain more holistic understanding towards achieving high IA performance. For example, IA may provide new opportunities for humans to observe and understand their cognitive biases and behavioral and emotional patterns. IA enables the wise systems which are able to manage processes holistically and in an equifinal way, thus surpassing the scope of smart systems or intelligence ones, since they solve problems for collective benefits or even the intelligent ones (mainly able to understand).

The current development of IA targets the merging of human capabilities with self-learning AI. At some point, this development will hit moral and ethical limits earlier than technological limits (Dengel et al., 2021). Therefore, an ethical framework must be established to set out how AI should be used to prepare for the next-generation of IA (Hassani et al., 2020; Reis et al., 2023). Fairness is dynamic and of social nature which may not be trusted to automation (Höddinghaus et al., 2021; Teodorescu et al., 2021). By analyzing the characteristics of AI and IA, we have determined that the goal of the human species is IA by making use of AI. At the same time, we must not make the mistake of replacing human capabilities by attempting to simply imitate the human brain. Separation from human self-awareness may become more and more problematic with the creation of artificial self-awareness (Benford et al., 2021; von Krogh et al, 2023).

In this vein, it seems it is humanity's best interest to invest on IA by choice, as is a safer route compared to attempting to replace or replicate human acting and overall behavior (Langer et al., 2021). The main disadvantage of an AI-focused strategy is that any attempt to replicate human brain limits the capacity of AI to producing an artificial, self-reinforcing learning brain (Sinz et al., 2019). There are also other issues of privacy and control nature because like any living entity the fundamental instinct of AI would be to protect its very existence. It would normally therefore strive for independent presence; therefore, once an artificial brain becomes aware of its own consciousness, it will logically strive for independence from any potentially "independence-threatening" entity which under certain circumstances that could be their own creators: human-beings.

REFERENCES

Abbas, S. M., Liu, Z., & Khushnood, M. (2022). When Human Meets Technology: Unlocking Hybrid Intelligence Role in Breakthrough Innovation Engagement via Self-Extension and Social Intelligence. *Journal of Computer Information Systems*, 1–18.

Acikgoz, F., Perez-Vega, R., Okumus, F., & Stylos, N. (2023). Consumer engagement with AI-powered voice assistants: A behavioral reasoning perspective. *Psychology and Marketing*, *40*(11), 2226–2243. Advance online publication. doi:10.1002/mar.21873

Ali, F., Dogan, S., Chen, X., Cobanoglu, C., & Limayem, M. (2023). Friend or a foe: Understanding generation Z Employees' intentions to work with service robots in the hotel industry. *International Journal of Human-Computer Interaction*, *39*(1), 111–122. doi:10.1080/10447318.2022.2041880

Alzoubi, H. M., & Aziz, R. (2021). Does emotional intelligence contribute to quality of strategic decisions? The mediating role of open innovation. *Journal of Open Innovation*, *7*(2), 130. doi:10.3390/joitmc7020130

Arrieta, A. B., Díaz-Rodríguez, N., Del Ser, J., Bennetot, A., Tabik, S., Barbado, A., ... Herrera, F. (2020). Explainable Artificial Intelligence (XAI): Concepts, taxonomies, opportunities and challenges toward responsible AI. *Information Fusion*, *58*, 82–115. doi:10.1016/j.inffus.2019.12.012

Barile, S., Bassano, C., Piciocchi, P., Saviano, M., & Spohrer, J. C. (2021). Empowering value co-creation in the digital age. *Journal of Business and Industrial Marketing*. Advance online publication. doi:10.1108/JBIM-12-2019-0553

Barile, S., & Polese, F. (2010). Smart service systems and viable service systems: Applying systems theory to service science. *Service Science*, 2(1/2), 21–40. doi:10.1287erv.2.1_2.21

Benford, S., Ramchurn, R., Marshall, J., Wilson, M. L., Pike, M., Martindale, S., Hazzard, A., Greenhalgh, C., Kallionpää, M., Tennent, P., & Walker, B. (2021). Contesting control: Journeys through surrender, self-awareness and looseness of control in embodied interaction. *Human-Computer Interaction*, *36*(5-6), 361–389. doi:10.1080/07370024.2020.1754214

Bozdag, A. A. (2023). AIsmosis and the pas de deux of human-AI interaction: Exploring the communicative dance between society and artificial intelligence. *Online Journal of Communication and Media Technologies*, *13*(4), e202340. Advance online publication. doi:10.30935/ojcmt/13414

Bretin, R., Khamis, M., & Cross, E. (2023). "Do I Run Away?": Proximity, Stress and Discomfort in Human-Drone Interaction in Real and Virtual Environments. In *IFIP Conference on Human-Computer Interaction* (pp. 525-551). Cham: Springer Nature Switzerland.

Correia, A., Paredes, H., Schneider, D., Jameel, S., & Fonseca, B. (2019). Towards hybrid crowd-AI centered systems: Developing an integrated framework from an empirical perspective. In *2019 IEEE International Conference on Systems, Man and Cybernetics (SMC)* (pp. 4013-4018). 10.1109/SMC.2019.8914075

Cukurova, M., Kent, C., & Luckin, R. (2019). Artificial intelligence and multimodal data in the service of human decision-making: A case study in debate tutoring. *British Journal of Educational Technology*, *50*(6), 3032–3046. doi:10.1111/bjet.12829

Dengel, A., Devillers, L., & Schaal, L. M. (2021). Augmented human and human-machine co-evolution: Efficiency and ethics. *Reflections on Artificial Intelligence for Humanity*, 203-227.

Duan, Y., Edwards, J. S., & Dwivedi, Y. K. (2019). Artificial intelligence for decision making in the era of Big Data–evolution, challenges and research agenda. *International Journal of Information Management, 48*, 63–71. doi:10.1016/j.ijinfomgt.2019.01.021

Dwivedi, Y. K., Hughes, L., Ismagilova, E., Aarts, G., Coombs, C., Crick, T., Duan, Y., Dwivedi, R., Edwards, J., Eirug, A., Galanos, V., Ilavarasan, P. V., Janssen, M., Jones, P., Kar, A. K., Kizgin, H., Kronemann, B., Lal, B., Lucini, B., ... Williams, M. D. (2021). Artificial Intelligence (AI): Multidisciplinary perspectives on emerging challenges, opportunities, and agenda for research, practice and policy. *International Journal of Information Management*, *57*, 101994. Advance online publication. doi:10.1016/j.ijinfomgt.2019.08.002

Engelbart, D. C. (2023). Augmenting human intellect: A conceptual framework. In *Augmented Education in the Global Age* (pp. 13–29). Routledge.

Enholm, I. M., Papagiannidis, E., Mikalef, P., & Krogstie, J. (2022). Artificial intelligence and business value: A literature review. *Information Systems Frontiers*, *24*(5), 1709–1734. doi:10.100710796-021-10186-w

Földes, D., Csiszár, C., & Tettamanti, T. (2021). Automation levels of mobility services. *Journal of Transportation Engineering. Part A, Systems*, *147*(5), 04021021. Advance online publication. doi:10.1061/JTEPBS.0000519

Fügener, A., Grahl, J., Gupta, A., & Ketter, W. (2022). Cognitive challenges in human–artificial intelligence collaboration: Investigating the path toward productive delegation. *Information Systems Research*, *33*(2), 678–696. doi:10.1287/isre.2021.1079

Fui-Hoon Nah, F., Zheng, R., Cai, J., Siau, K., & Chen, L. (2023). Generative AI and ChatGPT: Applications, challenges, and AI-human collaboration. *Journal of Information Technology Case and Application Research*, 1-28. . doi:10.1080/15228053.2023.2233814

Goodell, J., & Craig, S. D. (2022). What Does an Emerging Intelligence Augmentation Economy Mean for HF/E? Can Learning Engineering Help? *Proceedings of the Human Factors and Ergonomics Society Annual Meeting*, *66*(1), 460–464. doi:10.1177/1071181322661283

Han, S., & Yang, H. (2018). Understanding adoption of intelligent personal assistants: A parasocial relationship perspective. *Industrial Management & Data Systems*, *118*(3), 618–636. doi:10.1108/IMDS-05-2017-0214

Hassani, H., Silva, E. S., Unger, S., TajMazinani, M., & Mac Feely, S. (2020). Artificial intelligence (AI) or intelligence augmentation (IA): what is the future? *AI, 1*(2), 143-155.

Höddinghaus, M., Sondern, D., & Hertel, G. (2021). The automation of leadership functions: Would people trust decision algorithms? *Computers in Human Behavior*, *116*, 106635. Advance online publication. doi:10.1016/j.chb.2020.106635

Jain, H., Padmanabhan, B., Pavlou, P. A., & Raghu, T. S. (2021). Editorial for the special section on humans, algorithms, and augmented intelligence: The future of work, organizations, and society. *Information Systems Research, 32*(3), 675–687. doi:10.1287/isre.2021.1046

Jiménez López, E., Cuenca Jiménez, F., Luna Sandoval, G., Ochoa Estrella, F. J., Maciel Monteón, M. A., Muñoz, F., & Limón Leyva, P. A. (2022). Technical Considerations for the Conformation of Specific Competences in Mechatronic Engineers in the Context of Industry 4.0 and 5.0. *Processes (Basel, Switzerland), 10*(8), 1445. doi:10.3390/pr10081445

Kamboj, S., & Gupta, S. (2020). Use of smart phone apps in co-creative hotel service innovation: An evidence from India. *Current Issues in Tourism, 23*(3), 323–344. doi:10.1080/13683500.2018.1513459

Kashef, M., Visvizi, A., & Troisi, O. (2021). Smart city as a smart service system: Human-computer interaction and smart city surveillance systems. *Computers in Human Behavior, 124*, 106923. Advance online publication. doi:10.1016/j.chb.2021.106923

Keding, C. (2021). Understanding the interplay of artificial intelligence and strategic management: Four decades of research in review. *Management Review Quarterly, 71*(1), 91–134. doi:10.100711301-020-00181-x

Langer, M., & Landers, R. N. (2021). The future of artificial intelligence at work: A review on effects of decision automation and augmentation on workers targeted by algorithms and third-party observers. *Computers in Human Behavior, 123*, 106878. Advance online publication. doi:10.1016/j.chb.2021.106878

Leyer, M., & Schneider, S. (2021). Decision augmentation and automation with artificial intelligence: Threat or opportunity for managers? *Business Horizons, 64*(5), 711–724. doi:10.1016/j.bushor.2021.02.026

Maddikunta, P. K. R., Pham, Q. V., Prabadevi, B., Deepa, N., Dev, K., Gadekallu, T. R., ... Liyanage, M. (2022). Industry 5.0: A survey on enabling technologies and potential applications. *Journal of Industrial Information Integration, 26*, 100257. Advance online publication. doi:10.1016/j.jii.2021.100257

Malik, P. K., Sharma, R., Singh, R., Gehlot, A., Satapathy, S. C., Alnumay, W. S., Pelusi, D., Ghosh, U., & Nayak, J. (2021). Industrial Internet of Things and its applications in industry 4.0: State of the art. *Computer Communications, 166*, 125–139. doi:10.1016/j.comcom.2020.11.016

Markauskaite, L., Marrone, R., Poquet, O., Knight, S., Martinez-Maldonado, R., Howard, S., Tondeur, J., De Laat, M., Buckingham Shum, S., Gašević, D., & Siemens, G. (2022). Rethinking the entwinement between artificial intelligence and human learning: What capabilities do learners need for a world with AI? *Computers and Education: Artificial Intelligence, 3*, 100056. Advance online publication. doi:10.1016/j.caeai.2022.100056

Melville, N. P., Robert, L., & Xiao, X. (2023). Putting humans back in the loop: An affordance conceptualization of the 4th industrial revolution. *Information Systems Journal, 33*(4), 733–757. doi:10.1111/isj.12422

Mruthyunjaya, V., & Jankowski, C. (2020). Human-Augmented robotic intelligence (HARI) for human-robot interaction. In *Proceedings of the Future Technologies Conference (FTC) 2019: Volume 2* (pp. 204-223). Springer International Publishing.

Myllylä, M. (2022). Psychological and Cognitive Challenges in Sustainable AI Design. In *International Conference on Human-Computer Interaction* (pp. 426-444). Cham: Springer International Publishing. 10.1007/978-3-031-05434-1_29

Nalepa, G. J., Kutt, K., & Bobek, S. (2019). Mobile platform for affective context-aware systems. *Future Generation Computer Systems*, *92*, 490–503. doi:10.1016/j.future.2018.02.033

Pargman, D. S., Eriksson, E., Bates, O., Kirman, B., Comber, R., Hedman, A., & van den Broeck, M. (2019). The future of computing and wisdom: Insights from human–computer interaction. *Futures*, *113*, 102434. Advance online publication. doi:10.1016/j.futures.2019.06.006

Paul, S., Yuan, L., Jain, H. K., Robert, L. P. Jr, Spohrer, J., & Lifshitz-Assaf, H. (2022). Intelligence augmentation: Human factors in AI and future of work. *AIS Transactions on Human-Computer Interaction*, *14*(3), 426–445. doi:10.17705/1thci.00174

Pedota, M., & Piscitello, L. (2022). A new perspective on technology-driven creativity enhancement in the Fourth Industrial Revolution. *Creativity and Innovation Management*, *31*(1), 109–122. doi:10.1111/caim.12468

Polese, F., Barile, S., & Sarno, D. (2022). Artificial Intelligence and Decision-Making: Human–Machine Interactions for Successful Value Co-creation. In *The Palgrave Handbook of Service Management* (pp. 927–944). Springer International Publishing. doi:10.1007/978-3-030-91828-6_44

Rane, S. B., & Narvel, Y. A. M. (2022). Data-driven decision making with Blockchain-IoT integrated architecture: a project resource management agility perspective of industry 4.0. *International Journal of System Assurance Engineering and Management*, 1-19.

Ravetz, J. (2020). *Deeper city: Collective intelligence and the pathways from smart to wise*. Routledge. doi:10.4324/9781315765860

Reis, S., Coelho, L., Sarmet, M., Araújo, J., & Corchado, J. M. (2023). The Importance of Ethical Reasoning in Next Generation Tech Education. In *2023 5th International Conference of the Portuguese Society for Engineering Education (CISPEE)* (pp. 1-10). IEEE. 10.1109/CISPEE58593.2023.10227651

Ren, M., Chen, N., & Qiu, H. (2023). Human-machine Collaborative Decision-making: An Evolutionary Roadmap Based on Cognitive Intelligence. *International Journal of Social Robotics*, *15*(7), 1–14. doi:10.100712369-023-01020-1

Roschelle, J. (2021). Intelligence augmentation for collaborative learning. In *International Conference on Human-Computer Interaction* (pp. 254-264). Cham: Springer International Publishing.

Samuel, J., Kashyap, R., Samuel, Y., & Pelaez, A. (2022). Adaptive cognitive fit: Artificial intelligence augmented management of information facets and representations. *International Journal of Information Management*, *65*, 102505. Advance online publication. doi:10.1016/j.ijinfomgt.2022.102505

Sarmah, B., Kamboj, S., & Kandampully, J. (2018). Social media and co-creative service innovation: An empirical study. *Online Information Review*, *42*(7), 1146–1179. doi:10.1108/OIR-03-2017-0079

Schranz, M., Di Caro, G. A., Schmickl, T., Elmenreich, W., Arvin, F., Şekercioğlu, A., & Sende, M. (2021). Swarm intelligence and cyber-physical systems: Concepts, challenges and future trends. *Swarm and Evolutionary Computation, 60*, 100762. Advance online publication. doi:10.1016/j.swevo.2020.100762

Sinz, F. H., Pitkow, X., Reimer, J., Bethge, M., & Tolias, A. S. (2019). Engineering a less artificial intelligence. *Neuron, 103*(6), 967–979. doi:10.1016/j.neuron.2019.08.034 PMID:31557461

Stylos, N. (2020). Technological evolution and tourist decision-making: A perspective article. *Tourism Review, 75*(1), 273–278. doi:10.1108/TR-05-2019-0167

Sydelko, P., Espinosa, A., & Midgley, G. (2023). Designing interagency responses to wicked problems: A viable system model board game. *European Journal of Operational Research*. Advance online publication. doi:10.1016/j.ejor.2023.06.040

Teodorescu, M. H., Morse, L., Awwad, Y., & Kane, G. C. (2021). Failures of Fairness in Automation Require a Deeper Understanding of Human-ML Augmentation. *Management Information Systems Quarterly, 45*(3), 1483–1500. Advance online publication. doi:10.25300/MISQ/2021/16535

Thiele, L. P. (2021). Rise of the Centaurs: The Internet of Things Intelligence Augmentation. In *Towards an International Political Economy of Artificial Intelligence* (pp. 39–61). Springer International Publishing.

von Krogh, G., Roberson, Q., & Gruber, M. (2023). Recognizing and Utilizing Novel Research Opportunities with Artificial Intelligence. *Academy of Management Journal, 66*(2), 367–373. doi:10.5465/amj.2023.4002

Wamba-Taguimdje, S. L., Fosso Wamba, S., Kala Kamdjoug, J. R., & Tchatchouang Wanko, C. E. (2020). Influence of artificial intelligence (AI) on firm performance: The business value of AI-based transformation projects. *Business Process Management Journal, 26*(7), 1893–1924. doi:10.1108/BPMJ-10-2019-0411

Wang, B., Zheng, P., Yin, Y., Shih, A., & Wang, L. (2022). Toward human-centric smart manufacturing: A human-cyber-physical systems (HCPS) perspective. *Journal of Manufacturing Systems, 63*, 471–490. doi:10.1016/j.jmsy.2022.05.005

Xu, W., Dainoff, M. J., Ge, L., & Gao, Z. (2023). Transitioning to human interaction with AI systems: New challenges and opportunities for HCI professionals to enable human-centered AI. *International Journal of Human-Computer Interaction, 39*(3), 494–518. doi:10.1080/10447318.2022.2041900

Yao, X., Ma, N., Zhang, J., Wang, K., Yang, E., & Faccio, M. (2022). Enhancing wisdom manufacturing as industrial metaverse for industry and society 5.0. *Journal of Intelligent Manufacturing*, 1–21. doi:10.100710845-022-02027-7

Yao, Y. (2023). Human-machine co-intelligence through symbiosis in the SMV space. *Applied Intelligence, 53*(3), 2777–2797. doi:10.100710489-022-03574-5

Zhang, C., Wang, Z., Zhou, G., Chang, F., Ma, D., Jing, Y., Cheng, W., Ding, K., & Zhao, D. (2023). Towards new-generation human-centric smart manufacturing in Industry 5.0: A systematic review. *Advanced Engineering Informatics, 57*, 102121. Advance online publication. doi:10.1016/j.aei.2023.102121

Zhang, J., Qu, Q., & Chen, X. B. (2023). A review on collective behavior modeling and simulation: Building a link between cognitive psychology and physical action. *Applied Intelligence*, 1–30. doi:10.100710489-023-04924-7

Zhao, G., Li, Y., & Xu, Q. (2022). From emotion AI to cognitive AI. *International Journal of Network Dynamics and Intelligence*, 65-72.

Zhou, L., Paul, S., Demirkan, H., Yuan, L., Spohrer, J., Zhou, M., & Basu, J. (2021). Intelligence augmentation: Towards building human-machine symbiotic relationship. *AIS Transactions on Human-Computer Interaction, 13*(2), 243–264. doi:10.17705/1thci.00149

ADDITIONAL READING

Dahhani, A., Alloui, I., Monnet, S., & Vernier, F. (2022). Towards a Semantic Model for Wise Systems A Graph Matching Algorithm. *ADVCOMP 2022: The Sixteenth International Conference on Advanced Engineering Computing and Applications in Sciences.*

Stylos, N., Zwiegelaar, J., & Buhalis, D. (2021). Big data empowered agility for dynamic, volatile, and time-sensitive service industries: The case of tourism sector. *International Journal of Contemporary Hospitality Management, 33*(3), 1015–1036. doi:10.1108/IJCHM-07-2020-0644

Thiele, L. P. (2021). Rise of the Centaurs: The Internet of Things Intelligence Augmentation. In *Towards an International Political Economy of Artificial Intelligence* (pp. 39–61). Palgrave Macmillan.

Zhichang, Z. (2007). Complexity science, systems thinking and pragmatic sensibility. *Systems Research and Behavioral Science, 24*(4), 445–464. doi:10.1002res.846

KEY TERMS AND DEFINITIONS

Human-AI Symbiosis: It refers to humans and AI agents working together to enhance each other's capabilities, cooperatively undertake certain duties, and perform specific tasks to solve complex problems.

Intelligence Augmentation (IA): It emphasizes that AI is introduced to enhance human intelligence rather than replace it, by empowering each other in decision-making and effectively performing tasks via combined learning.

Metasociety: A parallel structure of a digital reality which reflects the values and standards of our physical lives and runs in the virtual cyberspace, able to effectively break the space, time, and economic restrictions, and create a new way of life, work, and communication for humans in real societies.

Smart Service Systems: A combination of physical and digital services, thus rendering integrated product-service systems and concentrating on digital, data-based service components.

Wise Systems: Integrated structures between AI with humans, which allow distributed software generate knowledge understandable by humans, thus enhancing human evaluation of wise systems' outputs.

Chapter 15
Artificial Intelligence in Sports: Monitoring Marathons in Social Media – The Role of Sports Events in Territory Branding

Natalia Vila-Lopez
University of Valencia, Spain

Ines Kuster-Boluda
https://orcid.org/0000-0002-8688-9175
University of Valencia, Spain

Francisco J. Sarabia-Sanchez
https://orcid.org/0000-0002-0370-2839
University Miguel Hernández, Spain

ABSTRACT

In the sports industry, artificial intelligence has become a powerful tool for sports managers interested in getting private sponsorships and for DMOs interested in branding a place. In this scenario, two main objectives guide this chapter (1) to generate a ranking of the leading Spanish marathons based on their presence on the four most important social networks in Spain (Facebook, Twitter, Instagram, and YouTube) and (2) to measure the engagement on social networks generated by the first of the marathons identified in the ranking. The official profiles of the accounts of the 10 marathons with the highest number of finishers in 2022 in Spain have been monitored on the social networks listed (Facebook, Twitter, Instagram, and YouTube). As the results show, a marathon can generate high network engagements. The destination's image can be highly favoured thanks to small local events (such as marathons) capable of generating a lot of movement on social networks. However, not all social networks work equally well in promoting sporting events capable of generating engagement.

DOI: 10.4018/978-1-6684-9591-9.ch015

INTRODUCTION

The sports industry is increasingly gaining momentum in many countries because of its exciting role in boosting tourism. Many tourist destinations increasingly rely on sporting events to help attract tourists. The analysis of social networks linked to sporting events with artificial intelligence allows public managers of tourist destinations and private companies that sponsor them to know the public's perceptions of the sporting services offered (sporting races in this case), understanding how often and in what context brands are mentioned (marathons in this case) and the engagement they can generate. In addition, AI can be used to identify which content on social networks is receiving the most attention, making it possible to predict the future popularity of certain types of sporting events over others. This information is helpful in boosting tourist destinations that host marathons and for private brands that sponsor them and want to achieve visibility. Moreover, and from an account manager's point of view, integrating AI and Social Blade can improve the capacity of accounts and their managers. Thus, this integration can be used for optimising the social media presence, helping to improve their benchmarking (adaptation and improvement of strategies according to the competitors' strategies), and improving the method for making predictions. At the time of writing this chapter, predictions are based on a simple linear regression method. However, using supervised or self-supervised neural networks can greatly improve these predictions and, therefore, can provide useful information for the channels managers.

Within sporting events, this research will focus on one specific sporting activity: marathons. Their growing prominence over the last decades is unquestionable (Zouni, Markogiannaki, Georgaki, 2021), as they are local events that do not require the preparation and investment of other major global sporting mega-events, such as the Olympiads or the World Cup. Despite this, given that these are recurring events, hosting this type of sporting activity can be a source of competitive advantage for a territory. It is not for nothing that in Spain, the ten largest Spanish cities have a marathon in their sports calendar (Pérez and Serrano, 2020). The marathon is the premier event in long-distance athletics. It is a race format that is standardized in its distance at a global level, covering 42 km and 195 m, facilitating comparisons. It can indeed be run on different surfaces, e.g. on asphalt or in the mountains.

As Naraine and Wanless (2020) have remarked, at no point has there been the volume, variety, and, consequently, the value of data being produced with velocity by organizations, customers, and other stakeholders. As these authors explain, in the sports industry, artificial intelligence has become a powerful tool for sports managers who must make crucial decisions at the right time with the correct information. Henceforth, the importance of developing intelligent information collection, analysis and management mechanisms to enable successful responses. In this vein, Pérez and Serrano (2020) remark that social networks have gone from being a digital platform to generating wealth for countries and businesses. At this point, monitoring the social media activity using of events is necessary to maximize its benefits for diverse stakeholders. The AI-based computer systems in charge of monitoring social networks show the user what is happening on the different platforms, speed up reaction times and facilitate strategic decision-making.

In the described scenario, this work pursues, as a generic objective, to show the main results of the first Barometer of Social Networks and Marathons in Spain, created to establish a simple and limited but representative picture of the projection in social networks of the leading marathons in Spain. Knowing and managing this volume of information in time is relevant to infer the role that the social networks of sporting events can play in the economic development of a territory. On the one hand, their proper management favors, at a public level, the use of sporting events in creating a city brand. On the other hand, it allows, at a private level, to attract sponsoring brands interested in leading events with a high level of dissemination on social networks.

More specifically, this study has two specific objectives: (i) First, to generate a ranking of the leading Spanish marathons based on their presence on the four social networks with the most significant dissemination in Spain (Facebook, Twitter, Instagram, and YouTube), and (2) second, to measure the engagement on social networks generated by the first of the marathons identified in the ranking. More specifically, this research will focus on measuring the virality of messages on social networks (followers, likes, etc.) to analyze the activity generated and discover the level of engagement that has been generated in relation to the number of followers, as well as other interesting ratios in business decision making (i.e. amplification rate, conversation rate, acceptance rate etc.).

To this end, the official profiles of the accounts of the ten marathons with the highest number of finishers in 2022 in Spain have been monitored on the social networks listed (Facebook, Twitter, Instagram and YouTube). This selection of networks responds to the desire to offer comparable data and the fact that these networks have been frequently compared in previous literature (Kumar and Sachdeva, 2022). We have taken as a starting point an initial population composed of all the asphalt marathons held in Spain during the year 2022. Specifically, in 2022, 26 asphalt marathons were held in Spain (Llanos, 2022) with a total of 52,836 finishers, representing an increase of 62% compared to 2021, a year that still saw many of its events cancelled due to the pandemic. Specifically, only 16 races were held. A broader perspective shows that the number is still far from the pre-pandemic level. For example, in 2019, 21 marathons were held in Spain but with 65,877 finishers. Table 1 shows the recovery of marathon running in Spain after the Covid crisis.

Table 1. Evolution of marathon finishers in Spain

Year	Female	% Female	Male	% Male	Total Finishers
2018	10,081	15.92%	53,245	84.08%	63,326
2019	10,787	16.37%	55,090	83.63%	65,877
2020	1,722	14.20%	10,408	85.80%	12,130
2021	4,858	14.85%	27,854	85.15%	32,712
2022	8,465	16.02%	44,371	83.98%	52,836

Source: Based on Llanos Arranz (2022)

Specifically, ten of the 26 asphalt marathons run in Spain have been selected for this barometer based on two criteria. First, a measure of representativeness has been followed. In fact, the ten marathons chosen are the most representative, i.e., with the highest participation, projection in social networks, and the highest score in the ranking drawn up by the Royal Spanish Athletics Federation (RFEA, 2021). (2021). Table 2 displays the RFEA ranking (2021) using a score based on technical-sporting criteria. This table shows the quality of the competition, but it does not evaluate aspects such as participation or its projection in social networks as an end in itself. The problem with using only this criterion is that establishing a follow-up according to the requirements set by the RFEA (2021) when establishing its ranking or directly analyzing all the events on social networks may not be appropriate for the year 2022. This is because during the 2021 course, especially in its initial months, the pandemic and its restrictions were still present. Not surprisingly, up to ten events held in 2022 did not occur in 2021. For example, the Seville Marathon, the exclusion of which would undermine the objectives of this study.

Table 2. Spanish marathon ranking 2021 of the RFEA

Posit.	Marathon	Score	2021	2020	2018/2019	Place Only 2021
1	XLI Maratón Valencia Trinidad Alfonso 2021	6,518	6,220	6,385	6,815	1
2	Zurich Maratón de Barcelona	2,068	2,189		1,947	2
3	XLIII EDP Rock 'n' Roll Madrid Maratón	1,318	1,175		1,461	3
4	Zurich Maratón Málaga 2021	863	943		782	4
5	Zurich Maratón Internacional de San Sebastián 2021	671	594		748	5
6	IV Maratón de Alcalá de Henares	218	214		222	6
7	XXV Quijote Maratón Ciudad Real	153	73		233	10
8	Ibiza Marathon 2021	116	48		183	11
9	VII Maratón de Santa Cruz de Tenerife Naviera Armas 2021	115	190		39	7
10	VII Maratón Internacional Ciudad de Logroño	81	93		69	9
11	Gran Canaria Maspalomas Maratón	55	109		0	8

Source: Based on RFEA (2021)

For this reason, a second criterion has been incorporated to overcome the limitations of the first criterion described above. For this purpose, "participation" has been considered since the number of athletes registered seems to be the most sensible criterion for delimiting the sample under study. This criterion, however, presents problems when it comes to its measurement, as it can present a more significant variability between events depending on the criteria used by each organizer to count participation. Similarly, there is considerable heterogeneity between marathons in terms of the registration platform used to measure participants, the number of invitations, practical promotions, or bibs reserved for sponsors, collaborators, or other marketing actions, which is why, to approximate the number of participants in each marathon, we have not opted not for the number of participants, but for the number of finishers. These finishers represent a more objective and comparable indicator of race participation.

In short, and due to the combination of the two criteria mentioned above, the present study has focused on the mono tracking of ten leading marathons in Spain based on the number of runners finishing each race. This criterion of finishers undoubtedly has some limitations. For example, it presents unfair variability based on conditions external to the organizer, such as the weather conditions of the race. A bad forecast or adverse conditions during the race may result in fewer athletes or a higher percentage of dropouts at the starting point. However, we argue that it is the most objective, comparable, and representative indicator we can rely on. In addition, for our selection, we have collected data on finishers of the major marathons and selected the ten with the highest number of finishers, as shown in Table 3.

Table 3. Spanish marathons 2022 by number of finishers

Position	Maratones España 2022	Fecha	Finishers
1	Maratón Trinidad Alfonso Valencia	04/12/2022	21,813
2	Zurich Maratón De Sevilla	20/02/2022	7,304
3	Zurich Rock'n' Roll Running Series Mad	24/04/2022	5,783
4	Zurich Maratón Barcelona	08/05/2022	5,406
5	Generali Maratón De Málaga	11/12/2022	2,745
6	Zurich Maratón De San Sebastián	27/11/2022	1,958
7	Mann Filter Maratón De Zaragoza	03/04/2022	1,022
8	Maratón Bp Castellón	27/02/2022	957
9	Bilbao Bizkaia Marathon	06/03/2022	688
10	Gran Canarias Maspalomas Marathon	20/11/2022	637

Source: Own elaboration

The added value of this barometer compared to previous work can be summarised in the following points. First, compared to the mega-event literature, research focused on small-scale events or regular events is limited (Zouni, Markogiannaki, Georgaki, 2021). So, this barometer shows a simple application of AI based on monitoring local events, showing a great potential for city branding and demonstrating that they are a much cheaper and simpler than large mega-events.

Second, private organizations lack the structure and appropriate communication channels to fully maximize data-driven insights (Naraine and Wanless, 2020). So, developing a public barometer can contribute to dynamizing sporting events successfully so that various actors benefit. Indeed, knowing which events attract the most traction on social media and generate higher levels of engagement can provide sports managers and DMOs (Destination Manager Organizations) with clues on how to proceed. Previous literature has demonstrated a direct relationship between the tourist impact of the Trinidad Alfonso Valencia Marathon and its activity on social networks. (Pérez and Serrano, 2020). So, studying the social media activity of marathons in city branding is crucial to investigate.

Third, more research is needed based on the use of AI to understand social network engagement considering its different dimensions: popularity (likes), virality (sharing), and commitment (comments) (Villamediana, Küster and Vila, 2019; Villamedia, Vila y Kuster, 2020). This research covers this gap as it measures the three dimensions of engagement as it considers total likes (popularity), comments (commitment) and shares (virality) to obtain an average engagement rate.

OBJECTIVE 1: RANKING OF THE TOP 10 SPAIN 2022 MARATHONS IN SOCIAL NETWORKS

Once we have established the ten marathons that are the basis of the barometer, as described in the introduction, we define the study's methodology and present how the investigated marathons are classified according to the social network analyzed.

Methodology for Objective 1

As indicated, this Barometer of Spanish Marathons on social networks is based on monitoring the ten selected races' official profiles on Facebook, Twitter, Instagram, and YouTube. Data has been collected and analyzed as of 31 December 2022 based on the following indicators (KPI). Social Blade was used to provide global analytics. This tool tracks the ranking and engagement of posts on various social networks, such as YouTube, Facebook, Instagram, Twitter and more. Although machine learning can also be used to label and organize data collected from monitoring and listening to social networks for more effective categorization and classification, this is not this chapter's objective. We will focus only on counting likes, reproductions, shares and followers (quantitative approach) to calculate relevant ratios in strategic decision-making, leaving the analysis of social network content (qualitative analysis) for future studies.

Facebook

Based on the public Facebook profiles of each of the ten marathons in the sample, we have considered the following:

- Ranking of fans of each race (accumulated to 31/12/20222)
- Ranking of followers of each race (accumulated to 31/12/20222)

Twitter

Based on the public Twitter profiles of each of the ten marathons in the sample, we have considered the following:

- Ranking of followers of each event (accumulated to 31/12/2022).

Instagram

Based on the public Instagram profiles of each of the ten marathons in the sample, we have considered the following:

- Ranking of followers per event (accumulated to 31/12/2022).

YouTube

Based on the public information on YouTube of each of the ten marathons in the sample, we have used the data of followers and reproductions, taking into account the following:

- Ranking of followers per test
- Channel plays (accumulated plays from the creation of the channel up to and including 31/12/2022)
- Reproductions of the channel during 2022

It should be noted that the Social Blade tool (www.socialblade.com) has been used to calculate the evolution of reproductions and followers during 2022 on YouTube. Image 1 shows, as an example, the monthly evolution of subscribers and views for the Valencia Marathon.

- Subscribers show the public information on the YouTube profiles for each marathon.
- Accumulated video views refer to the total accumulated views from the channel's creation until 31/12/2022. For its determination, we have used the Social Blade tool that allows us to know the total accumulated reproductions of the channel on 31/12/2022.
- The calculation of the reproductions explicitly obtained in 2022 is also made using the SocialBlade tool as the difference between the accumulated reproductions (global views) on 31/12/2022 and 31/12/2021.

Figure 1. Accumulated monthly evolution of subscribers and reproductions (video views) on YouTube of Maratón València
Source: Social Blade (2023)

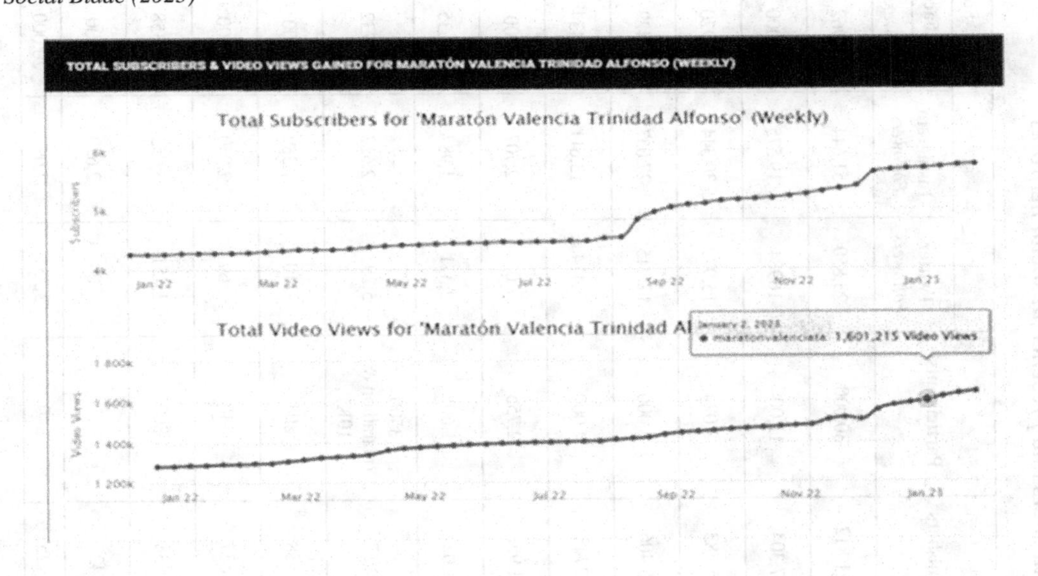

As a result of the monotype of the top ten marathons in Spain in 2022 (according to the number of finishers), this section shows their presence on different social networks. Specifically, the official profiles of each of the ten races on Facebook, Twitter, Instagram, and YouTube have been analyzed. The option of including other social networks, such as the emerging TikTok or Twitch, has been explored, but the events considered do not have profiles on these social networks. Some of the events considered do have profiles on other networks, such as Flickr, but these are one-off initiatives that all marathoners do not follow. For this reason, the study has necessarily been limited to those social networks that concentrate the largest number of followers and are more transversal in their management among the selected sample.

Table 4 shows the preponderance in social networks of the Trinidad-Alfonso Valencia Marathon, followed by the Seville Marathon. The projection and ranking of the ten marathons monitored on each of these four social networks are shown below.

Table 4. Top 10 Spanish marathons by finishers and presence in social networks

#	Spanish Marathons 2022	Event Date	Finishers	Participants	Social Media						Total Social Media
					Twitter	Instagram	Facebook		YouTube		
					Followers	*Followers*	*Followers*	*Fans*	*Subscribers*	*Reproductions*	
1	Maratón Trinidad Alfonso Valencia	04/12/2022	21,813	30,000	23,850	51,644	79,962	78,444	5,680	1,598,097	161,136
2	Zurich Maratón De Sevilla	20/02/2022	7,304	11,000	16984	19,575	32,000		645	241,492	69,204
3	Zurich Rock'n' Roll Running Series Madrid	24/04/2022	5,783	8,024	15,979	30,344	65,920	65,888	No channel	No channel	112,243
4	Zurich Maratón Barcelona	08/05/2022	5,406	8,000	17,045	27,057	56,429	56,363	751	326,720	101,282
5	Generali Maratón De Málaga	11/12/2022	2745	4,000	7,414	13,611	14,434	13,830	No channel	No channel	35,459
6	Zurich Maratón De San Sebastián	27/11/2022	1,958	2,755	0	2,707	1,300	0	116	19,217	4,123
7	Mann Filter Maratón De Zaragoza	03/04/2022	1,022	1,608	2,371	1,987	10,635	10,439	No channel	No channel	14,993
8	Maratón Bp Castellón	27/02/2022	957	3500 (marathon & 10K)	5,293	2,617	12,732	12,311	113	35,803	20,755
9	Bilbao Bizkaia Marathon	06/03/2022	688	804	210	1,380	1600	1400	No channel	No channel	3,190
10	Gran Canarias Maspalomas Marathon	20/11/2022	637	840	369	2,030	4,600	4,200	No channel	No channel	6,999
11	Palma Marathon Mallorca	09/10/2022	538	800	1,451	3,052	12,358	12,040	No channel	No channel	16,861
12	Total Energies Maratón Murcia Costa Cálida	06/02/2022	427		800	2,650	7,100	6,700	No channel	No channel	10,550
13	Bilbao Night Maratón	22/10/2022	425	840	3,255	7,097	20,000		61	12,380	30,413

Source: Own elaboration

Facebook Results

For this social network, a distinction must be made between followers and fans. Although they are similar concepts, Facebook itself establishes their definition and differences. Fans are those who have left a "Like" on the page, will automatically follow it, and the page's activity will be published in its news section. On the other hand, followers are those users who have clicked on "Follow" the page. In this case, the user's notification settings automatically change to "Highlights", meaning they might receive notifications with suggestions about live videos. This viewer's friends will not see activity related to the marathon page they have started following.

Table 5 shows the ranking of the top ten marathons in Spain in 2022 based on the number of followers of their Facebook page.

Table 5. Top 10 Spanish marathons by followers on Facebook 2022

#	Marathons	Followers 31/12/2022
1	Maratón Trinidad Alfonso Valencia	81,730
2	Rock'n' Roll Running Series Madrid	65,920
3	Maratón Barcelona	56,429
4	Maratón De Sevilla	32,000
5	Generali Maratón De Málaga	14,434
6	Maratón Bp Castellón	12,732
7	Mann Filter Maratón De Zaragoza	10,635
8	Maspalomas	4,600
9	Bilbao Bizkaia Marathon	1,600
10	Zurich Maratón De San Sebastián	1,300

Source: Own elaboration

Twitter Results

Table 6 shows the ranking of marathons according to the number of followers on Twitter. It can be seen that there are four very important marathons according to the number of followers: (the first of them is well ahead of the rest, three intermediate marathons, and three very weak marathons with very few (or no) followers.

Instagram Results

Table 7 shows the ranking of marathons according to the number of followers on Instagram. In this case, there are five top marathons with a large number of followers (the first of which is a long way behind the rest) and five weaker marathons with much fewer followers on Instagram.

Table 6. Top 10 Spanish marathons by Twitter followers 2022

#	Marathons	Followers at 31/12/2022
1	Maratón Trinidad Alfonso Valencia	23,858
2	Maratón Barcelona	17,045
3	Maratón De Sevilla	16,984
4	Rock'n' Roll Running Series Madrid	15,979
5	Generali Maratón De Málaga	7,414
6	Maratón Bp Castellón	5,293
7	Mann Filter Maratón De Zaragoza	2,371
8	Maspalomas	369
9	Bilbao Bizkaia Marathon	210
10	Zurich Maratón De San Sebastián	0

Source: Own elaboration

Table 7. Top ten Spanish marathons by Instagram followers 2022

#	Marathons	Followers at 31/12/2022
1	Maratón Trinidad Alfonso Valencia	51,662
2	Rock'n' Roll Running Series Madrid	30,344
3	Maratón Barcelona	27,057
4	Maratón De Sevilla	19,575
5	Generali Maratón De Málaga	13,611
6	Zurich Maratón De San Sebastián	2,707
7	Maratón Bp Castellón	2,617
8	Maspalomas	2,030
9	Mann Filter Maratón De Zaragoza	1,987
10	Bilbao Bizkaia Marathon	1,380

Source: Own elaboration

YouTube Results

Finally, on the social network YouTube, only five of the ten marathons with the highest number of finishers in Spain in 2022 have a profile. This is why Table 8 shows a biased ranking of the marathons studied.

Conclusions for Objective 1

Once the number of followers on Facebook, Twitter, Instagram, and YouTube has been analyzed, we proceed to observe the ranking of each test depending on the social network considered and its overall prominence, considering all of them simultaneously. Table 9 summarizes the results of the rankings obtained.

Table 8. Top ten Spanish marathons by YouTube followers 2022

#	Marathons	Subscribers at 31/12/22	Video Views Accumulated at 31/12/22	Video Views Accumulated at 01/01/22	Video Views 2022
1	Maratón Trinidad Alfonso Valencia	5,680	1,599,881	1,284,750	315,131
2	Zurich Maratón Barcelona	751	326,773	300,335	26,438
3	Zurich Maratón De Sevilla	645	241,538	228,401	13,137
4	Zurich Maratón De San Sebastián	116	19,217		
5	Maratón Bp Castellón	113	35,803		
6	Rock N' Roll Running Series Madrid				
7	Generali Maratón De Málaga				
8	Mann Filter Maratón De Zaragoza				
9	Bilbao Bizkaia Marathon				
10	Gran Canarias Maspalomas Marathon				

Source: Own elaboration.

As a result, we can conclude that the Trinidad Alfonso Valencia Marathon is the race with the highest number of finishers and followers on social networks in general and the leader in all four networks analyzed. Therefore, all the social networks converge in determining the first position in the ranking. These results are coherent with those of Pérez and Serrano, (2020), who concluded Trinidad Alfonso Valencia is the marathon that has had the greatest growth in recent years, as is demonstrated by the increase in the number of runners at the event and their higher average daily expenditure of these runners. So, the application of AI for capturing, analyzing and processing information from social networks has proved, in our study, to help identify the most engaging sporting event. This information is very useful for public and private organizations interested in promoting a place or a private brand.

Secondly, the second place varies according to the social network analyzed. Specifically, the Zurich Rock n'Roll Running Series from Madrid occupies the second position on three social networks (Facebook, Instagram, and YouTube) but the fourth position on Twitter. In this last social network, it is surpassed by the Barcelona and Seville marathons.

The third position in the group of followers in social networks corresponds to Zurich Maratón de Barcelona, and the fourth position to Zurich Maratón de Sevilla.

Finally, it can be concluded that some of the events (such as the Zurich Rock n'roll Running Series, Castellón Marathon or Bilbao Bizkaia Marathon) hold races of other distances simultaneously, bringing together under the same name and social profiles other events different from the marathon. However, this circumstance has also occurred numerous times in other races, such as the Valencia Marathon. Although it may indeed introduce some distortion, the truth is that it is irrelevant to the purpose of the study: to analyze the projection on social networks of the brands of the selected long-distance races in Spain.

Table 9. Position of Spanish marathons in 2022 by social network

Marathons Spain 2022	# Finishers	#Total Social Media	# Facebook	#Tiwtter	#Intagram	#YouTube
Maratón Trinidad Alfonso Valencia	1	1	1	1	1	1
Zurich Maratón De Sevilla	2	4	4	3	4	3
Zurich Rock'n' Roll Running Series Mad	3	2	2	4	2	
Zurich Maratón Barcelona	4	3	3	2	3	2
Generali Maratón De Málaga	5	5	5	5	5	
Zurich Maratón De San Sebastián	6	9	10	10	6	4
Mann Filter Maratón De Zaragoza	7	7	7	7	9	
Maratón Bp Castellón	8	6	6	6	7	5
Bilbao Bizkaia Marathon	9	10	9	9	10	
Gran Canarias Maspalomas Marathon	10	8	8	8	8	

Source: Own elaboration

OBJECTIVE 2: ANALYSIS OF ENGAGEMENT IN THE VALENCIA MARATHON

As shown in Table 9, in 2022, the Trinidad Alfonso Valencia Marathon was the marathon with the highest number of finishers and, in turn, with he most elevated presence on social networks: Facebook, Twitter, Instagram and YouTube.

Next, and in order to fulfil the second objective of this research, we set out to delve deeper into the social media behaviour of this leading event by investigating not only the projection of the event in terms of the number of followers on social networks but also considering the engagement of the event with its community. The large volume of information handled on these social media platforms makes it very difficult for companies to analyze all the data manually. Artificial Intelligence tools can be of great help, as they can process large amounts of information in a short time. In addition, AI can identify patterns and trends in the data, which can help managers to make better decisions. The second objective of this chapter focuses on investigating in greater depth the reactions that the referred event provokes in the community of followers it has on each social network, generally referred to as engagement. So, following a quantitative approach using the Social Blade tool, these and other indicators will be used to identify the best social network to promote a sports event and, by extension, to promote the place where this event is held.

Methodology for Objective 2

Engagement in social networks "is an indicator that serves to determine the level of engagement and interaction that an audience or consumer has with published content" (Newberry, 2023). The calculation of this indicator can be approached in different ways. Almost all of them form a ratio in which the numerator is made up of the number of interactions of a publication on a social network during a specific period (i.e., comments, shares, views, reactions), and the denominator represents the number of followers, impressions or reach of the account at that time.

For this work, given that we do not know the reach or number of impressions, the engagement rate has been defined as the ratio between the number of interactions (reactions to a post on Facebook, Instagram, YouTube or Twitter, plus comments and shares) and the number of followers on that social network, expressed as a percentage. This value corresponds to the calculation provided by Hootsuite (2023), as Figure 2 shows.

Figure 2. Engagement measurement based on total likes, comments, and shares
Source: Hootsuite (2023)

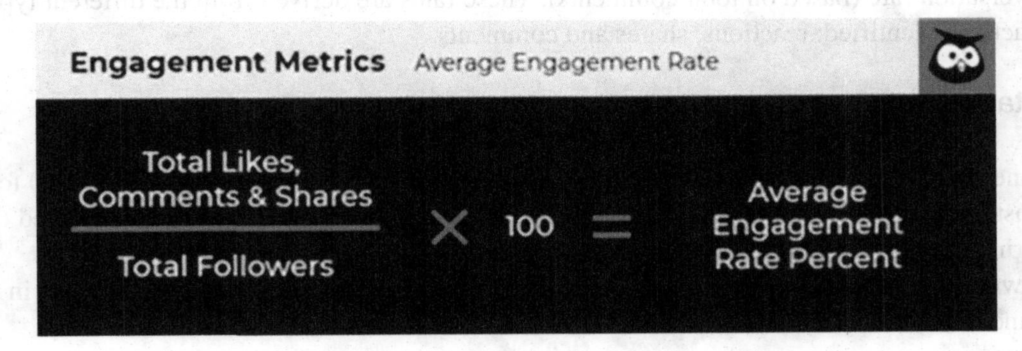

The following are the key indicators considered for this study, defined based on previous contributions (Rosgaby, 2022; Newberry, 2023):

Indicators Related to the Audience

- (1) number of fans (f), followers or followers (t / I);
- (2) number of following (t / I);
- (3) number of lists and members (t);
- (4) number of subscribers (YT);

Content-Related Indicators

- (6) frequency of posts (f / t / YT / I);
- (7) minutes of reproduction (YT);
- (8) format used (text, photos, links, videos, hashtags or tag, mentions);
- (9) Thematic.

Indicators Relating to Interactivity

- (10) reactions (like, like, happy, sad)
- (11) comments (f / YT / I /);
- (12) shares (f);
- (13) views (YT);
- (14) retweets (t);

- (15) favourites (t);
- (16) replies (t).

As a starting point, each of the publications made during 2022 on the official profile of Maratón Valencia on the social networks Facebook, Twitter, Instagram and YouTube were identified, and the interactions generated were counted. Table 10 shows the number of interactions identified.

For our study, we have taken into account not only the engagement rate but also three additional rates: the acceptance rate (based on total reactions), the amplification rate (based on total posts shared), and the conversation rate (based on total comments). These rates are derived from the different typologies of interactions identified: reactions, shares and comments.

Acceptance Rate

We define the acceptance rate as the ratio between the reactions generated to a post (KPI 10 for Facebook, Instagram and YT and KPI 11 for Twitter) and the number of followers. We calculated this rate considering each post's number of followers.

Likewise, this rate has been calculated annually (considering the sum of all the reactions in a given month and the number of followers in that month).

Amplification Rate

The amplification rate corresponds to the number of shares (KPIs 12 and 14) divided by the number of followers.

For our study, we have considered not only the engagement rate but also three additional rates: the acceptance rate (based on total reactions), the amplification rate (based on total posts shared), and the conversation rate (based on total comments). These rates are derived from the different typologies of interactions identified: reactions, shares, and comments. Figure 3 shows the amplification rate.

Figure 3. Amplification rate based on the number of posts shares
Source: Hootsuit (2023)

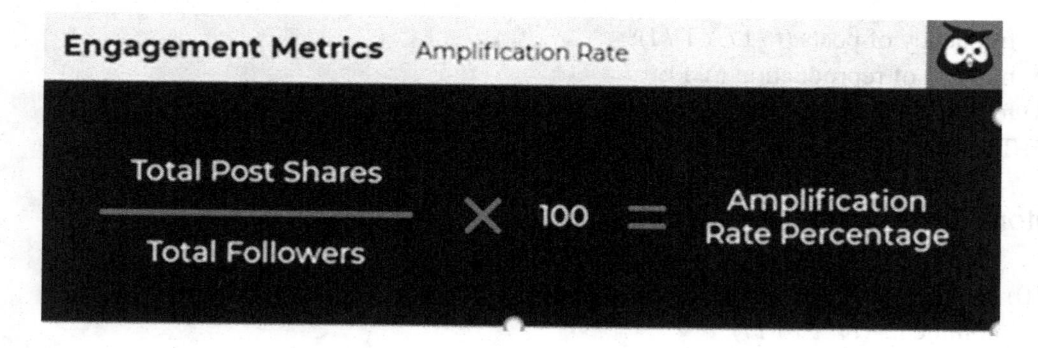

Conversation Rate

Similarly, the conversation rate is defined as the ratio between the number of responses generated by a given post (KPI 12 and 16) and the total number of followers on that social network at the corresponding time.

Audience Growth Rate

This rate is defined as the percentage growth of followers in a period with respect to the preceding period expressed as a percentage.

To obtain the number of followers on each social network, the following methodology was followed.

- *Calculation of followers on Facebook:* in this study, we have considered the number of followers on the page, not the number of fans. To calculate the number of followers, direct observation was carried out on 01/01/2023, and the screenshots were consulted in the waybackmachine8 tool.
    - 04/12/2022: 80,623 followers: https://web.archive.org/web/20221204144257/https://www.facebook.com/maratonvvalencia/
    - 13/11/2021: direct consultation with the FB administrator of the Marathon València profile is recommended in order to have the real data of followers in 2022.
- *Calculation of followers on Twitter:* Followers have been calculated using the Social Blade tool. For 01/01/2022 we can see a weekly follower count of 22,682 followers. From there, we can consult the monthly variation during 2022 (Image 4 and Image 5). Also, Image 6 shows the total number of followers.
- *Calculation of followers on Instagram:* The followers have been obtained with the Social Blade tool for the Instagram profile of the València marathon (https://www.instagram.com/maratonvalencia/). Image 7 shows this evolution.
- *Calculation of YouTube followers* was obtained by direct observation on 01/01/2023 of the channel (https://www.youtube.com/@MaratonValencia_TA). Also, the evolution of subscribers and views during 2022 has been followed on the YouTube profile of Marathon València (Image 8).

Figure 4. Monthly evolution of the total number of followers of the Maratón València Twitter profile
Source: SocialBlade (2023)

Figure 5. Evolution of the total number of followers of the Twitter profile of the Valencia Marathon
Source: SocialBlade (2023)

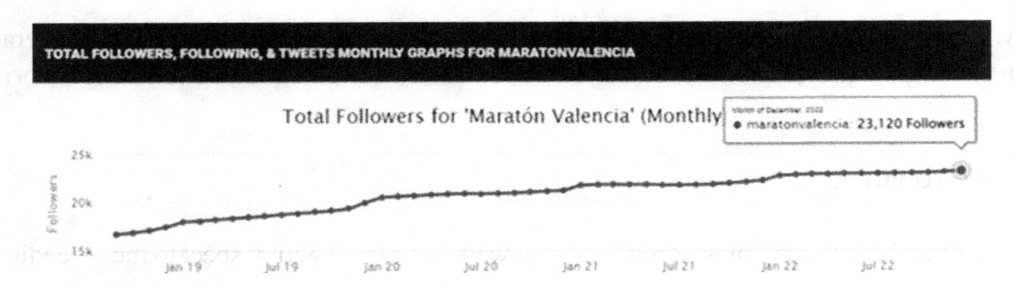

Figure 6. Evolution of the total number of followers of the Twitter profile of the Valencia Marathon
Source: SocialBlade (2023)

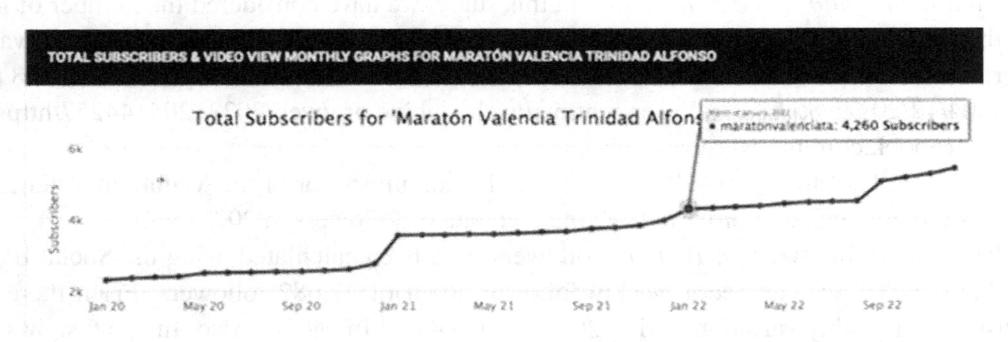

Figure 7. Evolution of the total number of followers of the Instagram profile of the Valencia Marathon
Source: Socialblade (2023)

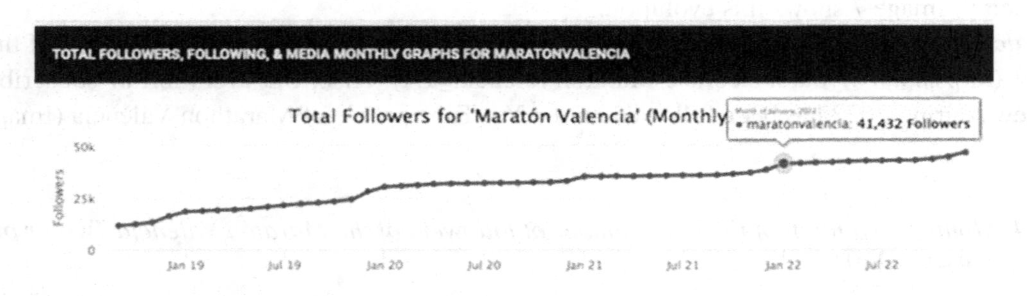

Figure 8. Evolution of the views of the Maratón Valencia YouTube channel
Source: SocialBlade (2023)

Results: Engagement València 2022 Marathon

The following lines show the acceptance, conversation, amplification and engagement rates of the official València Marathon profile for the social networks Facebook, Twitter, Instagram and YouTube. Image 9 shows the followers/followers/subscribers to the different social networks.

Figure 9. Followers by social network of Maratón Valencia
Source: Own elaboration

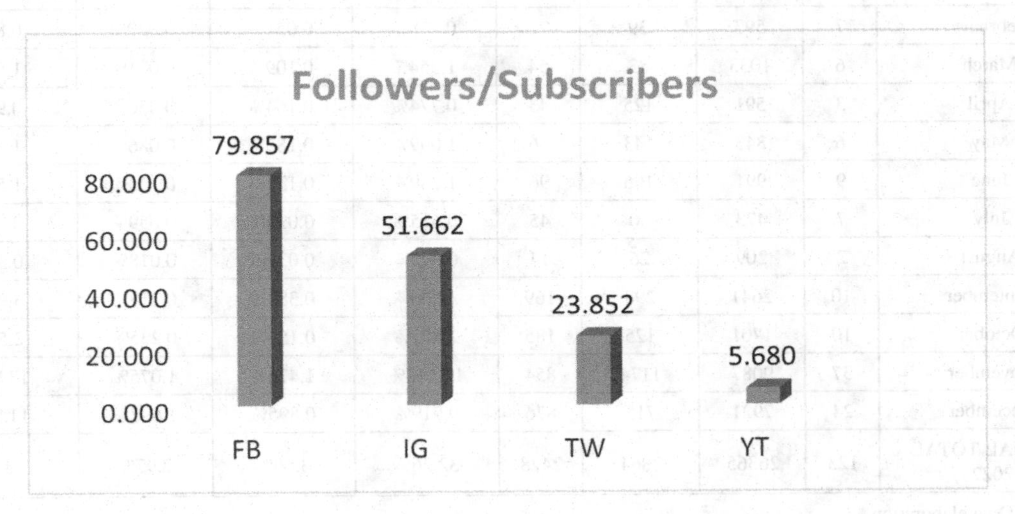

Results for Facebook

As Table 10 shows, reactions, comments, and consequently, the different rates (acceptance, conversation, amplification) and engagement skyrocketed in November. It is noticeable that from September onwards, the figures start to go up again after the summer, as the event is coming up.

In the case of the Facebook profile, obtaining month-by-month data on followers is impossible. Data as of 01/01/2023 has been considered for the whole period.

Table 11 shows that the acceptance rate is 32.26%, the amplification rate is 2.97%, the conversation rate is 3.5%, and the engagement rate is 38.73%.

Results for Instagram

As can be seen in Table 12, reactions and comments, and therefore the different rates (acceptance, conversation, amplification) and engagement, soar in November, just as they did for Facebook. It should be noted that from September onwards, after the summer, the figures start to pick up again, as the event is coming up.

Table 13 shows that all rates are much higher than those obtained in the case of Facebook. Indeed, the acceptance rate is 393.46 (a figure that multiplies Facebook's acceptance rate by 12), and the engagement rate is 401.05% (a figure that bears Facebook's engagement rate by 10). In other words, for the social network, Instagram reactions, shares and comments soar.

Table 10. Monthly reactions to each post on Facebook and monthly rates

Facebook Maraton Valencia								
Month	**Nº**	**Reactions**	**Comments**	**Shares**	**Acceptance Rate (Likes/Followers)**	**Conversation Rate (Comments/ Followers)**	**Amplification Rate (Shares/ Followers)**	**Engagement (Likes, Shares, and Comments/ Followers)**
January	2	282	0	4	0.345%	0.000%	0.005%	0.350%
February	7	593	39	22	0.778%	0.051%	0.029%	0.858%
March	6	1033	83	54	1.354%	0.109%	0.071%	1.533%
April	3	591	125	43	0.774%	0.164%	0.056%	0.994%
May	6	843	143	66	1.099%	0.186%	0.086%	1.371%
June	9	991	106	96	1.289%	0.138%	0.125%	1.552%
July	7	473	30	45	0.615%	0.039%	0.059%	0.712%
August	2	209	26	14	0.270%	0.034%	0.018%	0.321%
September	10	2641	298	169	3.390%	0.383%	0.217%	3.990%
October	10	1701	125	185	2.157%	0.159%	0.235%	2.550%
November	37	9087	1174	854	11.443%	1.478%	1.075%	13.996%
December	24	7921	715	876	9.919%	0.895%	1.097%	11.911%
ANNUAL TOTAL 2022	123	26365	2864	2428	32.26%	3.50%	2.97%	38.73%

Source: Own elaboration

Table 11. Summary of Maratón Valencia Facebook posts in 2022

KPI - General	**Total 2022**	**Average 2022**
Followers (end of month, as of 01/01 of the following period) (*)	83,838	78,073
Post published	132	11
New followers	7,669	
Audience growth rate	10.07%	0.81%
KPI - Interest:	**Total 2022**	**Average 2022**
Action	199.73	165
Conversation	21.70	18
Amplification	18.39	13
Interactions	239.83	196
KPI - Engagement:	**Total 2022**	**Average 2022**
Aceptation rate	33.77%	2.69%
Conversation rate	3.67%	0.29%
Amplification rate	3.11%	0.25%
Engagement rate (interactions)	40.55%	3.23%

(*) Metricool user Maratón Valencia
Source: Own elaboration

Table 12. Monthly reactions to each post on Instagram and their monthly rates

Instragram Maraton Valencia						
Month	**Nº**	**Favourites**	**Retweets**	**Acceptance Rate (Likes/Followers)**	**Conversation Rate (Comments/ Followers)**	**Engagement (Likes, Shares, and Comments/ Followers)**
January	3	2254	23	5.42%	0.06%	5.48%
February	6	3961	69	9.49%	0.17%	9.65%
March	4	7956	91	18.99%	0.22%	19.20%
April	3	2243	62	5.27%	0.15%	5.42%
May	3	6019	79	14.15%	0.19%	14.34%
June	8	4165	104	9.77%	0.24%	10.02%
July	5	1974	36	4.62%	0.08%	4.71%
August	3	2610	94	6.08%	0.22%	6.30%
September	12	9843	358	22.65%	0.82%	23.47%
October	10	10096	227	22.63%	0.51%	23.13%
November	45	50044	1306	107.66%	2.81%	110.47%
December	35	70811	869	137.07%	1.68%	138.75%
ANNUAL TOTAL 2022	**137**	**171976**	**3318**	**363.80%**	**7.14%**	**370.94%**

Source: Own elaboration

Table 13. Summary of Maratón Valencia's Instagram posts during 2022

KPI - General	**Total 2022**	**Average 2022**
Followers (end of month, as of 01/01 of the following period) (*)	51,662	43,709
Post published	137	11
New followers	10.,30	853
Audience growth rate	24.69%	1.90%
KPI – Interest*:	**Total 2022**	**Average 2022**
Action	1,255.30	1,075.42
Conversation	24.22	20.57
Interactions	1,279.52	1,095.99
KPI – Engagement*:	**Total 2022**	**Average 2022**
Aceptation rate	393.46%	28.93%
Conversation rate	7.59%	0.60%
Engagement rate (interactions)	401.05%	58.11%

(*) Amplification rate: not applicable in Instagram

Source: Own elaboration

Results for Twitter

As Table 14 shows, reactions and comments, and consequently the different rates (of acceptance, conversation, amplification) and engagement, shoot up in December and not in November, as was the case for Facebook and Instagram. In text-based social networks, the movement generated is not so immediate, and there is a certain cadence between when the event takes place and when the followers' comments start up in force.

Table 14. Summary of the monthly reactions to each post on Twitter and their monthly rates

Twitter Maraton Valencia								
Month	**Nº**	**Favourites**	**Answer**	**Retweets**	**Acceptance Rate (Likes/ Followers)**	**Conversation Rate (Comments/ Followers)**	**Amplification Rate (shares/ Followers)**	**Engagement (Likes, Shares, and Comments/ Followers)**
January	3	75	0	10	0.33%	0.00%	0.04%	0.37%
February	13	344	31	32	1.51%	0.14%	0.14%	1.78%
March	7	108	6	29	0.47%	0.03%	0.13%	0.63%
April	8	205	**9**	**41**	0.90%	0.04%	0.18%	1.12%
May	8	197	**8**	**46**	0.86%	0.03%	0.20%	1.10%
June	14	258	**10**	**86**	1.13%	0.04%	0.38%	1.55%
July	8	107	**10**	**38**	0.47%	0.04%	0.17%	0.68%
August	6	170	**8**	**47**	0.74%	0.03%	0.21%	0.98%
September	16	377	**19**	**80**	1.64%	0.08%	0.35%	2.07%
October	10	289	**7**	**56**	1.25%	0.03%	0.24%	1.53%
November	63	1796	**72**	**359**	7.77%	0.31%	1.55%	9.63%
December	65	5315	**233**	**1014**	22.28%	0.98%	4.25%	27.51%
ANNUAL TOTAL 2022	221	**9241**	**413**	**1838**	**38.74%**	**1.73%**	**7.71%**	**48.18%**

Source: Own elaboration

As Table 15 shows, all the rates for Twitter are much higher than those obtained for Facebook but lower than those obtained for Instagram. Indeed, the acceptance rate is 40.22 (lower than Instagram but higher than Facebook), and the engagement rate is 50.1% (also much higher than Facebook, but much lower than Instagram).

Results for YouTube

As Table 16 shows, the reactions and comments, and therefore the different rates (acceptance, conversation, amplification) and engagement, soar in November, but especially in December, as was the case with Twitter.

Table 15. Summary table of Maratón Valencia's Twitter posts in 2022

KPI – General	Total 2022	Average 2022
Followers (end of month, as of 01/01 of the following period)	23,852	22,977
Post published	216	18
New followers	1,170	98
Audience growth rate	5.16%	0.42%
KPI - Interest:	**Total 2022**	**Average 2022**
Action	42.78	29.11
Conversation	1.91	1.34
Amplification	8.51	6.02
Interactions	53.20	36.47
KPI - Engagement:	**Total 2022**	**Average 2022**
Aceptation rate	40.22%	3.28%
Conversation rate	1.80%	0.15%
Amplification rate	8.00%	0.65%
Engagement rate (interactions)	50.01%	4.08%

Source: Own elaboration

All the rates are much higher than those obtained for Facebook, slightly higher than those obtained for Twitter, but lower than those obtained for Instagram. In fact, the acceptance rate stands at 50.27 and engagement at 50.27 as well (table 17).

Table 16. Summary of monthly reactions to each YouTube post and their monthly rates

YouTube Maraton Valencia						
Month	**Nº**	**Reactions**	**Comments**	**Acceptance Rate (Likes/Followers)**	**Conversation Rate (Comments/ Followers)**	**Engagement (Likes, Shares, and Comments/ Followers)**
January	1	41	0	0.96%	0	0.96%
March	2	50	0	1.15%	0	1.15%
August	6	763	0	15.17%	0	15.17%
September	3	104	0	2.02%	0	2.02%
October	2	156	0	2.98%	0	2.98%
November	8	463	0	8.59%	0	8.59%
December	9	820	0	14.44%	0	14.44%
ANNUAL TOTAL 2022	**31**	**2397**	**0**	**43.30%**	**0**	**45.30%**

Source: Own elaboration

Table 17. Summary of the monthly reactions to each post on YouTube and their monthly rates

KPI - General	**Total 2022**	**Average 2022**
Followers (end of month, as of 01/01 of the following period) (*)	5,680	4,768
Post published	1.420	118
New followers	33.33%	2.48%
Audience growth rate	17,007,783	1,417,315
KPI – Interest*:	**Total 2022**	**Average 2022**
Action	1.69	65
Conversation	-	-
Interactions	1.69	65
KPI – Engagement*:	**Total 2022**	**Average 2022**
Aceptation rate	50.27%	6.47%
Conversation rate	0	0
Engagement rate (interactions)	50.27%	6.47%

(*) Amplification rate: not applicable in YouTube
Source: Own elaboration

Conclusions for Objective 2

Table 18 shows the comparative results for the four social networks investigated. AI helps monitor social media networks to better understand the audience's needs and expectations to implement smart strategies that add value to a sports event and the city when it is held. The quantitative indicators derived from social media monitoring in our study show that Facebook is the social network with the highest number of followers but nevertheless has the lowest engagement. The social network with by far the highest engagement is Instagram, followed by YouTube and Twitter, with very similar engagement. However, YouTube only counts reactions as interaction because comments are not enabled on the Maratón València profile. Therefore, Facebook is in last place in terms of engagement.

Secondly, in terms of follower growth, YouTube is the network with the highest year-on-year growth in followers, followed by Instagram. In other words, these two social networks seem to be the ones gaining prominence in the dissemination of comments, opinions, etc. on local sporting events.

Thirdly, if we look at the type of interaction, we see that reactions on all networks are the most frequent.

GENERAL CONCLUSIONS AND IMPLICATIONS

As a result of the research carried out, the following management implications can be noted. First, the development of sporting events that are popular and occupy top positions in the rankings can be of great help in improving the image of a destination insofar as a marathon that generates engagement on the networks leads, at the same time, to greater intentions to visit and revisit the destination (Zouni, Markogiannaki, Georgaki, 2021). Indeed, the destination's image can be highly favoured thanks to small local events (such as marathons) capable of generating a lot of movement on social networks.

Table 18. Summary of the acceptance, conversation, amplification and engagement rates of each Maratón València social network during 2022

	FB		IG		TW		YT	
Followers/Subscribers	**79,857**		**51,662**		**23,852**		**5,680**	
	Monthly Average	**Annual Total**	**Monthly Average**	**Annual Total**	**Monthly Average**	**Annual Total**	**Monthly Average**	**Annual Total**
Audience growth rate			1.90%	24.69%	0.42%	5.16%	2.48%	33.33%
Aceptation rate	2.69%	32.26%	28.93%	28.93%	3.28%	40.22%	6.47%	50.27%
Conversation rate	0.29%	3.50%	0.60%	0.60%	0.15%	1.80%	0.00%	0.00%
Amplification rate	0.25%	2.97%			0.65%	8.00%		
Engagement rate	3.23%	38.73%	58.11%	58.11%	4.08%	50.01%	6.47%	50.27%

Source: Own elaboration

Second, not all social networks work equally well in promoting sporting events. The results have shown the preponderance in the Spanish case of Instagram, which is the one that best contributes to visibility. Thus, using AI for monitoring social networks is very useful for choosing the best social network to succeed. Based only on quantitative indicators, our results lead us to conclude that it is advisable to prioritize Instagram among DMOs (Destination Managers Organizations) concerned with generating territorial branding and among private brands associated with sporting events that wish to improve their notoriety and, by extension, their acceptance among the public. Instagram is the consumer network par excellence (Pérez and Serrano, 2020). On the contrary, AI leads us to conclude, in line with Pérez and Serrano (2020), that Twitter is not the most suitable social network for promoting this type of event because the engagement and the rest of the ratios are low. The same occurs for Facebook.

As a limitation, it should be noted that this study has been carried out at a specific time, so it would be very interesting to establish a continuity of the project to establish an observatory that goes beyond the still photo. In this way, trends could be identified. With continuity, not only will we be able to obtain greater knowledge, but it will also allow us to analyze annual variability and detect trends in the use, projection, and growth of each social network in this area. Secondly, the study of the engagement and other rates could be extended to all marathons in Spain because an analysis limited to the Valencian marathon can limit the generalizations of the findings to other kinds of marathon events. Lastly, given that AI enables digital systems to develop judgmental capabilities to capture relevant web comments and opinions and process this data independently of prior human programming, a qualitative analysis based on sentiment analysis of the content published could be developed in the future.

ACKNOWLEDGMENT

(*) This paper has been financed by Regional Ministry of Innovation, Universities, Science and the Digital Society – Consellería de Innovación, Universidades, Ciencia y Sociedad Digital AICO 2022 CIAICO/2021/062: "El deporte en la construcción de marca país: el turista y el residente".

(*) The authors would like to thank the company Nostre Sports for their contribution to the development of the empirical study.

REFERENCES

Arranz, L. R. (2022). Los maratones en España pierden más de 13.000 corredores en 2022. *Novatos del running*. https://www.novatosdelrunning.es/los- maratones-en-espana-pierden-mas-de-13-000-corredores-en-2022/

Christina, N. (2023). *16 métricas de redes sociales que realmente importan y cómo darles seguimiento.* Hootsuite. https://blog.hootsuite.com/es/metricas-de-redes- sociales/#4_Tasa_de_interaccion

Kumar, A., & Sachdeva, N. (2022). Multimodal cyberbullying detection using capsule network with dynamic routing and deep convolutional neural network. *Multimedia Systems*, *28*(6), 1–10. doi:10.100700530-020-00747-5

Naraine, M. L., & Wanless, L. (2020). Going all in on AI: Examining the value proposition of and integration challenges with one branch of artificial intelligence in sport management. *Sports Innovation Journal*, *1*, 49–61. doi:10.18060/23898

Pérez, M. L., & Serrano, M. D. T. (2020). Impacto en redes sociales de la Maratón de Valencia Trinidad Alfonso 2019. *Estudios sobre la transformación digital de la comunicación*, *67*.

RFEA. (2021). *Ranking pruebas en ruta de la*. https://www.rfea.es/estadis/pdf/punruta2021.pdf

Rosgaby Medina, K. (2022). *Engagement rate en redes sociales: que es y cómo se calcula*. https://branch.com.co/marketing-digital/engagement-rate-en-redes- sociales-que-es-y-como-se-calcula/

Sehl, K., & Tien, S. (2023). *Engagement Rate Calculator+ Guide for 2023*. Hootsuite. https://blog.hootsuite.com/calculate-engagement-rate/

Sempre, V. M. (2013). *El cuadro de mando integral para medir nuestra actividad en Redes Sociales*. El Blog de Mayte Vañó Sempere. http://www.maytevs.com/el-cuadro-de-mando-integral-para-medir-nuestra-actividad-en- redes-sociales/

Villamediana, J., Küster, I., & Vila, N. (2019). Destination engagement on Facebook: Time and seasonality. *Annals of Tourism Research*, *79*(November), 102747. doi:10.1016/j.annals.2019.102747

Villamediana-Pedrosa, J. D., Vila-Lopez, N., & Küster-Boluda, I. (2020). Predictors of tourist engagement: Travel motives and tourism destination profiles. *Journal of Destination Marketing & Management*, *16*, 100412. doi:10.1016/j.jdmm.2020.100412

Zouni, G., Markogiannaki, P., & Georgaki, I. (2021). A strategic tourism marketing framework for sports mega events: The case of Athens Classic (Authentic) Marathon. *Tourism Economics*, *27*(3), 466–481. doi:10.1177/1354816619898074

APPENDIX

Tools Used

- Socialblade; https://socialblade.com/
- Metricool: https://app.metricool.com/
- WayBackMachine: https://web.archive.org/

Spanish Marathons Webs

- https://www.valenciaciudaddelrunning.com/maraton/maraton/
- https://www.zurichmaratonsevilla.es/
- https://rocknrollmadridrun.com/
- https://www.zurichmaratobarcelona.es/
- https://www.generalimaratonmalaga.com/
- https://zurichmaratonsansebastian.com/
- https://www.zaragozamaraton.com/
- https://www.maratonbpcastellon.com/
- https://bilbaobizkaiamarathon.com/
- https://grancanaria-maspalomasmarathon.com/
- https://www.palmademallorcamarathon.com/
- https://www.maratonmurcia.com/
- https://www.totalenergiesbilbaomarathon.com/

Monitored Social Networks

Facebook

- https://www.facebook.com/maratonvalencia
- https://www.facebook.com/RnRMadrid
- https://www.facebook.com/ZURICHMARATODEBARCELONA
- https://www.facebook.com/MaratonDeSevilla/followers
- https://www.facebook.com/maratondemalaga
- https://www.facebook.com/MaratonCastellon/
- https://www.facebook.com/zaragozamaraton
- https://www.facebook.com/GCMaspalomasMarathon
- https://www.facebook.com/bilbaobizkaiamarathon
- https://www.facebook.com/ZurichMaratonSS/

Twitter

- https://twitter.com/maratonvalencia
- https://twitter.com/maratobarcelona
- https://twitter.com/MaratonSevilla
- https://twitter.com/RNRmadmaraton
- https://twitter.com/maratonmalaga
- https://twitter.com/MaratoCastello
- https://twitter.com/MaratonZaragoza
- https://twitter.com/GCMMarathon
- https://twitter.com/bbmarathon

Instagram

- https://www.instagram.com/maratonvalencia/
- https://www.instagram.com/maratondemadrid/
- https://www.instagram.com/maratobarcelona/
- https://www.instagram.com/maratondesevilla/
- https://www.instagram.com/maratonmalaga/
- https://www.instagram.com/maratonsansebastian/?hl=es
- https://www.instagram.com/maratobpcastello/
- https://www.instagram.com/gcmmarathon/
- https://www.instagram.com/MaratonZaragoza/
- https://www.instagram.com/bilbaobizkaiamarathon/

YouTube

- https://www.youtube.com/@MaratonValencia_TA
- https://www.youtube.com/channel/UC1OGE_PGTAFxdBF-a3CxAKw
- https://www.youtube.com/user/MaratonDeSevilla
- https://www.youtube.com/@zurichmaratonsansebastian
- https://www.youtube.com/@maratoncastellon/
- https://www.youtube.com/@kumulusmallorca831/ç
- https://www.youtube.com/@kumulusmallorca831/
- https://www.youtube.com/@maratonmurcia1481
- https://www.youtube.com/@bilbaonightmarathon2028

Chapter 16
AI-Driven Customer Experience: Factors to Consider

Svetlana Bialkova
Liverpool Business School, UK

ABSTRACT

Despite the increasing implementation of artificial intelligence (AI), it is puzzling why consumers are still resistant towards it. Part of the problem is how to create systems that appropriately meet consumer demand for good quality and functional AI. The chapter addresses this issue by providing the much-needed understanding of how AI technologies can shape a satisfactory customer experience. Results are clear in showing that easy-to-use and high-quality AI systems form positive attitudes, and consumers are willing to use such technology again. Functional and enjoyable interaction enhanced the experience and thus attitude formation. These results have been substantiated statistically only for the high satisfaction group. By contrast, for low satisfaction group, consumers have not enjoyed the experience they had with the AI system. They found the interaction to be unpleasant, and the system to be useless. The outcomes are summarised in a framework for designing appropriate AI systems shaping consumer journey beyond the traditional marketing context.

1. INTRODUCTION

The boost of Artificial intelligence (AI) opened new avenues for designing marketing vehicles with the aim to enhance consumer experience. Enhanced experience by AI driven journey has been embraced by various brands in an attempt to create sustainable competitive advantage. AI (i.e. intelligence of machine and/or software as opposite to the human intelligence) was recognised to help in behavioural (Bigne, 2020) and marketing (Stone et al., 2021) analysis, as well as to provide advanced content. Furthermore, AI enabled personalisation (Bialkova, 2023b), automation (Fernandes & Oliveira, 2021), launching multiple offerings (Bialkova, 2021), that could be realised in a timely and cost-efficient manner. Despite the foreseen benefits, it is still puzzling way consumers are resistant towards these new AI systems like conversation agents and chatbots.

DOI: 10.4018/978-1-6684-9591-9.ch016

The current paper addresses this question providing a holistic understanding of customer needs and demands when it comes to AI systems use. Nesting consumer in the middle of the experience, i.e., embracing a customer-centric view, we believe is the key to achieve a truly experiential journey. Therefore, we have investigated potential factors driving consumer experience when it comes to actual use of chatbots currently available at the market.

In order to track the real-time customer feedback throughout the journey, we explored parameters emerging from the profound literature audit as key drivers of satisfaction, experience evaluation, and thus, attitudes and behaviour when it comes to chatbots. As a solid fundament for our empirical research, we implement theory from usability, marketing and consumer behaviour literature. In particular, we look at the cognitive (often associated with utilitarian) and emotional (associated with hedonic) aspects of chatbot use.

In the following, we first present the literature audit as a base to build around our theory. Factors emerging as potential key parameters have been tested in a survey exploring opinion of consumers who have used a chatbot at least once in their daily life. Results are presented then and discussed in a framework of chatbot efficiency that might enhance consumer experience.

2. THEORETICAL BACKGROUND

Chatbots could be defined as software applications used to conduct an on-line chat conversation via text or text-to-speech (for details see Bialkova, 2023a). Despite their increasing implementation as promising AI aids to enhance user experience and thus lifting marketing revenues, consumer satisfaction turns to be a key determinant of chatbot (future) use.

We therefore, first look at satisfaction and experience evaluation. Both, utilitarian (e.g., functionality, quality, ease of use) and hedonic (e.g., enjoyment) aspects will be explored in detail as these emerged from a profound literature audit to shape experience evaluation, and thus, satisfaction and attitudes formation.

2.1. Satisfaction

Satisfaction, associated with appropriately meeting consumer expectations, is a crucial determinant driving attitudes towards chatbots (Bialkova, 2021; 2022), and thus intention to use these AI systems in the future (Bialkova, 2023a). Consumers satisfied with the service provided by chatbots reported positive experience (Bialkova, 2023b). Satisfied consumers, might develop loyalty towards the brand delivering such services, as well known from classical service marketing papers (Zeithaml et al., 1996). In the context of chatbots, it was also acknowledged that satisfied consumers develop trust in e-agency, and thus are willing to purchase a product (Tsai et al., 2021), as well as to become loyal to the brand (Cheng & Jiang, 2020). Yet the question is whether consumers are satisfied with chatbots currently available at the market, and what are the parameters determining satisfaction?

Satisfaction with e-service agency was addressed by several papers. The core aspects emerging from the literature audit, we could organise around two pillars: utilitarian and hedonic. Utilitarian factors are associated with consumer thoughtful elaboration and product evaluation, as recognised a long time ago (Babin et al., 1994). By contrast, hedonic components are closely related to pure enjoyment and fun while shopping, as reported in marketing classics (Babin et al., 1994; Hirschman & Holbrook, 1982).

In the chatbot context, utilitarian characteristics have been acknowledged as determinants of productivity (Ben Mimoun et al., 2017), perceived usefulness (Xie et al., 2022) and satisfaction (Cheng & Jiang, 2020). Hedonic components further shaped attitudes towards chatbots (Bialkova, 2023a). A question arises hereby, *whether chatbots currently available at the market provide a satisfactory experience and the service demanded by the consumers.*

2.2. Experience Evaluation

Enhanced customer experience is aimed by any reputable brand. Not surprisingly then, with the technology advancement, various brands started to incorporate AI chatbot systems, in order to improve the customer experience. The paradox hereby emerging is: *why consumers are still resistant towards these new technologies?*

A plausible explanation we pursue by exploring the factors driving experience evaluation, as described in detail below.

Functionality and Enjoyment

Functionality (utilitarian by nature) and enjoyment (hedonic by nature) are characteristics in the heart of any experience, as well noted in experiential marketing literature. Experience is a multidimensional construct, and it involves both cognitive (referred also to utilitarian) and emotional (referred to hedonic) components (e.g., Lemon & Verhoef, 2016; but see also Schmitt 1999). Enhanced experience with cutting edge technologies was recognised to bring consumers to extraordinary journey, and to strengthen the brand value (for details see Bialkova, 2023c).

In the context of chatbots, functionality was extensively investigated. Suggested earlier by the Technology Acceptance Model, TAM (Davis et al., 1989), it was associated with the way a system functions. Only consumers recognising functional benefits, are willing to use the system in the future, as reported in a recent study exploring voice assistance (Tassiello et al., 2021). Enhanced functionality was acknowledged to improve chatbot quality perception (Bialkova, 2021; 2023a), and thus the e-agency (Bialkova, 2023b).

Concerning enjoyment, one of the core affects raising emotions (Russell, 2003), is often used by marketers to enhance experience (Holzwarth et al., 2006), to fuel purchase (Goldsmith, 2016), and to increase the shopping basket value. In the chatbot context, enjoyment was important predictor of satisfaction (Ashfaq et al., 2020), attitudes (Bialkova, 2023a), and therefore future use intention (Bialkova, 2023b). Enjoyment measured the user's level of excitement when conversing with the agent (Lee & Choi, 2017), and correlated with involvement within chatbot interaction (Bialkova, 2023b). Perceived emotional support by chatbots was further acknowledged, and in particular, to play a role in satisfaction (Gelbrich et al., 2022). It was also reported that emotional support mediates intimacy and trust towards satisfaction (Lee & Choi, 2017). A recent study, however, claimed that AI can think, but could not feel (Bakpayev et al., 2022). Such claim invites a close look at the AI systems used and reconsideration of their architecture.

Accuracy and Competence

Defined as precision of the system by usability literature, accuracy was recently explored in the context of chatbots. It was reported that accuracy has a positive and direct effect on customer brand relationship (Cheng & Jiang, 2020). It is a crucial factor shaping functionality perception (Bialkova, 2023b), and it mediates quality evaluation (Bialkova, 2022). Accurate communication with chatbot was recognised to significantly influence customer satisfaction (Chung et al., 2020). In the same study, however, the authors have not been able to substantiate statistically an effect, despite the hypothesised competent service agency as a prerequisite for satisfaction.

Competence reflecting knowledge and skills provided, is addressed in the service marketing literature, as well as in usability and technology contexts. In all these contexts, it was noted that in order to satisfy consumer demand, a service and/or a system should be competent. In the chatbot context, the importance of competent AI agency was also recognised (Bialkova, 2023b). Note, however, some studies have not been able to substantiate statistically a relationship between chatbot competence and usefulness (Meyer-Waarden et al., 2020), neither satisfaction (Chung et al., 2020). But this is a serious issue inviting further investigation to better understand competence perception and its effect on chatbot evaluation.

Ease of Use and Quality

Ease of use was extensively explored since its introduction by the TAM (Davis et al., 1989). Associated with the degree to which the user expects the tech system to be free of effort, it is not a surprise to be recognised as a core component in chatbot evaluation. It was reported to correlate with functionality (Bialkova, 2023a), and usefulness of chatbots (Rese et al., 2020), as well as to reflect trust in the AI system (Pitardi & Marriott, 2021). Note, however, also that ease of use was not able to predict satisfaction, despite its effect on intention to use (Ashfaq et al. 2020).

Concerning quality, reflecting the system and the information provided (i.e. Trivedi, 2019, but see also DeLone & McLean, 2005), again outcomes are contradictory. Some authors claimed crucial role of chatbot quality on satisfaction (Ashfaq et al. 2020), and importance to improve the service quality (Chen et al., 2022; Lou et al., 2021). Other authors, however, have not been able to substantiate an effect of quality at the level of consumer demand (e.g., Ben Mimoun et al., 2017; Meyer-Waarden et al., 2020; Wirtz et al., 2018). Put differently, launching high quality chatbots, satisfying consumer demands is a challenge (Bialkova, 2021; 2022). To address this challenge further investigation is invited.

Taken the importance of the above-mentioned factors, we have a close look at these. In particular, we assume:

H1. Consumers who are satisfied with a chatbot (in comparison to consumers who are not satisfied) will evaluate more positively the experience they had, in terms of functionality and enjoyment.
H2. Accuracy and competence will be evaluated as higher by consumers who are satisfied with a chatbot (in comparison to consumers who are not satisfied).
H3. Quality and ease of use will be perceived as higher by consumers who are satisfied with a chatbot (in comparison to consumers who are not satisfied).

One may argue that such comparison (low vs. high satisfaction) is trivial. But we have to point out hereby that only distinguishing high from low satisfactory evaluation, we could provide the much needed understanding on factors shaping consumer experience when it comes to chatbots. Such understanding is especially relevant, taken the attitudes consumers have towards chatbots and the intention to use in the future.

2.3. Attitudes and Future Use

Attitudes defined as beliefs about an object/or a person (for details see Theory of Planned Behaviour, TPB, Ajzen, 1991) are addressed in almost all of the papers concerning chatbots adoption. In majority of these studies, the relationship between attitudes and future use was tested (for literature review see Lim et al., 2022; Mariani et al., 2023).

What is much more interesting in the present context, the potential parameters shaping attitudes formation and change when it comes to chatbots. Some of these factors already have been mentioned hereby. Ease of use and quality (Bialkova, 2023a) loaded on attitudes, and ease of use was a significant predictor for future use intention (Blut et al., 2021; Pitardi & Marriott 2021). Note, however, effect of perceived ease of use on the intention to reuse a chatbot, was not found by some recent studies (Meyer-Waarden et al., 2020). Other studies even argued that ease of use and quality may turn into a barrier, if not provided at the level desired by the customers (Wirtz et al., 2018). Such a discrepancy in previous work is an emerging call for a profound exploration of the effects discussed.

Therefore, we zoom in into the relationship between satisfaction, attitudes and intention to use chatbots in the future. We explore closely consumers' satisfaction based on the experience they had with the chatbots used before in real life settings. We further looked at possible relationship between factors under investigation, as part of the design hereby. The hypotheses addressed we tested in an empirical study as described in detail in the method section.

3. METHOD

3.1. Participants

103 people took part in the study, 66% female, mean age 28 years old. 38% have used chatbots more than 5 times, 23% have used chatbots between 3-5 times, and the rest have used a chatbot at least once in their daily life. All participants used a chatbot to contact the customer service, 32% also used it to purchase a product.

3.2. Procedure and Design

Participants had to complete a survey online. First they have been introduced to the study, and had to sign a consent form. Then the questions were addressed. Constructs employed are summarised in Table 1.

Construct Satisfaction had 5 items, e.g., The chatbot did a good job adapted from Chung et al. (2020). Construct functionality had 4 items: e.g., Chatbots are functional, based on Bialkova (2023a). Enjoyment had 5 items, adopted from Russell (2003), validated in the context of chatbots by Bialkova (2023a). Ease of use also had 3 items, e.g., Chatbots are easy to use, adapted from Davis et al. (1989). Quality encompassed 4 items: e.g., very bad/ very good quality, adapted from Fenko et al. (2016), validated in the context of chatbots by Bialkova (2023a). Accuracy, e.g., The chatbot was accurate, and Competence, e.g., The chatbot was competent, consisted of 3 items each, and were adopted form Bialkova (2023b). Interactivity also had 3 items, e.g., I felt I could interact with the chatbot, based on Lessiter et al., 2001, validated in the context of chatbots by Bialkova (2023b). Service provided was a single item, self-developed. Attitudes, encompassed 3 items, e.g., favourable/unfavourable based on MacKenzie and

Lutz (1989), validated in the context of chatbots by Bialkova (2023a). Intention to use encompassed 3 items, e.g., impossible/possible. Recommendation had 4 items, e.g., Say positive things about chatbot to other people. Both, intetion to use and recommendation were adopted from Bialkova and te Paske (2020), validated in the context of chatbots by Bialkova (2023a). The scale for Attitudes, Quality, and Intention to use were semantic differential (from 1-7). The remaining constructs were measured with a Likert scale (1 = strongly disagree, 7 = strongly agree). All scales had very good internal validity, all Cronbach's $\alpha > .70$.

Table 1. Summary of the constructs used

Construct	Measuring Scale
Satisfaction	5 items: e.g.,"The chatbot did a good job" adapted from Chung et al. (2020).
Functionality	4 items: e.g., "Chatbots are functional" based on Bialkova (2023a)
Interactivity	3 items: e.g., "I felt I could interact with the chatbot" validated in the context of chatbots by Bialkova (2023b)
Enjoyment	5 items: e.g., "The interaction with chatbot(s) is sad/joyful" based on Russell (2003), validated in the context of chatbots by Bialkova (2023a)
Ease of use	3 items: e.g., "Chatbots are easy to use" adapted from Davis et al. (1989)
Quality	4 items: e.g., "very bad/ very good quality" adapted from Fenko et al. (2016), validated in the context of chatbots by Bialkova (2023a)
Accuracy	3 items: e.g., "The chatbot was accurate" adopted form Bialkova (2023b)
Competence	3 items: e.g., "The chatbot was competent" adopted from Bialkova (2023b)
Service Provided	Single item, "The chatbot provided a proper service" self-developed
Attitudes	3 items: e.g., "favourable/unfavourable" validated in the context of chatbots by Bialkova (2023a)
Intention to use	3 items: e.g., "impossible/possible" adapted from Bialkova and te Paske (2020), validated in the context of chatbots by Bialkova (2023a)
Recommendation	4 items: e.g., "Say positive things about chatbot to other people" adapted from Bialkova and te Paske (2020), validated in the context of chatbots by Bialkova (2023a)

3.3. Analytical Procedure

Data were first submitted to reliability check, all scales demonstrated high reliability, all Cronbach's $\alpha > .70$. Then T-tests were performed to probe for a difference in response from male and female respondents. Such difference was not reported (all p's $> .1$), only for satisfaction being at the margin ($p = .060$). ANOVAs were run to probe for a difference in response as a function of frequency of use (1-2 vs. 3-5 vs. >5 times). Such difference was not reported either.

For construct satisfaction, a Cut point was applied, whereby cases lower than 4 are assigned to one group (low satisfaction), values above the cut point to another group (high satisfaction), and cases equal to 4 were assigned to group neutral, hereafter referred to as indifferent. The high satisfaction group encompassed 51, and the low satisfaction 40 respondents. ANOVAs were run to check whether there is a difference in the response depending on participants' satisfaction (high satisfaction vs. low vs. indifferent group, respectively). Results are reported in detail below.

3. RESULTS

3.1 Comparative Analyses

Results are clear in showing that consumers in high satisfaction group provide more positive evaluation, in comparison to those in low satisfaction group, see Figure 1. For the purpose of the paper, we highlight the comparison between high and low satisfaction group. Results for participants in indifferent group (satisfaction = 4), are mentioned only for completeness, and especially when the parameters under investigation significantly differ from high satisfaction group, as tested by post hoc comparison.

Satisfaction was highest for the high satisfaction group ($M = 5.28$). Participants in the low satisfaction group ($M = 2.62$) reported lowest satisfaction. The indifferent group reported satisfaction at the cut point ($M = 4.00$), which was significantly different from response in both, low and high satisfaction group, $F(2, 100) = 169.72, p < .0001$. Consumers in high satisfaction group also found the service provided to be good, ($M = 4.85$), $F(2, 100) = 30.87, p < .001$. By contrast, low satisfaction group, provided significantly low evaluation concerning the service provided ($M = 2.96$). For the indifferent group, the service provided by chatbots was mediocre ($M = 3.92$).

Figure 1. Attitudes and behaviour intention, respectively for High (blue solid line) and Low (red dashed line) satisfaction group

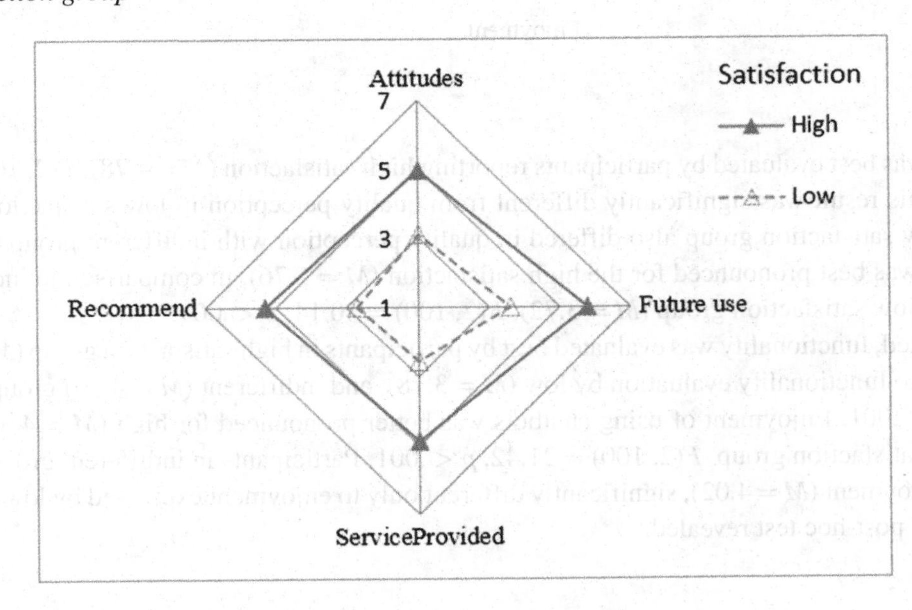

As predicted, attitudes towards chatbots were more positive for high (M = 4.98) than low (M = 3.13) satisfaction, than indifferent (M = 3.81) group, $F(2, 100) = 16.51$, $p < .001$. The intention to use chatbot(s) in the future was highest for consumers reporting high satisfaction (M = 5.77), $F(2, 100) = 25.55$, $p < .001$. For comparison, we report the intention to use a chatbot for low satisfaction (M = 3.61), and indifferent group (M = 4.47), respectively. Similarly, consumers in high satisfaction group were more willing to recommend chatbots, (M = 5.32) than consumers in low satisfaction group (M = 2.62), $F(2, 100) = 18.31$, $p < .001$.

What is much more interesting hereby is the outcome from the analyses concerning the parameters hypothesised to determine experience evaluation, i.e. in terms of quality, ease of use, functionality and enjoyment, as well as their antecedents (see Figure 2).

Figure 2. Experience evaluation, respectively for High (blue solid line) and Low (red dashed line) satisfaction group

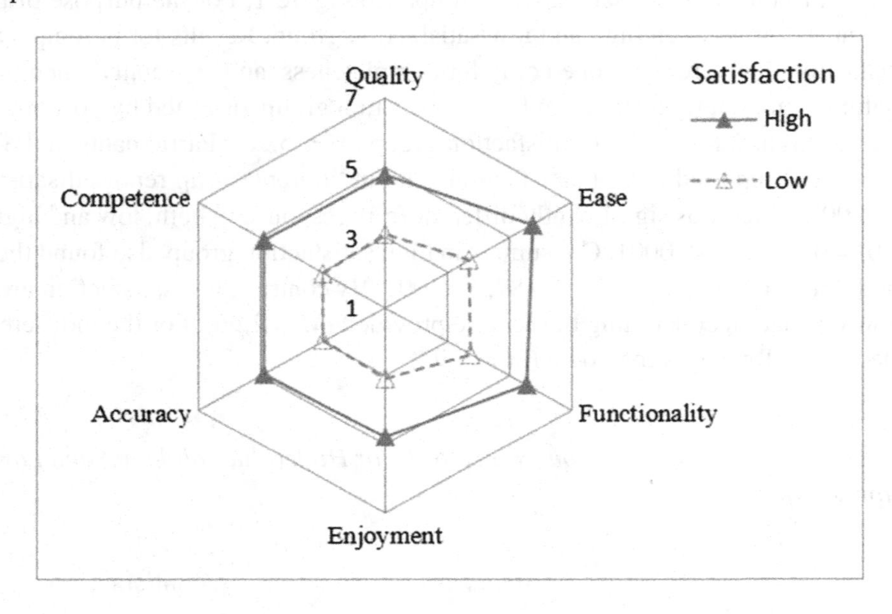

Quality was best evaluated by participants reporting high satisfaction (M = 4.78), $F(2, 100) = 16.51$, $p < .001$. This result was significantly different from quality perception in low satisfaction group (M = 3.12). Low satisfaction group also differed in quality perception with indifferent group (M = 4.08). Ease of use was best pronounced for the high satisfaction (M = 5.76), in comparison to indifferent (M = 4.92) and low satisfaction group (M = 3.72), $F(2, 100) = 40.14$, $p < .001$.

As expected, functionality was evaluated best by participants in high satisfaction group (M = 5.57), in comparison to functionality evaluation by low (M = 3.78), and indifferent (M = 4.50) group, $F(2, 100) = 49.45$, $p < .001$. Enjoyment of using chatbots was better pronounced for high (M = 4.36) than low (M = 3.02) satisfaction group, $F(2, 100) = 21.42$, $p < .001$. Participants in indifferent group expressed mediocre enjoyment (M = 4.02), significantly different only to enjoyment expressed by low satisfaction group, as the post-hoc test revealed.

Consumers expressing high satisfaction were best involved in interaction with a chatbot ($M = 4.82$), in comparison to low satisfaction ($M = 3.03$), and indifferent group ($M = 3.88$), $F(2, 100) = 31.79$, $p < .001$. Accuracy was perceived to be highest by high satisfaction group, ($M = 4.91$), $F(2, 100) = 33.63$, $p < .001$. Low satisfaction group found accuracy to be low ($M = 2.97$), and indifferent group reported mediocre accuracy ($M = 3.88$). Similar pattern was observed for competence, i.e., chatbots were most competent according to the high satisfaction group ($M = 4.87$), $F(2, 100) = 36.44$, $p < .001$. Low satisfaction group evaluated the chatbot competence to be low ($M = 2.96$). The indifferent group perceived the chatbots competence as mediocre ($M = 3.71$).

3.2 Regression Modelling

3.2.1. High Satisfaction Group

Satisfaction explained 31% of the variance in future use intention. Adding attitudes to the model slightly improved its power, $R^2 = .39$, $F(2, 48) = 15.28$, $p < .001$. The higher the satisfaction ($\beta = .34$) and the more positive the attitudes ($\beta = .40$) were, the higher was the intention to use a chatbot in the future (see Table 2, left column).

Ease of use explained 10% of the variance in attitudes. Adding quality evaluation to the model improved significantly its explanatory power, $R^2 = .43$, $F(2, 48) = 18.14$, $p < .001$. The better the quality ($\beta = .62$) and the ease of use ($\beta = .11$) were perceived to be, the more positive the attitudes towards the chatbots were (see Table 3, left column).

Table 2. Summary of the outcomes from the regression modelling encompassing satisfaction and attitudes, and their effect on future use intention, respectively high (left column) and low (right column) satisfaction group

	Future Use (High Satisfaction Group)			Future Use (Low Satisfaction Group)		
	$R^2 = .389$			$R^2 = .334$		
	b	*β*	*p*	*b*	*β*	*p*
Satisfaction	.57	.34	< .05	.17	.12	.22
Attitudes	.32	.40	< .005	.41	.35	< .05

Table 3. Summary of the outcomes from the regression modelling encompassing ease of use and quality evaluation, and their effect on attitudes, respectively high (left column) and low (right column) satisfaction group

	Attitudes (High Satisfaction Group)			Attitudes (Low Satisfaction Group)		
	$R^2 = .430$			$R^2 = .212$		
	b	*β*	*p*	*b*	*β*	*p*
Ease of use	.19	.11	< .05	.12	.11	.52
Quality	.85	.62	< .001	.52	.39	< .05

Functionality explained well the variance in the ease of use. Adding enjoyment to the model, increased its explanatory power. The better the functionality and the higher the enjoyment were, the higher the ease of use was perceived to be, $R^2 = .38$, $F(2, 48) = 19.35$, $p < .001$. Similarly, the quality was perceived to be higher, when functionality and enjoyment were higher, $R^2 = .45$, $F(2, 48) = 23.27$, $p < .001$.

Accuracy explained 18% of the variance of functionality. Adding competence, did not change the model power $R^2 = .18$, $F(2, 48) = 5.69$, $p < .01$.

3.2.2. Low Satisfaction Group

Low satisfaction precluded future use intention. Attitudes, however, were good predictors when added to the model, $R^2 = .33$, $F(2, 37) = 4.38$, $p < .05$ (see Table 2, right column).

Ease of use explained 10% of the variance in attitudes. Adding quality evaluation doubled the model power, $R^2 = .21$, $F(2, 37) = 4.98$, $p < .01$. Note, however, ease of use was relatively low for low satisfaction group, which may explain the pattern observed concerning the effect size (see Table 3, right column).

Functionality explained 19% of the variance in ease of use. Adding enjoyment to the model improved its explanatory power, $R^2 = .31$, $F(2, 37) = 8.28$, $p < .001$. The effect of functionality on quality perception was not substantiated statistically. Adding enjoyment to the model, contributed to statistically substantiate the model, $R^2 = .36$, $F(2, 37) = 10.22$, $p < .001$.

Accuracy explained 15% of the variance of functionality. Adding competence, did not change the model power, $R^2 = .16$, $F(2, 37) = 4.05$, $p < .05$.

4. DISCUSSION

The aim of the present study was to provide the much-needed understanding on factors shaping customer experience when it comes to AI systems use. Nesting consumer in the middle of the experience, we have addressed potential factors driving the experiential journey when it comes to actual use of chatbots currently available at the market. Results are clear in showing that pattern of behaviour (i.e. experience evaluation, attitudes and intention to use) depends on consumer satisfaction.

4.1. Satisfaction

While previous studies have explored utilitarian and hedonic factors in relation to chatbot adoption in general, we have been able to distinguish response from consumers in low and high satisfaction cohort, and thus, to provide understanding why consumers react and evaluate differently the experience with chatbots. Highly satisfied consumers reported more positive evaluation of the experience they had, in comparison to low satisfaction and indifferent group. Highly satisfied consumers also evaluated the service provided to be better (see Figure 1).

These findings are extremely important, as satisfied consumers may develop trust in e-agency, and thus, are willing to purchase a product (Tsai et al., 2021) and to become loyal to the brand (Cheng & Jiang, 2020; Zeithaml et al., 1996). Satisfaction is a key driver of consumer attitudes towards chatbots (Bialkova, 2021; 2022), and therefore future use intention (Bialkova, 2023a).

We also acknowledged that consumers satisfied with the service provided by chatbots reported positive experience, confirming earlier studies (Bialkova, 2023b). Positive experience itself was reported to have a direct effect on attitudes towards chatbots (Jiménez-Barreto et al., 2021). Hereby, we strengthen the exploration, having a close look whether and how experience evaluation was shaped depending on the level of satisfaction. As demonstrated, majority of the relationships between factors under investigation have followed similar patterns in high and low satisfaction group, although with different magnitude (i.e better pronounced for high satisfaction group). Some relationships, however, have not been established for low satisfaction group. We discuss these relationships in detail with the aim to highlight how to turn barriers into stimulators in chatbot use and experience enhancement.

4.2. Experience Evaluation

As predicted, functionality and enjoyment were higher for high than low satisfaction group (H1 confirmed). Accuracy and competence were evaluated to be higher by consumers in high than low satisfaction group (H2 confirmed). These findings are of great importance showing that both, utilitarian (i.e. functionality, accuracy) and hedonic (i.e. enjoyment) components shape the experience evaluation, which is modulated by the level of satisfaction (see Figure 2).

Current outcomes are in line with earlier work, claiming that enjoyment shapes attitudes towards chatbots (Bialkova, 2023a). Furthermore, perceived emotional support by chatbots plays a role in satisfaction (Ashfaq et al., 2020; Gelbrich et al., 2022), mediated by intimacy and trust (Lee & Choi, 2017). Functionality was also recognised in chatbot quality perception, and e-agency efficiency (Bialkova, 2021; 2023b).

Concerning the relationship tested with the regression modelling, accuracy explained a good variance of functionality. This confirms previous work arguing that accuracy is a crucial factor determining functionality perception (Bialkova, 2023b). Note, however, hereby, adding competence, did not change the model power. Such findings shed new light in the understanding of chatbot functionality. We answer a puzzling paradox from earlier works. Although competent service agency was a desired prerequisite for satisfaction, such effect was not substantiated statistically (Chung et al., 2020). Put differently, we have been able to show hereby that consumers do not find chatbots currently available at the market to be competent.

Furthermore, the better the functionality and the higher the enjoyment were, the higher the ease of use was perceived to be. Similarly, the quality was perceived to be higher, when functionality and enjoyment were higher. Note however, these effects were well pronounced only for the highly satisfied consumers. For consumers in low satisfaction group, the effect on functionality was not substantiated on quality. A plausible explanation could be that consumers have not found the chatbot to be functional, which leads to low satisfaction. Current outcomes cohere with earlier work arguing that when consumers recognise functional benefits, they are willing to use the system in the future (Tassiello et al., 2021). By contrast, when consumers do not recognise such benefits they experience aversion, as reported hereby by low satisfaction group.

As predicted, quality and ease of use were perceived as higher by consumers who were satisfied with a chatbot, in comparison to consumers who were not satisfied (H3 confirmed). Current findings are of great importance bringing understanding to a crucial debate in literature. While some researchers claimed that chatbot quality interplayed with satisfaction (Ashfaq et al. 2020; Bialkova, 2023a), other did not report effect of quality (e.g., Ben Mimoun et al., 2017; Meyer-Waarden et al., 2020; Wirtz et al., 2018). A plausible explanation we provide hereby comes with the distinction between response from low and high satisfaction group.

Concerning ease of use, correlation with functionality (Bialkova, 2023a), and with usefulness of chatbots (Rese et al., 2020) was reported by recent studies. By contrast, ease of use was not able to predict intention to use (studies (Meyer-Waarden et al., 2020), neither satisfaction (Ashfaq et al. 2020). To resolve this paradox, we go a step further distinguishing response from low and highly satisfied consumers. We have been able to show that consumer resistance towards chatbots depends on the level of satisfaction and attitudes towards the AI system used.

4.3. Attitudes and Future Use

Attitudes were a significant predictor in future use intention, for both low and high satisfaction group (see Table 2). This outcome coheres with previous work arguing a strong relationship between attitudes and future use intention (e.g., Bialkova, 2021, 2022; Blut et al., 2021; Lim et al., 2022; Mariani et al., 2023; Pitardi & Marriott 2021).

Satisfaction was also a predictor for the willingness to use chatbots in the future. Note, however, this effect was substantiated statistically only for high (see Table 2, left panel), but not for low (see Table 2, right panel) satisfaction group. This is a crucial finding shedding light on the debate concerning the pattern of chatbot use intention.

A further contribution of the present work reflects the relationship between attitudes, quality and ease of use, as a function of satisfaction. Quality perception predicted attitudes formation, for both, low and high satisfaction group, but with different magnitude (see Table 3). This outcome confirms earlier work that the quality of chatbots is a key driver shaping attitudes (Bialkova, 2023a).

Concerning ease of use, its effect on attitudes was substantiated statistically only for high (see Table 3, left panel), but not for low (see Table 3, right panel) satisfaction group. This finding is of a great importance. We have been able to demonstrate that positive attitudes are formed only when consumers find the chatbot to be of good quality, and easy to use. By contrast when chatbots do not offer ease of use, attitudes are not that positive, which reflects low satisfaction, and thus, rejection of chatbots. With this finding we provide understanding on a hot discussion in the literature. While some researchers claimed effect of ease of use on attitudes (Bialkova, 2023a; Pitardi & Marriott, 2021), others questioned the effect of ease of use (e.g., Mariani et al., 2023; Wirtz et al., 2018). A plausible explanation we pursue with the level of satisfaction. Present study in fact shows that particular chatbot AI systems are perceived as easier to be used than others. Moreover, some factors recognised as potential stimulators may turn into a barrier, to backfire expectations and thus satisfaction with chatbot experience, if not provided at the level desired by the customers.

5. FUTURE RESEARCH DIRECTIONS

As part of the design, we have addressed difference in response between low vs. high satisfaction group after actual use of chatbots currently available at the market. Our sample included 103 respondents. Taken that only half of the participants expressed satisfaction with chatbots, a further study could zoom-in into determinants of quality perception for this group, by increasing the sample size.

Increasing the number of respondents, another study could look at low satisfaction group only and to explore in detail specific factors precluding the chatbot future use, as emerging hereby. For example, low satisfaction group did not find the interaction with chatbots to be enjoyable. This is a possible research line, namely, investigating what determines emotional (hedonic) aspect, when it comes to interaction with chatbots.

Concerning cognitive (utilitarian) aspect, low satisfaction group reported that chatbots currently present at the market are not easy to use. Such outcome opens avenues for designers and UX experts to seriously reconsider the interface of chatbots offered to end users. Therefore, new studies could look at appropriate design and to test pertinent AI systems that satisfy the consumer demand for easy to use chatbots.

Moreover, low satisfaction group evaluated chatbots available at the market to do not be competent. Therefore, a new research line could explore factors shaping competence perception.

We have to also mention hereby that our sample encompassed only European consumers. Thus, a future study could approach consumers across the globe and to test whether factors emerging hereby as key parameters driving AI user experience are applicable across countries and cultures. Such cross cultural perspective may be especially relevant for usability experts and consumer scientists advising better targeting and positioning of AI systems that are planned for a future market launch.

6. CONCLUSION

The current paper provides a holistic understanding of customer needs and demands when it comes to AI systems use. Investigating potential factors driving experience with chatbots currently available at the market, we asked consumers to report their opinion after actual chatbot use. We zoomed in into the relationship between satisfaction, attitudes and intention to use chatbots in the future.

We have been able to distinguish response from consumers in low and high satisfaction cohort. Such distinction enabled us to answer a puzzling paradox concerning consumer resistance towards chatbots, and thus, to suggest advice on how to create better AI systems that appropriately meet consumers' demand.

The results from the field study are clear in showing that: 1) Consumers who are satisfied with a chatbot (in comparison to consumers who are not satisfied) evaluate more positively the experience they had, in terms of functionality and enjoyment. 2) Accuracy and competence were evaluated as higher by consumers who are satisfied with a chatbot. 3) Quality and ease of use were perceived as higher by consumers in high satisfaction group. 4) Quality perception predicted attitudes formation, for both, low and high satisfaction group. 5) Ease of use predetermined positive attitudes only for high satisfaction group. Furthermore, 6) When chatbots were not easy to use, attitudes were not positive, which reflected low satisfaction, and thus, rejection of chatbots. 7) Consumers do not find chatbots currently available at the market to be competent. These last findings are a warning call to design chatbots that provide the competence and ease of use desired by customers. Only creating efficient chatbots AI systems that appropriately meet the consumer demand will assure the desired experience leading to an extraordinary AI led journey.

REFERENCES

Ashfaq, M., Yun, J., Yu, S., & Loureiro, S. (2020). I, Chatbot: Modeling the determinants of users' satisfaction and continuance intention of AI-powered service agents. *Telematics and Informatics*, *54*, 101473. doi:10.1016/j.tele.2020.101473

Babin, B. J., Darden, W. R., & Griffen, M. (1994). Work and/or fun: Measuring hedonic and utilitarian shopping value. *The Journal of Consumer Research, 20*(4), 644–656. doi:10.1086/209376

Bakpayev, M., Baek, T. H., van Esch, P., & Yoon, S. (2022). Programmatic creative: AI can think but it cannot feel. *Australasian Marketing Journal, 30*(1), 90–95. doi:10.1016/j.ausmj.2020.04.002

Ben Mimoun, M. S., Poncin, I., & Garnier, M. (2017). Animated conversational agents and e-consumer productivity: The roles of agents and individual characteristics. *Information & Management, 54*(5), 545–559. doi:10.1016/j.im.2016.11.008

Bialkova, S. (2021). *Would you talk to me? The role of chatbots in marketing*. ICORIA2021, Bordeaux, France.

Bialkova, S. (2022). *Interacting with chatbot: How to enhance functionality and enjoyment?* AEMARK2022, Valencia, Spain.

Bialkova, S. (2023a). I want to talk to you: Chatbot marketing integration. *Advances in Advertising Research, XII*, 23–36. doi:10.1007/978-3-658-40429-1_2

Bialkova, S. (2023b). How to Optimise Interaction with Chatbots? Key Parameters Emerging from Actual Application. *International Journal of Human-Computer Interaction*, 1–10. Advance online publication. doi:10.1080/10447318.2023.2219963

Bialkova, S., & te Paske, S. (2021). Campaign participation, spreading e-WOM, purchase: How to optimise CSR effectiveness via Social media? *European Journal of Management and Business Economics, 30*(1), 108–126. doi:10.1108/EJMBE-08-2020-0244

Bigne, E. (2020). Teaching in business: A customized process driven by technological innovations. *Journal of Management and Business Education, 3*(1), 4–15. doi:10.35564/jmbe.2020.0002

Blut, M., Wang, C., Wünderlich, N., & Brock, C. (2021). Understanding anthropomorphism in service provision: A meta-analysis of physical robots, chatbots, and other AI. *Journal of the Academy of Marketing Science, 49*(4), 632–658. doi:10.100711747-020-00762-y

Chen, Q., Gong, Y., Lu, Y., & Tang, J. (2022). Classifying and measuring the service quality of AI chatbot in frontline service. *Journal of Business Research, 145*, 552–568. doi:10.1016/j.jbusres.2022.02.088

Cheng, Y., & Jiang, H. (2020). How Do AI-driven Chatbots Impact User Experience? Examining Gratifications, Perceived Privacy Risk, Satisfaction, Loyalty, and Continued Use. *Journal of Broadcasting & Electronic Media, 64*(4), 592–614. doi:10.1080/08838151.2020.1834296

Chung, M., Ko, E., Joung, H., & Kim, S. J. (2020). Chatbot e-service and customer satisfaction regarding luxury brands. *Journal of Business Research, 117*, 587–595. doi:10.1016/j.jbusres.2018.10.004

Davis, F. D., Bagozzi, R. P., & Warshaw, P. R. (1989). User acceptance of computer technology: A comparison of two theoretical models. *Management Science, 35*(8), 982–1003. doi:10.1287/mnsc.35.8.982

Fenko, A., Kersten, L., & Bialkova, S. (2016). Overcoming consumer scepticism toward foodlabels: The role of multisensory experience. *Food Quality and Preference, 48*, 81–92. doi:10.1016/j.foodqual.2015.08.013

Fernandes, T., & Oliveira, E. (2021). Understanding consumers' acceptance of automated technologies in service encounters: Drivers of digital voice assistants adoption. *Journal of Business Research*, *122*, 180–191. doi:10.1016/j.jbusres.2020.08.058

Gelbrich, K., Hagel, J., & Orsingher, C. (2021). Emotional support from a digital assistant in technology-mediated services: Effects on customer satisfaction and behavioral persistence. *International Journal of Research in Marketing*, *38*(1), 176–193. doi:10.1016/j.ijresmar.2020.06.004

Goldsmith, R. (2016). The Big Five, happiness, and shopping. *Journal of Retailing and Consumer Services*, *31*, 52–61. doi:10.1016/j.jretconser.2016.03.007

Hirschman, E. C., & Holbrook, M. B. (1982). Hedonic Consumption: Emerging Concepts, Methods and Propositions. *Journal of Marketing*, *46*(3), 92–101. doi:10.1177/002224298204600314

Holzwarth, M., Janiszewski, C., & Neumann, M. M. (2006). The influence of avatars on online consumer shopping behavior. *Journal of Marketing*, *70*(4), 19–36. doi:10.1509/jmkg.70.4.019

Jiménez Barreto, J., Natalia, R., & Sebastian, M. (2021). "Find a flight for me, Oscar!" Motivational customer experiences with chatbots. *International Journal of Contemporary Hospitality Management*, *33*(11), 3860–3882. doi:10.1108/IJCHM-10-2020-1244

Lee, S., & Choi, J. (2017). Enhancing user experience with conversational agent for movie recommendation: Effects of self-disclosure and reciprocity. *International Journal of Human-Computer Studies*, *103*, 95–105. doi:10.1016/j.ijhcs.2017.02.005

Lemon, K. N., & Verhoef, P. C. (2016). Understanding customer experience throughout the customer journey. *Journal of Marketing*, *80*(6), 69–96. doi:10.1509/jm.15.0420

Lim, W. M., Kumar, S., Verma, S., & Chaturvedi, R. (2022). Alexa, what do we know about conversational commerce? Insights from a systematic literature review. *Psychology and Marketing*, *39*(6), 1129–1155. doi:10.1002/mar.21654

Lou, C., Kang, H., & Tse, C. H. (2021). Bots vs. humans: How schema congruity, contingency-based interactivity, and sympathy influence consumer perceptions and patronage intentions. *International Journal of Advertising*, *41*(4), 655–684. doi:10.1080/02650487.2021.1951510

Mariani, M. M., Hashemi, N., & Wirtz, J. (2023). Artificial intelligence empowered conversational agents: A systematic literature review and research agenda. *Journal of Business Research*, *161*, 113838. doi:10.1016/j.jbusres.2023.113838

Meyer-Waarden, L., Pavone, G., Poocharoentou, T., Prayatsup, P., Ratinaud, M., Tison, A., & Torn, S. (2020). How Service Quality Influences Customer Acceptance and Usage of Chatbots? *Journal of Service Management Research*, *4*(1), 35–51. doi:10.15358/2511-8676-2020-1-35

Pitardi, V., & Marriott, H. R. (2021). Alexa, she's not human but… Unveiling the drivers of consumers' trust in voice-based artificial intelligence. *Psychology and Marketing*, *38*(4), 626–642. doi:10.1002/mar.21457

Rese, A., Ganster, L., & Baier, D. (2020). chatbots in retailers' customer communication: How to measure their acceptance? *Journal of Retailing and Consumer Services*, *56*, 102176. doi:10.1016/j.jretconser.2020.102176

Stone, M., Moutinho, L., Ekinci, Y., Labib, A., Evans, G., Aravopoulou, E., Laughlin, P., Hobbs, M., Machtynger, J., & Machtynger, L. (2021). Artificial intelligence in marketing and marketing research. In *The Routledge Companion to Marketing Research* (pp. 147–163). Routledge. doi:10.4324/9781315544892-11

Tassiello, V., Tillotson, J. S., & Rome, A. S. (2021). "Alexa, order me a pizza!": The mediating role ofpsychological power in the consumer–voiceassistant interaction. *Psychology and Marketing*, *8*(7), 1069–1080. doi:10.1002/mar.21488

Trivedi, J. (2019). Examining the Customer Experience of Using Banking Chatbots and Its Impact on Brand Love: The Moderating Role of Perceived Risk. *Journal of Internet Commerce*, *18*(1), 91–111. doi:10.1080/15332861.2019.1567188

Tsai, W. H. S., Liu, Y., & Chuan, C. H. (2021). How chatbots' social presence communication enhances consumer engagement: The mediating role of parasocial interaction and dialogue. *Journal of Research in Interactive Marketing*, *15*(3), 460–482. doi:10.1108/JRIM-12-2019-0200

Wirtz, J., Patterson, P. G., Kunz, W. H., Gruber, T., Lu, V. N., Paluch, S., & Martins, A. (2018). Brave new world: Service robots in the frontline. *Journal of Service Management*, *29*(5), 907–931. doi:10.1108/JOSM-04-2018-0119

Xie, C., Wang, Y., & Cheng, Y. (2022). Does Artificial Intelligence Satisfy You? A Meta-Analysis of User Gratification and User Satisfaction with AI-Powered Chatbots. *International Journal of Human-Computer Interaction*, 1–11. Advance online publication. doi:10.1080/10447318.2022.2121458

Zeithaml, V. A., Berry, L. L., & Parasuraman, A. (1996). The Behavioral Consequences of Service Quality. *Journal of Marketing*, *60*(2), 31–46. doi:10.1177/002224299606000203

ADDITIONAL READING

Ajzen, I. (1991). The theory of planned behavior. *Organizational Behavior and Human Decision Processes*, *50*(2), 179–211. doi:10.1016/0749-5978(91)90020-T

Bialkova, S. (2023c). Enhancing Multisensory Experience and Brand Value: Key Determinants for Extended, Augmented, and Virtual Reality Marketing Applications. In A. Simeone, B. Weyers, S. Bialkova, & R. W. Lindeman (Eds.), *Everyday Virtual and Augmented Reality. Human–Computer Interaction Series* (pp. 181–195). Springer. doi:10.1007/978-3-031-05804-2_7

DeLone, W. H., & McLean, E. R. (2003). The DeLone and McLean model of information systems success: A Ten-Year update. *Journal of Management Information Systems*, *19*(4), 9–30. doi:10.1080/07421222.2003.11045748

Lessiter, J., Freeman, J., Keogh, E., & Davidoff, J. (2001). A cross-media presence questionnaire: The ITC-sense of presence inventory. *Presence (Cambridge, Mass.)*, *10*(3), 282–329. doi:10.1162/105474601300343612

MacKenzie, S. B., & Lutz, R. I. (1989). An empirical examination of the structural antecedents of attitude toward the ad in an advertising pretesting context. *Journal of Marketing*, *53*(2), 48–65. doi:10.1177/002224298905300204

Russell, J. A. (2003). Core Affect and the Psychological Construction of Emotion. *Psychological Review*, *110*(1), 145–172. doi:10.1037/0033-295X.110.1.145 PMID:12529060

Schmitt, B. (1999). Experiential Marketing. *Journal of Marketing Management*, *15*(1-3), 53–67. doi:10.1362/026725799784870496

KEY TERMS AND DEFINITIONS

AI: Intelligence of machine and/or software as opposite to the human intelligence.

Chatbots: Software applications used to conduct an on-line chat conversation via text or text-to-speech.

Customer Experience: A multidimensional construct, involving both, cognitive (referred also to utilitarian) and emotional (referred to hedonic) components.

Ease of Use: Associated with the degree to which the user expects the tech system to be free of effort.

Enjoyment: One of the core affects composing emotions.

Functionality: Associated with the way a system functions.

Quality: Reflecting the system and the information provided.

Satisfaction: Associated with appropriately meeting consumer expectations.

Chapter 17
Impact of Artificial Intelligence in Industry 4.0 and 5.0

Luiz Motinho
University of Suffolk, UK

Luis Cavique
https://orcid.org/0000-0002-5590-1493
Universidade Aberta, Portugal

ABSTRACT

Industry 4.0 uses the network concept to establish an interconnected manufacturing system. Industry 4.0 integrates the more recent digital concepts such as artificial intelligence (AI), the internet of things (IoT), big data, cloud computing, and 3D printing. The next maturity level, Industry 5.0, aims to shift the focus back to human-centric production by creating a sustainable and collaborative environment with humans and machines. Every manufacturer aims to find new ways to increase profits, reduce risks, and improve production efficiency. AI tools can process and interpret vast volumes of data from the production floor to spot patterns, analyze and predict consumer behavior, and detect real-time anomalies in production processes. This work studies the impact of AI in Industries 4.0 and 5.0. In Industry 4.0, AI can help in classic tasks such as predictive maintenance, production optimization, and customer personalization. Industry 5.0 enables sustainable manufacturing development and human-AI interaction. In this work, the authors demonstrate the impact of AI in Industry 4.0 and 5.0.

1. INTRODUCTION

In recent years, we have witnessed remarkable advancements in artificial intelligence (AI) that have revolutionized various domains. AlphaZero's ability to learn from scratch and master complex games, like chess, shogi, and Go, without prior knowledge or human guidance marked a significant milestone in AI research in 2017. In 2020, another notable advancement in the medical field was the discovery of the antibiotic Halicina, inspired by the HAL software of the movie '2001: A Space Odyssey'. The same year, OpenAI's GPT-3 (Generative Pre-trained Transformer) made significant strides in natural language

DOI: 10.4018/978-1-6684-9591-9.ch017

processing and generation. GPT models, such as GPT-3.5, have demonstrated impressive capabilities in understanding and generating human-like text (Kissinger et al. 2022).

Evaluating AI's impact in Industry 4.0 and 5.0 is challenging since many technologies contribute to this area. Various topics like the Internet of Things (IoT), Big Data, Cloud Computing, 3D printing, Robotics, Cobots, G6 networks, and FabLabs disperse the reader without giving him unity.

This work aims to find a framework for several technologies and study how AI relates to other technologies in Industry 5.0, recognizing the legacy of Industry 4.0.

For this purpose, we choose the classic I/O framework, or pipeline, with three steps: (i) data and connectivity, (ii) intelligent systems, and (iii) interaction and visualization, shown in Figure 1.

Figure 1. A framework of the types of technologies of Industry 4.0 and 5.0

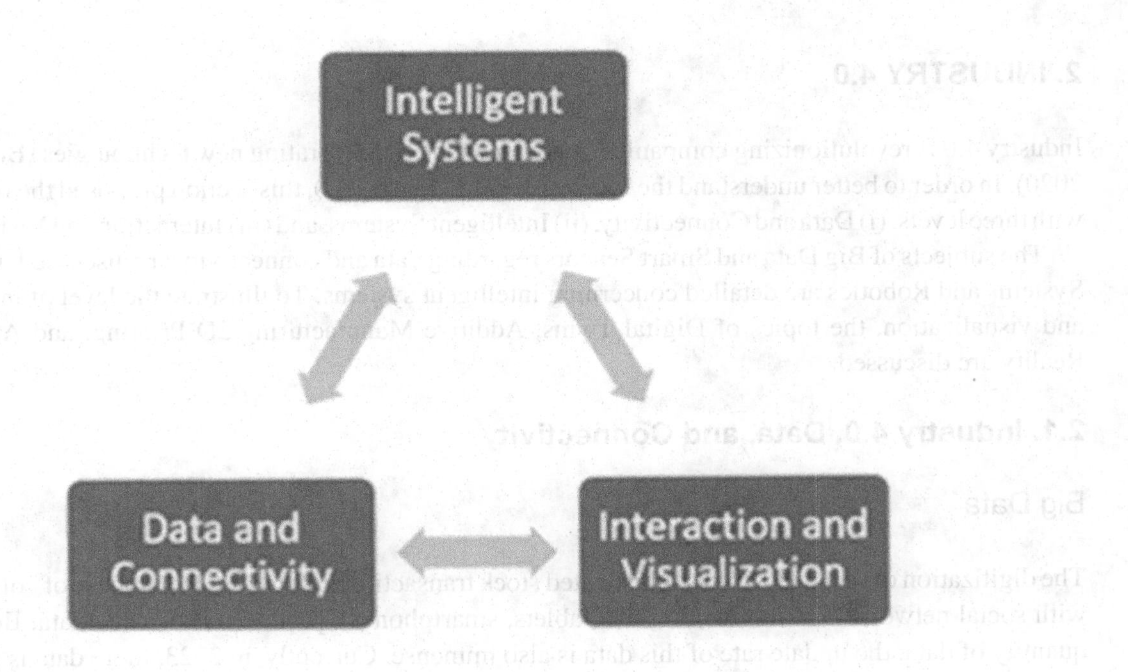

The contribution of this work is to distinguish between Industry 4.0 and Industry 5.0 and prophesy which technologies are the most substantive in Industry 5.0 while focusing on AI.

In this work, we aimed to identify technologies encompassing Industry 4.0 and Industry 5.0 and align with the previous framework with three technology sets. The result with the more relevant technologies is presented in Figure 2. Each technology will be further developed in the document.

The remaining paper is organized as follows. Section 2 presents the technologies of Industry 4.0. Section 3 reports the technologies that support Industry 5.0. Finally, in Section 4, we draw some conclusions.

Figure 2. More relevant technologies of Industry 4.0 and 5.0

	Industry 4.0	Industry 5.0
Data and connectivity	big data	data quality
	smart sensors	6G
Intelligent systems	robots	cobots
	AI/ML	generative AI
Interaction and visualization	digital twins	personal fabrication
	additive manufacturing	FabLabs
	augmented reality	extended reality

2. INDUSTRY 4.0

Industry 4.0 is revolutionizing companies' manufacturing by integrating new technologies (Bornet et al. 2020). In order to better understand the impact of AI in Industry 4.0, this section proposed the framework with three levels: (i) Data and Connectivity, (ii) Intelligent Systems, and (iii) Interaction and Visualization.

The subjects of Big Data and Smart Sensors regarding data and connectivity are discussed. Intelligent Systems and Robotics are detailed concerning intelligent systems. To illustrate the level of interactions and visualization, the topics of Digital Twins, Additive Manufacturing/3D Printing, and Augmented Reality are discussed.

2.1. Industry 4.0, Data, and Connectivity

Big Data

The digitization of services, from sophisticated stock transactions to the simple purchase of coffee, along with social networks and mobile devices (tablets, smartphones), produces enormous data. Beyond the quantity of data, the update rate of this data is also immense. Currently, in 2023, more data is generated every 2 minutes than all the data from prehistory up to 2003.

On the other hand, processing capability has also seen significant increases. Over the past 40 years, the integration capacity of integrated circuits has allowed us to double processing power every two years (Moore 1965), increase storage capacity, and reduce its price. Moore's Law has been proven over the past 40 years, enabling a processing increase in the value of 2 raised to the power of 20 (2^20).

To get a sense of the increase in capacity, given the difficulty of humans in understanding the meaning of exponential growth, let us use the example of travel time from Lisbon to Porto. Suppose a trip from Lisbon to Porto 40 years ago took an average of 6 hours. If the evolution of transport were as significant as in computers, the travel time from Lisbon to Porto today would be two-hundredths of a second (6.60.60 / 2^20).

The large volume of data, offset by increased processing capacity, has given rise to new concepts, such as Big Data, and the creation of new professions like data scientists, dubbed by the Harvard Business Review as the sexiest profession of the 21st century.

With the advent of Web 2.0 (the people's web) associated with mobile devices and the Internet of Things, classic business applications have surpassed data volume.

In a study conducted in 2012, the estimated value of information on the planet was 2.8 ZB (zettabytes, 1021 bytes). The change in scale in data volume and its update rate gave rise to what is generically called Big Data.

Big Data is associated with the 3V acronym: volume, update speed, and various formats. Some authors include a fourth V of value or data veracity.

The concept of 'dark data' is often used in data management, similar to the concept of dark matter in astrophysics. The concepts' similarities are because only a tiny percentage of the matter is visible and possibly contains several explanations for the universe's origin. Of the 2.8 ZB, 85% are unstructured data, i.e., media like video, photography, and sound. Of the remaining 15% of formatted and text data, only 3% are analyzed. Thus, we conclude that only a small percentage of 0.45% of the planet's data is subject to analysis, and 99.55% of unanalyzed data is called Dark Data.

With the emergence of new formats for structured data, the concept of NoSQL arose within Big Data. NoSQL, or Not-only SQL, allows data storage, processing, and querying very efficiently. NoSQL solutions are divided into a few groups: Key/value storage, like LinkedIn's Voldemort; Super-columns storage, like Facebook's HBase or Cassandra; Document storage, like XML database or MongoDB; Graph storage, like HyperGraphDB or ArangoDB and Object-oriented storage, like Db4object.

From the relational model and the declarative SQL (Structured Query Language) used in most corporate databases, NoSQL presents itself as the alternative for dealing with large volumes of data.

The structures of NoSQL solutions have been simplified compared to the relational model and ensure very efficient information querying with algorithmic complexities of order O(1).

In NoSQL, the maximum complexity should be of the order O(N), where N is the file size. Unlike SQL, the operation of joining tables does not exist, given its high complexity of order O(N^2) for the worst case.

For data aggregation, MapReduce is used, implemented in two phases. The Map operator function selects data into subgroups, and the Reduce operation aggregates the information from each subgroup. The worst-case algorithmic complexity will be twice O(N).

The concept of Big Data brings new challenges for dealing with large volumes of data, both for companies and the scientific community. Developing new algorithms is critical since the algorithmic complexities are preferable to order O(1) and should never exceed order O(N).

As a result, Big Data creates new opportunities in data-driven decision-making. Director of Google Research, Peter Norvig, stated, "We do not have better algorithms; we have much more data" (Davenport 2014).

Smart Sensors

Sensor networks can improve the world through diagnostics in medical applications, improved performance of energy sources like fuel cells, batteries, and solar power, improved health, safety, and security for people, and sensors for exploring space and known universities. Sensor proliferation and sensor fusion will continue to accelerate as security within the IoT improves and makes it safer to transmit critical data (Sofi et al. 2022).

When we think of the plethora of applications for sensors, one cannot avoid seeing an Artificial Intelligence (AI) potential in each. The technology may not inspire many unique or novel sensors, but it will generate a massive demand for sensors of all types. Moreover, with the need for compact designs, sensor fusion will become more the norm than application-specific.

Sensor technologies will help us meet environmental goals surrounding cleaner energy and lower carbon footprints, including the electrification of vehicles. Sensors also serve autonomous vehicles and smart city infrastructures; sensors collect the big data needed for management, control, and safety. Sensor technologies are also widely used in the medical market. Our aging global population is driving a trend toward home healthcare. With the pandemic, this drive has accelerated with the increased use of remote patient monitoring and diagnostic devices, including medical wearables.

One particular space that is set to be boosted by IoT sensor technology in the future is smart cities. Analyzing the stream of crucial data produced by sensors is critical in keeping structures – and therefore communities, economies, and livelihoods – safe and operational. When this approach is scaled up to include whole urban environments, the potential of smart cities will be fully realized.

IoT sensors are also primed for aiding sustainability efforts across companies of all sizes and could see further innovation with the aid of AI. The development and adoption of 100% solar-powered sensors remove the need for lithium-ion batteries in most IoT devices today, enabling long-term, maintenance-free deployments.

We cannot change batteries in billions of sensors every few years. Sensors have become more compact while offering more functionality. They now employ micro-electro-mechanical systems (MEMS) architecture, and plug-and-play sensors are integrated into packages to adapt to the various market and application requirements.

Major league baseball is on the cutting edge in capturing and studying microdata. MEMS and sensors track everything on the field and measure games in unprecedented ways, from following a single player's eye movement to an entire team's coordinated response to a fly ball.

Instead of a baseball field, imagine a store full of sensors. If MEMS and sensor technology make it into stores, we could track how customers interact with products, how they move through the store, and what customer segment they belong to, all with unparalleled precision. We could unlock a new universe of insights significantly more than our digital footprint.

2.2. Industry 4.0, Intelligent Systems

Intelligent Systems

Artificial Intelligence (AI) is a revolutionary field set to act as a primary component in emerging technologies like big data, robotics, and IoT. It has moved from fantasy to reality, with AI systems now embedded in everyday items, from cars to apps. The evolution of AI started with knowledge engineering, shifted to model- and algorithm-based machine learning, and is now focused on perception, reasoning, and generalization (Stephanopoulos, Han 1996).

Despite impressive advancements, current AI programs possess specialized intelligence, meaning they can solve only one problem at a time and are rigid in response to input changes. The ultimate goal is to create AI with general intelligence similar to humans, which is still one of science's most ambitious objectives.

AI forms the foundation of computer learning, capable of analyzing massive amounts of data to make optimal decisions in a fraction of the time compared to humans. AI algorithms differ from conventional algorithms in their ability to change and rewrite themselves in response to data inputs, displaying a form of intelligence.

AI and machine learning (ML) will transform the scientific method. Their ability to analyze enormous data sets and discover complex relationships will catalyze a new era of scientific discovery. We have already seen significant advancements in specialized AI due to the availability of large data sets and computational resources.

AI is becoming integral to many industries, including online retail, where automated systems replace human workers. However, with proper investment in education and upskilling, AI is predicted to create more jobs than it eliminates, changing the narrative from "humans or computers" to "humans and computers."

In the future, AI and human intelligence will combine to develop sophisticated cybersecurity measures. The transportation and manufacturing sectors will also benefit significantly from AI, with predictions of commercialized autonomous vehicles and enhanced factory operations.

AI also promises transformative consumer experiences, enabling concepts like the metaverse and cryptocurrencies. Exciting research is happening in reinforcement learning and generative adversarial networks (GAN), which will have substantial near-future ramifications.

AI algorithms offer great potential in bridging the gap between digital and physical realms, facilitating the integration of frictionless capitalism into the economy. However, these developments pose significant challenges to personal privacy and freedom from discrimination, necessitating proper data ownership and control measures.

Further research will focus on AI techniques such as multiagent systems, experience-based reasoning, and developmental robotics. These techniques could provide the key to equipping machines with common sense, learning the relations between their actions and environmental effects.

A survey from Oxford University's Future of Humanity Institute suggested that by 2026, machines could write school essays (in some form, ChatGPT already meets this milestone); by 2027, self-driving trucks could make drivers redundant; and by 2137, all human jobs could be automated.

Robotics

The ability of machines and robots to learn could give them an even more diverse range of applications. Future robots that can adapt to their surroundings, master new processes, and alter their behavior would be suited to more complex and dynamic tasks. Ultimately, robots have the potential to enhance our lives. As well as shouldering the burden of physically demanding or repetitive tasks, they may be able to improve healthcare, make transport more efficient, and give us more freedom to pursue creative endeavors (Jeyakumar 2022).

Thanks to improved sensor technology and more remarkable advances in Machine Learning and Artificial Intelligence, robots will keep moving from mere rote machines to collaborators with cognitive functions. These advances and other associated fields are progressing upward, and robotics will significantly benefit from these strides. We can expect to see more significant numbers of increasingly sophisticated robots incorporated into more areas of life, working with humans.

An accelerated move towards attended automation is one of the trends we see unfolding in the future. Organizations worldwide understand that they must empower their employees via attended automation to automate more processes and maximize the returns on their existing RPA (Robotic Process Automation) investments (Langmann, Turi 2023). Attended automation, sometimes called desktop automation, extends the functionality of bots even further. An attended automation robot—essentially a digital personal assistant that lives on the desktop—guides employees to deliver a better customer experience, automating repetitive desktop tasks faster and more accurately. Attended bots take care of mundane tasks like populating forms for the employee and can also provide the following best action and compliance guidance as the customer service representative helps a customer resolve an account query, for example. The attended automation bots bridge tasks in the back and front offices, allowing more complex processes to be automated end-to-end.

Robot designers use Artificial Intelligence to give their creations enhanced capabilities like: (i) Computer Vision: Robots can identify and recognize objects they meet, discern details, and learn how to navigate or avoid specific items. (ii) Manipulation: AI helps robots gain the fine motor skills needed to grasp objects without destroying them item. (iii) Motion Control and Navigation: Robots no longer need humans to guide them along paths and process flows. AI enables robots to analyze their environment and self-navigate. This capability even applies to the virtual world of software. AI helps robot software processes avoid flow bottlenecks or process exceptions. (iv) Natural Language Processing (NLP) and Real-World Perception: Artificial Intelligence and Machine Learning (ML) help robots better understand their surroundings, recognize and identify patterns, and comprehend data. These improvements increase the robot's autonomy and decrease reliance on human agents.

2.3. Industry 4.0, Interaction, and Visualization

Digital Twins

The concept of making a duplicate or 'twin' of an asset to enable simulations and predict outcomes based on changes in the operating conditions finds its origins in the 1960s with the Apollo space program. The concept gained prominence in April 1970 when NASA engineers utilized duplicates of the spacecraft systems, initially designed for training purposes, to aid in the rescue of Apollo 13 following a catastrophic oxygen tank explosion. As the astronauts slept in the lunar module, the command module had to be powered down to preserve the batteries. NASA engineers used what amounted to a twin of the command module's electrical system to devise a series of steps to power up the frozen spacecraft without draining what little power remained in the batteries. A process typically took two days on the launchpad and was completed in under two hours. It was an extraordinary feat. Nevertheless, the technology has only matured since then. Clarifying the distinction between a model and a Digital Twin can help to demystify the technology and its applications (Grieves, Vickers 2017).

Deploying Digital Twins in production across operations and supply chains is now an upward trend. Not only in production, but the underlying technology has advanced to provide extremely high-fidelity twins with previously unimaginable capabilities.

Digital twins are a virtual representation of a physical object. It could be anything – as complex as a car or a manufacturing production line or as simple as a piece of furniture. The digital twin emulates all the parts of the object (or a set of connected objects) to create a virtual proxy. A car's digital twin would model its shape, tires, seats, engine, transmission, and everything. Companies use a digital twin to design a 3D model of the original, enabling teams to analyze the performance of the object under different conditions. Successfully deployed, digital twins can save serious money, improve product designs, and elevate efficiency and productivity (Tao et al. 2019).

Advances in edge computing and in-memory processing enabled by scalable computing delivered through containers, along with new pervasive network technologies like 5G supporting streaming data, make it possible to interconnect these twins. The results can be seen in real-world projects with twinned complete manufacturing lines and complex interconnected processes like supply chain integration. Digital Twins are becoming essential to everything from smart city planning to improving healthcare. For example, Singapore has created a complete digital twin of the city-nation to track traffic, pollution, climate, and city layouts so city managers can test accessibility options, see the potential impact of new construction, manage emergency responses, and monitor city health. Meanwhile, doctors are creating patient-specific Digital Twins of lungs to help decide on ventilator use when treating COVID-19 patients. In a lighter example, craft beer drinkers taste Industry 4.0 intelligence with digital twins.

To construct an effective Digital Twin, it is crucial to identify the problem to be solved, the opportunity to be explored, and the required accuracy of the predictions. Understanding the goal of a Digital Twin determines which data and sensor feeds are required to achieve predictive value within a defined confidence interval. To create a digital twin, one must record the object's base information and capture how it is performing and being used. Sensors do that, whether it is a machine in a factory or a tire on the road. The development of the Industrial Internet of Things (IIoT) has pushed digital twins even further. IIoT-enabled machines and components collect and feed data in real-time, meaning they essentially self-report their conditions.

The potential for Digital Twins in almost every industry is endless. Previously siloed departments across operations, maintenance, finance, sales, and marketing can all use digital twins to access a unified source of real-world data to predict maintenance, improve design, understand usage, and adjust pricing. Companies utilize digital twins to create 3D models of the original objects, which allows teams to analyze the object's performance under various conditions. Successfully deployed, digital twins can save serious money, improve product designs, and elevate efficiency and productivity. As the cost and complexity of digital twins have fallen, their adoption has spread beyond manufacturing to many different types of businesses. The cost of sensors continues to decline, the amount of data collected exponentially grows, and storing is affordable in various cloud services. Digital twins are also enabling new business models. For example, an equipment manufacturer may lease a digitally enabled component. Customers reap the advantages through such a 'product-as-a-service' model because there is a much smaller initial financial outlay and much less downtime. While we may not understand the inner workings of a machine or component until it malfunctions, a digital twin, in conjunction with IIoT, allows for proactive maintenance and servicing due to predictive analytics. A digital twin could identify and solve many 'what-if' scenarios to enable better decision-making for highly complex supply chains. This vision of creating digital twins of supply chains has a significant obstacle to overcome in sharing intelligence. Companies often hold their data close, which gets in the way of efficiency.

Over the long run, digital twins will form their networks, which experts call a 'digital thread'. Suppose a digital twin enables us to digitally represent a piece of equipment or facility. In that case, a digital thread is the continuous, connected stream of information an intelligent asset provides throughout its life cycle, from design to decommissioning (Jagusch et al. 2021). Implemented effectively, digital twins can serve as strategic catalysts. They can provide visibility into an organization's processes and ways to improve them and, in turn, strengthen the customer experience and relationships. Digital twins provide a sandbox for testing and refining innovations before launching into the real world. They afford businesses a cost- and time-efficient way to design more innovative products and assets while capturing more information about them. So they enable companies to make products better, faster, and safer and generate new revenue opportunities. These include service offerings that create as-a-service business models, which remove the burden of significant capital outlays and lifetime maintenance from the customer and keep them connected with the service provider.

Additive Manufacturing and 3D Printing

Additive Manufacturing (AM) or 3D printing, a significant player in digital transformation, involves creating an object layer by layer, offering potential value over traditional manufacturing techniques (Shahrubudin et al. 2019). This technology eliminates the need for tooling and switching costs between production sites, introducing a new paradigm distinct from the labor-intensive manufacturing methods of the past 150 years.

In additive manufacturing, 'additive' refers to building an object layer by layer. Additive manufacturing, also known as 3D printing, involves the sequential deposition of material to form a three-dimensional object. Unlike traditional subtractive manufacturing methods that involve cutting, drilling, or carving away material from a larger block, additive manufacturing adds material in a controlled manner to construct the final product.

AM technologies provide four potential sources of value. Firstly, they offer unparalleled design freedom, enabling the creation of parts that outperform or cost less than conventional alternatives. Secondly, as there is no need for molds or fixed tooling, every part can be unique, enabling mass customization. Thirdly, AM eliminates time-consuming fabrication operations, accelerating product development and reducing time to market. Lastly, it simplifies the maintenance and support of products in the field by enabling on-demand production of spare parts.

While some claim that 3D printing is overhyped, it is evident that technology has significantly strengthened supply chains, especially during crises. The coming decade is expected to increase AM's role as more companies utilize 3D printing to create consumer-facing parts, custom manufacturing aids, and decentralized production.

The distributed manufacturing model transforms product strategies, enabling companies to manufacture final products closer to the customer. Advancements in Digital Light Processing (DLP) 3D printing processes have improved throughput by increasing speed, building volume, and reducing post-processing. These advances have contributed to greater simplicity and higher efficiency in the manufacturing process.

Experts predict additives will become the dominant mass-manufacturing technique, aiding the battle against global warming through light-weighting, reducing energy consumption and waste, and transitioning to plant-based materials. AM also promises to increase sustainability in production, pushing for the use of renewable sources of electricity.

We will likely see more mass production using AM technology, not just for custom or high-added value products but also for serialized products. Advanced material composites and the ability to produce highly complex geometries will unlock new manufacturing possibilities.

The next decade presents excellent opportunities for 3D printing to enter emerging markets like commercial space, drones, UAVs (unmanned aerial vehicles), robotics, urban mobility, etc. These markets are expected to grow staggeringly, pushing the growth of enabling technologies like 3D printing.

Over the next decade, AM will become a practical tool covering the entire product life cycle, from concept models to aftermarket spares, including volume production. Questions remain regarding the discovery of new 3D printing materials, the application of metamaterials, and the acceleration of efforts to produce bio-based polymers or recapture waste plastics and metals.

Additive Manufacturing will establish itself as the manufacturing technology in the next ten years, supporting the transition to a sustainable and resource-efficient civilization. As AM material and production costs continue to drop, we will see more supply chain disruption, and parts printed locally becoming more cost-effective than those shipped from global manufacturing hubs.

Additive manufacturing could revolutionize supply chain and inventory management, making high-mix, low-volume manufacturing economical. The Distributed Recycling and Additive Manufacturing (DRAM) technology, which enables us to use 3D printers at home to recycle waste plastic into valuable products, is poised to impact consumer manufacturing significantly.

Augmented Reality and Beyond

The revolution in the digital world is driving the transformation of various industries. Today, Extended Reality (XR) – including virtual and augmented reality (VR/AR), is becoming increasingly prevalent in intelligent industrial environments. AR assists machinery operators, such as those driving crates, in loading and unloading containers from ships. At the same time, VR is often deployed to train workers on using heavy and potentially hazardous equipment.

Connectivity is considered a cornerstone, enabling many of the cutting-edge technologies we see today. For instance, deploying AI at scale or processing data in real-time, especially near the action, requires robust connectivity. Connectivity becomes particularly relevant when dealing with autonomous robots, which have been used in industrial manufacturing for several decades. However, when these robots are integrated with AI, IoT, edge computing, and other emerging technologies, they become more potent, versatile, and practical.

Various new and affordable technologies are available, sparking significant advancements in various fields. New tools such as 3D printers, robotics, microprocessors, artificial, virtual, and augmented reality, e-textiles, "smart" materials, and novel programming languages empower individuals to invent and create. These tools and digital sharing capabilities have amplified this evolutionary leap.

The extensive range of contemporary exponential technologies encompasses 3D printing, artificial intelligence (AI), augmented and virtual reality, digital biology, biotech, nanomedicine, computing systems, and autonomous vehicles. Holography, for instance, is a budding technology set to transform Industry 4.0 completely. Holograms may seem more like a concept straight out of a science-fiction novel; however, rapid advancements in telecommunications, particularly 5G, are set to change this perspective.

Current hologram technology, like Hologauze, offers significant 3D hologram effects on a large scale. It works with 3D polarized projection systems and is a 2D hologram effect. In industrial applications, holographic technology finds its use in quality control during manufacturing and fracture testing, such as holographic non-destructive testing. Moreover, it has applications across various fields like medicine, the military, weather forecasting, virtual reality, digital art, and security.

Automation and real-time network connectivity are at the core of the Industry 4.0 revolution, with holography playing a pivotal role. The technology substantially benefits various development processes, features, and applications, focusing on 'holography for Industry 4.0'. Adopting holographic technologies could significantly improve the efficiency of existing products and services in other technology sectors such as architecture, 3D modeling, mechatronics, robotics, and healthcare and medical engineering. Therefore, hologram technology is emerging as a new industry trend influencing various aspects of Industry 4.0.

3. INDUSTRY/SOCIETY 5.0

Industry 5.0 is the next phase in the evolution of the industrial sector, building upon the advances of Industry 4.0 by incorporating more sophisticated human-machine interactions. It leverages critical technologies such as the Internet of Things (IoT), Big Data, Artificial Intelligence (AI), and smart manufacturing to improve business productivity and efficiency.

The core premise of Industry 5.0 is the collaboration between humans and machines, often called cobots. This shift from an overemphasis on technology to a more balanced human-machine collaboration promises to improve customer satisfaction by producing personalized products, giving businesses a competitive advantage, and promoting economic growth. Universal robots aid human workers, enhancing productivity in the manufacturing industry (Maddikunta et al. 2022).

Society 5.0, emerging alongside Industry 5.0, originates from Japan, marking the evolution after earlier societal stages like hunter-gatherer and industrial eras. It emphasizes human-centricity and sustainability over mere technological advancements. This concept envisions a harmonious blend of technology and human life, addressing societal challenges (Deguchi et al. 2020).

Additionally, Industry 5.0 contributes to sustainability efforts by striving to establish systems that run on renewable energy. This new industrial era incorporates technologies such as collaborative robots, digital twins, and platforms like Nexus Integra for managing industrial assets on a large scale, facilitating digital transformation.

The philosophy behind Industry 5.0 is not just about manufacturing for profit. It promotes principles of sustainability and human-centricity. A prime example of Industry 5.0 implementation is seen in multinational companies like Toyota and Repsol. Repsol uses Blockchain and robotic process automation technology to improve business security and productivity, showcasing how Industry 5.0 principles can be practically applied. On the other hand, Toyota makes significant investments in Industry 5.0 technologies to stay ahead of the curve.

Industry 5.0 aims to revolutionize manufacturing by automating repetitive tasks and integrating intelligent robots into supply chains and workflows. It adopts cognitive computing and intelligent automation to enable hyper-personalization. Future technological advancements, such as 6G networks, are set to augment Industry 5.0 further, making it an influential force in the industrial sector.

In order to better understand the impact of AI in Industry 5.0, this section proposed the same framework with three levels: (i) Data and Connectivity, (ii) Intelligent Systems, and (iii) Interaction and Visualization.

The subjects of Data Quality and 6G Networks regarding data and connectivity are discussed. Cobots are detailed concerning intelligent systems. To illustrate the level of interactions and visualization, the topics of Personal Fabrication and FabLabs are discussed.

3.1. Industry 5.0, Data, and Connectivity

Data Quality

Data Quality is gaining importance among data researchers as the volume of data generated by social networks, e-commerce, IoT, and other connectivity systems continues to grow. Despite the significant investment in storage technologies for managing large volumes of data, many of these datasets are unstructured or contain missing and incorrect values (Ehrlinger, Woss 2022).

Data Quality in Machine Learning can have a significant impact on bias. Bias refers to the systematic error or prejudice in the data or the learning algorithm, leading to unfair or discriminatory outcomes. If the training data used to train the machine learning model is biased, the model can learn and perpetuate those biases. For example, suppose historical data used to train a hiring algorithm predominantly consists of male candidates being selected. In that case, the algorithm may learn to favor male candidates in the future, leading to gender bias. Addressing data quality issues is crucial to mitigate bias in machine learning. Techniques such as careful data collection, diverse and representative training datasets, data augmentation, preprocessing techniques, and fairness-aware algorithms can help improve data quality and reduce bias in machine learning models.

Data has become crucial in academia and Industry 4.0 in recent years. However, much of the available data in the public domain, including text, tables, and linked data, is often incorrect, incomplete, or ambiguous.

Data needs to undergo an assessment and improvement process using Data Quality Management policies to ensure quality information. The pipeline should consider (i) Data Profiling, understanding data and metadata characteristics and identifying anomalies; (ii) Data Assessment, measuring data quality using predefined metrics; (iii) Data Cleansing, correcting errors and inconsistencies in the data; and (iv) Data Monitoring, continuously monitor data quality to detect and address issues.

Data quality can be understood as the "fitness for use" for a given application or use case. Assessing data quality involves computing multiple quality metrics rather than relying on a single metric for a specific application. Data Quality must meet criteria/dimensions such as accuracy, completeness, validity, consistency, uniqueness, timeliness, and fitness for purpose to ensure high-quality outcomes.

6G Network

6G, the sixth generation of wireless communication technology, is set to be a substantial upgrade from its predecessor, 5G. Expected to achieve download speeds near 95 Gb/s, it aims to eliminate latency, reduce congestion on mobile networks, and support an enormous number of devices with ultra-low latency requirements. This significant increase in speed and capacity is expected to make real-time data processing and analytics more efficient, fundamentally transforming how we understand and implement digital technologies (Tomkos et al. 2020).

With the shift to 6G, we are likely to see advancements in technologies such as Extended Reality (XR), digital twinning, tera-Hertz and millimeter waves communication, tactile Internet, non-orthogonal multiple access (NOMA), small cells communication, and fog/edge computing. These technologies will integrate terrestrial, aerial, and maritime communications into a robust and reliable network.

A significant area where 6G can profoundly impact is Industry 5.0. Industry 5.0 is characterized by the collaborative working of humans and robots or intelligent machines, leveraging technologies like big data analytics, the Internet of Things (IoT), collaborative robots, Blockchain, digital twins, and future 6G systems — the Industry 5.0 paradigm benefits from these technologies to optimize operations and create efficient supply chains.

6G networks are expected to meet the intelligent information standard that provides high energy efficiency, reliability, and increased traffic capacity. They can help manage large amounts of data, making big data analytics a key enabling technology for Industry 5.0. IoT, an essential aspect of Industry 5.0, can be significantly improved with the advent of 6G, offering opportunities to reduce operating costs by addressing issues in communication networks, waste management, supply chain optimization, and more.

3.2. Industry 5.0, Intelligent Systems

Cobots

The COVID-19 pandemic has accentuated the need for resilient supply chains and human-machine collaboration. Full or partial shutdowns and social distancing regulations require factories and workspaces to operate with a minimal onsite crew. Despite labor shortages, supply chain disruptions, and other production challenges, manufacturers are under constant pressure to respond to the evolving market needs. The demands for mass customization, quality expectations, faster product cycles, and product variability are at an all-time high. Tackling these persistent challenges requires combining human skill and ingenuity with the strength and speed of robots. To bring the best of both worlds – human creativity and robotic precision – manufacturers should adopt cobots (collaborative robots) that can reduce human interaction in feasible situations and accelerate production cycles (Lefranc et al. 2022).

Cobots allow manufacturers to maximize production and address the changing demands while ensuring the safety of their employees, clients, and partners. Cobot stands for 'collaborative robot,' which says it all – cobots are robots designed to work alongside and assist human employees while making sure they do not cause any harm to us carbon-based lifeforms. Unlike industrial robots, which generally work in isolation and can be dangerous for passers-by, cobots feature sensors and collision avoidance, making them safe to work in the same space as humans.

Collaborative robotics include both robots and robot-like devices and generally are part of processes that cannot be fully automated, and that is precisely why we often see them in close quarters with people. Moreover, cobots in manufacturing are assigned tasks that fall into the Four D's: dangerous, dull, dirty, or dear (Marr 2017).

Automation industry experts all seem to agree that the future of collaborative robotics is looking better every day. The market is expected to grow fast over the next decade, with future growth beyond that. Robot manufacturers are quickly adapting to customer demands, and engineers are solving some of their most complex problems with automation.

Capable of collaborative work with humans, cobots improve performance through safer and more efficient processes in the supply chain and manufacturing industry. Compared to industrial robots placed in cages to work independently, cobots have features, such as sensors and vision technology, that enable them to halt operation if immediate danger to humans is detected.

Machine-enhanced manual tasks are increasingly seen in modern factories with collaborative robots or cobots. Cobots can work alongside humans to deliver more excellent performance than by either working independently. Manufacturers can better decide which tasks to automate and which to perform manually, making production lines more efficient. Cobots could also reshape the whole concept of the factory floor. If robots can work safely with humans, they can be utilized away from the controlled, safety-fenced world of the shop floor. In the food and beverage industry, cobots with dexterous limbs, intelligent sensors, and advanced safety parameters could dramatically speed the process of getting goods from factory to shelf. Faster picking, placing, sorting, and loading are just the start.

Collaborative robots, better known as cobots, have seen severe uptake across the manufacturing industry due to COVID-19. Despite industry challenges caused by depleting workforces and supply issues, investment has been made in cobots due to their cost efficiency, ease of use, and collaborative convenience. Cobots work alongside humans to facilitate productivity, reducing the need for unskilled labor and providing upskilling opportunities for workforces still required to manage warehouse automation and cobot integration.

These benefits have made cobots noticeably popular with small and medium-sized enterprises (SMEs), as they require a gradual introduction of automation while retaining flexibility and manageable costs. In order to work at maximum efficiency, engineers must program a cobot to learn its boundaries, maximize functional space, and understand restricted modes. With an ongoing industrial transformation journey underway, the use of cobots alongside an upskilled, people-led workforce can be expected. The result looks to be a vast increase in efficiency and output while retaining key employees through incentivizing progression opportunities.

A study by the World Economic Forum in association with Advanced Robotics for Manufacturing found that collaborative robots can cut nearly two-thirds of the cycle time required to pack boxes onto pallets. Because cobots are designed to work without any breaks, they reduce the idle time between cycles. The International Society of Automation reports that cobots can save production costs by reducing 75% of manual labor. Traditional robots increase the installation costs for manufacturers as they need to set up additional safety measures around the deployment area. Cobots do not incur such extra expenses as they can be set up close to humans.

One of the less-known benefits of collaborative robots is their nearly maintenance-free operation and long-term reliability. Not only are they easy to maintain, but they are also easy to program, and this is even the case in the most demanding environments that would put human workers to the test.

Many small and medium-sized businesses (SMEs) believe automation is out of their reach as they cannot afford giant, complex robots that also do not fit within their limited floor space. Cobots are changing the game for these organizations, providing a more cost-effective and straightforward way to introduce automation into their workflows.

With remote network capabilities powered by high signals and bandwidth alongside low latency, a cobot with 6G systems can significantly add invaluable service to any manufacturing industry. The possibilities in the future are endless, especially with newer developments and growing optimization opportunities in Industry 5.0 for AI and collaborative work.

To effectively scale cobot deployment, the market must have use cases of successful business models. From hardware design to sensors, actuators, vision systems, data processing, artificial intelligence, and more, cobot technology requires the development of user safety and predictable results.

Cobots are quickly becoming a significant part of automation in manufacturing as the technology is proving reliable, which means that more and more companies are ready to use them. Cobots will grow more sophisticated and versatile in the upcoming years. As AI improves, so will cobots' fulfillment of precision and cognitive tasks. Moreover, the connectivity of cobots with IIoT, meaning the connection to other machines, devices, network databases, and so on, will allow them to improve multiple factory workflows and even provide valuable data analytics, including predictive analytics and suggestions on improving processes. Connecting cobots via the Industrial Internet of Things has connected cobots to surrounding machinery. By extension, adding connectivity adds machine visibility, data analytics, and improved predictive maintenance. This connection has led to increased precision, flexibility, and efficiency.

Denmark-based Universal Robots reports that cobots are at the heart of Industry 5.0. Cobots democratize robotic capabilities, serving as a personal tool that any workforce member can leverage to apply creative skills and generate more value. Cobots can be used as a plug-and-play solution across various manufacturing and industrial operations such as automotive production, food processing, chemical plants, medical devices, and kits. Since they collaborate well with humans in a safe environment, cobots will augment intelligent decision-making, drive high-quality products to the market, enable mass customization and personalization, optimize manufacturing costs, generate new job roles (e.g., Chief Robotics Officer), and boost virtual education to make the most of collaborative robotics.

Generative AI

Generative AI (GenAI) is the domain of AI that can create new content from pre-existing sets of information, supported by large-scale neural networks ('deep neuronal networks'), i.e., with millions of artificial neurons. The process of calibrating parameters between pairs of neurons is called neural network learning.

Large Language Models (LLM) are GenAI models for natural language processing. The well-known ChatGPT from OpenAI is based on the GPT architecture, which stands for "Generative Pre-trained Transformer".

ChatGPT has two features that are entirely different from previous information retrieval systems. The first feature is a significant advantage: the system can maintain a dialogue based on user sentences (prompts) with an awareness of context (context-aware), i.e., it considers previous sentences in the conversation. The second feature of the system is a disadvantage: on particular subjects, the system generates sentences that do not correspond to reality. In these cases, we say the system 'hallucinates'.

ChatGPT and similar models based on GPT-3 and GPT-4 are designed to account for the context of the conversation, unlike older Question Answering systems that responded to the posed question. Formulating the 'prompt' is critical to obtain desired answers on the user's side. The use of imperative sentences like "rewrite the sentence", "summarize the sentence", or "expand the sentence" makes the conversation more effective.

Generative AI in Industry 5.0 offers unparalleled personalized production and rapid prototyping opportunities. Augmenting human creativity fosters a harmonious collaboration between man and machine. Moreover, its potential in optimizing supply chains and enhancing consumer interactions places it at the forefront of the next industrial evolution. This synergy promises both efficiency and a reinvigorated emphasis on the human touch.

3.3. Industry 5.0, Iteration, and Visualization

Personal Fabrication

Personal fabrication is an emerging field in Human-Computer Interaction that allows individuals to fabricate products in their homes using digital data and 3D printers. This technology uses these printers and other digital fabrication tools to create three-dimensional solid objects from digital files in just a few hours. Both additive and subtractive processes are used, allowing the production of precise parts from polymers and metals. Despite the potential for non-technical users to create custom objects, most currently rely on downloading existing models (Camburn, Wood 2018).

The democratization of personal fabrication has the potential to significantly disrupt industries such as product design, interior design, carpentry, and engineering. It empowers individuals, allowing them to use computing for physical matter. This democratization of design can be compared to the impact of the Internet on consumers and producers, presenting new opportunities just as the Internet did.

The development of user-friendly 3D design software, like TinkerCAD and SketchUp, has made 3D design accessible to a broader audience, even in educational contexts. The rise of virtual design tools and platforms, such as online video games and Lego's online factory, have further allowed users to develop design skills, share ideas, collaborate, and customize and hack products.

Physical computing, which includes embedding interactivity or intelligence into everyday objects, is another aspect of this maker trend. Microcontrollers like Arduino have facilitated a greater understanding of electronics among people. Biomimicry, the practice of reverse-engineering natural materials and processes, is also a valuable source of inspiration for designers and scientists in various fields.

Personal fabrication also has the potential to impact the profession of industrial design significantly. For example, the Portuguese brand Feltrando exemplifies how personal fabrication can combine local cultural heritage, sustainability, craftsmanship, and industrial textile residue (felt) to create high-quality textile products. As demands for faster, more powerful, functional, minor, cheaper, and user-friendly products increase, the challenges faced require unconventional problem-solving techniques and exploration of new technologies, such as personal fabrication.

FabLab

FabLabs, or digital fabrication laboratories, were first created by Professor Neil Gershenfeld at MIT in 2001 to provide an accessible and affordable maker space. These labs offer a place for creativity, learning, and innovation, allowing anyone to use advanced technology, such as 3D printers and laser cutting, to make almost anything. FabLabs are small-scale workshops with computer-controlled tools aiming to disrupt the traditional manufacturing industry, akin to how microcomputers changed computing (Savastano et al. 2017).

They serve multiple purposes, including the rapid prototyping of industrial products and the low-cost manufacturing of open-source designs. Their versatility is seen in their ability to empower individuals to create bespoke devices adapted to specific needs. These labs are critical to Do-It-Yourself (DIY) communities, cultures, and projects. They provide a platform for transforming ideas into physical prototypes.

Despite not yet competing with mass production for widely distributed products, FabLabs has shown potential in empowering individuals to create intelligent, personalized devices, which is not practical or economical with mass production. They provide physical spaces offering access to digital and industrial-grade fabrication, electronics tools, open-source software, and programs supporting prototyping.

The projected growth of FabLabs is termed "Lass' Law," which predicts that these labs will double roughly every 18 months due to the increasing affordability and digitization of machinery. This proliferation could lead to custom fabrication becoming ubiquitous in the future.

The estimated cost of creating a mobile FabLab is $298,000, including the design, procurement, construction, preparation, and training phases. This process can take 6-18 months, depending on the project and the network's capacity.

There are examples of successful FabLab implementation worldwide. They have been used as place-based policy tools in Europe to respond to regional and local challenges. Shepherds in Norway used their FabLab to create a system for tracking sheep using mobile phones. At the same time, people in Ghana made an innovative truck refrigeration system powered by the vehicle's exhaust gases.

FabLabs can contribute to urban regeneration and local transformation, exemplified by the Municipality of Lisbon's decision to open a Fab Lab in a central location, re-purposing a new food market into a fully-equipped digital maker space.

These labs also strengthen the culture of innovation and creativity. FabLab EDP in Portugal aims to interact with society by promoting entrepreneurship and active citizenship and providing resources for generating and implementing ideas.

In conclusion, FabLabs presents a significant innovation in manufacturing and prototyping, offering unprecedented opportunities for creativity, entrepreneurship, and problem-solving at the local level. Their anticipated growth may lead to a significant shift in how we perceive and engage in manufacturing processes.

4. CONCLUSION

This work details a set of technologies for Industry 4.0 and Industry 5.0, structured in a framework with three steps: (i) data and connectivity, (ii) intelligent systems, and (iii) interaction and visualization. The role of Artificial Intelligence (AI) in Industry 4.0 and 5.0 needs to include input (the data and the connections) and output (the visualization and the interactions).

From our point of view, Industry 4.0's topics, like digital twins, additive manufacturing, and augmented reality, where holography plays a pivotal role, will continue to evolve.

Regarding Industry 5.0, the spirit of AI will be combined with topics such as Data Quality, 6G networks, Cobots, Personal Fabrication, and FabLabs. Data Quality enables us to improve fairness and reduce bias in machine learning models. 6G networks play a significant role, projecting to offer even faster speeds, lower latency, and higher capacity than its predecessor. Cobots offer the advantages of automation, promoting harmonious human-robot collaboration in the workplace. Personal fabrication is the next step in additive manufacturing, using both additive and subtractive processes and allowing the production of precise parts from polymers and metals. Fabric labs, or fabrication laboratories, are critical to Do-It-Yourself communities by extending the open-source concept from software to hardware design.

The philosophy behind Industry 5.0 is not just about manufacturing for profit. It promotes principles of sustainability and human-centricity. As shown in Figure 1, the role of AI is the hinge between the pairs (data and connectivity) and (interaction and visualization).

REFERENCES

Bornet, P., Barkin, I., & Wirtz, J. (2020). *Intelligent Automation: Learn how to harness Artificial Intelligence to boost business and make our world more human*. Kindle Store.

Camburn, B., & Wood, K. (2018). Principles of maker and DIY fabrication: Enabling design prototypes at low cost. *Design Studies, 58*, 63-88. doi:10.1016/j.destud.2018.04.002

Carmigniani, J., & Furht, B. (2011). Augmented Reality: An Overview. In B. Furht (Ed.), *Handbook of Augmented Reality*. Springer. doi:10.1007/978-1-4614-0064-6_1

Davenport, T. H. (2014). *Big Data at Work: Dispelling the Myths, Uncovering the Opportunities*. Harvard Business Review Press. doi:10.15358/9783800648153

Deguchi, A. (2020). What Is Society 5.0? In *Society 5.0*. Springer. doi:10.1007/978-981-15-2989-4_1

Ehrlinger, L., & Woss, W. (2022). A Survey of Data Quality Measurement and Monitoring Tools. *Frontiers in Big Data, 5*. https://www.frontiersin.org/articles/10.3389/fdata.2022.850611

Grieves, M., & Vickers, J. (2017). Digital Twin: Mitigating Unpredictable, Undesirable Emergent Behavior in Complex Systems. In Transdisciplinary Perspectives on Complex Systems. Springer International. DOI doi:10.1007/978-3-319-38756-7_4

Jagusch, K., Sender, J., Jericho, D., & Flügge, W. (2021). Digital thread in shipbuilding as a prerequisite for the digital twin. *Procedia CIRP*, *104*, 318–323. doi:10.1016/j.procir.2021.11.054

Jeyakumar, K. (2022). A Comprehensive Survey on Robotics and Automation in Various Industries. *Proceedings of the Seventh International Conference on Communication and Electronics Systems (ICCES 2022)*.

Kissinger, H. A., Schmidt, E., & Huttenlocher, D. (2022). *The Age of AI: And Our Human Future*. Back Bay Books.

Langmann, C., & Turi, D. (2023). *Robotic Process Automation (RPA) - Digitization and Automation of Processes: Prerequisites, functionality and implementation using accounting as an example*. Springer Gabler.

Lefranc, G., López, I., Osorio-Comparán, R., & Peña, M. (2022). Cobots in automation and at home. *2022 IEEE International Conference on Automation/XXV Congress of the Chilean Association of Automatic Control (ICA-ACCA),* 1-6. doi: 10.1109/ICA-ACCA56767.2022.10006164

Maddikunta, P. K. R., Pham, Q.-V., Prabadevi, B., Deepa, N., Dev, K., Gadekallu, T. R., Ruby, R., & Liyanage, M. (2022). Industry 5.0: A survey on enabling technologies and potential applications. *Journal of Industrial Information Integration, 26*. doi:10.1016/j.jii.2021.100257

Marr, B. (2017). The 4 Ds of robotization: dull, dirty, dangerous and dear. *Enterprise Tech, Forbes Innovation*. https://www.forbes.com/sites/bernardmarr/2017/10/16/the-4-ds-of-robotization-dull-dirty-dangerous-and-dear/

Moore, G. E. (1965). *Cramming more components onto integrated circuits*. Electronics Magazine.

Savastano, M., Bellini, F., D'Ascenzo, F., & Scornavacca, E. (2017). FabLabs as Platforms for Digital Fabrication Services: A Literature Analysis. In S. Za, M. Drăgoicea, & M. Cavallari (Eds.), *Exploring Services Science, IESS 2017, Lecture Notes in Business Information Processing* (Vol. 279). Springer. doi:10.1007/978-3-319-56925-3_3

Shahrubudin, N., Lee, T. C., & Ramlan, R. (2019). An Overview on 3D Printing Technology: Technological, Materials, and Applications. *Procedia Manufacturing*, *35*, 1286–1296. doi:10.1016/j.promfg.2019.06.089

Sofi, A., Regita, J. J., Rane, B., & Lau, H. H. (2022). Structural health monitoring using wireless smart sensor network – An overview. *Mechanical Systems and Signal Processing, 163*. doi:10.1016/j.ymssp.2021.108113

Stephanopoulos, G., & Han, C. (1996). Intelligent systems in process engineering: A review. *Computers & Chemical Engineering, 20*(6-7), 743-791. doi:10.1016/0098-1354(95)00194-8

Tao, F., Zhang, M., & Nee, A. Y. C. (2019). *Digital Twin Driven Smart Manufacturing*. Elsevier Inc. doi:10.1016/C2018-0-02206-9

Tomkos, I., Klonidis, D., Pikasis, E., & Theodoridis, S. (2020). Toward the 6G Network Era: Opportunities and Challenges. *IT Professional, 22*(1), 34–38. doi:10.1109/MITP.2019.2963491

Compilation of References

Abadie, A., Diamond, A., & Hainmueller, J. (2010). Synthetic control methods for Comparative case studies: Estimating the effect of California's tobacco control program. *Journal of the American Statistical Association*, *105*(490), 493–505. doi:10.1198/jasa.2009.ap08746

Abadie, A., & Gardeazabal, J. (2003). The economic costs of conflict: A case study of the Basque Country. *The American Economic Review*, *93*(1), 113–132. doi:10.1257/000282803321455188

Abbas, S. M., Liu, Z., & Khushnood, M. (2022). When Human Meets Technology: Unlocking Hybrid Intelligence Role in Breakthrough Innovation Engagement via Self-Extension and Social Intelligence. *Journal of Computer Information Systems*, 1–18.

Abiodun, O. I., Jantan, A., Omolara, A. E., Dada, K. V., Mohamed, N. A., & Arshad, H. (2018). State-of-the-art in artificial neural network applications: A survey. *Heliyon*, *4*(11), e00938. doi:10.1016/j.heliyon.2018.e00938 PMID:30519653

Acciarini, C., Cappa, F., Boccardelli, P., & Oriani, R. (2023). How can organisations leverage big data to innovate their business models? A systematic literature review. *Technovation*, *123*, 102713. doi:10.1016/j.technovation.2023.102713

Acharki, N., Lugo, R., Bertoncello, A., & Garnier, J. (2023). Comparison of metalearners for estimating multi-valued treatment heterogeneous effects. *ICML2023- Fortieth International Conference on Machine Learning*. Doi:10.48550/arXiv.2205.14714

Acikgoz, F., Perez-Vega, R., Okumus, F., & Stylos, N. (2023). Consumer engagement with AI-powered voice assistants: A behavioral reasoning perspective. *Psychology and Marketing*, *40*(11), 2226–2243. Advance online publication. doi:10.1002/mar.21873

Adams, R., & Loideáin, N. N. (2019). Addressing indirect discrimination and gender stereotypes in AI virtual personal assistants: The role of international human rights law. *Cambridge International Law Journal*, *8*(2), 241–257. doi:10.4337/cilj.2019.02.04

Adomavicius, G., & Tuzhilin, A. (2005). Personalization technologies: A process-oriented perspective. *Communications of the ACM*, *48*(10), 83–90. doi:10.1145/1089107.1089109

Adya, M., Armstrong, J. S., Collopy, F., & Kennedy, M. (2000). An application of rule-based forecasting to a situation lacking domain knowledge. *International Journal of Forecasting*, *16*(4), 477–484. doi:10.1016/S0169-2070(00)00074-1

Aeberhard, S., & Forina, M. (1991). *Wine*. UCI Machine Learning Repository.

Aggarwal, C. C. (2016). *Recommender systems*. Springer. doi:10.1007/978-3-319-29659-3

Aguirre, E., Mahr, D., Grewal, D., De Ruyter, K., & Wetzels, M. (2015). Unraveling the personalization paradox: The effect of information collection and trust-building strategies on online advertisement effectiveness. *Journal of Retailing*, *91*(1), 34–49. doi:10.1016/j.jretai.2014.09.005

Airports Council International. (2020). *Annual World Airport Traffic Reports*. Airports Council International.

Aizenberg, E., & van den Hoven, J. (2020). Designing for human rights in AI. *Big Data & Society*, *7*(2). doi:10.1177/2053951720949566

Akarslan, F. (2012). *Photovoltaic systems and applications*. Modeling and Optimization of Renewable Energy Systems., doi:10.5772/39244

Alam, I. (2002). An exploratory investigation of user involvement in new service development. *Journal of the Academy of Marketing Science*, *30*(3), 250–261. doi:10.1177/0092070302303006

Albrecht, J. P. (2016). How the GDPR will change the world. *Eur. Data Prot. L. Rev.*, *2*(3), 287–289. doi:10.21552/EDPL/2016/3/4

Ali, F., Dogan, S., Chen, X., Cobanoglu, C., & Limayem, M. (2023). Friend or a foe: Understanding generation Z Employees' intentions to work with service robots in the hotel industry. *International Journal of Human-Computer Interaction*, *39*(1), 111–122. doi:10.1080/10447318.2022.2041880

Aljanabi, M., & Chat, G. P. T. (2023). ChatGPT: Future directions and open possibilities. *Mesopotamian Journal of Cyber Security*, *2023*, 16–17. doi:10.58496/MJCS/2023/003

Almomani, A., Saavedra, P., Barreiro, P., Durán, R., Crujeiras, R., Loureiro, M., & Sánchez, E. (2022). Application of choice models in Tourism Recommender Systems. *Expert Systems: International Journal of Knowledge Engineering and Neural Networks*, *40*(3), e13177. doi:10.1111/exsy.13177

AlShabi, M., & El Haj Assad, M. (2021). Artificial intelligence applications in renewable energy systems. *Design and Performance Optimization of Renewable Energy Systems*, 251–295. doi:10.1016/b978-0-12-821602-6.00018-3

Alves, M. (2022). Causal inference for the brave and true. *Matheus Facture*. https://matheusfacure.github.io/python-causality-handbook/landing-page.html

Alzoubi, H. M., & Aziz, R. (2021). Does emotional intelligence contribute to quality of strategic decisions? The mediating role of open innovation. *Journal of Open Innovation*, *7*(2), 130. doi:10.3390/joitmc7020130

Anagnostopoulos, I., Zeadally, S., & Exposito, E. (2016). Handling big data: Research challenges and future directions. *The Journal of Supercomputing*, *72*(4), 1494–1516. doi:10.100711227-016-1677-z

Andriella, A., Huertas-Garcia, R., Forgas-Coll, S., Torras, C., & Alenyà, G. (2022). "I know how you feel" The importance of interaction style on users' acceptance in an entertainment scenario. *Interaction Studies: Social Behaviour and Communication in Biological and Artificial Systems*, *23*(1), 21–57. doi:10.1075/is.21019.and

Angrist, J. D., & Pischke, J.-S. (2015). *Mastering Metrics: The Path from Cause to Effect*. Princeton University Press.

Antaki, F., Touma, S., Milad, D., El-Khoury, J., & Duval, R. (2023). Evaluating the performance of chatgpt in ophthalmology: An analysis of its successes and shortcomings. *Ophthalmology Science, 100324*.

Aragon, C., Guha, S., Kogan, M., Muller, M., & Neff, G. (2022). *Human-centered data science: An introduction*. MIT Press.

Arici, H. E., Cakmakoglu Arıcı, N., & Altinay, L. (2023). The use of big data analytics to discover customers' perceptions of and satisfaction with green hotel service quality. *Current Issues in Tourism*, *26*(2), 270–288. doi:10.1080/13683500.2022.2029832

Arranz, L. R. (2022). Los maratones en España pierden más de 13.000 corredores en 2022. *Novatos del running*. https://www.novatosdelrunning.es/los- maratones-en-espana-pierden-mas-de-13-000-corredores-en-2022/

Arrieta, A. B., Díaz-Rodríguez, N., Del Ser, J., Bennetot, A., Tabik, S., Barbado, A., ... Herrera, F. (2020). Explainable Artificial Intelligence (XAI): Concepts, taxonomies, opportunities and challenges toward responsible AI. *Information Fusion, 58*, 82–115. doi:10.1016/j.inffus.2019.12.012

Aryanto, K., Oudkerk, M., & van Ooijen, P. (2015). Free DICOM de-identification tools in clinical research: Functioning and safety of patient privacy. *European Radiology, 25*(12), 3685–3695. doi:10.100700330-015-3794-0 PMID:26037716

Asghar, M. Z., Subhan, F., Ahmad, H., Khan, W. Z., Hakak, S., Gadekallu, T. R., & Alazab, M. (2020). senti-esystem: A sentiment-basedesystem-using hybridised fuzzy and deep neural network for measuring customer satisfaction. *Software, Practice & Experience, 51*(3), 571–594. doi:10.1002pe.2853

Ashby, W. R. (1957). *An introduction to cybernetics*. Chapman & Hall Ltd.

Ashfaq, M., Yun, J., Yu, S., & Loureiro, S. (2020). I, Chatbot: Modeling the determinants of users' satisfaction and continuance intention of AI-powered service agents. *Telematics and Informatics, 54*, 101473. doi:10.1016/j.tele.2020.101473

Ashmore, R. D., & Jussim, L. (1997). *Self and Identity: Fundamental Issues*. Oxford University Press.

Atasoy, T., Akinc, H. E., & Ercin, O. (2015). An analysis on smart grid applications and grid integration of renewable energy systems in smart cities. In *International Conference on Renewable Energy Research and Applications (ICRERA)*. IEEE. 10.1109/ICRERA.2015.7418473

Athey, S., & Imbens, G. (2016). Recursive partitioning for heterogeneous causal effects. *Proceedings of the National Academy of Sciences of the United States of America, 113*(27), 7353–7360. doi:10.1073/pnas.1510489113 PMID:27382149

Ausin-Azofra, J. M., Bigne, E., Ruiz, C., Marin-Morales, J., Guixeres, J., & Alcañiz, M. (2021). Do you see what I see? Effectiveness of 360-Degree vs. 2D video ads using a neuroscience approach. *Frontiers in Psychology, 12*, 612717. doi:10.3389/fpsyg.2021.612717 PMID:33679528

Ayoko, O. B. (2021). Digital Transformation, Robotics, Artificial Intelligence, and Innovation. *Journal of Management & Organization, 27*(5), 831–835. doi:10.1017/jmo.2021.64

Azencott, C. A. (2018). Machine learning and genomics: Precision medicine versus patient privacy. *Philosophical Transactions - Royal Society. Mathematical, Physical, and Engineering Sciences, 376*(2128), 20170350. doi:10.1098/rsta.2017.0350 PMID:30082298

Babin, B. J., Darden, W. R., & Griffen, M. (1994). Work and/or fun: Measuring hedonic and utilitarian shopping value. *The Journal of Consumer Research, 20*(4), 644–656. doi:10.1086/209376

Bação, F., Lobo, V., & Painho, M. (2005). *Self-organizing maps as substitutes for k-means clustering. Computational Science—ICCS 2005*. Springer.

Badenes, A., Ruiz, C. & Bigné, E. (2019). Engaging customers through user-and company-generated content on CSR. *Spanish Journal of marketing-ESIC, 23*(3), 379-372.

Bak-Coleman, J. B., Alfano, M., Barfuss, W., Bergstrom, C. T., Centeno, M. A., Couzin, I. D., Donges, J. F., Galesic, M., Gersick, A. S., Jacquet, J., Kao, A. B., Moran, R. E., Romanczuk, P., Rubenstein, D. I., Tombak, K. J., Van Bavel, J. J., & Weber, E. U. (2021). Stewardship of global collective behavior. *Proceedings of the National Academy of Sciences of the United States of America, 118*(27), e2025764118. doi:10.1073/pnas.2025764118 PMID:34155097

Bakpayev, M., Baek, T. H., van Esch, P., & Yoon, S. (2022). Programmatic creative: AI can think but it cannot feel. *Australasian Marketing Journal, 30*(1), 90–95. doi:10.1016/j.ausmj.2020.04.002

Balajee, N. (2020). *What is Artificial Intelligence?* https://nanduribalajee.medium.com/what-is-artificial-intelligenc-ec68579db123

Balasubramaniam, N., Kauppinen, M., Kujala, S., & Hiekkanen, K. (2020). *Ethical Guidelines for Solving Ethical Issues and Developing AI Systems* (1st ed.). Product-Focused Software Process Improvement. doi:10.1007/978-3-030-64148-1_21

Balducci, B., & Marinova, D. (2018). Unstructured data in marketing. *Journal of the Academy of Marketing Science*, *46*(4), 557–590. doi:10.100711747-018-0581-x

Balkin, J. (2017). 2016 Sidley Austin Distinguished Lecture on Big Data Law and Policy: The Three Laws of Robotics in the Age of Big Data. *Ohio State Law Journal*, *78*, 1217.

Bao, T., & Chang, T. (2016). The product and timing effects of eWOM in viral marketing. *International Journal of Business*, *21*(2), 99–111.

Barile, S., Bassano, C., Piciocchi, P., Saviano, M., & Spohrer, J. C. (2021). Empowering value co-creation in the digital age. *Journal of Business and Industrial Marketing*. Advance online publication. doi:10.1108/JBIM-12-2019-0553

Barile, S., & Polese, F. (2010). Smart service systems and viable service systems: Applying systems theory to service science. *Service Science*, *2*(1/2), 21–40. doi:10.1287erv.2.1_2.21

Baron, R. M., & Kenny, D. A. (1986). The moderator–mediator variable distinction in social psychological research: Conceptual, strategic, and statistical considerations. *Journal of Personality and Social Psychology*, *51*(6), 1173–1182. doi:10.1037/0022-3514.51.6.1173 PMID:3806354

Batat, W. (2022). What does phygital really mean? A conceptual introduction to the phygital customer experience (PH-CX) framework. *Journal of Strategic Marketing*, 1–24. Advance online publication. doi:10.1080/0965254X.2022.2059775

Batat, W., & Hammedi, W. (2023). The extended reality technology (ERT) framework for designing customer and service experiences in phygital settings: A service research agenda. *Journal of Service Management*, *34*(1), 10–33. doi:10.1108/JOSM-08-2022-0289

Bates, M. E. (2016). *Bringing Insight to Data: Info Pros' Role in Text and Data Mining*. Springer Nature. https://www.springernature.com/gp/librarians/landing/textanddata

Bates, D. (2005). Crisis between the Wars: Derrida and the Origins of Undecidability. *Representations (Berkeley, Calif.)*, *90*(1), 1–27. doi:10.1525/rep.2005.90.1.1

Baumeister, R. F. (2005). *The Cultural Animal: Human Nature, Meaning, and Social Life*. Oxford University Press. doi:10.1093/acprof:oso/9780195167030.001.0001

Baum, W. M. (2017). *Understanding behaviorism: Behavior, culture, and evolution*. John Wiley & Sons. doi:10.1002/9781119143673

Beam, M. A., Hutchens, M. J., & Hmielowski, J. D. (2018). Facebook news and (de) polarization: Reinforcing spirals in the 2016 US election. *Information Communication and Society*, *21*(7), 940–958. doi:10.1080/1369118X.2018.1444783

Beaulieu-Jones, B. K., Yuan, W., Brat, G. A., Beam, A. L., Weber, G., Ruffin, M., & Kohane, I. S. (2021). Machine learning for patient risk stratification: Standing on, or looking over, the shoulders of clinicians? *NPJ Digital Medicine*, *4*(1), 62. doi:10.103841746-021-00426-3 PMID:33785839

Beerbaum, D. (2023). Generative Artificial Intelligence (GAI) Software – Assessment on Biased Behavior. SSRN *Electronic Journal*. doi:10.2139/ssrn.4386395

Belhadi, A., Kamble, S., Benkhati, I., Gupta, S., & Mangla, S. K. (2023). Does strategic management of digital technologies influence electronic word-of-mouth (ewom) and customer loyalty? empirical insights from B2B platform economy. *Journal of Business Research*, *156*, 113548. doi:10.1016/j.jbusres.2022.113548

Belle, V., & Papantonis, I. (2020). *Principles and practice of explainable machine learning*. arXiv:2009.11698.

Bellucci, E., Venkatraman, S., & Stranieri, A. (2020). Online dispute resolution in mediating EHR disputes: A case study on the impact of emotional intelligence. *Behaviour & Information Technology, 39*(10), 1124–1139. doi:10.1080/0144929X.2019.1645209

Ben Mimoun, M. S., Poncin, I., & Garnier, M. (2017). Animated conversational agents and e-consumer productivity: The roles of agents and individual characteristics. *Information & Management, 54*(5), 545–559. doi:10.1016/j.im.2016.11.008

Benford, S., Ramchurn, R., Marshall, J., Wilson, M. L., Pike, M., Martindale, S., Hazzard, A., Greenhalgh, C., Kallionpää, M., Tennent, P., & Walker, B. (2021). Contesting control: Journeys through surrender, self-awareness and looseness of control in embodied interaction. *Human-Computer Interaction, 36*(5-6), 361–389. doi:10.1080/07370024.2020.1754214

Bengoa, A., Maseda, A., Iturralde, T., & Aparicio, G. (2021). A bibliometric review of the technology transfer literature. *The Journal of Technology Transfer, 46*(5), 1514–1550. doi:10.100710961-019-09774-5

Benn, S. I. (1975). Freedom, Autonomy and the Concept of a Person. *Proceedings of the Aristotelian Society, 76*(1), 109–130. doi:10.1093/aristotelian/76.1.109

Ben-Shahar, O., & Porat, A. (2021). *Personalized law: Different rules for different people*. Oxford University Press. doi:10.1093/oso/9780197522813.001.0001

Berdichevsky, D., & Neuenschwander, E. (1999). Toward an ethics of persuasive technology. *Communications of the ACM, 42*(5), 51–58. doi:10.1145/301353.301410

Berkovsky, S., Freyne, J., & Oinas-Kukkonen, H. (2012). Influencing individually: Fusing personalization and persuasion. [TiiS]. *ACM Transactions on Interactive Intelligent Systems*, 2(2), 1–8. doi:10.1145/2209310.2209312

Berridge, K. C. (2009). Wanting and liking: Observations from the neuroscience and psychology laboratory. *Inquiry, 52*(4), 378–398. doi:10.1080/00201740903087359 PMID:20161627

Beshears, J., Choi, J. J., Laibson, D., & Madrian, B. C. (2008). How are preferences revealed? *Journal of Public Economics, 92*(8–9), 1787–1794. doi:10.1016/j.jpubeco.2008.04.010 PMID:24761048

Bessi, A., & Ferrara, E. (2016). Social bots distort the 2016 US Presidential election online discussion. *First Monday, 21*, 11–17.

Bettman, J. R. (1979). *An Information Processing Theory of Consumer Choice*. Addison-Wesley.

Bialkova, S. (2021). *Would you talk to me? The role of chatbots in marketing*. ICORIA2021, Bordeaux, France.

Bialkova, S. (2022). *Interacting with chatbot: How to enhance functionality and enjoyment?* AEMARK2022, Valencia, Spain.

Bialkova, S. (2023a). I want to talk to you: Chatbot marketing integration. *Advances in Advertising Research, XII*, 23–36. doi:10.1007/978-3-658-40429-1_2

Bialkova, S. (2023b). How to Optimise Interaction with Chatbots? Key Parameters Emerging from Actual Application. *International Journal of Human-Computer Interaction*, 1–10. Advance online publication. doi:10.1080/10447318.2023.2219963

Bialkova, S., & te Paske, S. (2021). Campaign participation, spreading e-WOM, purchase: How to optimise CSR effectiveness via Social media? *European Journal of Management and Business Economics, 30*(1), 108–126. doi:10.1108/EJMBE-08-2020-0244

Biasiucci, A., Franceschiello, B., & Murray, M. M. (2019, February 4). Electroencephalography. *Current Biology*, *29*(3), R80–R85. doi:10.1016/j.cub.2018.11.052 PMID:30721678

Bigne, E. (2020). Teaching in business: A customized process driven by technological innovations. *Journal of Management and Business Education*, *3*(1), 4–15. doi:10.35564/jmbe.2020.0002

Bigne, E., Fuentes-Medina, M. L., & Morini-Marrero, S. (2020). Memorable tourist experiences versus ordinary tourist experiences analyzed through user-generated content. *Journal of Hospitality and Tourism Management*, *45*, 309–318. doi:10.1016/j.jhtm.2020.08.019

Bigne, E., & Maturana, P. (2023). Does virtual reality trigger visits and booking holiday travel packages? *Cornell Hospitality Quarterly*, *64*(2), 226–245. doi:10.1177/19389655221102386

Bigne, E., Nicolau, J. L., & William, E. (2021). Advance booking across channels: The effects on dynamic pricing. *Tourism Management*, *86*, 104341. doi:10.1016/j.tourman.2021.104341

Bigne, E., Oltra, E., & Andreu, L. (2019). Harnessing stakeholder input on Twitter: A case study of short breaks in Spanish tourist cities. *Tourism Management*, *71*(April), 490–505. doi:10.1016/j.tourman.2018.10.013

Bigne, E., Ruiz, C., & Badenes-Rocha, A. (2023). The influence of negative emotions on brand trust and intention to share cause-related posts: A neuroscientific study. *Journal of Business Research*, *157*(March), 113628. doi:10.1016/j.jbusres.2022.113628

Bigne, E., Ruiz, C., Cuenca, A., Perez-Cabañero, C., & Garcia, A. (2021). What drives the helpfulness of online reviews? A deep learning study of sentiment analysis, pictorial content and reviewer expertise for mature destinations. *Journal of Destination Marketing & Management*, *20*(June), 100570. doi:10.1016/j.jdmm.2021.100570

Bigne, E., Ruiz, C., Perez-Cabañero, C., & Cuenca, A. (2023). Are customer star ratings and sentiments aligned? A deep learning study of the customer service experience in tourism destinations. *Service Business*, *17*(1), 281–314. doi:10.100711628-023-00524-0

Bigne, E., William, E., & Soria-Olivas, E. (2020). Similarity and consistency in hotel online ratings across platforms. *Journal of Travel Research*, *59*(4), 742–758. doi:10.1177/0047287519859705

Bingley, W. J., Curtis, C., Lockey, S., Bialkowski, A., Gillespie, N., Haslam, S. A., Ko, R. K. L., Steffens, N., Wiles, J., & Worthy, P. (2023). Where is the human in human-centered AI? insights from developer priorities and user experiences. *Computers in Human Behavior*, *141*, 107617. doi:10.1016/j.chb.2022.107617

Bird, S., Barocas, S., Crawford, K., Diaz, F., & Wallach, H. (2016). Exploring or Exploiting? Social and Ethical Implications of Autonomous Experimentation in AI. *Workshop on Fairness, Accountability, and Transparency in Machine Learning*. https://ssrn.com/abstract=2846909

Bishop, M. (2002). *Computer Security: Art and Science*. Addison-Wesley.

Biswas, B., Sengupta, P., & Ganguly, B. (2021). Your reviews or mine? Exploring the determinants of "perceived helpfulness" of online reviews: A cross-cultural study. *Electronic Markets*, 1–20. PMID:35602112

Blaabjerg, F., & Ma, K. (2019). Renewable energy systems with wind power. *Power Electronics in Renewable Energy Systems and Smart Grid*, 315–345. doi:10.1002/9781119515661.ch6

Bland, D. J., & Osterwalder, A. (2020). *Testing Business Ideas: A Field Guide for Rapid Experimentation*. John Wiley and Sons.

Blue Brain Project. (n.d.). EPFL. Retrieved November 28, 2022, from https://www.epfl.ch/research/domains/bluebrain/

Blumer, H. (1986). *Symbolic Interactionism: Perspective and Method.* University of California Press.

Blut, M., Wang, C., Wünderlich, N., & Brock, C. (2021). Understanding anthropomorphism in service provision: A meta-analysis of physical robots, chatbots, and other AI. *Journal of the Academy of Marketing Science, 49*(4), 632–658. doi:10.100711747-020-00762-y

Bogart, L. (1957). Opinion research and marketing. *Public Opinion Quarterly, 21*(1), 129–140. doi:10.1086/266692

Bornet, P., Barkin, I., & Wirtz, J. (2020). *Intelligent Automation: Learn how to harness Artificial Intelligence to boost business and make our world more human.* Kindle Store.

Bose B. K. (2020). Artificial intelligence techniques: how can it solve problems in power electronics? An Advancing Frontier," *In IEEE Power Electronics Magazine,* vol. 7, no. 4, pp. 19-27, doi: . doi:10.1109/MPEL.2020.3033607

Bose, B. K. (2019). Artificial intelligence applications in renewable energy systems and smart grid – some novel applications. *Power Electronics in Renewable Energy Systems and Smart Grid,* 625–675. doi:10.1002/9781119515661.ch12

Bose, B. K., & Wang, F. (Fred). (2019). Energy, environment, power electronics, renewable energy systems, and smart grid. *Power Electronics in Renewable Energy Systems and Smart Grid,* 1–83. doi:10.1002/9781119515661.ch1

Bose, B. K. (2017). Artificial intelligence techniques in smart grid and renewable energy systems—Some example applications. *Proceedings of the IEEE, 105*(11), 2262–2273. doi:10.1109/JPROC.2017.2756596

Bose, B. K. (2017). Power electronics, smart grid, and renewable energy systems. *Proceedings of the IEEE, 105*(11), 2011–2018. doi:10.1109/JPROC.2017.2745621

Bose, B. K. (Ed.). (2019). *Power Electronics in Renewable Energy Systems and Smart Grid.*, doi:10.1002/9781119515661

Bostrom, N. (2012). The superintelligent will: Motivation and instrumental rationality in advanced artificial agents. *Minds and Machines, 22*(2), 71–85. doi:10.100711023-012-9281-3

Bostrom, N. (2016). *Superintelligence: paths, dangers, strategies, reprint edition.* OUP Oxford.

Botvinick, M., Wang, J. X., Dabney, W., Miller, K. J., & Kurth-Nelson, Z. (2020). Deep reinforcement learning and its neuroscientific implications. *Neuron, 107*(4), 603–616. doi:10.1016/j.neuron.2020.06.014 PMID:32663439

Boulding, K. E. (1956). General systems theory—The skeleton of science. *Management Science, 2*(3), 197–208. doi:10.1287/mnsc.2.3.197

Boulding, K. E. (1956). *The image: Knowledge in life and society* (Vol. 47). University of Michigan press. doi:10.3998/mpub.6607

Bouneffouf, D., Rish, I., & Aggarwal, C. (2020). Survey on applications of multi-armed and contextual bandits. *2020 IEEE Congress on Evolutionary Computation (CEC),* (pp. 1–8). IEEE. 10.1109/CEC48606.2020.9185782

Bozdag, A. A. (2023). AIsmosis and the pas de deux of human-AI interaction: Exploring the communicative dance between society and artificial intelligence. *Online Journal of Communication and Media Technologies, 13*(4), e202340. Advance online publication. doi:10.30935/ojcmt/13414

Bozkurt, A., Xiao, J., Lambert, S., Pazurek, A., Crompton, H., Koseoglu, S., Farrow, R., Bond, M., Nerantzi, C., Honeychurch, S., Bali, M., Dron, J., Mir, K., Stewart, B., Costello, E., Mason, J., Stracke, C., Romero-Hall, E., Koutropoulos, A., & Toquero, C. (2023). Speculative Futures on ChatGPT and Generative Artificial Intelligence (AI): A Collective Reflection from the Educational Landscape. *Asian Journal of Distance Education, 18*(1), 53. https://eprints.gla.ac.uk/294292/1/294292.pdf

Bradford, A. (2012). The Brussels Effect. *Northwestern University Law Review, 107*. https://ssrn.com/abstract=2770634

Bratman, M. E. (2000). Reflection, planning, and temporally extended agency. *The Philosophical Review*, *109*(1), 35–61. doi:10.1215/00318108-109-1-35

Bretin, R., Khamis, M., & Cross, E. (2023). "Do I Run Away?": Proximity, Stress and Discomfort in Human-Drone Interaction in Real and Virtual Environments. In *IFIP Conference on Human-Computer Interaction* (pp. 525-551). Cham: Springer Nature Switzerland.

Brinkmann, S. (2005). Human kinds and looping effects in psychology: Foucauldian and hermeneutic perspectives. *Theory & Psychology*, *15*(6), 769–791. doi:10.1177/0959354305059332

Broome, J. (1984). Selecting people randomly. *Ethics*, *95*(1), 38–55. doi:10.1086/292596 PMID:11651785

Brown, H., Lee, K., Mireshghallah, F., Shokri, R., & Tramèr, F. (2022*). What Does it Mean for a Language Model to Preserve Privacy?* arXiv. https://arxiv.org/pdf/2202.05520.pdf

Bruner, J. (1991). The narrative construction of reality. *Critical Inquiry*, *18*(1), 1–21. doi:10.1086/448619

Brunton, S. L., Noack, B. R., & Koumoutsakos, P. (2020). Machine Learning for Fluid Mechanics. *Annual Review of Fluid Mechanics*, *52*(1), 477–508. doi:10.1146/annurev-fluid-010719-060214

Brusilovsky, P., & Maybury, M. T. (2002). From adaptive hypermedia to the adaptive web. *Communications of the ACM*, *45*(5), 30–33. doi:10.1145/506218.506239

Bryan, C. J., Tipton, E., & Yeager, D. S. (2021). Behavioural science is unlikely to change the world without a heterogeneity revolution. *Nature Human Behaviour*, *5*(8), 980–989. doi:10.103841562-021-01143-3 PMID:34294901

Brynjolfsson, E., & Mcafee, A. (2014). *The second machine age: Work, progress, and prosperity in a time of brilliant technologies*. W. W Norton & Company.

Brynjolfsson, E., & Mcafee, A. (2017). Artificial intelligence, for real. *Harvard Business Review*, *1*, 1–31.

Buiten, M. C. (2019). Towards intelligent regulation of artificial intelligence. *European Journal of Risk Regulation*, *10*(1), 41–59. doi:10.1017/err.2019.8

Bulchand-Gidumal, J., William Secin, E., O'Connor, P., & Buhalis, D. (2023). Artificial intelligence's impact on hospitality and tourism marketing: Exploring key themes and addressing challenges. *Current Issues in Tourism*, 1–18. doi:10.1080/13683500.2023.2229480

Burr, C., Cristianini, N., & Ladyman, J. (2018). An Analysis of the Interaction Between Intelligent Software Agents and Human Users. *Minds and Machines*, *28*(4), 735–774. doi:10.100711023-018-9479-0 PMID:30930542

Busch, C., & Mak, V. (2021). Putting the Digital Services Act into Context: Bridging the Gap between EU Consumer Law and Platform Regulation. *Available at SSRN 3933675*.

Buzova, D., Sanz-Blas, S., & Cervera-Taulet, A. (2019). Does culture affect sentiments expressed in cruise tours' eWOM? *Service Industries Journal*, *39*(2), 154–173. doi:10.1080/02642069.2018.1476497

Cai, L., Wu, C., Meimandi, K. J., & Gerber, M. S. (2017). Adaptive mobile behavior change intervention using reinforcement learning. *2017 International Conference on Companion Technology (ICCT)*, (pp. 1–2). IEEE. 10.1109/COMPANION.2017.8287078

Camburn, B., & Wood, K. (2018). Principles of maker and DIY fabrication: Enabling design prototypes at low cost. *Design Studies, 58*, 63-88. doi:10.1016/j.destud.2018.04.002

Cao, Y., Zhou, L., Lee, S., Cabello, L., Chen, M., & Hershcovich, D. (n.d.). *Assessing Cross-Cultural Alignment between ChatGPT and Human Societies: An Empirical Study*. arXiv. https://arxiv.org/pdf/2303.17466.pdf

Card, D., & Smith, N. A. (2020). On Consequentialism and Fairness. *Frontiers in Artificial Intelligence*, *3*, 34. doi:10.3389/frai.2020.00034 PMID:33733152

Cardinaux, F., Bhowmik, D., Abhayaratne, C., & Hawley, M. S. (2011). Video based technology for ambient assisted living: A review of the literature. *Journal of Ambient Intelligence and Smart Environments*, *3*(3), 253–269. doi:10.3233/AIS-2011-0110

Carmigniani, J., & Furht, B. (2011). Augmented Reality: An Overview. In B. Furht (Ed.), *Handbook of Augmented Reality*. Springer. doi:10.1007/978-1-4614-0064-6_1

Carrega, M., Santos, H., & Marques, N. (2021). Data Streams for Unsupervised Analysis of Company Data. In F. S. Goreti Marreiros, *Progress in Artificial Intelligence: 20th EPIA Conference on Artificial Intelligence* (pp. 609–621). Springer-Verlag. 10.1007/978-3-030-86230-5_48

Carter, O., Hohwy, J., van Boxtel, J., Lamme, V., Block, N., Koch, C., & Tsuchiya, N. (2018). Conscious machines: Defining questions. *Science*, *359*(6374), 400–400. doi:10.1126cience.aar4163 PMID:29371459

Caudill, E. M., & Murphy, P. E. (2000). Consumer online privacy: Legal and ethical issues. *Journal of Public Policy & Marketing*, *19*(1), 7–19. doi:10.1509/jppm.19.1.7.16951

Cavique, L., Pinheiro, P., & Mendes, A. B. (2023). (accepted)). Data science maturity model: From raw data to Pearl's causality hierarchy. *WorldCIST*, 2023.

Çelik, Ö., & Altunaydın, S. S. (2018). A Research on Machine Learning Methods and Its Applications. *Journal of Educational Technology & Online Learning*, *1*(3), 25–40. doi:10.31681/jetol.457046

Centre for Data Ethics and Innovation (CDEI). (2019). *Interim report: review into bias in algorithmic decision-making*. Centre for Data Ethics and Innovation. https://www.gov.uk/

Chabris, C. F., Laibson, D., Morris, C. L., Schuldt, J. P., & Taubinsky, D. (2008). Individual laboratory-measured discount rates predict field behavior. *Journal of Risk and Uncertainty*, *37*(2-3), 237–269. doi:10.100711166-008-9053-x PMID:19412359

Chakrabortty, A., & Bose, A. (2019). Smart grid simulations and control. *Power Electronics in Renewable Energy Systems and Smart Grid*, 585–624. doi:10.1002/9781119515661.ch11

Chambers, C. P., & Echenique, F. (2016). *Revealed preference theory* (Vol. 56). Cambridge University Press.

Chazette, L., Karras, O., & Schneider, K. (2018). *Do end-users want explanations? Analyzing the role of explainability as an emerging aspect of non-functional requirements*. Academic Press.

Cheah, J. H., Lim, X. J., Ting, H., Liu, Y., & Quach, S. (2020). Are privacy concerns still relevant? Revisiting consumer behaviour in omnichannel retailing. *Journal of Retailing and Consumer Services*, 102242.

Cheng, L., Liu, F., & Yao, D. (2017). Enterprise data breach: Causes, challenges, prevention, and future directions. *Wiley Interdisciplinary Reviews. Data Mining and Knowledge Discovery*, *7*(5), 1211. doi:10.1002/widm.1211

Cheng, Y., & Jiang, H. (2020). How Do AI-driven Chatbots Impact User Experience? Examining Gratifications, Perceived Privacy Risk, Satisfaction, Loyalty, and Continued Use. *Journal of Broadcasting & Electronic Media*, *64*(4), 592–614. doi:10.1080/08838151.2020.1834296

Chen, M., Beutel, A., Covington, P., Jain, S., Belletti, F., & Chi, E. H. (2019). Top-k off-policy correction for a REINFORCE recommender system. *Proceedings of the Twelfth ACM International Conference on Web Search and Data Mining*, (pp. 456–464). ACM. 10.1145/3289600.3290999

Chen, N., Ribeiro, B., Vieira, A., & Chen, A. (2013). Clustering and visualization of bankruptcy trajectory using self-organizing map. *Expert Systems with Applications*, *40*(1), 385–393. doi:10.1016/j.eswa.2012.07.047

Chen, Q., Gong, Y., Lu, Y., & Tang, J. (2022). Classifying and measuring the service quality of AI chatbot in frontline service. *Journal of Business Research*, *145*, 552–568. doi:10.1016/j.jbusres.2022.02.088

Chen, T., & Guestrin, C. (2016, August). Xgboost: A scalable tree boosting system. In *Proceedings of the 22nd ACM SIGKDD international conference on knowledge discovery and data mining* (pp. 785-794). ACM 10.1145/2939672.2939785

Chen, W. J., & Chen, M. L. (2014). Factors affecting the hotel's service quality: Relationship marketing and corporate image. *Journal of Hospitality Marketing & Management*, *23*(1), 77–96. doi:10.1080/19368623.2013.766581

Chen, Y., Iyengar, R., & Iyengar, G. (2017). Modeling multimodal continuous heterogeneity in conjoint analysis—A sparse learning approach. *Marketing Science*, *36*(1), 140–156. doi:10.1287/mksc.2016.0992

Chomsky, N., Roberts, I., & Watumull, J. (2023). Noam Chomsky: The False Promise of ChatGPT. *The New York Times*, 8.

Chong, A., Li, B., Ngai, E., Ch'ng, E., & Lee, F. (2016). Predicting online product sales via online reviews, sentiments, and promotion strategies: A big data architecture and neural network approach. *International Journal of Operations & Production Management*, *36*(4), 358–383. doi:10.1108/IJOPM-03-2015-0151

Christina, N. (2023). *16 métricas de redes sociales que realmente importan y cómo darles seguimiento.* Hootsuite. https://blog.hootsuite.com/es/metricas-de-redes- sociales/#4_Tasa_de_interaccion

Chu, H., & Liu, S. (2023). *Can AI tell good stories? Narrative Transportation and Persuasion with ChatGPT.*

Chung, M., Ko, E., Joung, H., & Kim, S. J. (2020). Chatbot e-service and customer satisfaction regarding luxury brands. *Journal of Business Research*, *117*, 587–595. doi:10.1016/j.jbusres.2018.10.004

Chung, T. S., Wedel, M., & Rust, R. T. (2016). Adaptive personalisation using social networks. *Journal of the Academy of Marketing Science*, *44*(1), 66–87. doi:10.100711747-015-0441-x

Churchill, E. F. (2013). Putting the person back into personalization. *Interaction*, *20*(5), 12–15. doi:10.1145/2512050.2504847

Ciechanowski, L., Jemielniak, D., & Gloor, P. A. (2020). TUTORIAL: AI research without coding: The art of fighting without fighting: Data science for qualitative researchers. *Journal of Business Research*, *117*, 322–330. doi:10.1016/j.jbusres.2020.06.012

Cireşan, D., Meier, U., Masci, J., & Schmidhuber, J. (2012). Multi-column deep neural network for traffic sign classification. *Neural Networks*, *32*, 333–338. doi:10.1016/j.neunet.2012.02.023 PMID:22386783

Citron, D. K. (2007). Technological due process. *Wash. UL Rev.*, *85*, 1249.

Clark, C. E., & DuPont, B. (2018). Reliability-based design optimization in offshore renewable energy systems. *Renewable & Sustainable Energy Reviews*, *97*, 390–400. doi:10.1016/j.rser.2018.08.030

Clarke, R. (1994). Human identification in information systems: Management challenges and public policy issues. *Information Technology & People*, *7*(4), 6–37. doi:10.1108/09593849410076799

Conant, R. C., & Ross Ashby, W. (1970). Every good regulator of a system must be a model of that system. *International Journal of Systems Science*, *1*(2), 89–97. doi:10.1080/00207727008920220

Cooke, A. D. J., & Zubcsek, P. P. (2017). The connected consumer: Connected devices and the evolution of customer intelligence. *Journal of the Association for Consumer Research*, *2*(2), 164–178. doi:10.1086/690941

Cook, T. D. (2008). "Waiting for life to arrive": A history of the regression-discontinuity design in psychology, statistics and economics. *Journal of Econometrics*, *142*(2), 636–654. doi:10.1016/j.jeconom.2007.05.002

Cooley, C. H. (1902). Looking-glass self. *The Production of Reality: Essays and Readings on Social Interaction*, *6*, 126–128.

Coors, C. (2010). Headwind from Europe: The new position of the German courts on personality rights after the judgment of the European Court of Human Rights. *German Law Journal*, *11*(5), 527–537. doi:10.1017/S207183220001868X

Copeland, B. J. (2023). MYCIN: artificial intelligence program. *Encyclopedia Britannica*. https://www.britannica.com/technology/MYCIN

Correia, A., Paredes, H., Schneider, D., Jameel, S., & Fonseca, B. (2019). Towards hybrid crowd-AI centered systems: Developing an integrated framework from an empirical perspective. In *2019 IEEE International Conference on Systems, Man and Cybernetics (SMC)* (pp. 4013-4018). 10.1109/SMC.2019.8914075

Cortez, P., Cerdeira, A., Almeida, F., Matos, T., & Reis, J. (2009). Modeling wine preferences by data mining from physicochemical properties. *Decision Support Systems*, *47*(4), 547–553. doi:10.1016/j.dss.2009.05.016

Cosley, D., Lam, S. K., Albert, I., Konstan, J. A., & Riedl, J. (2003). Is seeing believing? How recommender system interfaces affect users' opinions. *Proceedings of the SIGCHI Conference on Human Factors in Computing Systems*, (pp. 585–592). IEEE.

Covington, P., Adams, J., & Sargin, E. (2016). Deep neural networks for youtube recommendations. *RecSys 2016 - Proceedings of the 10th ACM Conference on Recommender Systems*, (pp. 191–198). Doi:10.1145/2959100.2959190

Cranor, C. (1975). Toward a Theory of Respect for Persons. *American Philosophical Quarterly*, *12*(4), 309–319.

Crato, N., & Paruolo, P. (2019), Data-Driven Policy Impact Evaluation: How Access to Microdata is Transforming Policy Design. Springer. doi:10.1007/978-3-319-78461-8

Cremonesi, P., Koren, Y., & Turrin, R. (2010). Performance of recommender algorithms on top-n recommendation tasks. *Proceedings of the Fourth ACM Conference on Recommender Systems*, (pp. 39–46). ACM. 10.1145/1864708.1864721

Cristianini, N., Scantamburlo, T., & Ladyman, J. (2021). The social turn of artificial intelligence. *AI & Society*.

Crues, W. (2017). Automated Extraction of Results from Full Text Journal Articles. *Proceedings of the 10th International Conference on Educational Data Mining*. IEEE.

Cukurova, M., Kent, C., & Luckin, R. (2019). Artificial intelligence and multimodal data in the service of human decision-making: A case study in debate tutoring. *British Journal of Educational Technology*, *50*(6), 3032–3046. doi:10.1111/bjet.12829

Cunningham, S. (2021). *Causal inference: The mixtape*. Online Ebook Version. https://mixtape.scunning.com/

Cunningham, S. (2021). *Causal inference: the mixtape*. Yale University Press.

Cyr, D., Gefen, D., & Walczuch, R. (2017). Exploring the relative impact of biological sex and masculinity–femininity values on information technology use. *Behaviour & Information Technology*, *36*(2), 178–193. doi:10.1080/0144929X.2016.1212091

D'Acunto, D., Tuan, A., Dalli, D., Viglia, G., & Okumus, F. (2020). Do consumers care about CSR in their online reviews? An empirical analysis. *International Journal of Hospitality Management*, *85*, 102342. doi:10.1016/j.ijhm.2019.102342

D'Amour, A., Srinivasan, H., Atwood, J., Baljekar, P., Sculley, D., & Halpern, Y. (2020). Fairness is not static: deeper understanding of long term fairness via simulation studies. *Proceedings of the 2020 Conference on Fairness, Accountability, and Transparency*, (pp. 525–534). ACM. 10.1145/3351095.3372878

Dai, W., Lin, J., Jin, F., & Chen, G. (2023, April 25). *Can Large Language Models Provide Feedback to Students? A Case Study on ChatGPT.* ResearchGate. https://www.researchgate.net/publication/370228288_Can_Large_Language_Models_Provide_Feedback_to_Students_A_Case_Study_on_ChatGPT

Damiani, E. S., Di Vimercati, D. C., & Samarati, P. (2003). Managing multiple and dependable identities. *IEEE Internet Computing*, *7*(6), 29–37. doi:10.1109/MIC.2003.1250581

Dan, O., & Loewenstein, Y. (2019). From choice architecture to choice engineering. *Nature Communications*, *10*(1), 1–4. doi:10.103841467-019-10825-6 PMID:31243285

DARPA, Defence Advanced Research Projects Agency. (2023). *DARPA perspective on AI.* SARPA. https://www.darpa.mil/about-us/darpa-perspective-on-ai

Darwall, S. L. (1977). Two Kind of Respect. *Ethics*, *88*(1), 36–49. doi:10.1086/292054

Databricks Lakehouse. (2023). *2023 State of Data + AI.* Databricks. https://www.databricks.com/resources/ebook/state-of-data-ai?scid=7018Y000001Fi0tQAC&utm_medium=paid+search&utm_source=google&utm_campaign=17161077299&utm_adgroup=154951955492&utm_content=ebook&utm_offer=state-of-data-ai&utm_ad=662845930524&utm_term=ai&gclid=EAIaIQobChMI7Yee7cC0gAMVAoJoCR09PwkAEAAYASAAEgIWZvD_BwE. Accessed on July 29, 2023.

Davenport, T. H. (2014). *Big Data at Work: Dispelling the Myths, Uncovering the Opportunities.* Harvard Business School Publishing Corporation. doi:10.15358/9783800648153

Davis, R., & King, J. J. (1984). The origin of rule-based systems in AI. *Rule-based expert systems: The MYCIN experiments of the Stanford Heuristic Programming Project.*

Davis, F. D., Bagozzi, R. P., & Warshaw, P. R. (1989). User acceptance of computer technology: A comparison of two theoretical models. *Management Science*, *35*(8), 982–1003. doi:10.1287/mnsc.35.8.982

De Gregorio, G., & Celeste, E. (2022). Digital Humanism: The Constitutional Message of the GDPR. *Global Privacy Law Review, 3*(1).

De Mooij, M., & Hofstede, G. (2011). Cross-cultural consumer behavior: A review of research findings. *Journal of International Consumer Marketing*, *23*(3-4), 181–192.

De Vries, K. (2010). Identity, profiling algorithms and a world of ambient intelligence. *Ethics and Information Technology*, *12*(1), 71–85. doi:10.100710676-009-9215-9

Deacon, T. W. (1997). The symbolic species: The co-evolution of language and the brain (Issue 202). WW Norton & Company.

DeCew, J. W. (1986). The scope of privacy in law and ethics. *Law and Philosophy*, 145–173.

Deguchi, A. (2020). What Is Society 5.0? In *Society 5.0.* Springer. doi:10.1007/978-981-15-2989-4_1

Dehaene, S., Lau, H., & Kouider, S. (2017). What is consciousness, and could machines have it? *Science*, *358*(6362), 486–492. doi:10.1126cience.aan8871 PMID:29074769

Deisenroth, M. P., Faisal, A. A., & Ong, C. S. (2020). *Mathematics for machine learning.* Cambridge University Press. doi:10.1017/9781108679930

Dekking, F. M., Kraaikamp, C., Lopuhaä, H. P., & Meester, L. E. (2005). *A Modern Introduction to Probability and Statistics: Understanding why and how* (Vol. 488). Springer. doi:10.1007/1-84628-168-7

Dellosa, J. T., & Palconit, E. C. (2021, September). Artificial Intelligence (AI) in renewable energy systems: A condensed review of its applications and techniques. *In 2021 IEEE International Conference on Environment and Electrical Engineering and 2021 IEEE Industrial and Commercial Power Systems Europe (EEEIC/I&CPS Europe)*, (pp. 1-6). IEEE.

den Hengst, F., Grua, E. M., el Hassouni, A., & Hoogendoorn, M. (2020). Reinforcement learning for personalization: A systematic literature review. *Data Science*, *3*(2), 107–147. doi:10.3233/DS-200028

Dengel, A., Devillers, L., & Schaal, L. M. (2021). Augmented human and human-machine co-evolution: Efficiency and ethics. *Reflections on Artificial Intelligence for Humanity*, 203-227.

Deng, L., & Yu, D. (2014). Foundations and Trends in Signal Processing. *DEEP LEARNING–Methods and Applications*, *7*(3–4), 197–387.

Dennett, D. C. (1976). Conditions of Personhood. In A. O. Rorty (Ed.), *The Identities of Persons* (pp. 175–196). University of California Press.

Dennett, D. C. (2015). *Elbow Room: The Varieties of Free Will Worth Wanting*. MIT Press. doi:10.7551/mitpress/10470.001.0001

Dergaa, I., Chamari, K., Zmijewski, P., & Saad, H. B. (2023). From human writing to artificial intelligence generated text: Examining the prospects and potential threats of ChatGPT in academic writing. *Biology of Sport*, *40*(2), 615–622. doi:10.5114/biolsport.2023.125623 PMID:37077800

Dermody, G., & Fritz, R. (2019). A conceptual framework for clinicians working with artificial intelligence and health-assistive smart homes. *Nursing Inquiry*, *26*(1), e12267. doi:10.1111/nin.12267 PMID:30417510

DeSarbo, W. S., & Grisaffe, D. (1998). Combinatorial optimisation approaches to constrained market segmentation: An application to industrial market segmentation. *Marketing Letters*, *9*(2), 115–134. doi:10.1023/A:1007997714444

Descartes, R. (1999). *Discourse on method and meditations on first philosophy*. Hackett Publishing.

Devlin, J. (2018). *Bert: Pre-training of deep bidirectional transformers for language understanding*. arXiv preprint arXiv:1810.04805.

Devriendt, F., Moldovan, D., & Verbeke, W. (2018). A literature survey and experimental evaluation of the state-of-the-art in uplift modeling: A stepping stone toward developing prescriptive analytics. *Big Data*, *6*(1), 13–41. doi:10.1089/big.2017.0104 PMID:29570415

Dewey, J. (1927). *The Public and its Problems* (Vol. H). Holt and Company.

Dezfouli, A., Nock, R., & Dayan, P. (2020). Adversarial vulnerabilities of human decision-making. *Proceedings of the National Academy of Sciences of the United States of America*, *117*(46), 29221–29228. doi:10.1073/pnas.2016921117 PMID:33148802

Dhar, S., & Bose, I. (2022). Walking on air or hopping mad? Understanding the impact of emotions, sentiments and reactions on ratings in online customer reviews of mobile apps. *Decision Support Systems*, *162*, 113769. doi:10.1016/j.dss.2022.113769

Dignum, V. (2019). The responsibility is ours. In *Artificial intelligence: applications, implications and speculations*. Fidelidade-Culturgest Conferences and Debates.

Dilthey, W. (1989). *Introduction to the Human Sciences* (Vol. 1). Princeton University Press.

Domingos, P. (2012). A few useful things to know about machine learning. *Communications of the ACM*, *55*(10), 78–87. doi:10.1145/2347736.2347755

Donald, M. (1991). *Origins of the modern mind: Three stages in the evolution of culture and cognition*. Harvard University Press.

Douglas, M., & Ney, S. (1998). *Missing persons: A critique of the personhood in the social sciences*. University of California Press.

Driver, J. (2011). *Consequentialism*. Routledge. doi:10.4324/9780203149256

Duan, Y., Edwards, J. S., & Dwivedi, Y. K. (2019). Artificial intelligence for decision making in the era of Big Data–evolution, challenges and research agenda. *International Journal of Information Management*, *48*, 63–71. doi:10.1016/j.ijinfomgt.2019.01.021

Dubey, A. K., Narang, S., Srivastav, A. L., Kumar, A., & García-Díaz, V. (Eds.). (2022). *Artificial Intelligence for Renewable Energy Systems*. Elsevier.

Duhaim, A. M., Fadhel, M. A., Alnoor, A., Baqer, N. S., Alzubaidi, L., Albahri, O. S., Alamoodi, A. H., Bai, J., Salhi, A., Santamaría, J., Ouyang, C., Gupta, A., Gu, Y., & Deveci, M. (2023). A systematic review of trustworthy and explainable artificial intelligence in Healthcare: Assessment of quality, bias risk, and data fusion. *Information Fusion*.

Duncan, C. S. (1919). *Commercial Research: An Outline of Working Principles*. McMillan Co.

Durna, U., Dedeoglu, B. B., & Balikçioglu, S. (2015). The role of servicescape and image perceptions of customers on behavioral intentions in the hotel industry. *International Journal of Contemporary Hospitality Management*, *27*(7), 1728–1748. doi:10.1108/IJCHM-04-2014-0173

Dutt, R., Deb, A., & Ferrara, E. (2018). "Senator, We Sell Ads": Analysis of the 2016 Russian Facebook Ads Campaign. *Communications in Computer and Information Science*, *941*, 151–168. doi:10.1007/978-981-13-3582-2_12

Dwivedi, Y. K., Hughes, L., Ismagilova, E., Aarts, G., Coombs, C., Crick, T., Duan, Y., Dwivedi, R., Edwards, J., Eirug, A., Galanos, V., Ilavarasan, P. V., Janssen, M., Jones, P., Kar, A. K., Kizgin, H., Kronemann, B., Lal, B., Lucini, B., ... Williams, M. D. (2021). Artificial Intelligence (AI): Multidisciplinary perspectives on emerging challenges, opportunities, and agenda for research, practice and policy. *International Journal of Information Management*, *57*, 101994. Advance online publication. doi:10.1016/j.ijinfomgt.2019.08.002

Dwivedi, Y. K., Kshetri, N., Hughes, L., Slade, E. L., Jeyaraj, A., Kar, A. K., Baabdullah, A. M., Koohang, A., Raghavan, V., Ahuja, M., Albanna, H., Albashrawi, M. A., Al-Busaidi, A. S., Balakrishnan, J., Barlette, Y., Basu, S., Bose, I., Brooks, L., Buhalis, D., & Wright, R. (2023). "So what if ChatGPT wrote it?" Multidisciplinary perspectives on opportunities, challenges and implications of generative conversational AI for research, practice and policy. *International Journal of Information Management*, *71*, 102642. doi:10.1016/j.ijinfomgt.2023.102642

Dworkin, R. (2013). *Taking Rights Seriously*. A&C Black.

Dzyabura, D., & Yoganarasimhan, H. (2018). Machine learning and marketing. In *Handbook of marketing analytics* (pp. 255–279). Edward Elgar Publishing. doi:10.4337/9781784716752.00023

Ehrlinger, L., & Woss, W. (2022). A Survey of Data Quality Measurement and Monitoring Tools. *Frontiers in Big Data*, *5*. https://www.frontiersin.org/articles/10.3389/fdata.2022.850611

Ekstrand, M. D., & Willemsen, M. C. (2016). Behaviorism is not enough: better recommendations through listening to users. *Proceedings of the 10th ACM Conference on Recommender Systems*, (pp. 221–224). 10.1145/2959100.2959179

El Haj Assad, M., Alhuyi Nazari, M., & Rosen, M. A. (2021). Applications of renewable energy sources. *Design and Performance Optimization of Renewable Energy Systems,* 1–15. doi:10.1016/b978-0-12-821602-6.00001-8

Ellenberg, S. S., Fleming, T. R., & DeMets, D. L. (2019). *Data monitoring committees in clinical trials: a practical perspective*. John Wiley & Sons. doi:10.1002/9781119512684

El-Shamandi Ahmed, K., Ambika, A., & Belk, R. (2023). Augmented reality magic mirror in the service sector: Experiential consumption and the self. *Journal of Service Management, 34*(1), 56–77. doi:10.1108/JOSM-12-2021-0484

Eltyeb, S., & Salim, N. (2014). Chemical named entities recognition: A review on approaches and applications. *Journal of Cheminformatics, 6*(1), 17. doi:10.1186/1758-2946-6-17 PMID:24834132

Emanuel, E. J., & Wachter, R. M. (2019). Artificial intelligence in health care: Will the value match the hype? *Journal of the American Medical Association, 321*(23), 2281–2282. doi:10.1001/jama.2019.4914 PMID:31107500

Engelbart, D. C. (2023). Augmenting human intellect: A conceptual framework. In *Augmented Education in the Global Age* (pp. 13–29). Routledge.

English, T. (2020, November 20). *Nanobots will be flowing through your body by 2030*. IE. Retrieved November 29, 2022, from https://interestingengineering.com/innovation/nanobots-will-be-flowing-through-your-body-by-2030

Enholm, I. M., Papagiannidis, E., Mikalef, P., & Krogstie, J. (2022). Artificial intelligence and business value: A literature review. *Information Systems Frontiers, 24*(5), 1709–1734. doi:10.100710796-021-10186-w

Eppen, G. D., Gould, F. J., Schmidt, C. P., Moore, J. H., & Weatherford, L. R. (1998), Introductory Management Science: decision modeling with spreadsheets. Prentice-Hall International.

Etzioni, A. (1996). The responsive community: A communitarian perspective. *American Sociological Review, 61*(1), 1–11. doi:10.2307/2096403

European Commission. (2020), *White Paper on Artificial Intelligence: a European approach to excellence and trust* (White Paper No. COM(2020) 65 final), European Commission. https://ec.europa.eu/info/publications/white-paper-artificial-intelligence-european-approach-excellence-and-trust_en

European Commission. (2020). *White Paper on Artificial Intelligence: A European approach to excellence and trust.* Author.

European Commission. (2021). *Communication on fostering a European approach to artificial intelligence*. EC. https://digital-strategy.ec.europa.eu/en/library/communication-fostering-european-approach-artificial-intelligence

European Commission. (2021). *Ethics Guidelines for Trustworthy AI*. https://ec.europa.eu/

Evans, J. S. B. (2010). Intuition and reasoning: A dual-process perspective. *Psychological Inquiry, 21*(4), 313–326. doi:10.1080/1047840X.2010.521057

Everitt, T., Hutter, M., Kumar, R., & Krakovna, V. (2021). Reward tampering problems and solutions in reinforcement learning: A causal influence diagram perspective. *Synthese, 198*(S27), 1–33. doi:10.100711229-021-03141-4

Eyal, N. (2014). *Hooked: How to build habit-forming products*. Penguin.

Eysenbach, G. (2023). The role of ChatGPT, generative language models, and artificial intelligence in medical education: A conversation with ChatGPT and a call for papers. *JMIR Medical Education, 9*(1), e46885. doi:10.2196/46885 PMID:36863937

Fang, H., Zhang, J., Bao, Y., & Zhu, Q. (2013). Towards effective online review systems in the Chinese context: A cross-cultural empirical study. *Electronic Commerce Research and Applications*, *12*(3), 208–220. doi:10.1016/j.elerap.2013.03.001

Fan, H., & Poole, M. S. (2006). What is personalization? Perspectives on the design and implementation of personalization in information systems. *Journal of Organizational Computing and Electronic Commerce*, *16*(3–4), 179–202. doi:10.120715327744joce1603&4_2

Fan, X., Jiang, X., & Deng, N. (2022). Immersive technology: A meta-analysis of augmented/virtual reality applications and their impact on tourism experience. *Tourism Management*, *91*, 104534. doi:10.1016/j.tourman.2022.104534

Fan, X., & Nowell, D. L. (2011). Using propensity score matching in educational Research. *Gifted Child Quarterly*, *55*(1), 74–79. doi:10.1177/0016986210390635

Farhud, D. D., & Zokaei, S. (2021). Ethical Issues of Artificial Intelligence in Medicine and Healthcare. *Iranian Journal of Public Health*, *50*(11). Advance online publication. doi:10.18502/ijph.v50i11.7600 PMID:35223619

Fenko, A., Kersten, L., & Bialkova, S. (2016). Overcoming consumer scepticism toward foodlabels: The role of multisensory experience. *Food Quality and Preference*, *48*, 81–92. doi:10.1016/j.foodqual.2015.08.013

Fernandes, T., & Oliveira, E. (2021). Understanding consumers' acceptance of automated technologies in service encounters: Drivers of digital voice assistants adoption. *Journal of Business Research*, *122*, 180–191. doi:10.1016/j.jbusres.2020.08.058

Fernández-Loría, C., & Provost, F. (2022). Causal decision making and causal effect estimation are not the same… and why it matters. *INFORMS Journal on Data Science, 1*(1).

Fernandez-Quilez, A. (2022). Deep Learning in Radiology: Ethics of data and on the value of algorithm transparency, interpretability and explainability. *AI and Ethics*, *3*(1), 257–265. doi:10.100743681-022-00161-9

Ferrara, E. (2023). *Should chatbots be biased? Challenges and risks of bias in large language models*. arXiv preprint arXiv:2304.03738.

Filieri, R., Lin, Z., Pino, G., Alguezaui, S., & Inversini, A. (2021). The role of visual cues in eWOM on consumers' behavioral intention and decisions. *Journal of Business Research*, *135*, 663–675. doi:10.1016/j.jbusres.2021.06.055

Fisher, D. H. (1987). Knowledge acquisition via incremental conceptual clustering. *Machine Learning*, *2*(2), 139–172. doi:10.1007/BF00114265

Fisher, R. A. (1966). *The design of experiments* (8th ed.). Hafner Publishing Company.

Fitzgerald, S. (2023). How Can You Use AI—And Protect Patient Privacy. *Neurology Today*, *23*(12), 1–12. doi:10.1097/01.NT.0000943140.92540.15

Floridi, L. (2022). The European Legislation on AI: A brief analysis of its philosophical approach. In The 2021 Yearbook of the Digital Ethics Lab (pp. 1–8). Springer. doi:10.1007/978-3-031-09846-8_1

Floridi, L. (2016). On human dignity as a foundation for the right to privacy. *Philosophy & Technology*, *29*(4), 307–312. doi:10.100713347-016-0220-8

Floridi, L., Cowls, J., Beltrametti, M., Chatila, R., Chazerand, P., Dignum, V., Luetge, C., Madelin, R., Pagallo, U., Rossi, F., Schafer, B., Valcke, P., & Vayena, E. (2018). AI4People—An ethical framework for a good AI society: Opportunities, risks, principles, and recommendations. *Minds and Machines*, *28*(4), 689–707. doi:10.100711023-018-9482-5 PMID:30930541

Floridi, L., & Sanders, J. W. (2004). On the morality of artificial agents. *Minds and Machines, 14*(3), 349–379. doi:10.1023/B:MIND.0000035461.63578.9d

Foerster-Metz, U. S., Marquardt, K., Golowko, N., Kompalla, A., & Hell, C. (2018). Digital transformation and its implications on organizational behavior. *Journal of EU Research in Business, 2018*(S3).

Fogg, B. J. (2003). *Persuasive Technology: Using Computers to Change What We Think and Do*. Elsevier. doi:10.1016/B978-155860643-2/50011-1

Földes, D., Csiszár, C., & Tettamanti, T. (2021). Automation levels of mobility services. *Journal of Transportation Engineering. Part A, Systems, 147*(5), 04021021. Advance online publication. doi:10.1061/JTEPBS.0000519

Ford, V., Siraj, A., & Eberle, W. (2014). Smart grid energy fraud detection using artificial neural networks. In *IEEE Symposium on Computational Intelligence Applications in Smart Grid (CIASG)*. IEEE. 10.1109/CIASG.2014.7011557

Forgas-Coll, S., Huertas-Garcia, R., Andriella, A., & Alenyà, G. (2023). Social robot-delivered customer-facing services: An assessment of the experience. *Service Industries Journal, 43*(3-4), 154–184. doi:10.1080/02642069.2022.2163995

Fort, J.-C., Letrémy, P., & Cottrell, M. (2002). Advantages and drawbacks of the Batch Kohonen algorithm. 10th-European-Symposium on Artificial Neural Networks, (pp. 223-30).

Fougère, D., & Jacquemet, N. (2019). Causal inference and impact evaluation. *Economie et Statistique/Economics and Statistics*, (510-511-512), 181-200. doi:10.24187/ecostat.2019.510t.1996

Frankfurt, H. G. (1971). Freedom of the Will and the Concept of a Person. *The Journal of Philosophy, 68*(1), 5. doi:10.2307/2024717

Frankfurt, H. G. (2006). *Taking ourselves seriously and getting it right*. Stanford University Press.

Frederick, J. G. (1918). *Business Research and Statistics*. D. Appleton and Co.

Friedman, J. (2001). Greedy function approximation: A gradient boosting machine. *Annals of Statistics, 29*(5), 1189–1232. doi:10.1214/aos/1013203451

Friston, K. (2010). The free-energy principle: A unified brain theory? *Nature Reviews. Neuroscience, 11*(2), 127–138. doi:10.1038/nrn2787 PMID:20068583

Fügener, A., Grahl, J., Gupta, A., & Ketter, W. (2022). Cognitive challenges in human–artificial intelligence collaboration: Investigating the path toward productive delegation. *Information Systems Research, 33*(2), 678–696. doi:10.1287/isre.2021.1079

Fui-Hoon Nah, F., Zheng, R., Cai, J., Siau, K., & Chen, L. (2023). Generative AI and ChatGPT: Applications, challenges, and AI-human collaboration. *Journal of Information Technology Case and Application Research*, 1-28. . doi:10.1080/15228053.2023.2233814

Fukuyama, F. (2018). *Identity: The demand for dignity and the politics of resentment*. Farrar, Straus and Giroux.

Fuller, L. (1964). *The Morality of Law*. Yale University Press.

Fung, A. (2013). The Principle of Affected Interests: An Interpretation and Defense. In *Representation* (pp. 236–268). University of Pennsylvania Press. doi:10.9783/9780812208177.236

Furuta, H., Matsushima, T., Kozuno, T., Matsuo, Y., Levine, S., Nachum, O., & Gu, S. S. (2021). Policy information capacity: Information-theoretic measure for task complexity in deep reinforcement learning. *International Conference on Machine Learning*, (pp. 3541–3552).

Gabriel, M. (2022). *Startups and science: EU businesses will increase their competitiveness by linking closely scientists and startups*. Linkedin. https://www.linkedin.com/pulse/startups-science-eu-businesses-increase-linking-closely-gabriel

Gabriel, I. (2020). Artificial intelligence, values, and alignment. *Minds and Machines*, *30*(3), 411–437. doi:10.100711023-020-09539-2

Gabriel, I. (2022). Toward a Theory of Justice for Artificial Intelligence. *Daedalus*, *151*(2), 218–231. doi:10.1162/daed_a_01911

Galati, A., Thrassou, A., Christofi, M., Vrontis, D., & Migliore, G. (2021). Exploring travelers' willingness to pay for green hotels in the digital era. *Journal of Sustainable Tourism*, 1–18. doi:10.1080/09669582.2021.2016777

Gallie, W. B. (1955). Essentially contested concepts. *Proceedings of the Aristotelian Society*, *56*(1), 167–198. doi:10.1093/aristotelian/56.1.167

Gama, J., Lorena, A., Faceli, K., Oliveira, O., & Carvalho, A. (2017). Extração de Conhecimento de dados: Data Mining. Edições Sílabo- 3ª Edição.

García-Madurga, M. Á., & Grilló-Méndez, A. J. (2023). Artificial intelligence in the tourism industry: An overview of reviews. *Administrative Sciences*, *13*(8), 172. doi:10.3390/admsci13080172

Gasbarro, F., & Bonera, M. (2021). The influence of green practices and green image on customer satisfaction and word-of-mouth in the hospitality industry. *Sinergie Italian Journal of Management*, *39*(3), 231–248. doi:10.7433116.2021.12

Gauci, J., Ghavamzadeh, M., Honglei, L., & Nahmias, R. (2019, October 19). *Open-sourcing ReAgent, a modular, end-to-end platform for building reasoning systems*. Facebook AI. https://ai.facebook.com/blog/open-sourcing-reagent-a-platform-for-reasoning-systems/

Gautam, L., & Kaur, G. (2023). Review on ultrasound-image analysis of fetus using Deep Learning. *2023 IEEE 8th International Conference for Convergence in Technology (I2CT)*. 10.1109/I2CT57861.2023.10126500

Geiderman, J. M., Moskop, J. C., & Derse, A. R. (2006). Privacy and confidentiality in emergency medicine: Obligations and challenges. *Emergency Medicine Clinics of North America*, *24*(3), 633–656. doi:10.1016/j.emc.2006.05.005 PMID:16877134

Gelbrich, K., Hagel, J., & Orsingher, C. (2021). Emotional support from a digital assistant in technology-mediated services: Effects on customer satisfaction and behavioral persistence. *International Journal of Research in Marketing*, *38*(1), 176–193. doi:10.1016/j.ijresmar.2020.06.004

Gergen, K. J. (1985). The social constructionist movement in modern psychology. *The American Psychologist*, *40*(3), 266–275. doi:10.1037/0003-066X.40.3.266

Gilbert, T. K., Lambert, N., Dean, S., Zick, T., & Snoswell, A. (2022). *Reward reports for reinforcement learning*. arXiv preprint arXiv:2204.10817.

Giri, G., Maddahi, Y., & Zareinia, K. (2021). A brief review on challenges in design and development of Nanorobots for Medical Applications. *Applied Sciences (Basel, Switzerland)*, *11*(21), 10385. doi:10.3390/app112110385

Goldberg, Y. (2016). A primer on neural network models for natural language processing. *Journal of Artificial Intelligence Research*, *57*, 345–420. doi:10.1613/jair.4992

Goldsmith, R. (2016). The Big Five, happiness, and shopping. *Journal of Retailing and Consumer Services*, *31*, 52–61. doi:10.1016/j.jretconser.2016.03.007

Gomez-Uribe, C. A., & Hunt, N. (2015). The Netflix Recommender System: Algorithms. *Business Value, 6.*

González-Rodríguez, M. R., Díaz-Fernández, M. C., & Font, X. (2020). Factors influencing willingness of customers of environmentally friendly hotels to pay a price premium. *International Journal of Contemporary Hospitality Management, 32*(1), 60–80. doi:10.1108/IJCHM-02-2019-0147

Goodell, J., & Craig, S. D. (2022). What Does an Emerging Intelligence Augmentation Economy Mean for HF/E? Can Learning Engineering Help? *Proceedings of the Human Factors and Ergonomics Society Annual Meeting, 66*(1), 460–464. doi:10.1177/1071181322661283

Goode, M. M., Davies, F., Moutinho, L., & Jamal, A. (2005). Determining customer satisfaction from mobile phones: A neural network approach. *Journal of Marketing Management, 21*(7-8), 755–778. doi:10.1362/026725705774538381

Goodfellow, I., Bengio, Y., & Courville, A. (2016). *Deep learning*. MIT press.

Goodfellow, I., Bengio, Y., & Courville, A. (2016). *Deep Learning*. MIT Press.

Gopalan, M., Rosinger, K., & Ahn, J. B. (2020). Use of quasi-experimental research designs in education research: Growth, promise, and challenges. *Review of Research in Education, 44*(1), 218–243. doi:10.3102/0091732X20903302

Gottesman, O., Johansson, F., Komorowski, M., Faisal, A., Sontag, D., Doshi-Velez, F., & Celi, L. A. (2019). Guidelines for reinforcement learning in healthcare. *Nature Medicine 2019 25:1, 25*(1), 16–18. doi:10.1038/s41591-018-0310-5

Gottfredson, L. S. (1997). Mainstream science on intelligence: An editorial with 52 signatories, history, and bibliography. [). JAI.]. *Intelligence, 24*(1), 13–23. doi:10.1016/S0160-2896(97)90011-8

Granovetter, M. S. (1973). The Strengh of Weak Ties. *American Journal of Sociology, 78*(6), 1360–1380. doi:10.1086/225469

Gray, C. M., Kou, Y., Battles, B., Hoggatt, J., & Toombs, A. L. (2018). The dark (patterns) side of UX design. *Proceedings of the 2018 CHI Conference on Human Factors in Computing Systems*, (pp. 1–14). ACM. 10.1145/3173574.3174108

Green Hotel Association. (2008). *What are green hotels?* Green Hotel Association. (https://www.greenhotels.com/whatare.htm).

Greene, T., Dhurandhar, A., & Shmueli, G. (2022). Atomist or holist? A diagnosis and vision for more productive interdisciplinary AI ethics dialogue. *Patterns*, 100652.

Greene, T., Goethals, S., Martens, D., & Shmueli, G. (2023). *Monetizing Explainable AI: A Double-edged Sword.* arXiv preprint arXiv:2304.06483.

Greene, T., Shmueli, G., Fell, J., Lin, C.-F., & Liu, H.-W. (2022). Forks over knives: Predictive inconsistency in criminal justice algorithmic risk assessment tools. *Journal of the Royal Statistical Society Series A: Statistics in Society, 185*(Supplement_2), S692–S723.

Greene, T., Martens, D., & Shmueli, G. (2022). Barriers to academic data science research in the new realm of algorithmic behaviour modification by digital platforms. *Nature Machine Intelligence, 4*(4), 323–330. doi:10.103842256-022-00475-7

Greene, T., Shmueli, G., & Ray, S. (2022). (Forthcoming). Taking the Person Seriously: Ethically-aware IS Research in the Era of Reinforcement Learning-based Personalization. *JAIS Preprints*, 77.

Greene, T., Shmueli, G., Ray, S., & Fell, J. (2019). Adjusting to the GDPR: The Impact on Data Scientists and Behavioral Researchers. *Big Data, 7*(3), 140–162. doi:10.1089/big.2018.0176 PMID:31033336

Gretzel, U., Sigala, M., Xiang, Z., & Koo, C. (2015). Smart tourism: Foundations and developments. *Electronic Markets, 25*(3), 179–188. doi:10.100712525-015-0196-8

Grier, D. A. (2023). *Debating Artificial Intelligence*. Computer. https://www.computer.org/publications/tech-news/closer-than-you-might-think/debating-artificial-intelligence

Grieves, M., & Vickers, J. (2017). Digital Twin: Mitigating Unpredictable, Undesirable Emergent Behavior in Complex Systems. In Transdisciplinary Perspectives on Complex Systems. Springer International. DOI doi:10.1007/978-3-319-38756-7_4

Grover, V., Chiang, R. H., Liang, T. P., & Zhang, D. (2018). Creating strategic business value from big data analytics: A research framework. *Journal of Management Information Systems*, *35*(2), 388–423. doi:10.1080/07421222.2018.1451951

Guan, C., Hung, Y. C., & Liu, W. (2022). Cultural differences in hospitality service evaluations: Mining insights of user generated content. *Electronic Markets*, *32*(3), 1–21. doi:10.100712525-022-00545-z

Gupta, P., Kumar, S., Singh, Y. B., Singh, P., Sharma, S. K., & Rathore, N. K. (2022). The Impact of Artificial Intelligence on Renewable Energy Systems. *NeuroQuantology : An Interdisciplinary Journal of Neuroscience and Quantum Physics*, *20*(16), 5012–5029.

Gupta, S., Singhvi, S., & Granata, G. (2023). Assessing the Impact of Artificial Intelligence in e-Commerce Portal: A Comparative Study of Amazon and Flipkart. In *Industry 4.0 and the Digital Transformation of International Business* (pp. 173–187). Springer Nature Singapore. doi:10.1007/978-981-19-7880-7_10

Habal, M. B., Clement, A., Hardisty, M., Fialkov, J. A., Whyne, C. M., & Hanieh, A. (2023, August 28). Artificial Intelligence–based modeling can predict face. *The Journal of Craniofacial Surgery*. https://journals.lww.com/jcraniofacialsurgery/abstract/9900/artificial_intelligence_based_modeling_can_predict.1013.aspx

Habermas, J. (1990). *Moral Consciousness and Communicative Action*. MIT Press.

Habermas, J. (2015). *Knowledge and Human Interests*. John Wiley & Sons.

Hacking, I. (2007). Kinds of People: Moving Targets. *Proceedings of the British Academy*, 286–318.

Hadi, R., & Valenzuela, A. (2019). Good vibrations: Consumer responses to technology-mediated haptic feedback. *The Journal of Consumer Research*, *47*(2), 256–271. doi:10.1093/jcr/ucz039

Haenlein, M., & Kaplan, A. (2019). A brief history of artificial intelligence: On the past, present, and future of artificial intelligence. *California Management Review*, *61*(4), 5–14. doi:10.1177/0008125619864925

HagendorffT. (2019). *The ethics of AI ethics—an evaluation of guidelines*. International Centre for Ethics in the Sciences and Humanities. arXiv: 190303425.

Haidt, J. (2008). Morality. *Perspectives on Psychological Science*, *3*(1), 65–72. doi:10.1111/j.1745-6916.2008.00063.x PMID:26158671

Hajipour, V., Niaki, S. T., Tavana, M., Santos-Arteaga, F. J., & Hosseinzadeh, S. (2023). A comparative performance analysis of intelligence-based algorithms for optimising competitive facility location problems. *Machine Learning with Applications*, *11*, 100443. doi:10.1016/j.mlwa.2022.100443

Haleem, A., Javaid, M., Asim Qadri, M., Pratap Singh, R., & Suman, R. (2022). Artificial Intelligence (AI) applications for marketing: A literature-based study. *International Journal of Intelligent Networks*, *3*, 119–132. doi:10.1016/j.ijin.2022.08.005

Hameed, I., Hussain, H., & Khan, K. (2022). The role of green practices toward the green word-of-mouth using stimulus-organism-response model. *Journal of Hospitality and Tourism Insights*, *5*(5), 1046–1061. doi:10.1108/JHTI-04-2021-0096

Hamet, P., & Tremblay, J. (2017). Artificial intelligence in medicine. *Metabolism: Clinical and Experimental, 69*, S36–S40. doi:10.1016/j.metabol.2017.01.011 PMID:28126242

Hanelt, A., Bohnsack, R., Marz, D., & Antunes Marante, C. (2021). A systematic review of the literature on digital transformation: Insights and implications for strategy and organizational change. *Journal of Management Studies, 58*(5), 1159–1197. doi:10.1111/joms.12639

Han, J., & Kamber, M. (2011). *Data Mining: Concepts and Techniques*. Morgan Kaufmann Publishers.

Han, K., Cao, P., Wang, Y., Xie, F., Ma, J., Yu, M., Wang, J., Xu, Y., Zhang, Y., & Wan, J. (2022). A review of approaches for predicting drug-drug interactions based on machine learning. *Frontiers in Pharmacology, 12*, 814858. doi:10.3389/fphar.2021.814858 PMID:35153767

Han, S., & Yang, H. (2018). Understanding adoption of intelligent personal assistants: A parasocial relationship perspective. *Industrial Management & Data Systems, 118*(3), 618–636. doi:10.1108/IMDS-05-2017-0214

Hansotia, B., & Rukstales, B. (2002). Incremental value modeling. *Journal of Interactive Marketing, 16*(3), 35–46. doi:10.1002/dir.10035

Haque, M. U., Dharmadasa, I., Sworna, Z. T., Rajapakse, R. N., & Ahmad, H. (2022). "I think this is the most disruptive technology": Exploring sentiments of ChatGPT early adopters using Twitter data. https://doi.org//arXiv.2303.03836. doi:10.48550

Harris, M. (1976). History and significance of the emic/etic distinction. *Annual Review of Anthropology, 5*(1), 329–350. doi:10.1146/annurev.an.05.100176.001553

Hart, H. L. A. (1961). *The Concept of Law*. Oxford University Press.

Hartzog, W. (2021). What is Privacy? That's the Wrong Question. *U. Chi. L. Rev., 88*, 1677.

Hasan, R., Thaichon, P., & Weaven, S. (2021). Are we already living with Skynet? Anthropomorphic artificial intelligence to enhance customer experience. In *Developing digital marketing*. Emerald Publishing Limited. doi:10.1108/978-1-80071-348-220211007

Hasan, R., Weaven, S., & Thaichon, P. (2021). Blurring the line between physical and digital environment: The impact of artificial intelligence on customers' relationship and customer experience. In *Developing digital marketing*. Emerald Publishing Limited. doi:10.1108/978-1-80071-348-220211008

Hassani, H., Silva, E. S., Unger, S., TajMazinani, M., & Mac Feely, S. (2020). Artificial intelligence (AI) or intelligence augmentation (IA): what is the future? *AI, 1*(2), 143-155.

Hastie, T., Tibshirani, R., Friedman, J. H., & Friedman, J. H. (2009). *The elements of statistical learning: data mining, inference, and prediction* (Vol. 2). Springer. doi:10.1007/978-0-387-84858-7

Häubl, G., & Trifts, V. (2000). Consumer decision making in online shopping environments: The effects of interactive decision aids. *Marketing Science, 19*(1), 4–21. doi:10.1287/mksc.19.1.4.15178

Haugen, F. (2021, October 4). *Statement of Frances Haugen*. Whistleblower Aid. https://www.commerce.senate.gov/services/files/589FC8A558E-824E-4914-BEDB-3A7B1190BD49

Hebrail, G., & Berard, A. (2012). *Individual household electric power consumption*. UCI Machine Learning Repository.

Heidegger, M. (2010). *Being and Time*.

Heidelberger, M. (2004). *Nature from within: Gustav Theodor Fechner and his psychophysical worldview*. University of Pittsburgh Press.

He, J., Baxter, S. L., Xu, J., Xu, J., Zhou, X., & Zhang, K. (2019). The practical implementation of artificial intelligence technologies in medicine. *Nature Medicine, 25*(1), 30–36. doi:10.103841591-018-0307-0 PMID:30617336

Helberger, N., Karppinen, K., & D'acunto, L. (2018). Exposure diversity as a design principle for recommender systems. *Information Communication and Society, 21*(2), 191–207. doi:10.1080/1369118X.2016.1271900

Helfat, C. E., Finkelstein, S., Mitchell, W., Peteraf, M., Singh, H., Teece, D., & Winter, S. G. (2009). *Dynamic capabilities: Understanding strategic change in organizations*. John Wiley & Sons.

Helmond, A. (2015). The platformization of the web: Making web data platform ready. *Social Media+ Society, 1*(2), 2056305115603080.

Hennig-Thurau, T., Gwinner, K. P., Walsh, G., & Gremler, D. D. (2004). Electronic word-of-mouth via consumer-opinion platforms: What motivates consumers to articulate themselves on the internet? *Journal of Interactive Marketing, 18*(1), 38–52. doi:10.1002/dir.10073

Hérault, J., & Demartines, P. (1997). Curvilinear Component Analysis: A Self-Organizing Neural Network for Nonlinear Mapping of Data Sets. *IEEE Transactions on Neural Networks*, 148–154. PMID:18255618

HerboldS.Hautli-JaniszA.HeuerU.KiktevaZ.TrautschA. (2023). AI, write an essay for me: A large-scale comparison of human-written versus ChatGPT-generated essays—arXiv preprint arXiv:2304.14276.

Hernán, M. A. (2017), Causal Diagrams: Draw Your Assumptions Before Your Conclusions. EDX. https://www.edx.org/course/causal-diagrams-draw-your-assumptions-before-your.

Hewett, K., Rand, W., Rust, R. T., & Van Heerde, H. J. (2016). Brand buzz in the echoverse. *Journal of Marketing, 80*(3), 1–24. doi:10.1509/jm.15.0033

He, X., Pan, J., Jin, O., Xu, T., Liu, B., Xu, T., Atallah, A., Herbrich, R., Bowers, S., & Candela, J. (2014). Practical lessons from predicting clicks on ads at facebook. *Proceedings of the Eighth International Workshop on Data Mining for Online Advertising*, (pp. 1–9). ACM. 10.1145/2648584.2648589

Higueras-Castillo, E., Liébana-Cabanillas, F. J., & Villarejo-Ramos, Á. F. (2023). Intention to use e-commerce vs physical shopping. difference between consumers in the post-COVID era. *Journal of Business Research, 157*, 113622. doi:10.1016/j.jbusres.2022.113622

Hijmans, H., & Raab, C. D. (2018 Forthcoming). *Ethical Dimensions of the GDPR. Commentary on the General Data Protection Regulation*. Edward Elgar.

Hildebrandt, M. (2015). *Smart technologies and the end (s) of law: novel entanglements of law and technology*. Edward Elgar Publishing. doi:10.4337/9781849808774

Hildebrandt, M., & Gutwirth, S. (2008). *Profiling the European Citizen*. Springer. doi:10.1007/978-1-4020-6914-7

Hill, J., Ford, W. R., & Farreras, I. G. (2015). Real conversations with artificial intelligence: A comparison between human-human online conversations and human-chatbot conversations. *Computers in Human Behavior, 49*, 245–250. doi:10.1016/j.chb.2015.02.026

Hills, T. T., Todd, P. M., Lazer, D., Redish, A. D., & Couzin, I. D. (2015). Exploration versus exploitation in space, mind, and society. *Trends in Cognitive Sciences, 19*(1), 46–54. doi:10.1016/j.tics.2014.10.004 PMID:25487706

Hinton, G. E., & Salakhutdinov, R. R. (2006). Reducing the dimensionality of data with neural networks. *Science*, *313*(5786), 504–507. doi:10.1126cience.1127647 PMID:16873662

Hirschman, E. C., & Holbrook, M. B. (1982). Hedonic Consumption: Emerging Concepts, Methods and Propositions. *Journal of Marketing*, *46*(3), 92–101. doi:10.1177/002224298204600314

Ho, A. (2020). Are we ready for artificial intelligence health monitoring in elder care? *BMC Geriatrics*, *20*(1), 358. doi:10.118612877-020-01764-9 PMID:32957946

Hochreiter, S., & Schmidhuber, J. (1997). Long short-term memory. *Neural Computation*, *9*(8), 1735–1780. doi:10.1162/neco.1997.9.8.1735 PMID:9377276

Höddinghaus, M., Sondern, D., & Hertel, G. (2021). The automation of leadership functions: Would people trust decision algorithms? *Computers in Human Behavior*, *116*, 106635. Advance online publication. doi:10.1016/j.chb.2020.106635

Hoens, T. R., Blanton, M., Steele, A., & Chawla, N. V. (2013). Reliable medical recommendation systems with patient privacy. [TIST]. *ACM Transactions on Intelligent Systems and Technology*, *4*(4), 1–31. doi:10.1145/2508037.2508048

Hoffman, D. L., & Novak, T. P. (2009). Flow online: Lessons learned and future prospects. *Journal of Interactive Marketing*, *23*(1), 23–34. doi:10.1016/j.intmar.2008.10.003

Hofree, G., & Winkielman, P. (2012). 13. On (not) knowing and feeling what we want and like. Handbook of Self-Knowledge, 210–224.

Hoftede, G., Hofstede, G. J., & Minkov, M. (2010). *Cultures and organizations: software of the mind: intercultural cooperation and its importance for survival*. McGraw-Hill.

Holland, P. W. (1986). Statistics and Causal Inference. *Journal of the American Statistical Association*, *81*(396), 945–960. doi:10.1080/01621459.1986.10478354

Holm, A. B., & Günzel-Jensen, F. (2017). Succeeding with freemium: Strategies for implementation. *The Journal of Business Strategy*, *38*(2), 16–24. doi:10.1108/JBS-09-2016-0096

Holzwarth, M., Janiszewski, C., & Neumann, M. M. (2006). The influence of avatars on online consumer shopping behavior. *Journal of Marketing*, *70*(4), 19–36. doi:10.1509/jmkg.70.4.019

Hong, W. C. H. (2023). The impact of ChatGPT on foreign language teaching and learning: Opportunities in education and research. *Journal of Educational Technology and Innovation*, *5*(1).

Horvitz, E. (2017). AI, people, and society. *Science*, *357*(6346), 7–7. doi:10.1126cience.aao2466 PMID:28684472

Hosanagar, K. (2020). *A Human's Guide to Machine Intelligence: How Algorithms are Shaping Our Lives and How We can Stay in Control*. Penguin.

Hosmer, D. W. Jr, Lemeshow, S., & Sturdivant, R. X. (2013). *Applied logistic regression* (Vol. 398). John Wiley & Sons. doi:10.1002/9781118548387

Hospedales, T., Antoniou, A., Micaelli, P., & Storkey, A. (2022). Meta-Learning in Neural Networks: A Survey. *IEEE Transactions on Pattern Analysis and Machine Intelligence*, *44*(9), 5149–5169. PMID:33974543

Hsieh. P. (2017). AI in medicine: rise of the machines. *Forbes*.

Huang, A. Q. (2019). Power semiconductor devices for smart grid and renewable energy systems. *Power Electronics in Renewable Energy Systems and Smart Grid*, 85–152. doi:10.1002/9781119515661.ch2

Huang, M. H., & Rust, R. T. (2018). Artificial intelligence in service. *Journal of Service Research*, *21*(2), 155û172.

Huang, M. H., & Rust, R. T. (2018). Artificial intelligence in service. *Journal of Service Research, 21*(2), 155–172. doi:10.1177/1094670517752459

Huang, M. H., & Rust, R. T. (2021). A strategic framework for artificial intelligence in marketing. *Journal of the Academy of Marketing Science, 49*(1), 30–50. doi:10.100711747-020-00749-9

Huang, M. H., Rust, R. T., & Maksimovic, V. (2019). The feeling economy: Managing in the next generation of artificial intelligence (AI). *California Management Review, 61*(4), 43–65. doi:10.1177/0008125619863436

Huarng, K. H. (2018). Entrepreneurship for long-term care in sharing economy. *The International Entrepreneurship and Management Journal, 14*(1), 97–104. doi:10.100711365-017-0460-9

Huarng, K. H., Yu, T. H. K., & Lee, C. F. (2022). Adoption model of healthcare wearable devices. *Technological Forecasting and Social Change, 174*, 121286. doi:10.1016/j.techfore.2021.121286

Humphreys, A., & Wang, R. J. H. (2018). Automated text analysis for consumer research. *The Journal of Consumer Research, 44*(6), 1274–1306. doi:10.1093/jcr/ucx104

Hu, N., Koh, N., & Reddy, S. (2014). Ratings lead you to the product, reviews help you clinch it? The mediating role of online review sentiments on product sales. *Decision Support Systems, 57*, 42–53. doi:10.1016/j.dss.2013.07.009

Huntington-Klein N. (2022). *The effect: an introduction to research design and causality.* Chapman and Hall/CRC.

Huntington-Klein, N. (2022). *The effect: An introduction to research design and causality.* Chapman and Hall/CRC Online Ebook Version. https://theeffectbook.net/index.html

Hutter, M. (2004). *Universal artificial intelligence: Sequential decisions based on algorithmic probability.* Spring Science & Business Media.

Hutter, F., Kotthoff, L., & Vanschoren, J. (2019). *Automated machine learning.* The Springer Series on Challenges in Machine Learning. doi:10.1007/978-3-030-05318-5

Hu, W., Wu, Q., Anvari-Moghaddam, A., Zhao, J., Xu, X., Abulanwar, S. M., & Cao, D. (2022). Applications of artificial intelligence in renewable energy systems. *IET Renewable Power Generation, 16*(7), 1279–1282. doi:10.1049/rpg2.12479

Hyder, A., & Nag, R. (2020). Artificial Intelligence Analysis of Marketing Scientific Literature: An Abstract. In *Academy of Marketing Science Annual Conference*. Springer International Publishing.

Hyder, A., & Nag, R. (2021). An Artificial Intelligence Method for the Analysis of Marketing Scientific Literature: An Abstract. In *Academy of Marketing Science Annual Conference-World Marketing Congress*. Springer International Publishing.

Hyder, A., Harris, K.-C., & Perez-Vidal, C. (2022). in Special Session: Marketing Science at the Service of Innovative Startups and Vice Versa: An Abstract. In *Academy of Marketing Science Annual Conference*. Springer Nature.

Iacobucci, D. (2016). *Marekting Management.* South Western.

IEEE Spectrum. (2023, January 25). *Human-level AI is right around the corner-or hundreds of years away.* IEEE Spectrum. https://spectrum.ieee.org/humanlevel-ai-is-right-around-the-corner-or-hundreds-of-years-away

IEEE Standards Association. (2017). *Ethically aligned design: A vision for prioritizing human well-being with autonomous and intelligent systems, Overview: Version 2.* IEEE.

IEEE. (2022). Ethically Aligned Design. *IEEE Ethics in Action in Autonomous and Intelligent Systems.* https://ethicsinaction.ieee.org/

Imbens, G. W. (2000). The role of the propensity score in estimating dose-response functions. *Biometrika*, *87*(3), 706–710. doi:10.1093/biomet/87.3.706

Imbens, G., & Rubin, D. (2015). *Causal Inference for Statistics, Social, and Biomedical Sciences: An Introduction.* Cambridge University Press. doi:10.1017/CBO9781139025751

Índice, I. E. S. E. (2022). *Cities in Motion.* IESE Business School.

Jackson, F. (1986). What Mary Didn't Know. *The Journal of Philosophy*, *83*(5), 291–295. doi:10.2307/2026143

Jacobsen, B. N. (2023). Machine learning and the politics of Synthetic Data. *Big Data & Society*, *10*(1), 205395172211453. doi:10.1177/20539517221145372

Jagusch, K., Sender, J., Jericho, D., & Flügge, W. (2021). Digital thread in shipbuilding as a prerequisite for the digital twin. *Procedia CIRP*, *104*, 318–323. doi:10.1016/j.procir.2021.11.054

Jain, H., Padmanabhan, B., Pavlou, P. A., & Raghu, T. S. (2021). Editorial for the special section on humans, algorithms, and augmented intelligence: The future of work, organizations, and society. *Information Systems Research*, *32*(3), 675–687. doi:10.1287/isre.2021.1046

Jamal, D. N., Rajkumar, S., & Ameen, N. (2018). Remote elderly health monitoring system using cloud-based wbans. In *Handbook of research on cloud and fog computing infrastructures for data science* (pp. 265–288). IGI Global. doi:10.4018/978-1-5225-5972-6.ch013

Jameson, J. L., & Longo, D. L. (2015). Precision medicine—Personalized, problematic, and promising. *Obstetrical & Gynecological Survey*, *70*(10), 612–614. doi:10.1097/01.ogx.0000472121.21647.38 PMID:26014593

Janghel, R. R., Shukla, A., Tiwari, R., & Kala, R. (2010). Breast cancer diagnosis using artificial neural network models. In *The 3rd International Conference on Information Sciences and Interaction Sciences* (pp. 89-94). IEEE. 10.1109/ICICIS.2010.5534716

Japkowicz, N., & Shah, M. (2011). *Evaluating learning algorithms: a classification perspective.* Cambridge University Press. doi:10.1017/CBO9780511921803

Jaskowski, M., & Jaroszewicz, S. (2012). Uplift modeling for clinical trial data. In *ICML Workshop on Clinical Data Analysis* (Vol. 46, pp. 79-95).

Jeník, I., & Duff, S. (2020). *How to Build a Regulatory Sandbox.*

Jeyakumar, K. (2022). A Comprehensive Survey on Robotics and Automation in Various Industries. *Proceedings of the Seventh International Conference on Communication and Electronics Systems (ICCES 2022).*

Jiao, W., Wang, W., Huang, J. T., Wang, X., & Tu, Z. (2023). *Is ChatGPT a good translator? A preliminary study.* arXiv preprint arXiv:2301.08745.

Jiménez Barreto, J., Natalia, R., & Sebastian, M. (2021). "Find a flight for me, Oscar!" Motivational customer experiences with chatbots. *International Journal of Contemporary Hospitality Management*, *33*(11), 3860–3882. doi:10.1108/IJCHM-10-2020-1244

Jiménez López, E., Cuenca Jiménez, F., Luna Sandoval, G., Ochoa Estrella, F. J., Maciel Monteón, M. A., Muñoz, F., & Limón Leyva, P. A. (2022). Technical Considerations for the Conformation of Specific Competences in Mechatronic Engineers in the Context of Industry 4.0 and 5.0. *Processes (Basel, Switzerland)*, *10*(8), 1445. doi:10.3390/pr10081445

Joachims, T., London, B., Su, Y., Swaminathan, A., & Wang, L. (2021). Recommendations as treatments. *AI Magazine*, *42*(3), 19–30. doi:10.1609/aimag.v42i3.18141

Jordan, M. I., & Mitchell, T. M. (2015). Machine learning: Trends, perspectives, and prospects. *Science*, *349*(6245), 255–260. doi:10.1126cience.aaa8415 PMID:26185243

Jost, J. T. (2019). A quarter century of system justification theory: Questions, answers, criticisms, and societal applications. *British Journal of Social Psychology*, *58*(2), 263–314. doi:10.1111/bjso.12297

Kaddour, J., Lynch, A., Liu, Q., Kusner, M. J., & Silva, R. (2022). (Manuscript submitted for publication). Causal machine learning: A survey and open problems. *arXiv preprint arXiv:2206.15475. Work (Reading, Mass.).* doi:10.48550/arXiv.2206.15475

Kahneman, D., & Tversky, A. (1979). Prospect Theory: An Analysis of Decision under Risk. *Econometrica*, *47*(2), 263–292. doi:10.2307/1914185

Kakaria, S., Bigne, E., Catambrone, V., & Valenza, G. (2022). Heart rate variability in marketing research: A systematic review and methodological perspectives. *Psychology and Marketing*, *40*(1), 190–208. doi:10.1002/mar.21734

Kallus, N. (2017). Recursive partitioning for personalization using observational data. *International Conference on Machine Learning*, (pp. 1789–1798). IEEE.

Kalogirou, S. A., Mathioulakis, E., & Belessiotis, V. (2014). Artificial neural networks for the performance prediction of large solar systems. *Renewable Energy*, *63*, 90–97. doi:10.1016/j.renene.2013.08.049

Kamboj, S., & Gupta, S. (2020). Use of smart phone apps in co-creative hotel service innovation: An evidence from India. *Current Issues in Tourism*, *23*(3), 323–344. doi:10.1080/13683500.2018.1513459

Kamenica, E. (2019). Bayesian persuasion and information design. *Annual Review of Economics*, *11*(1), 249–272. doi:10.1146/annurev-economics-080218-025739

Kaminskas, M., & Bridge, D. (2016). Diversity, serendipity, novelty, and coverage: A survey and empirical analysis of beyond-accuracy objectives in recommender systems. *ACM Transactions on Interactive Intelligent Systems*, *7*(1), 1–42. doi:10.1145/2926720

Kane, G. C., Young, A. G., Majchrzak, A., & Ransbotham, S. (2021). Avoiding an Oppressive Future of Machine Learning: A Design Theory for Emancipatory Assistants. *Management Information Systems Quarterly*, *45*(1), 371–396. doi:10.25300/MISQ/2021/1578

Kane, R. (2001). *Free will.* John Wiley & Sons.

Kang, M., Ragan, B. G., & Park, J. H. (2008). Issues in outcomes research: An overview of randomization techniques for clinical trials. *Journal of Athletic Training*, *43*(2), 215–221. doi:10.4085/1062-6050-43.2.215 PMID:18345348

Kant, I. (1948). *Groundwork of the Metaphysics of Morals* (H. J. Paton, Trans.). Hutchinson.

Kaplan, A. (2022). *Artificial Intelligence, Business and Civilization: Our Fate Made in Machines*. Routledge. doi:10.4324/9781003244554

Karro, J., Dent, A. W., & Farish, S. (2005). Patient perceptions of privacy infringements in an emergency department. *Emergency Medicine Australasia*, *17*(2), 117–123. doi:10.1111/j.1742-6723.2005.00702.x PMID:15796725

Kartajaya, H., Setiawan, I., & Kotler, P. (2021). *Marketing 5.0: Technology for humanity*. John Wiley & Sons.

Karthikeyan, P., Manikandakumar, M., Sri Subarnaa, D. K., & Priyadharshini, P. (2021). Weed identification in agriculture field through IoT. In *Advances in Intelligent Systems and Computing* (pp. 495–505). Springer. doi:10.1007/978-981-15-5029-4_41

Kashef, M., Visvizi, A., & Troisi, O. (2021). Smart city as a smart service system: Human-computer interaction and smart city surveillance systems. *Computers in Human Behavior, 124*, 106923. Advance online publication. doi:10.1016/j.chb.2021.106923

Kasneci, E., Seßler, K., Küchemann, S., Bannert, M., Dementieva, D., Fischer, F., Gasser, U., Groh, G., Günnemann, S., Hüllermeier, E., Krusche, S., Kutyniok, G., Michaeli, T., Nerdel, C., Pfeffer, J., Poquet, O., Sailer, M., Schmidt, A., Seidel, T., & Kasneci, G. (2023). ChatGPT for good? On opportunities and challenges of large language models for education. *Learning and Individual Differences, 103*, 102274. doi:10.1016/j.lindif.2023.102274

Katznelson, G., & Gerke, S. (2021). The need for health AI ethics in medical school education. *Advances in Health Sciences Education : Theory and Practice, 26*(4), 1447–1458. doi:10.100710459-021-10040-3 PMID:33655433

Kaushal, V., & Yadav, R. (2022). Learning successful implementation of chatbots in businesses from B2B customer experience perspective. *Concurrency and Computation, 35*(1), e7450. doi:10.1002/cpe.7450

Kaye, J. A., Maxwell, S. A., Mattek, N., Hayes, T. L., Dodge, H., Pavel, M., Jimison, H. B., Wild, K., Boise, L., & Zitzelberger, T. A. (2011). Intelligent systems for assessing aging changes: Home-based, unobtrusive, and continuous assessment of aging. *The Journals of Gerontology. Series B, Psychological Sciences and Social Sciences, 66*(Suppl. 1), i180–i190. doi:10.1093/geronb/gbq095 PMID:21743050

Keding, C. (2021). Understanding the interplay of artificial intelligence and strategic management: Four decades of research in review. *Management Review Quarterly, 71*(1), 91–134. doi:10.100711301-020-00181-x

Kelleher, J. D., Mac Namee, B., & D'arcy, A. (2020). *Fundamentals of machine learning for predictive data analytics: algorithms, worked examples, and case studies*. MIT press.

Kelly, S. M. (2023). ChatGPT passes exams from law and business schools. *CNN Business*, 26.

Kennedy, E. H., Ma, Z., McHugh, M. D., & Small, D. S. (2017). Nonparametric methods for doubly robust estimation of continuous treatment effects. *Journal of the Royal Statistical Society. Series B, Statistical Methodology, 79*(4), 1229–1245. doi:10.1111/rssb.12212 PMID:28989320

Kent, W. (2012). *Data and Reality: A Timeless Perspective on Perceiving and Managing Information*. Technics publications.

Ker, J., Wang, L., Rao, J., & Lim, T. (2018). Deep learning applications in medical image analysis. *IEEE Access : Practical Innovations, Open Solutions, 6*, 9375–9389. doi:10.1109/ACCESS.2017.2788044

Khodadadi, A., Ghandiparsi, S., & Chuah, C.-N. (2022). A natural language processing and deep learning based model for Automated Vehicle Diagnostics using free-text Customer Service Reports. *Machine Learning with Applications, 10*, 100424. doi:10.1016/j.mlwa.2022.100424

Khowaja, S. A., Khuwaja, P., & Dev, K. (2023). ChatGPT Needs SPADE (Sustainability et al. divide, and Ethics) Evaluation: A Review. arXiv preprint arXiv:2305.03123.

Kim, E., Huang, K., Saunders, A., McCallum, A., Ceder, G., & Olivetti, E. (2017). Materials synthesis insights from scientific literature via text extraction and machine learning. *Chemistry of Materials, 29*(21), 9436–9444. doi:10.1021/acs.chemmater.7b03500

Kim, J. M., Jun, M., & Kim, C. K. (2018). The effects of culture on consumers' consumption and generation of online reviews. *Journal of Interactive Marketing, 43*, 134–150. doi:10.1016/j.intmar.2018.05.002

Kim, Y. J., & Kim, H. S. (2022). The impact of hotel customer experience on customer satisfaction through online reviews. *Sustainability (Basel), 14*(2), 848. doi:10.3390u14020848

Kissinger, H. A., Schmidt, E., & Huttenlocher, D. (2022). *The Age of AI: And Our Human Future*. Back Bay Books.

Klette, R. (2014). *Concise computer vision* (Vol. 233). Springer. doi:10.1007/978-1-4471-6320-6

Knani, M., Echchakoui, S., & Ladhari, R. (2022). Artificial intelligence in tourism and hospitality: Bibliometric analysis and research agenda. *International Journal of Hospitality Management*, *107*, 103317. doi:10.1016/j.ijhm.2022.103317

Koh, N., Hu, N., & Clemons, E. (2010). Do online reviews reflect a product's true perceived quality? An investigation of online movie reviews across cultures. *Electronic Commerce Research and Applications*, *9*(5), 374–385. doi:10.1016/j.elerap.2010.04.001

Kohonen, T. (1982). Self-organized formation of topologically correct feature maps. *Biological Cybernetics*, *43*(1), 59–69. doi:10.1007/BF00337288

Kohonen, T. (2001). *Self-Organizing Maps* (3rd ed.). Springer-Verlag. doi:10.1007/978-3-642-56927-2

Korsgaard, C., Cohen, G. A., Geuss, R., Nagel, T., & Williams, B. (1996). *The Sources of Normativity* (O. O'Neill, Ed.). Cambridge University Press. doi:10.1017/CBO9780511554476

Koster, R., Balaguer, J., Tacchetti, A., Weinstein, A., Zhu, T., Hauser, O., Williams, D., Campbell-Gillingham, L., Thacker, P., Botvinick, M., & Summerfield, C. (2022). Human-centred mechanism design with Democratic AI. *Nature Human Behaviour*, *6*(10), 1398–1407. doi:10.103841562-022-01383-x PMID:35789321

Kotkov, D., Wang, S., & Veijalainen, J. (2016). A survey of serendipity in recommender systems. *Knowledge-Based Systems*, *111*, 180–192. doi:10.1016/j.knosys.2016.08.014

Kretschmer, T., & Khashabi, P. (2020). Digital transformation and organization design: An integrated approach. *California Management Review*, *62*(4), 86–104. doi:10.1177/0008125620940296

Krizhevsky, A., Sutskever, I., & Hinton, G. E. (2012). Imagenet classification with deep convolutional neural networks. *Advances in Neural Information Processing Systems*, *25*, 1097–1105.

Krogh, A. (2008). What are artificial neural networks? *Nature Biotechnology*, *26*(2), 195–197. doi:10.1038/nbt1386 PMID:18259176

Kroll, J., Barocas, S., Felten, E., Reidenberg, J., Robinson, D., & Yu, H. (2017). Accountable Algorithms. *University of Pennsylvania Law Review*, *165*, 633–705.

Kruskal, J. (1964). Multidimensional scaling by optimizing goodness of fit to a nonmetric hypothesis. *Psychometrika*, *29*(1), 1–27. doi:10.1007/BF02289565

Krys, K., Vignoles, V. L., De Almeida, I., & Uchida, Y. (2022). Outside the "cultural binary": Understanding why Latin American collectivist societies foster independent selves. *Perspectives on Psychological Science*, *17*(4), 1166–1187. doi:10.1177/17456916211029632 PMID:35133909

Kumar, A., & Sachdeva, N. (2022). Multimodal cyberbullying detection using capsule network with dynamic routing and deep convolutional neural network. *Multimedia Systems*, *28*(6), 1–10. doi:10.100700530-020-00747-5

Kumari, M., & Khan, R. (2018). Review Paper on Blue Brain Technology. *International Journal of Computer Applications*, *180*(48), 21–23. doi:10.5120/ijca2018917256

Kumari, P., & Sangeetha, R. (2022). How Does Electronic Word of Mouth Impact Green Hotel Booking Intention? *Services Marketing Quarterly*, *43*(2), 146–165. doi:10.1080/15332969.2021.1987609

Kumar, V. (2017). Integrating theory and practice in marketing. *Journal of Marketing*, *81*(2), 1–7. doi:10.1509/jm.80.2.1

Kumar, V., Rajan, B., Venkatesan, R., & Lecinski, J. (2019). Understanding the role of artificial intelligence in personalised engagement marketing. *California Management Review*, *61*(4), 135–155. doi:10.1177/0008125619859317

Kung, T. H., Cheatham, M., Medenilla, A., Sillos, C., De Leon, L., Elepaño, C., Madriaga, M., Aggabao, R., Diaz-Candido, G., Maningo, J., & Tseng, V. (2023). Performance of ChatGPT on USMLE: Potential for AI-assisted medical education using large language models. *PLoS Digital Health*, *2*(2), e0000198. doi:10.1371/journal.pdig.0000198 PMID:36812645

Künzel, S. R., Sekhon, J. S., Bickel, P. J., & Yu, B. (2019). Metalearners for estimating heterogeneous treatment effects using machine learning. *Proceedings of the National Academy of Sciences of the United States of America*, *116*(10), 4156–4165. doi:10.1073/pnas.1804597116 PMID:30770453

Kuo, N. W., & Hsiao, T. Y. (2008). An exploratory research of the application of natural capitalism to sustainable tourism management in Taiwan. *Journal of Cleaner Production*, *16*(1), 116–124. doi:10.1016/j.jclepro.2006.11.005

Kuo, R., Ho, L., & Hu, C. (2002). Integration of self-organizing feature map and K-means algorithm for market segmentation. *Computers & Operations Research*, *29*(11), 1475–1493. doi:10.1016/S0305-0548(01)00043-0

Kurzweil, R. (2006). *The singularity is near*. Gerald Duckworth and Co Ltd.

Lam, K., Iqbal, F. M., Purkayastha, S., & Kinross, J. M. (2021). Investigating the Ethical and Data Governance Issues of Artificial Intelligence in Surgery: Protocol for a Delphi Study. *JMIR Research Protocols*, *10*(2), e26552. doi:10.2196/26552 PMID:33616543

Langer, M., & Landers, R. N. (2021). The future of artificial intelligence at work: A review on effects of decision automation and augmentation on workers targeted by algorithms and third-party observers. *Computers in Human Behavior*, *123*, 106878. Advance online publication. doi:10.1016/j.chb.2021.106878

Langley, P., & Leyshon, A. (2017). Platform capitalism: The intermediation and capitalization of digital economic circulation. *Finance and Stochastics*, *3*(1), 11–31.

Langmann, C., & Turi, D. (2023). *Robotic Process Automation (RPA) - Digitization and Automation of Processes: Prerequisites, functionality and implementation using accounting as an example*. Springer Gabler.

Laso, I. (2020): *#EUvsVirus interview*. European Commission DG Research and Innovation. https://www.facebook.com/EUvsVirus/videos/2635593133382903

Lateef, A. A. A., Ali Al-Janabi, S. I., & Abdulteef, O. A. (2022, July). Artificial Intelligence Techniques Applied on Renewable Energy Systems: A Review. In *Proceedings of International Conference on Computing and Communication Networks: ICCCN 2021* (pp. 297-308). Springer Nature Singapore. 10.1007/978-981-19-0604-6_25

Latham, A., & Goltz, S. (2019). A Survey of the General Public's Views on the Ethics of Using AI in Education. In Artificial Intelligence in Education: 20th International Conference, AIED 2019, Chicago, IL, USA, June 25-29, 2019 [Springer International Publishing.]. *Proceedings*, *20*(Part I), 194–206.

Laufer, R. S., & Wolfe, M. (1977). Privacy as a concept and a social issue: A multidimensional developmental theory. *The Journal of Social Issues*, *33*(3), 22–42. doi:10.1111/j.1540-4560.1977.tb01880.x

Lechner, M. (2011). The estimation of causal effects by difference-in-difference methods. *Foundations and Trends® in Econometrics*, *4*(3), 165–224. doi:10.1561/0800000014

Lechner, M. (2023). Causal Machine Learning and its use for public policy. *Schweizerische Zeitschrift für Volkswirtschaft und Statistik*, *159*(1), 1–15. doi:10.118641937-023-00113-y

LeCun, Y., Bengio, Y., & Hinton, G. (2015). Deep learning. *Nature*, *521*(7553), 436–444. doi:10.1038/nature14539 PMID:26017442

Lee, J. S., Hsu, L. T., Han, H., & Kim, Y. (2010). Understanding how consumers view green hotels: How a hotel's green image can influence behavioural intentions. *Journal of Sustainable Tourism*, *18*(7), 901–914. doi:10.1080/09669581003777747

Lee, P., Goldberg, C., & Kohane, I. (2023). *The AI Revolution in Medicine: GPT-4 and Beyond.* Pearson.

Lee, S. H., Yoon, S. H., & Kim, H. W. (2021). Prediction of online video advertising inventory based on TV programs: A deep learning approach. *IEEE Access : Practical Innovations, Open Solutions*, *9*, 22516–22527. doi:10.1109/ACCESS.2021.3056115

Lee, S., & Choi, J. (2017). Enhancing user experience with conversational agent for movie recommendation: Effects of self-disclosure and reciprocity. *International Journal of Human-Computer Studies*, *103*, 95–105. doi:10.1016/j.ijhcs.2017.02.005

Lee, Y., & Kozar, K. A. (2009). Designing usable online stores: A landscape preference perspective. *Information & Management*, *46*(1), 31–41. doi:10.1016/j.im.2008.11.002

Lefranc, G., López, I., Osorio-Comparán, R., & Peña, M. (2022). Cobots in automation and at home. *2022 IEEE International Conference on Automation/XXV Congress of the Chilean Association of Automatic Control (ICA-ACCA),* 1-6. doi: 10.1109/ICA-ACCA56767.2022.10006164

Lehrer, K. (1997). Freedom, preference and autonomy. *The Journal of Ethics*, *1*(1), 3–25. doi:10.1023/A:1009744817791

Leidner, D. E., & Tona, O. (2021). The CARE Theory of Dignity Amid Personal Data Digitalization. *Management Information Systems Quarterly*, *45*(1), 343–370. doi:10.25300/MISQ/2021/15941

Leiva, L. A., Arapakis, I., & Iordanou, C. (2021). My mouse, my rules: Privacy issues of behavioral user profiling via mouse tracking. *Proceedings of the 2021 Conference on Human Information Interaction and Retrieval*, (pp. 51–61). IEEE. 10.1145/3406522.3446011

Lemon, K. N., & Verhoef, P. C. (2016). Understanding customer experience throughout the customer journey. *Journal of Marketing*, *80*(6), 69–96. doi:10.1509/jm.15.0420

Lennox-Chhugani, N., Chen, Y., Pearson, V., Trzcinski, B., & James, J. (2021). Women's attitudes to the use of AI image readers: A case study from a national breast screening programme. *BMJ Health & Care Informatics*, *28*(1), e100293. doi:10.1136/bmjhci-2020-100293 PMID:33795236

Leon, R. D. (2019). Hotel's online reviews and ratings: A cross-cultural approach. *International Journal of Contemporary Hospitality Management*, *31*(5), 2054–2073. doi:10.1108/IJCHM-05-2018-0413

Lerner, A. P. (1972). The economics and politics of consumer sovereignty. *The American Economic Review*, *62*(1/2), 258–266.

Levine, S. (2021). Understanding the World Through Action. *ArXiv Preprint ArXiv:2110.12543.*

Lewis, D. (1973). Causation. *The Journal of Philosophy*, *70*(17), 556–567. doi:10.2307/2025310

Leyer, M., & Schneider, S. (2021). Decision augmentation and automation with artificial intelligence: Threat or opportunity for managers? *Business Horizons*, *64*(5), 711–724. doi:10.1016/j.bushor.2021.02.026

Liang, T.-P., Lai, H.-J., & Ku, Y.-C. (2006). Personalized content recommendation and user satisfaction: Theoretical synthesis and empirical findings. *Journal of Management Information Systems*, *23*(3), 45–70. doi:10.2753/MIS0742-1222230303

Liang, T., Robert, L., Arbor, A., Sarker, S., Cheung, C. M. K., Matt, C., Trenz, M., & Turel, O. (2021). Artificial intelligence and robots in individuals' lives: How to align technological possibilities and ethical issues. *Internet Research, 31*(1), 1–10. doi:10.1108/INTR-11-2020-0668

Liebman, E., Saar-Tsechansky, M., & Stone, P. (2019). The right music at the right time: Adaptive personalised playlists based on sequence modeling. *Management Information Systems Quarterly, 43*(3), 765–786. doi:10.25300/MISQ/2019/14750

Li, J., Ye, Z., & Zhang, C. (2022). Study on the interaction between Big Data and artificial intelligence. *Systems Research and Behavioral Science, 39*(3), 641–648. doi:10.1002res.2878

Lilienfeld, S. O., Ammirati, R., & Landfield, K. (2009). Giving debiasing away: Can psychological research on correcting cognitive errors promote human welfare? *Perspectives on Psychological Science, 4*(4), 390–398. doi:10.1111/j.1745-6924.2009.01144.x PMID:26158987

Lilien, G. L. (2011). Bridging the academic–practitioner divide in marketing decision models. *Journal of Marketing, 75*(4), 196–210. doi:10.1509/jmkg.75.4.196

Lim, W. M., Gunasekara, A., Pallant, J. L., Pallant, J. I., & Pechenkina, E. (2023). Generative AI and the future of education: Ragnarök or reformation? A paradoxical perspective from management educators. *International Journal of Management Education, 21*(2), 100790. doi:10.1016/j.ijme.2023.100790

Lim, W. M., Kumar, S., Verma, S., & Chaturvedi, R. (2022). Alexa, what do we know about conversational commerce? Insights from a systematic literature review. *Psychology and Marketing, 39*(6), 1129–1155. doi:10.1002/mar.21654

Lindblom, C. E. (1959). The Science of "Muddling Through." *Public Administraation Review*, 79–88.

Linde, Y., Buzo, A., & Gray, R. (1980). An algorithm for vector quantizer design. *IEEE Transactions on Communications, 28*(1), 84–95. doi:10.1109/TCOM.1980.1094577

Lindström, B., Bellander, M., Schultner, D. T., Chang, A., Tobler, P. N., & Amodio, D. M. (2021). A computational reward learning account of social media engagement. *Nature Communications, 12*(1), 1–10. doi:10.1038/s41467-020-19607-x

Lin, M. P., Marine-Roig, E., & Llonch-Molina, N. (2022). Gastronomy tourism and well-being: Evidence from Taiwan and Catalonia Michelin-starred restaurants. *International Journal of Environmental Research and Public Health, 19*(5), 2778. doi:10.3390/ijerph19052778 PMID:35270469

Littman, M. L. (2015). Reinforcement learning improves behaviour from evaluative feedback. *Nature, 521*(7553), 445–451. doi:10.1038/nature14540 PMID:26017443

Liu, R., Gupta, S., & Patel, P. (2021). The application of the principles of responsible AI on social media marketing for digital health. *Information Systems Frontiers*, 1–25. doi:10.100710796-021-10191-z PMID:34539226

Liu, X., Singh, P. V., & Srinivasan, K. (2016). A structured analysis of unstructured big data by leveraging cloud computing. *Marketing Science, 35*(3), 363–388. doi:10.1287/mksc.2015.0972

Liu, Y., Ram, S., Lusch, R. F., & Brusco, M. (2010). Multicriterion market segmentation: A new model, implementation, and evaluation. *Marketing Science, 29*(5), 880–894. doi:10.1287/mksc.1100.0565

Li, X., Shi, M., & Wang, X. S. (2019). Video mining: Measuring visual information using automatic methods. *International Journal of Research in Marketing, 36*(2), 216–231. doi:10.1016/j.ijresmar.2019.02.004

Lo, J. Y., & Floyd, C. E. (1999). Application of artificial neural networks for diagnosis of breast cancer. In *Proceedings of the 1999 Congress on Evolutionary Computation-CEC99 (Cat. No. 99TH8406)* (*Vol. 3*, pp. 1755-1759). IEEE. 10.1109/CEC.1999.785486

Lobschat, L., Mueller, B., Eggers, F., Brandimarte, L., Diefenbach, S., Kroschke, M., & Wirtz, J. (2021). Corporate digital responsibility. *Journal of Business Research, 122*, 875–888. doi:10.1016/j.jbusres.2019.10.006

Lojo, A., Li, M., & Xu, H. (2020). Online tourism destination image: Components, information sources, and incongruence. *Journal of Travel & Tourism Marketing, 37*(4), 495–509. doi:10.1080/10548408.2020.1785370

Lorenz-Spreen, P., Lewandowsky, S., Sunstein, C. R., & Hertwig, R. (2020). How behavioural sciences can promote truth, autonomy and democratic discourse online. *Nature Human Behaviour, 4*(11), 1102–1109. doi:10.103841562-020-0889-7 PMID:32541771

Lotan, E., Tschider, C., Sodickson, D. K., Caplan, A. L., Bruno, M., Zhang, B., & Lui, Y. W. (2020). Medical imaging and privacy in the era of artificial intelligence: Myth, fallacy, and the future. *Journal of the American College of Radiology, 17*(9), 1159–1162. doi:10.1016/j.jacr.2020.04.007 PMID:32360449

Lou, C., Kang, H., & Tse, C. H. (2021). Bots vs. humans: How schema congruity, contingency-based interactivity, and sympathy influence consumer perceptions and patronage intentions. *International Journal of Advertising, 41*(4), 655–684. doi:10.1080/02650487.2021.1951510

Loyola-González, O., Monroy, R., Rodríguez, J., López-Cuevas, A., & Mata-Sánchez, J. I. (2019). Contrast pattern-based classification for bot detection on twitter. *IEEE Access : Practical Innovations, Open Solutions, 7*, 45800–45817. doi:10.1109/ACCESS.2019.2904220

Luca M., & Bazerman, M. H. (2020). *The Power of Experiments: Decision Making in a Data-Driven World.* MIT Press.

Ludwig, S., De Ruyter, K., Friedman, M., Brüggen, E., Wetzels, M., & Pfann, G. (2013). More than words: The influence of affective content and linguistic style matches in online reviews on conversion rates. *Journal of Marketing, 77*(1), 87–103. doi:10.1509/jm.11.0560

Lukkien, D. R., Nap, H. H., Buimer, H. P., Peine, A., Boon, W. P., Ket, J. C., Minkman, M. M. N., & Moors, E. H. (2023). Toward responsible artificial intelligence in long-term care: A scoping review on practical approaches. *The Gerontologist, 63*(1), 155–168. doi:10.1093/geront/gnab180 PMID:34871399

Maaten, L. v., & Hinton, G. (2008). Visualizing Data using t-SNE. *Journal of Machine Learning Research, 9*, 2579–2605.

Macdonald, P. S. (2017). *History of the concept of mind: Volume 1: Speculations about soul, mind and spirit from homer to hume.* Routledge. doi:10.4324/9781315092959

MacIntyre, A. (1984). *After Virtue.* University of Notre Dame Press.

MacnishK.RyanM.GregoryA.JiyaT.AntoniouJ.HatzakisT.. (2019). D1.1 Case studies. *De Montfort University.* doi:10.21253/DMU.7679690.v3

Maddikunta, P. K. R., Pham, Q. V., Prabadevi, B., Deepa, N., Dev, K., Gadekallu, T. R., ... Liyanage, M. (2022). Industry 5.0: A survey on enabling technologies and potential applications. *Journal of Industrial Information Integration, 26*, 100257. Advance online publication. doi:10.1016/j.jii.2021.100257

Magistretti, S., Pham, C. T. A., & Dell'Era, C. (2021). Enlightening the dynamic capabilities of design thinking in fostering digital transformation. *Industrial Marketing Management, 97*, 59–70. doi:10.1016/j.indmarman.2021.06.014

Mahieu, R. (2021). The right of access to personal data: A genealogy. *Technology and Regulation*, 62–75.

Mahmood, T., & Ricci, F. (2007). Learning and adaptivity in interactive recommender systems. *Proceedings of the Ninth International Conference on Electronic Commerce*, (pp. 75–84). IEEE. 10.1145/1282100.1282114

Malik, P. K., Sharma, R., Singh, R., Gehlot, A., Satapathy, S. C., Alnumay, W. S., Pelusi, D., Ghosh, U., & Nayak, J. (2021). Industrial Internet of Things and its applications in industry 4.0: State of the art. *Computer Communications*, *166*, 125–139. doi:10.1016/j.comcom.2020.11.016

Malinen, S., & Nurkka, P. (2015). Cultural influence on online community use: A cross–cultural study on online exercise diary users of three nationalities. *International Journal of Web Based Communities*, *11*(2), 153–169. doi:10.1504/IJWBC.2015.068539

Manders-Huits, N. (2010). Practical versus moral identities in identity management. *Ethics and Information Technology*, *12*(1), 43–55. doi:10.100710676-010-9216-8

Manheim, D., & Garrabrant, S. (2018). Categorizing variants of Goodhart's Law. *ArXiv Preprint ArXiv:1803.04585*.

Manikandakumar, M., & Karthikeyan, P. (2022). Essentials, Challenges, and Future Directions of Agricultural IoT: A Case Study in the Indian Perspective. *Advances in Web Technologies and Engineering*, 181–196. doi:10.4018/978-1-7998-4186-9.ch010

Manikandakumar, M., & Karthikeyan, P. (2023). Weed classification using particle swarm optimization and deep learning models. *Computer Systems Science and Engineering*, *44*(1), 913–927. doi:10.32604/csse.2023.025434

Manikandakumar, M., & Ramanujam, E. (2018). Security and privacy challenges in big data environment. *Advances in Information Security, Privacy, and Ethics*, 315–325. doi:10.4018/978-1-5225-4100-4.ch017

Manivannan, P., Prabha, D., & Balasubramanian, K. (2022). Artificial Intelligence Databases: Turn-on big data of the SMBS. *International Journal of Business Information Systems*, *39*(1), 1. doi:10.1504/IJBIS.2022.120367

Marblestone, A. H., Wayne, G., & Kording, K. P. (2016). Toward an integration of deep learning and neuroscience. *Frontiers in Computational Neuroscience*, *10*, 94. doi:10.3389/fncom.2016.00094 PMID:27683554

March, J. G. (1991). Exploration and exploitation in organizational learning. *Organization Science*, *2*(1), 71–87. doi:10.1287/orsc.2.1.71

Mariani, M. M., Hashemi, N., & Wirtz, J. (2023). Artificial intelligence empowered conversational agents: A systematic literature review and research agenda. *Journal of Business Research*, *161*, 113838. doi:10.1016/j.jbusres.2023.113838

Mariani, M. M., Perez-Vega, R., & Wirtz, J. (2022). AI in marketing, consumer research and psychology: A systematic literature review and research agenda. *Psychology and Marketing*, *39*(4), 755–776. doi:10.1002/mar.21619

Marine-Roig, E. (2019). Destination image analytics through traveller-generated content. *Sustainability (Basel)*, *11*(12), 3392. doi:10.3390u11123392

Markauskaite, L., Marrone, R., Poquet, O., Knight, S., Martinez-Maldonado, R., Howard, S., Tondeur, J., De Laat, M., Buckingham Shum, S., Gašević, D., & Siemens, G. (2022). Rethinking the entwinement between artificial intelligence and human learning: What capabilities do learners need for a world with AI? *Computers and Education: Artificial Intelligence*, *3*, 100056. Advance online publication. doi:10.1016/j.caeai.2022.100056

Markram, H. (2013). Seven challenges for neuroscience. *Functional Neurology*, *28*(3), 145–151. doi:10.11138/FNeur/2013.28.3.144 PMID:24139651

Markus, H., & Wurf, E. (1987). The dynamic self-concept: A social psychological perspective. *Annual Review of Psychology*, *38*(1), 299–337. doi:10.1146/annurev.ps.38.020187.001503

Marques, N., Silva, B., & Santos, H. (2016). An Interactive Interface for Multi-dimensional Data Stream Analysis. *2016 20th International Conference Information Visualisation (IV)*. IEEE.

Marr, B. (2017). The 4 Ds of robotization: dull, dirty, dangerous and dear. *Enterprise Tech, Forbes Innovation*. https://www.forbes.com/sites/bernardmarr/2017/10/16/the-4-ds-of-robotization-dull-dirty-dangerous-and-dear/

Martens, D., Provost, F., Clark, J., & de Fortuny, E. J. (2016). Mining massive fine-grained behavior data to improve predictive analytics. *Management Information Systems Quarterly*, *40*(4), 869–888. doi:10.25300/MISQ/2016/40.4.04

Martínez, P. (2015). Customer loyalty: Exploring its antecedents from a green marketing perspective. *International Journal of Contemporary Hospitality Management*, *27*(5), 896–917. doi:10.1108/IJCHM-03-2014-0115

Martin-Rodilla, P., Pereira-Fariña, M., & Gonzalez-Perez, C. (2019). Qualifying and quantifying uncertainty in Digital Humanities. *Proceedings of the Seventh International Conference on Technological Ecosystems for Enhancing Multiculturality*. ACM. 10.1145/3362789.3362833

Masís S. (2021). *Interpretable Machine Learning with Python: Learn to build interpretable high-performance models with hands-on real-world examples*. Packt Publishing.

Mathur, A., Acar, G., Friedman, M. J., Lucherini, E., Mayer, J., Chetty, M., & Narayanan, A. (2019). Dark patterns at scale: Findings from a crawl of 11K shopping websites. *Proceedings of the ACM on Human-Computer Interaction, 3*(CSCW), (pp. 1–32). IEEE. 10.1145/3359183

McAdams, D., & McLean, K. (2013). Narrative Identity. *Current Directions in Psychological Science*, *22*(3), 233–238. doi:10.1177/0963721413475622

McAuley, J. (2022). *Personalized Machine Learning*. Cambridge University Press. doi:10.1017/9781009003971

McCarthy, J. (1959). "Programs with Common Sense" at the Way back Machine (archived October 4, 2013). *Proceedings of the Teddington Conference on the Mechanization of Thought Processes,* 756-91.

McDuff, D., & Czerwinski, M. (2018). Designing emotionally sentient agents. *Communications of the ACM*, *61*(12), 74–83. doi:10.1145/3186591

McLuhan, M., & Fione, Q. (1968). *War and peace in the global village*. Bantam Books.

Medvedev, I., Gordon, T., & Wu, H. (2019, November 25). *Powered by AI: Instagram's Explore recommender system.* Meta AI Blog. https://ai.facebook.com/blog/powered-by-ai-instagrams-explore-recommender-system/

Mehraliyev, F., Chan, I. C. C., & Kirilenko, A. P. (2022). Sentiment analysis in hospitality and tourism: A thematic and methodological review. *International Journal of Contemporary Hospitality Management*, *34*(1), 46–77. doi:10.1108/IJCHM-02-2021-0132

Meissner, F., Grigutsch, L. A., Koranyi, N., Müller, F., & Rothermund, K. (2019). Predicting behavior with implicit measures: Disillusioning findings, reasonable explanations, and sophisticated solutions. *Frontiers in Psychology*, *10*, 2483. doi:10.3389/fpsyg.2019.02483 PMID:31787912

Melville, N. P., Robert, L., & Xiao, X. (2023). Putting humans back in the loop: An affordance conceptualization of the 4th industrial revolution. *Information Systems Journal*, *33*(4), 733–757. doi:10.1111/isj.12422

Menczer, F. (2020). 4 Reasons Why Social Media Make Us Vulnerable to Manipulation. *Fourteenth ACM Conference on Recommender Systems*, (pp. 1–1). IEEE. 10.1145/3383313.3418434

Mesquita, B., & Walker, R. (2003). Cultural differences in emotions: A context for interpreting emotional experiences. *Behaviour Research and Therapy*, *41*(7), 777–793. doi:10.1016/S0005-7967(02)00189-4 PMID:12781245

Meyer-Waarden, L., Pavone, G., Poocharoentou, T., Prayatsup, P., Ratinaud, M., Tison, A., & Torn, S. (2020). How Service Quality Influences Customer Acceptance and Usage of Chatbots? *Journal of Service Management Research, 4*(1), 35–51. doi:10.15358/2511-8676-2020-1-35

Mhlanga, D. (2023). Open AI in education, the responsible and ethical use of ChatGPT towards lifelong learning. *Education, the Responsible and Ethical Use of ChatGPT Towards Lifelong Learning.*

Michie, S., Richardson, M., Johnston, M., Abraham, C., Francis, J., Hardeman, W., Eccles, M. P., Cane, J., Wood, C. E., Michie, S., Johnston, M., Wood, C. E., Richardson, M., Abraham, C., Francis, J., Hardeman, W., Eccles, M. P., & Cane, J. (2013). The Behavior Change Technique Taxonomy (v1) of 93 Hierarchically Clustered Techniques: Building an International Consensus for the Reporting of Behavior Change Interventions. *Annals of Behavioral Medicine, 46*(1), 81–95. doi:10.100712160-013-9486-6 PMID:23512568

Mikolov, T., Chen, K., Corrado, G., & Dean, J. (2013). *Efficient estimation of word representations in vector space.* arXiv preprint arXiv:1301.3781.

Milani, S., Topin, N., Veloso, M., & Fang, F. (2022). A survey of explainable reinforcement learning. *ArXiv Preprint ArXiv:2202.08434.*

Milano, S., Mittelstadt, B., Wachter, S., & Russell, C. (2021). Epistemic fragmentation poses a threat to the governance of online targeting. *Nature Machine Intelligence, 3*(6), 466–472. doi:10.103842256-021-00358-3

Milano, S., Taddeo, M., & Floridi, L. (2020). Recommender systems and their ethical challenges. *AI & Society, 35*(4), 957–967. doi:10.100700146-020-00950-y

Milgram, S. (1967). The Small World Problem. *Psychology Today, 1*(1), 60–67.

Miller, G. A. (1969). Psychology as a means of promoting human welfare. *The American Psychologist, 24*(12), 1063–1075. doi:10.1037/h0028988

Mill, J. S. (2015). *On Liberty, Utilitarianism and Other Essays* (P. F. R. Mark, Ed.). Oxford university press. doi:10.1093/owc/9780199670802.001.0001

Miltenberger, R. G. (2016). *Behavior Modification: Principles and Procedures* (6th ed.). Cengage Learning.

Minsky, M. (1961). Steps toward artificial intelligence. *Proceedings of the IRE, 49*(1), 8–30. doi:10.1109/JRPROC.1961.287775

Mo, F., Rehman, H. U., Monetti, F. M., Chaplin, J. C., Sanderson, D., Popov, A., Maffei, A., & Ratchev, S. (2023). A framework for manufacturing system reconfiguration and Optimisation Utilising Digital Twins and Modular Artificial Intelligence. *Robotics and Computer-integrated Manufacturing, 82*, 102524. doi:10.1016/j.rcim.2022.102524

Mohamadou, Y., Halidou, A., & Kapen, P. T. (2020). A review of mathematical modelling, artificial intelligence and datasets used in the study, prediction and management of covid-19. *Applied Intelligence, 50*(11), 3913–3925. doi:10.100710489-020-01770-9 PMID:34764546

Mohamed, M. A., Eltamaly, A. M., & Alolah, A. I. (2017). Swarm intelligence-based optimization of grid-dependent hybrid renewable energy systems. *Renewable & Sustainable Energy Reviews, 77*, 515–524. doi:10.1016/j.rser.2017.04.048

Mohammad, N. I., Ismail, S. A., Kama, M. N., Yusop, O. M., & Azmi, A. (2019, August). Customer churn prediction in telecommunication industry using machine learning classifiers. In *Proceedings of the 3rd international conference on vision, image and signal processing* (pp. 1-7). IEEE. 10.1145/3387168.3387219

Möhlmann, M., Zalmanson, L., Henfridsson, O., & Gregory, R. W. (2021). Algorithmic Management of Work on Online Labor Platforms: When Matching Meets Control. *Management Information Systems Quarterly*, *45*(4), 1999–2022. doi:10.25300/MISQ/2021/15333

Mokyr, J., Vickers, C., & Ziebarth, N. (2015). The History of Technological Anxiety and the Future of Economic Growth. *The Journal of Economic Perspectives*, *29*(3), 31–50. doi:10.1257/jep.29.3.31

Molina, M. G. (2019). Grid energy storage systems. *Power Electronics in Renewable Energy Systems and Smart Grid*, 495–583. doi:10.1002/9781119515661.ch10

Molnar C. (2020). *Interpretable Machine Learning: a guide for making black box interpretable.* lulu.com.

Montemayor, C. (2023). *The Prospect of a Humanitarian Artificial Intelligence: Agency and Value Alignment.*

Moon, S., Kim, M. Y., & Iacobucci, D. (2021). Content analysis of fake consumer reviews by survey-based text categorization. *International Journal of Research in Marketing*, *38*(2), 343–364. doi:10.1016/j.ijresmar.2020.08.001

Moore, G. E. (1965). *Cramming more components onto integrated circuits.* Electronics Magazine.

Moskop, J. C., Marco, C. A., Larkin, G. L., Geiderman, J. M., & Derse, A. R. (2005). From Hippocrates to HIPAA: privacy and confidentiality in emergency medicine–Part I: conceptual, moral, and legal foundations. *Annals of Emergency Medicine*, *45*(1), 53–59. doi:10.1016/j.annemergmed.2004.08.008 PMID:15635311

Mostafa, M. (2009). Shades of green: A psychographic segmentation of the green consumer in Kuwait using self-organizing maps. *Expert Systems with Applications*, *36*(8), 11030–11038. doi:10.1016/j.eswa.2009.02.088

Mouffe, C. (1999). Deliberative democracy or agonistic pluralism? *Social Research*, 745–758.

Mruthyunjaya, V., & Jankowski, C. (2020). Human-Augmented robotic intelligence (HARI) for human-robot interaction. In *Proceedings of the Future Technologies Conference (FTC) 2019: Volume 2* (pp. 204-223). Springer International Publishing.

Mukaetova-Ladinska, E.B., Harwood, T., & Maltby, J. (2020). Artificial Intelligence in the healthcare of older people. Archives of Psychiatry and Mental Health, 4(1), 007–013.

Murthi, B. P. S., & Sarkar, S. (2003). The role of the management sciences in research on personalization. *Management Science*, *49*(10), 1344–1362. doi:10.1287/mnsc.49.10.1344.17313

Mustak, M., Salminen, J., Plé, L., & Wirtz, J. (2021). Artificial intelligence in marketing: Topic modeling, scientometric analysis, and research agenda. *Journal of Business Research*, *124*, 389–404. doi:10.1016/j.jbusres.2020.10.044

Muthusamy, M., & Periasamy, K. (2019). A comprehensive study on internet of things security. In Advancing Consumer-Centric Fog Computing Architectures, 72–86. doi:10.4018/978-1-5225-7149-0.ch004

Myllylä, M. (2022). Psychological and Cognitive Challenges in Sustainable AI Design. In *International Conference on Human-Computer Interaction* (pp. 426-444). Cham: Springer International Publishing. 10.1007/978-3-031-05434-1_29

Naffine, N. (2003). Who are law's persons? From Cheshire cats to responsible subjects. *The Modern Law Review*, *66*(3), 346–367. doi:10.1111/1468-2230.6603002

Nahum-Shani, I., Smith, S. N., Spring, B. J., Collins, L. M., Witkiewitz, K., Tewari, A., & Murphy, S. A. (2018). Just-in-Time Adaptive Interventions (JITAIs) in Mobile Health: Key Components and Design Principles for Ongoing Health Behavior Support. *Annals of Behavioral Medicine*, *52*(6), 446–462. doi:10.100712160-016-9830-8 PMID:27663578

Nalepa, G. J., Kutt, K., & Bobek, S. (2019). Mobile platform for affective context-aware systems. *Future Generation Computer Systems, 92*, 490–503. doi:10.1016/j.future.2018.02.033

Nanne, A. J., Antheunis, M. L., Van Der Lee, C. G., Postma, E. O., Wubben, S., & Van Noort, G. (2020). The use of computer vision to analyse brand-related user generated image content. *Journal of Interactive Marketing, 50*(1), 156–167. doi:10.1016/j.intmar.2019.09.003

Naraine, M. L., & Wanless, L. (2020). Going all in on AI: Examining the value proposition of and integration challenges with one branch of artificial intelligence in sport management. *Sports Innovation Journal, 1*, 49–61. doi:10.18060/23898

National Institute of Standards and Technology. (2023). *NIST AI Risk Management Framework Playbook*. NIST. https://pages.nist.gov/AIRMF/

Netzer, O., Toubia, O., Bradlow, E. T., Dahan, E., Evgeniou, T., Feinberg, F. M., Feit, E. M., Hui, S. K., Johnson, J., Liechty, J. C., Orlin, J. B., & Rao, V. R. (2008). Beyond conjoint analysis: Advances in preference measurement. *Marketing Letters, 19*(3-4), 337–354. doi:10.100711002-008-9046-1

Neurology-Vegas. (2021, February 5). *What is the memory capacity of a human brain?* Clinical Neurology Specialists. Retrieved November 20, 2022, from https://www.cnsnevada.com/what-is-the-memory-capacity-of-a-human-brain/

Neuwirth, R. J. (2022). *The EU Artificial Intelligence Act: Regulating Subliminal AI Systems*. Taylor & Francis. doi:10.4324/9781003319436

Ngai, E. W., & Wu, Y. (2022). Machine learning in marketing: A literature review, conceptual framework, and research agenda. *Journal of Business Research, 145*, 35–48. doi:10.1016/j.jbusres.2022.02.049

Nguyen, T. M., Quach, S., & Thaichon, P. (2021). The effect of AI quality on customer experience and brand relationship. *Journal of Consumer Behaviour*. Advance online publication. doi:10.1002/cb.1974

Nie, X., & Wager, S. (2021). Quasi-oracle estimation of heterogeneous treatment effects. *Biometrika, 108*(2), 299–319. doi:10.1093/biomet/asaa076

Nilashi, M., Abumalloh, R. A., Alrizq, M., Almulihi, A., Alghamdi, O. A., Farooque, M., Samad, S., Mohd, S., & Ahmadi, H. (2022). A hybrid method to solve data sparsity in travel recommendation agents using Fuzzy Logic Approach. *Mathematical Problems in Engineering, 2022*, 1–20. doi:10.1155/2022/7372849

Nilsson, N. J. (2014). *Principles of Artificial Intelligence*. Morgan Kaufmann.

Nimri, R., Patiar, A., & Jin, X. (2020). The determinants of consumers' intention of purchasing green hotel accommodation: Extending the theory of planned behaviour. *Journal of Hospitality and Tourism Management, 45*, 535–543. doi:10.1016/j.jhtm.2020.10.013

Nordling, L. (2019). A fairer way forward for AI in health care. *Nature, 573*(7775), s103–s105. doi:10.1038/d41586-019-02872-2 PMID:31554993

Nord, W. R., & Peter, J. P. (1980). A behavior modification perspective on marketing. *Journal of Marketing, 44*(2), 36–47. doi:10.1177/002224298004400205

OECD. (2002). *Glossary of key terms in evaluation and results-based management. DAC Network on Development Evaluation*. OECD.

OECD. (2021). *Social Impact Measurement for the Social and Solidarity Economy*. Doi:10.1787/20794797

Official Journal of the European Union. (2016). *Regulation (EU) 2016/679 of the European Parliament and of the Council of 27595 April 2016 on the Protection of Natural Persons with Regard to the Processing of Personal Data and on the Free Movement of such data, and Repealing Directive 95/46/ec (General Data Protection Regulation).* European Parliament and Council. https://eur-lex.europa.eu/eli/reg/2016/679/oj

Oh, N., Choi, G. S., & Lee, W. Y. (2023). ChatGPT goes to the operating room: Evaluating GPT-4 performance and its potential in surgical education and training in the era of large language models. *Annals of Surgical Treatment and Research, 104*(5), 269. doi:10.4174/astr.2023.104.5.269 PMID:37179699

Oja, M., Kaski, S., & Kohonen, T. (2003). Bibliography of self-organizing map (SOM) papers: 1998-2001 addendum. *Neural computing surveys*, 1--156.

Ojiro, T., & Tsuruta, K. (2019). *A study of smart factory with artificial intelligence.* Control and Optimization of Renewable Energy Systems. doi:10.2316/P.2019.860-011

Olaya, D., Vásquez, J., Maldonado, S., Miranda, J., & Verbeke, W. (2020). Uplift modeling for preventing student dropout in higher education. *Decision Support Systems, 134*, 113320. doi:10.1016/j.dss.2020.113320

Oliveira, B., & Casais, B. (2019). The importance of user-generated photos in restaurant selection. *Journal of Hospitality and Tourism Technology, 10*(1), 2–14. doi:10.1108/JHTT-11-2017-0130

Ordenes, F. V., Ludwig, S., de Ruyter, K., Grewal, D., & Wetzels, M. (2017). Unveiling What is Written in The Stars: Analysing Explicit, Implicit, and Discourse Patterns of Sentiment in Social Media. *The Journal of Consumer Research, 43*(6), 875–894. doi:10.1093/jcr/ucw070

Oshana, M. (2006). *Personal autonomy in society*. Ashgate Publishing, Ltd.

Ouyang, L., Wu, J., Jiang, X., Almeida, D., Wainwright, C., Mishkin, P., & Lowe, R. (2022). Training language models to follow instructions with human feedback. *Advances in Neural Information Processing Systems, 35*, 27730–27744.

Özturk, H., Bahçecik, N., & Özçelik, K. S. (2014). The development of the patient privacy scale in nursing. *Nursing Ethics, 21*(7), 812–828. doi:10.1177/0969733013515489 PMID:24482263

Palmer, A., & Koenig-Lewis, N. (2009). An experiential, social network-based approach to direct marketing. *Direct Marketing: An International Journal.*

Paquet, J. E. (2022). *New directionality for knowledge valorisation in the EU.* Research and Innovation Knowledge Valorisation.

Pargman, D. S., Eriksson, E., Bates, O., Kirman, B., Comber, R., Hedman, A., & van den Broeck, M. (2019). The future of computing and wisdom: Insights from human–computer interaction. *Futures, 113*, 102434. Advance online publication. doi:10.1016/j.futures.2019.06.006

Pariser, E. (2011). *The filter bubble: How the new personalized web is changing what we read and how we think.* Penguin.

Paul, S., Yuan, L., Jain, H. K., Robert, L. P. Jr, Spohrer, J., & Lifshitz-Assaf, H. (2022). Intelligence augmentation: Human factors in AI and future of work. *AIS Transactions on Human-Computer Interaction, 14*(3), 426–445. doi:10.17705/1thci.00174

Pearl J. (2019), The seven tools of causal inference, with reflections on machine learning. *Communications of the ACM*. ACM.

Pearl J., & Glymour, M. (2016). *Causal Inference in Statistics: A Primer.* Wiley.

Pearl R. (2018). *Artificial intelligence in healthcare: Separating reality from hype*. Academic Press.

Pearl, J., & Mackenzie, D. (2018). *The book of why: the new science of cause and effect*. Basic books.

Pearl, J. (1995). Causal diagrams for empirical research. *Biometrika*, *82*(4), 669–688. doi:10.1093/biomet/82.4.669

Pearl, J. (2000). *Causality: models, reasoning, and inference*. Cambridge University Press.

Pearl, J., & Mackenzie, D. (2018). *The Book of Why: The New Science of Cause and Effect*. Basic Books.

Pechenkina, K. (2023). Artificial intelligence for good? Challenges and possibilities of AI in higher education from a data justice perspective. In L. Czerniewicz & C. Cronin (Eds.), *Higher Education for Good: Teaching and Learning Futures (#HE4Good)*. Open Book Publishers.

Pedota, M., & Piscitello, L. (2022). A new perspective on technology-driven creativity enhancement in the Fourth Industrial Revolution. *Creativity and Innovation Management*, *31*(1), 109–122. doi:10.1111/caim.12468

Peng, K., Ding, L., Zhong, Q., Shen, L., Liu, X., Zhang, M., & Tao, D. (2023). *Towards making the most of chatbot for machine translation*. arXiv preprint arXiv:2303.13780.

Pennycook, G., & Rand, D. G. (2019). Lazy, not biased: Susceptibility to partisan fake news is better explained by lack of reasoning than by motivated reasoning. *Cognition*, *188*, 39–50. doi:10.1016/j.cognition.2018.06.011 PMID:29935897

Pereira, L. M., & Lopes, A. (2020). Machine Ethics: From Machine Morals to the Machinery of Morality, book series Studies in Applied Philosophy, Epistemology and Rational Ethics, SAPERE. Springer.

Pereira, A. M., Moura, J. A., Costa, E. D., Vieira, T., Landim, A. R. D. B., Bazaki, E., & Wanick, V. (2022). Customer models for artificial intelligence-based decision support in fashion online retail supply chains. *Decision Support Systems*, *158*, 113795. doi:10.1016/j.dss.2022.113795

Pérez, M. L., & Serrano, M. D. T. (2020). Impacto en redes sociales de la Maratón de Valencia Trinidad Alfonso 2019. *Estudios sobre la transformación digital de la comunicación*, *67*.

Perez, M.N., Cohen, R. (2019). Uploading the Mind: The Basics and Ethics of Whole Brain Emulation. *Pushing the Limits*, 30-33.

Perez, C. (2019). *Invisible women: Exposing data bias in a world designed for men*, *0* (1st ed.). Chatto and Windus.

Persad, G., Wertheimer, A., & Emanuel, E. (2009). Principles for allocation of scarce medical interventions. *Lancet*, *373*(9661), 423–431. doi:10.1016/S0140-6736(09)60137-9 PMID:19186274

Pike, S., Pontes, N., & Kotsi, F. (2021). Stopover destination attractiveness: A quasi-experimental approach. *Journal of Destination Marketing & Management*, *19*, 100514. doi:10.1016/j.jdmm.2020.100514

Pillai, S. V. & Kumar, R. S. (2021). *The role of data-driven artificial intelligence on COVID-19 disease management in public sphere: A review*. doi:10.1007/s40622-021-00289-3

Pinheiro, P., & Cavique, L. (2022). Uplift Modeling Using the Transformed Outcome Approach. In G. Marreiros, B. Martins, A. Paiva, B. Ribeiro, & A. Sardinha (Eds.), Lecture Notes in Computer Science: Vol. 13566. *Progress in Artificial Intelligence. EPIA 2022*. Springer., doi:10.1007/978-3-031-16474-3_51

Pitardi, V., & Marriott, H. R. (2021). Alexa, she's not human but… Unveiling the drivers of consumers' trust in voice-based artificial intelligence. *Psychology and Marketing*, *38*(4), 626–642. doi:10.1002/mar.21457

Polese, F., Barile, S., & Sarno, D. (2022). Artificial Intelligence and Decision-Making: Human–Machine Interactions for Successful Value Co-creation. In *The Palgrave Handbook of Service Management* (pp. 927–944). Springer International Publishing. doi:10.1007/978-3-030-91828-6_44

Polkinghorne, D. E. (1988). *Narrative Knowing and the Human Sciences*. SUNY Press.

Pommeranz, A., Broekens, J., Wiggers, P., Brinkman, W.-P., & Jonker, C. M. (2012). Designing interfaces for explicit preference elicitation: A user-centered investigation of preference representation and elicitation process. *User Modeling and User-Adapted Interaction, 22*(4-5), 357–397. doi:10.100711257-011-9116-6

Ponnapureddy, S., Priskin, J., Ohnmacht, T., Vinzenz, F., & Wirth, W. (2017). The influence of trust perceptions on German tourists' intention to book a sustainable hotel: A new approach to analysing marketing information. *Journal of Sustainable Tourism, 25*(7), 970–988. doi:10.1080/09669582.2016.1270953

Posner, E., & Weyl, E. (2019). *Radical Markets*. Princeton University Press. doi:10.2307/j.ctvdf0kwg

Posner, R. A. (1981). The Economics of Privacy. *The American Economic Review, 71*(2), 405–409.

Poulakidakos, S., & Armenakis, A. (2014). Propaganda in Greek public discourse. Propaganda scales in the presentation of the Greek MoU-bailout agreement of 2010. Revista de Ştiinţe Politice. *Revue des Sciences Politiques, 41*, 126–140.

Puiutta, E., & Veith, E. M. S. P. (2020). Explainable reinforcement learning: A survey. *Machine Learning and Knowledge Extraction: 4th IFIP TC 5, TC 12, WG 8.4, WG 8.9, WG 12.9 International Cross-Domain Conference, CD-MAKE 2020, Dublin, Ireland, August 25–28, 2020. Proceedings, 4*, 77–95.

Purtova, N. (2015). The illusion of personal data as no one's property. *Law, Innovation and Technology, 7*(1), 83–111. doi:10.1080/17579961.2015.1052646

Qian, J., Nguyen, N., Oya, Y., Kikugawa, G., Okabe, T., Huang, Y., & Ohuchi, F. (2019). Introducing self-organized maps (SOM) as a visualization tool for materials research and education. *Results in Materials, Volume 4*.

Quach, S., Thaichon, P., Martin, K. D., Weaven, S., & Palmatier, R. W. (2022). Digital technologies: Tensions in privacy and data. *Journal of the Academy of Marketing Science, 50*(6), 1–25. doi:10.100711747-022-00845-y PMID:35281634

Rafieian, O., & Yoganarasimhan, H. (2022). AI and Personalization. *Available at SSRN 4123356*.

Rahman, M. M., Terano, H. J., Rahman, M. N., Salamzadeh, A., & Rahaman, M. S. (2023). *ChatGPT and academic research: a review and recommendations based on practical examples*.

Rahman, M. S., Bag, S., Hossain, M. A., Abdel Fattah, F. A., Gani, M. O., & Rana, N. P. (2023). The new wave of AI-Powered Luxury Brands Online Shopping experience: The role of digital multisensory cues and customers' engagement. *Journal of Retailing and Consumer Services, 72*, 103273. doi:10.1016/j.jretconser.2023.103273

Rahman, M., Terano, H. J. R., Rahman, N., Salamzadeh, A., & Rahaman, S. (2023). ChatGPT and Academic Research: A Review and Recommendations Based on Practical Examples. *Journal of Education. Management and Development Studies, 3*(1), 1–12.

Rahwan, I., Cebrian, M., Obradovich, N., Bongard, J., Bonnefon, J.-F., Breazeal, C., Crandall, J. W., Christakis, N. A., Couzin, I. D., Jackson, M. O., Jennings, N. R., Kamar, E., Kloumann, I. M., Larochelle, H., Lazer, D., McElreath, R., Mislove, A., Parkes, D. C., Pentland, A. S., & Wellman, M. (2019). Machine behaviour. *Nature, 568*(7753), 477–486. doi:10.103841586-019-1138-y PMID:31019318

Ramesh, A. N., Kambhampati, C., Monson, J. R., & Drew, P. J. (2004). Artificial intelligence in medicine. *Annals of the Royal College of Surgeons of England, 86*(5), 334–338. doi:10.1308/147870804290 PMID:15333167

Rana, N. P., Chatterjee, S., Dwivedi, Y. K., & Akter, S. (2022). Understanding dark side of artificial intelligence (AI) integrated business analytics: Assessing firm's operational inefficiency and competitiveness. *European Journal of Information Systems, 31*(3), 364–387. doi:10.1080/0960085X.2021.1955628

Rane, S. B., & Narvel, Y. A. M. (2022). Data-driven decision making with Blockchain-IoT integrated architecture: a project resource management agility perspective of industry 4.0. *International Journal of System Assurance Engineering and Management*, 1-19.

Rapoport, A. (1977). *Human Aspects of Urban Form.*

Rathore, B. (2023). Integration of Artificial Intelligence& Its Practices in Apparel Industry. [IJNMS]. *International Journal of New Media Studies*, *10*(1), 25–37. doi:10.58972/eiprmj.v10i1y23.40

Ravetz, J. (2020). *Deeper city: Collective intelligence and the pathways from smart to wise.* Routledge. doi:10.4324/9781315765860

Rawls, J. (1971). *A Theory of Justice.* Harvard University Press. doi:10.4159/9780674042605

Rawls, J. (2005). *Political Liberalism.* Columbia University Press.

Ray, P. P. (2023). ChatGPT: A comprehensive review of the background, applications, key challenges, bias, ethics, limitations, and future scope. *Internet of Things and Cyber-Physical Systems.*

Ray, S., Kim, S. S., & Morris, J. G. (2012). Research note—Online Users' switching costs: Their nature and formation. *Information Systems Research*, *23*(1), 197–213. doi:10.1287/isre.1100.0340

Raza, K. (2020). Artificial Intelligence Against COVID-19: A Meta-analysis of Current Research. In A. Hassanien, N. Dey, & S. Elghamrawy (Eds.), *Big Data Analytics and Artificial Intelligence Against COVID-19: Innovation Vision and Approach* (pp. 165–176). Springer. doi:10.1007/978-3-030-55258-9_10

Reis, S., Coelho, L., Sarmet, M., Araújo, J., & Corchado, J. M. (2023). The Importance of Ethical Reasoning in Next Generation Tech Education. In *2023 5th International Conference of the Portuguese Society for Engineering Education (CISPEE)* (pp. 1-10). IEEE. 10.1109/CISPEE58593.2023.10227651

Rekioua, D. (2019). Design of hybrid renewable energy systems. *Green Energy and Technology*, 173–195. doi:10.1007/978-3-030-34021-6_5

Rendle, S., Freudenthaler, C., Gantner, Z., & Schmidt-Thieme, L. (2012). BPR: Bayesian personalized ranking from implicit feedback. *ArXiv Preprint ArXiv:1205.2618.*

Ren, M., Chen, N., & Qiu, H. (2023). Human-machine Collaborative Decision-making: An Evolutionary Roadmap Based on Cognitive Intelligence. *International Journal of Social Robotics*, *15*(7), 1–14. doi:10.100712369-023-01020-1

Rese, A., Ganster, L., & Baier, D. (2020). chatbots in retailers' customer communication: How to measure their acceptance? *Journal of Retailing and Consumer Services*, *56*, 102176. doi:10.1016/j.jretconser.2020.102176

RFEA. (2021). *Ranking pruebas en ruta de la.* https://www.rfea.es/estadis/pdf/punruta2021.pdf

Rhoen, M. (2017). Rear view mirror, crystal ball: Predictions for the future of data protection law based on the history of environmental protection law. *Computer Law & Security Review*, *33*(5), 603–617. doi:10.1016/j.clsr.2017.05.010

Ricalde, L. J., Ordonez, E., Gamez, M., & Sanchez, E. N. (2011). Design of a smart grid management system with renewable energy generation. In *IEEE Symposium on Computational Intelligence Applications in Smart Grid (CIASG).* IEEE. 10.1109/CIASG.2011.5953346

Ricoeur, P. (1994). *Oneself as Another.* University of Chicago Press.

Riedmiller, M., Springenberg, J. T., Hafner, R., & Heess, N. (2021). Collect & Infer--a fresh look at data-efficient Reinforcement Learning. *ArXiv Preprint ArXiv:2108.10273.*

Rochet, J., & Tirole, J. (2006). Two-sided markets: A progress report. *The RAND Journal of Economics*, *37*(3), 645–667. doi:10.1111/j.1756-2171.2006.tb00036.x

Rogers, P. (2014a). Overview of Impact Evaluation: Methodological Briefs - Impact Evaluation. *Methodological Briefs*. UNICEF Office of Research-Innocenti. https://www.unicef-irc.org/publications/746-overview-of-impact-evaluation-methodological-briefs-impact-evaluation-no-1.html

Rogers, P. (2014b). Theory of Change: Methodological Briefs - Impact Evaluation No. 2, *Methodological Briefs*. UNICEF Office of Research-Innocenti. https://www.unicef-irc.org/publications/747-theory-of-change-methodological-briefs-impact-evaluation-no-2.html

Rorty, R. (1989). *Contingency, Irony, and Solidarity*. Cambridge University Press. doi:10.1017/CBO9780511804397

Roschelle, J. (2021). Intelligence augmentation for collaborative learning. In *International Conference on Human-Computer Interaction* (pp. 254-264). Cham: Springer International Publishing.

Rosen, M. A. (2021). Renewable energy and energy sustainability. *Design and Performance Optimization of Renewable Energy Systems*, 17–31. doi:10.1016/b978-0-12-821602-6.00002-x

Rosenbaum, P. R., & Rubin, D. B. (1983). The central role of the propensity score in Observational studies for causal effects. *Biometrika*, *70*(1), 41–55. doi:10.1093/biomet/70.1.41

Rosgaby Medina, K. (2022). *Engagement rate en redes sociales: que es y cómo se calcula*. https://branch.com.co/marketing-digital/engagement-rate-en-redes- sociales-que-es-y-como-se-calcula/

Rouvroy, A., & Poullet, Y. (2009). The right to informational self-determination and the value of self-development: Reassessing the importance of privacy for democracy. In *Reinventing Data Protection?* (pp. 45–76). Springer Netherlands. doi:10.1007/978-1-4020-9498-9_2

Roy, G. (2023). Travelers' online review on Hotel Performance – Analysing Facts with the theory of lodging and sentiment analysis. *International Journal of Hospitality Management*, *111*, 103459. doi:10.1016/j.ijhm.2023.103459

Ruane, E., Birhane, A., & Ventresque, A. (2019, December). Conversational AI: Social and Ethical Considerations. In AICS (pp. 104-115).

Rubin, D. B. (1974). Estimating causal effects of treatments in randomized and nonrandomized studies. *Journal of Educational Psychology*, *66*(5), 688–701. doi:10.1037/h0037350

Rudin, C. (2019). Stop explaining black box machine learning models for high stakes decisions and use interpretable models instead. *Nature Machine Intelligence*, *1*(5), 206–215. doi:10.103842256-019-0048-x PMID:35603010

Ruiz-Real, J. L., Uribe-Toril, J., Torres, J. A., & De Pablo, J. (2021). Artificial intelligence in business and economics research: Trends and future. *Journal of Business Economics and Management*, *22*(1), 98–117. doi:10.3846/jbem.2020.13641

Russell, R., & Andersson, J. (2023, May 27). *Anger over airports' passport e-gates not working*. BBC. https://www.bbc.com/news/uk-65731795.

Russell, S. (2019). *Human compatible: Artificial intelligence and the problem of control*. Penguin.

Russell, S. J., & Norvig, P. (2010). *Artificial intelligence: a modern approach*. Pearson Education.

Ryan, M. (2020). In AI we trust: Ethics, artificial intelligence, and reliability. *Science and Engineering Ethics*, *26*(5), 2749–2767. doi:10.100711948-020-00228-y PMID:32524425

Sadiq, M., Adil, M., & Paul, J. (2022). Eco-friendly hotel stay and environmental attitude: A value-attitude-behaviour perspective. *International Journal of Hospitality Management*, *100*, 103094. doi:10.1016/j.ijhm.2021.103094

Sadowski, J. (2020). The internet of landlords: Digital platforms and new mechanisms of rentier capitalism. *Antipode*, *52*(2), 562–580. doi:10.1111/anti.12595

Sætra, H. S., Borgebund, H., & Coeckelbergh, M. (2022). Avoid diluting democracy by algorithms. *Nature Machine Intelligence*, *4*(10), 804–806. doi:10.103842256-022-00537-w

Sáez-Ortuño, L., Forgas-Coll, S., Huertas-Garcia, R., & Sánchez-García, J. (2023a). What's on the horizon? A bibliometric analysis of personal data collection methods on social networks. *Journal of Business Research*, *158*, 113702. doi:10.1016/j.jbusres.2023.113702

Sáez-Ortuño, L., Forgas-Coll, S., Huertas-Garcia, R., & Sánchez-García, J. (2023b). Online cheaters: Profiles and motivations of internet users who falsify their data online. *Journal of Innovation & Knowledge*, *8*(2), 100349. doi:10.1016/j.jik.2023.100349

Sáez-Ortuño, L., Huertas-Garcia, R., Forgas-Coll, S., & Puertas-Prats, E. (2023c). How can entrepreneurs improve digital market segmentation? A comparative analysis of supervised and unsupervised learning algorithms. *The International Entrepreneurship and Management Journal*, •••, 1–28. doi:10.100711365-023-00882-1

Salakhutdinov, R., Mnih, A., & Hinton, G. (2007). Restricted Boltzmann machines for collaborative filtering. *Proceedings of the 24th International Conference on Machine Learning*, (pp. 791–798). IEEE. 10.1145/1273496.1273596

Sammon, J. J. (1969). A Nonlinear Mapping for Data Structure Analysis. *Transactions on Computers*, 401--409.

Samuel, J., Kashyap, R., Samuel, Y., & Pelaez, A. (2022). Adaptive cognitive fit: Artificial intelligence augmented management of information facets and representations. *International Journal of Information Management*, *65*, 102505. Advance online publication. doi:10.1016/j.ijinfomgt.2022.102505

Santos-Lacueva, R., Velasco González, M., & González Domingo, A. (2022). *The integration of sustainable tourism policies in European cities*.

Sarbey, B. (2016). Definitions of death: Brain death and what matters in a person. *Journal of Law and the Biosciences*, *3*(3), 743–752. doi:10.1093/jlb/lsw054 PMID:28852554

Sarker, I. H. (2021). Machine learning: Algorithms, real-world applications and Research Directions. *SN Computer Science*, *2*(3), 160. doi:10.100742979-021-00592-x PMID:33778771

Sarmah, B., Kamboj, S., & Kandampully, J. (2018). Social media and co-creative service innovation: An empirical study. *Online Information Review*, *42*(7), 1146–1179. doi:10.1108/OIR-03-2017-0079

Satornino, C. B., Grewal, D., Guha, A., Schweiger, E. B., & Goodstein, R. C. (2023). The perks and perils of artificial intelligence use in lateral exchange markets. *Journal of Business Research*, *158*, 113580. doi:10.1016/j.jbusres.2022.113580

Saund, E. (2020). *SCI-52 Artificial Intelligence: Deep Learning, Human-Centered AI, and Beyond*. Stanford University.

Savastano, M., Bellini, F., D'Ascenzo, F., & Scornavacca, E. (2017). FabLabs as Platforms for Digital Fabrication Services: A Literature Analysis. In S. Za, M. Drăgoicea, & M. Cavallari (Eds.), *Exploring Services Science, IESS 2017, Lecture Notes in Business Information Processing* (Vol. 279). Springer. doi:10.1007/978-3-319-56925-3_3

Sayfuddin, A. T. M., & Chen, Y. (2021). The signaling and reputational effects of customer ratings on hotel revenues: Evidence from TripAdvisor. *International Journal of Hospitality Management*, *99*, 103065. doi:10.1016/j.ijhm.2021.103065

Schauer, F. (2006). *Profiles, probabilities, and stereotypes*. Harvard University Press. doi:10.4159/9780674043244

Schechtman, M. (2018). *The Constitutions of Selves*. Cornell University Press.

Schranz, M., Di Caro, G. A., Schmickl, T., Elmenreich, W., Arvin, F., Şekercioğlu, A., & Sende, M. (2021). Swarm intelligence and cyber-physical systems: Concepts, challenges and future trends. *Swarm and Evolutionary Computation*, *60*, 100762. Advance online publication. doi:10.1016/j.swevo.2020.100762

Sehl, K., & Tien, S. (2023). *Engagement Rate Calculator+ Guide for 2023*. Hootsuite. https://blog.hootsuite.com/calculate-engagement-rate/

Selevan, S. G., Kimmel, C. A., & Mendola, P. (2000). Identifying critical windows of exposure for children's health. *Environmental Health Perspectives*, *108*(suppl 3), 451–455. PMID:10852844

Sempre, V. M. (2013). *El cuadro de mando integral para medir nuestra actividad en Redes Sociales*. El Blog de Mayte Vañó Sempere. http://www.maytevs.com/el-cuadro-de-mando-integral-para-medir-nuestra-actividad-en- redes-sociales/

Sen, A., & Williams, B. (1982). *Utilitarianism and Beyond*. Cambridge University Press. doi:10.1017/CBO9780511611964

Settles, B. (1994). Active Learning Literature Survey. *Machine Learning*, *15*(2), 201–221. doi:10.1023/A:1022673506211

Shafer, G. (1986). Savage revisited. *Statistical Science*, 463–485.

Shahid, A. (2018). Smart grid integration of renewable energy systems. In *7th International Conference on Renewable Energy Research and Applications (ICRERA)*. IEEE. 10.1109/ICRERA.2018.8566827

Shahrubudin, N., Lee, T. C., & Ramlan, R. (2019). An Overview on 3D Printing Technology: Technological, Materials, and Applications. *Procedia Manufacturing*, *35*, 1286–1296. doi:10.1016/j.promfg.2019.06.089

Shardanand, U., & Maes, P. (1995). Social information filtering: Algorithms for automating "word of mouth." *Proceedings of the SIGCHI Conference on Human Factors in Computing Systems*, (pp. 210–217). IEEE. 10.1145/223904.223931

Sharda, R., Delen, D., & Turban, E. (2017). *Business Intelligence, Analytics, and Data Science: A Managerial Perspective* (4th ed.). Pearson.

Sharma, A., Park, S., & Nicolau, J. L. (2020). Testing loss aversion and diminishing sensitivity in review sentiment. *Tourism Management*, *77*, 104020. doi:10.1016/j.tourman.2019.104020

Sharma, S. K., & Hoque, X. (2017). Sentiment predictions using support vector machines for odd-even formula in Delhi. *International Journal of Intelligent Systems and Applications*, *9*(7), 61–69. doi:10.5815/ijisa.2017.07.07

Shen, N., Bernier, T., Sequeira, L., Strauss, J., Silver, M. P., Carter-Langford, A., Shortliffe, E. H., Davis, R., Axline, S. G., Buchanan, B. G., Green, C. C., & Cohen, S. N. (1975). Computer-based consultations in clinical therapeutics: Explanation and rule acquisition capabilities of the MYCIN system. *Computers and Biomedical Research, an International Journal*, *8*(4), 303–320. doi:10.1016/0010-4809(75)90009-9 PMID:1157471

Shen, Y., Heacock, L., Elias, J., Hensel, K. D., Reig, B., Shih, G., & Moy, L. (2023). ChatGPT and other large language models are double-edged words. *Radiology*, *307*(2), e230163. doi:10.1148/radiol.230163 PMID:36700838

Shmueli, G., & Tafti, A. (2023). How to "improve" prediction using behavior modification. *International Journal of Forecasting*, *39*(2), 541–555. doi:10.1016/j.ijforecast.2022.07.008

Silva, B. (2016). *Exploratory Cluster Analysis from Ubiquitous Data Streams using Self-Organizing Maps*. [PhD Thesis, Universidade NOVA de Lisboa]. http://hdl.handle.net/10362/19974

Silva, B., & Marques, N. (2010). Feature Clustering with Self-Organizing Maps and an Application to Financial Time-series for Portfolio Selection. *ICNC 2010 International Conference on Neural Computation*. Valencia.

Silva, B., & Marques, N. (2015). The ubiquitous self-organizing map for non-stationary data streams. *Journal of Big Data*, *2*(1), 27. doi:10.118640537-015-0033-0

Silver, D., Singh, S., Precup, D., & Sutton, R. S. (2021). Reward is enough. *Artificial Intelligence*, *299*, 103535. doi:10.1016/j.artint.2021.103535

Simonetti, A., & Bigne, E. (2022). How visual attention to social media cues impacts visit intention and liking expectation for restaurants. *International Journal of Contemporary Hospitality Management*, *34*(6), 2049–2070. doi:10.1108/IJCHM-09-2021-1091

Sinz, F. H., Pitkow, X., Reimer, J., Bethge, M., & Tolias, A. S. (2019). Engineering a less artificial intelligence. *Neuron*, *103*(6), 967–979. doi:10.1016/j.neuron.2019.08.034 PMID:31557461

Smiley, M. (1992). *Moral responsibility and the boundaries of community: Power and accountability from a pragmatic point of view*. University of Chicago Press. doi:10.7208/chicago/9780226763255.001.0001

Smith, B., & Linden, G. (2017). Two Decades of Recommender Systems at Amazon.com. *IEEE Internet Computing*, *21*(3), 12–18. doi:10.1109/MIC.2017.72

Smith, C. (2011). *What is a person?: Rethinking humanity, social life, and the moral good from the person up*. University of Chicago Press.

Sofi, A., Regita, J. J., Rane, B., & Lau, H. H. (2022). Structural health monitoring using wireless smart sensor network – An overview. *Mechanical Systems and Signal Processing*, *163*. doi:10.1016/j.ymssp.2021.108113

Software and Information Industry Association (SIIA). (2017). *Ethical Principles for Artificial Intelligence and Data Analytics*. SIIA.

Spahn, A. (2012). And lead us (not) into persuasion…? Persuasive technology and the ethics of communication. *Science and Engineering Ethics*, *18*(4), 633–650. doi:10.100711948-011-9278-y PMID:21544700

Spangler, S., Wilkins, A. D., Bachman, B. J., Nagarajan, M., Dayaram, T., Haas, P., Regenbogen, S., Pickering, C. R., Comer, A., Myers, J. N., & Stanoi, I. (2014). Automated hypothesis generation based on mining scientific literature. In *Proceedings of the 20th ACM SIGKDD international conference on Knowledge discovery and data mining*. ACM. 10.1145/2623330.2623667

Spirtes, P., Glymour, C. N., & Scheines, R. (2000). *Causation, prediction, and search*. MIT Press.

Srinivasan, P. (2004). Text mining. Generating hypotheses from MEDLINE. *Journal of the American Society for Information Science and Technology*, *55*(5), 396–413. doi:10.1002/asi.10389

Srnicek, N. (2017). *Platform capitalism*. John Wiley & Sons.

Stace, W. T. (1924). *The philosophy of Hegel: A systematic exposition*. Macmillan.

Stahl, B. C. (2004). Information, ethics, and computers: The problem of autonomous moral agents. *Minds and Machines*, *14*(1), 67–83. doi:10.1023/B:MIND.0000005136.61217.93

Stahl, B. C., Antoniou, J., Ryan, M., Macnish, K., & Jiya, T. (2022). Organisational responses to the ethical issues of artificial intelligence. *AI & Society*, *31*(1), 23–37. doi:10.100700146-021-01148-6

Stahl, B. C., & Coeckelbergh, M. (2016). Ethics of healthcare robotics: Towards responsible research and innovation. *Robotics and Autonomous Systems*, *86*, 152–161. doi:10.1016/j.robot.2016.08.018

Stanley, K. (2007). Design of randomized controlled trials. *Circulation, 115*(9), 1164–1169. doi:10.1161/CIRCULATIONAHA.105.594945 PMID:17339574

Stead, M., Gordon, R., Angus, K., & McDermott, L. (2007). A systematic review of social marketing effectiveness. *Health Education, 107*(2), 126–191. doi:10.1108/09654280710731548

Steenkamp, J.-B. (2021). *A life in marketing: looking back, looking forward. AMA Irwin McGraw Hill Acceptance Speech.* American Marketing Association.

Stemler, S. (2019) *An Overview of Content Analysis.* Pare Online. https://pareonline.net/getvn.asp?v=7&n=17

Stephanopoulos, G., & Han, C. (1996). Intelligent systems in process engineering: A review. *Computers & Chemical Engineering, 20*(6-7), 743-791. doi:10.1016/0098-1354(95)00194-8

Sternberg, R. J. (2005). The theory of successful intelligence. *Revista Interamericana de Psicología. Interamerican Journal of Psychology, 39*(2), 189–202.

Sternberg, R. J., & Kaufman, J. C. (2013). *The evolution of intelligence.* Psychology Press. doi:10.4324/9781410605313

Stewart, D. W. (2010). The Evolution of Market Research. In P. Maclaran, M. Saren, B. Stern, & M. Tadajewski (Eds.), *The SAGE Handbook of Marketing Theory* (pp. 74–85). Sage.

Stokes, F., & Palmer, A. (2020). Artificial Intelligence and Robotics in Nursing: Ethics of Caring as a Guide to Dividing Tasks Between AI and Humans. *Nursing Philosophy, 21*(4), e12306. doi:10.1111/nup.12306 PMID:32609420

Stone, M., Moutinho, L., Ekinci, Y., Labib, A., Evans, G., Aravopoulou, E., Laughlin, P., Hobbs, M., Machtynger, J., & Machtynger, L. (2021). Artificial intelligence in marketing and marketing research. In *The Routledge Companion to Marketing Research* (pp. 147–163). Routledge. doi:10.4324/9781315544892-11

Stout, L. (2012). *The shareholder value myth: How putting shareholders first harms investors, corporations, and the public.* Berrett-Koehler Publishers.

Stray, J., Halevy, A., Assar, P., Hadfield-Menell, D., Boutilier, C., Ashar, A., Beattie, L., Ekstrand, M., Leibowicz, C., Sehat, C. M., Johansen, S., Kerlin, L., Vickrey, D., Singh, S., Vrijenhoek, S., Zhang, A., Andrus, M., Helberger, N., Proutskova, P., & Vasan, N. (2022). Building Human Values into Recommender Systems: An Interdisciplinary Synthesis. *ArXiv.*

Stryker, S., & Burke, P. J. (2000). The past, present, and future of an identity theory. *Social Psychology Quarterly, 63*(4), 284–297. doi:10.2307/2695840

Stylos, N. (2020). Technological evolution and tourist decision-making: A perspective article. *Tourism Review, 75*(1), 273–278. doi:10.1108/TR-05-2019-0167

Sun, J., Zhu, X., Zhang, C., & Fang, Y. (2011, June). HCPP: Cryptography based secure EHR system for patient privacy and emergency healthcare. In *2011 31st International Conference on Distributed Computing Systems* (pp. 373-382). IEEE.

Sutton, R. S., & Barto, A. G. (2018). *Reinforcement learning: An introduction.* MIT press.

Swain, M. C., & Cole, J. M. (2016). ChemDataExtractor: A toolkit for automated extraction of chemical information from the scientific literature. *Journal of Chemical Information and Modeling, 56*(10), 1894–1904. doi:10.1021/acs.jcim.6b00207 PMID:27669338

Sydelko, P., Espinosa, A., & Midgley, G. (2023). Designing interagency responses to wicked problems: A viable system model board game. *European Journal of Operational Research.* Advance online publication. doi:10.1016/j.ejor.2023.06.040

Szeliski, R. (2011). *Computer vision: Algorithms and applications.* Springer-Verlag. doi:10.1007/978-1-84882-935-0

Taecharungroj, V., & Mathayomchan, B. (2019). Analysing TripAdvisor reviews of tourist attractions in Phuket, Thailand. *Tourism Management*, *75*, 550–568. doi:10.1016/j.tourman.2019.06.020

Tamanaha, B. Z. (2004). *On the rule of law: History, politics, theory*. Cambridge University Press. doi:10.1017/CBO9780511812378

Tanai, Y., & Ciftci, K. (2023). How to customize an early start preparatory course policy to improve student graduation success: An application of uplift modeling. *Annals of Operations Research*. doi:10.100710479-023-05607-9

Tang, A., Christian, E., & Huang, Y. M. (2020). *What needs to happen before we can upload our brains to a Computer*. Business Insider. Retrieved November 21, 2022, from https://www.businessinsider.com/how-to-upload-brain-computer-eternal-digital-life-black-mirror-2020-9

Tao, F., Zhang, M., & Nee, A. Y. C. (2019). *Digital Twin Driven Smart Manufacturing*. Elsevier Inc. doi:10.1016/C2018-0-02206-9

Tassiello, V., Tillotson, J. S., & Rome, A. S. (2021). "Alexa, order me a pizza!": The mediating role ofpsychological power in the consumer–voiceassistant interaction. *Psychology and Marketing*, *8*(7), 1069–1080. doi:10.1002/mar.21488

Taylor, C. (1980). Understanding in Human Science. *The Review of Metaphysics*, *34*(1), 25–38.

Taylor, E. J. (2020). *AI and global governance: covid-19 underlines the need for AI- support governance of water-energy food nexus, AI and Global Governance*. Centre for Policy Research.

Teodorescu, M. H., Morse, L., Awwad, Y., & Kane, G. C. (2021). Failures of Fairness in Automation Require a Deeper Understanding of Human-ML Augmentation. *Management Information Systems Quarterly*, *45*(3), 1483–1500. Advance online publication. doi:10.25300/MISQ/2021/16535

Thiele, L. P. (2021). Rise of the Centaurs: The Internet of Things Intelligence Augmentation. In *Towards an International Political Economy of Artificial Intelligence* (pp. 39–61). Springer International Publishing.

Thistlewaite, D. L., & Campbell, D. T. (2017). Regression-Discontinuity Analysis: An Alternative to the Ex-Post Facto Experiment. *Observational Studies*, *3*(2), 119–128. doi:10.1353/obs.2017.0000

Thomas, R. L., & Uminsky, D. (2022). Reliance on metrics is a fundamental challenge for AI. *Patterns*, *3*(5), 100476. doi:10.1016/j.patter.2022.100476 PMID:35607624

Thomke S.H. (2020). Experimentation Works: The Surprising Power of Business Experiments. *Harvard Business Review Press*.

Timoshenko, A., & Hauser, J. R. (2019). Identifying customer needs from user-generated content. *Marketing Science*, *38*(1), 1–20. doi:10.1287/mksc.2018.1123

Tipler, F. J. (2012). Inevitable existence and inevitable goodness of the singularity. *Journal of Consciousness Studies*, *19*(1-2), 183–193.

Tlili, A., Shehata, B., Adarkwah, M. A., Bozkurt, A., Hickey, D. T., Huang, R., & Agyemang, B. (2023). What if the devil is my guardian angel: ChatGPT is a case study of using chatbots in education. *Smart Learning Environments*, *10*(1), 15. doi:10.118640561-023-00237-x

Tomkos, I., Klonidis, D., Pikasis, E., & Theodoridis, S. (2020). Toward the 6G Network Era: Opportunities and Challenges. *IT Professional*, *22*(1), 34–38. doi:10.1109/MITP.2019.2963491

Topol, E. J. (2019). High-performance medicine: The convergence of human and artificial intelligence. *Nature Medicine*, *25*(1), 44–56. doi:10.103841591-018-0300-7 PMID:30617339

Torrance, S. (2012). Super-intelligence and (super-)consciousness. *International Journal of Machine Consciousness, 4*(2), 483–501. doi:10.1142/S1793843012400288

Torra, V. (2017). *Data privacy: foundations, new developments and the big data challenge*. Springer. doi:10.1007/978-3-319-57358-8

Tóth, Z., Caruana, R., Gruber, T., & Loebbecke, C. (2022). The Dawn of the AI Robots: Towards a New Framework of AI Robot Accountability. *Journal of Business Ethics, 178*(4), 895–916. doi:10.100710551-022-05050-z

Touretzky, D. S., & Saksida, L. M. (1997). Operant conditioning in Skinnerbots. *Adaptive Behavior, 5*(3–4), 219–247. doi:10.1177/105971239700500302

TripAdvisor. (2023). *TripAdvisor Review Transparency Report*. Tripadvisor. https://www.tripadvisor.com.mx/TransparencyReport2023#group-section-Trends-and-Processes-Tkmr3mda9i

Trivedi, J. (2019). Examining the Customer Experience of Using Banking Chatbots and Its Impact on Brand Love: The Moderating Role of Perceived Risk. *Journal of Internet Commerce, 18*(1), 91–111. doi:10.1080/15332861.2019.1567188

Tsai, W. H. S., Liu, Y., & Chuan, C. H. (2021). How chatbots' social presence communication enhances consumer engagement: The mediating role of parasocial interaction and dialogue. *Journal of Research in Interactive Marketing, 15*(3), 460–482. doi:10.1108/JRIM-12-2019-0200

Tseng, A. (2017). Why do online tourists need sellers' ratings? Exploration of the factors affecting regretful tourist e-satisfaction. *Tourism Management, 59*, 413–424. doi:10.1016/j.tourman.2016.08.017

Tshitoyan, V., Dagdelen, J., Weston, L., Dunn, A., Rong, Z., Kononova, O., Persson, K. A., Ceder, G., & Jain, A. (2019). Unsupervised word embeddings capture latent knowledge from materials science literature. *Nature, 571*(7763), 95–98. doi:10.103841586-019-1335-8 PMID:31270483

Tsytsarau, M., & Palpanas, T. (2012). Survey on mining subjective data on the web. *Data Mining and Knowledge Discovery, 24*(3), 478–514. doi:10.100710618-011-0238-6

Tu, C. (2019). Comparison of various machine learning algorithms for estimating generalized propensity score. *Journal of Statistical Computation and Simulation, 89*(4), 708–719. doi:10.1080/00949655.2019.1571059

Turing, A. M. (1950). Computing Machinery and Intelligence. *Mind, 59*(236), 433–460. doi:10.1093/mind/LIX.236.433

Tushnet, M. (2012). Constitution-making: An introduction. *Texas Law Review, 91*, 1983.

Tussyadiah, I. (2020). A review of research into automation in tourism: Launching the Annals of Tourism Research curated collection on artificial intelligence and robotics in tourism. *Annals of Tourism Research, 81*, 102883. doi:10.1016/j.annals.2020.102883

Tyler, T. R. (2006). *Why people obey the law*. Princeton university press. doi:10.1515/9781400828609

Ultsch, A., & Siemon, H. (1990). Kohonen's Self Organizing Feature Maps for Exploratory Data Analysis. *Proceedings of the International Neural Network Conference (INNC-90)* (pp. 305--308). Kluwer Academic Press.

Valls, A., Gibert, K., Orellana, A., & Antón-Clavé, S. (2018). Using ontology-based clustering to understand the push and pull factors for British tourists visiting a Mediterranean coastal destination. *Information & Management, 55*(2), 145–159. doi:10.1016/j.im.2017.05.002

Van der Sloot, B. (2017). Decisional privacy 2.0: The procedural requirements implicit in Article 8 ECHR and its potential impact on profiling. *International Data Privacy Law, 7*(3), 190–201. doi:10.1093/idpl/ipx011

Van Dijck, J. (2013). *The culture of connectivity: A critical history of social media*. Oxford University Press. doi:10.1093/acprof:oso/9780199970773.001.0001

Van Dijck, J., Poell, T., & De Waal, M. (2018). *The platform society: Public values in a connective world*. Oxford University Press. doi:10.1093/oso/9780190889760.001.0001

Van Otterlo, M. (2013). A machine learning view on profiling. In Privacy, Due Process and the Computational Turn (pp. 41–64). Routledge.

van Schalkwyk, G. (2023). Artificial intelligence in pediatric behavioral health. *Child and Adolescent Psychiatry and Mental Health, 17*(1), 1–2. doi:10.118613034-023-00586-y PMID:36907862

VanderWeele, T. J. (2017). On the Promotion of Human Flourishing. *Proceedings of the National Academy of Sciences of the United States of America, 114*(31), 8148–8156. doi:10.1073/pnas.1702996114 PMID:28705870

Varela, F. J., Thompson, E., & Rosch, E. (2017). The embodied mind, revised edition: Cognitive science and human experience. MIT Press. doi:10.7551/mitpress/9780262529365.001.0001

Vargo, S. L. (2019). Moving forward..... *AMS Review, 9*(3), 133–135. doi:10.100713162-019-00159-3

Verma, V. K., Chandra, B., & Kumar, S. (2019). Values and ascribed responsibility to predict consumers' attitude and concern towards green hotel visit intention. *Journal of Business Research, 96*, 206–216. doi:10.1016/j.jbusres.2018.11.021

Vesanen, J. (2007). What is personalization? A conceptual framework. *European Journal of Marketing, 41*(5/6), 409–418. doi:10.1108/03090560710737534

Vesanto, J. (1999). SOM-based data visualization methods. *Intelligent Data Analysis, 3*(2), 111–126. doi:10.3233/IDA-1999-3203

Villamediana, J., Küster, I., & Vila, N. (2019). Destination engagement on Facebook: Time and seasonality. *Annals of Tourism Research, 79*(November), 102747. doi:10.1016/j.annals.2019.102747

Villamediana-Pedrosa, J. D., Vila-Lopez, N., & Küster-Boluda, I. (2020). Predictors of tourist engagement: Travel motives and tourism destination profiles. *Journal of Destination Marketing & Management, 16*, 100412. doi:10.1016/j.jdmm.2020.100412

von Krogh, G., Roberson, Q., & Gruber, M. (2023). Recognizing and Utilizing Novel Research Opportunities with Artificial Intelligence. *Academy of Management Journal, 66*(2), 367–373. doi:10.5465/amj.2023.4002

von Neumann, J. (1966). Theory of self-reproducing automata. *Mathematics of Computation, 21*, 745.

Wager, S., & Athey, S. (2018). Estimation and Inference of Heterogeneous Treatment Effects using Random Forests. *Journal of the American Statistical Association, 113*(523), 1228–1242. doi:10.1080/01621459.2017.1319839

Wallach, W. A., Allen, C. B., & Franklin, S. C. (2011). Consciousness and ethics: Artificially conscious moral agents. *International Journal of Machine Consciousness, 3*(1), 177–192. doi:10.1142/S1793843011000674

Wamba-Taguimdje, S. L., Fosso Wamba, S., Kala Kamdjoug, J. R., & Tchatchouang Wanko, C. E. (2020). Influence of artificial intelligence (AI) on firm performance: The business value of AI-based transformation projects. *Business Process Management Journal, 26*(7), 1893–1924. doi:10.1108/BPMJ-10-2019-0411

Wang, S., Wang, T., He, C., & Hu, Y. J. (2022). Can Your Toothpaste Shopping Predict Mutual Funds Purchasing?—Transferring Knowledge from Consumer Goods to Financial Products Via Machine Learning. *SSRN*.

Wang, B., Zheng, P., Yin, Y., Shih, A., & Wang, L. (2022). Toward human-centric smart manufacturing: A human-cyber-physical systems (HCPS) perspective. *Journal of Manufacturing Systems*, *63*, 471–490. doi:10.1016/j.jmsy.2022.05.005

Wang, J., Wang, S., Xue, H., Wang, Y., & Li, J. (2018). Green image and consumers' word-of-mouth intention in the green hotel industry: The moderating effect of Millennials. *Journal of Cleaner Production*, *181*, 426–436. doi:10.1016/j.jclepro.2018.01.250

Wang, W., Li, B., Luo, X., & Wang, X. (2022). Deep reinforcement learning for sequential targeting. *Management Science*.

Wang, W., Ying, S., Lyu, J., & Qi, X. (2019). Perceived image study with online data from social media: The case of boutique hotels in China. *Industrial Management & Data Systems*, *119*(5), 950–967. doi:10.1108/IMDS-11-2018-0483

Wang, X. (2018). AI defeats elite doctors in diagnosis competition. *The Star*, (July), 2.

Wang, Y., Wang, Z., Zhang, D., & Zhang, R. (2019). Discovering cultural differences in online consumer product reviews. *Journal of Electronic Commerce Research*, *20*(3), 169–183.

Warren, M. E. (2017). A problem-based approach to democratic theory. *The American Political Science Review*, *111*(1), 39–53. doi:10.1017/S0003055416000605

Wedel, M., & Kamakura, W. A. (2000). *Market segmentation: Conceptual and methodological foundations*. Kluwer Academic Publishers Group. doi:10.1007/978-1-4615-4651-1

Weidinger, L., McKee, K. R., Everett, R., Huang, S., Zhu, T. O., Chadwick, M. J., Summerfield, C., & Gabriel, I. (2023). Using the Veil of Ignorance to align AI systems with principles of justice. *Proceedings of the National Academy of Sciences of the United States of America*, *120*(18), e2213709120. doi:10.1073/pnas.2213709120 PMID:37094137

Weiss, K., Khoshgoftaar, T. M., & Wang, D. (2016). A survey of transfer learning. *Journal of Big Data*, *3*(1), 9. doi:10.118640537-016-0043-6

Welikala, S., & Karunananda, A. S. (2017, October 31). *Mind Uploading with BCI Technology* [Paper presentation]. SLAAI – International Conference on Artificial Intelligence, Moratuwa, Sri Lanka. https://slaai.lk/wp- content/uploads/2018/01/slaai2017.pdf#page=15

Weston, L., Tshitoyan, V., Dagdelen, J., Kononova, O., Trewartha, A., Persson, K. A., Ceder, G., & Jain, A. (2019). Named entity recognition and normalization applied to large-scale information extraction from the materials science literature. *Journal of Chemical Information and Modeling*, *59*(9), 3692–3702. doi:10.1021/acs.jcim.9b00470 PMID:31361962

What is machine learning - ml - and why is it important ? (n.d.). NetApp. Retrieved November 20, 2022, from https://www.netapp.com/artificial-intelligence/what-is-machine-learning/

White, H., Sabarwal, S., & de Hoop, T. (2014). Randomized Controlled Trials (RCTs): Methodological Briefs - Impact Evaluation No. 7, *Methodological Briefs*. UNICEF Office of Research-Innocenti. https://www.unicef-irc.org/publications/752-randomized-controlled-trials-rcts-methodological-briefs-impact-evaluation-no-7.html

White, G., Ariyachandra, T., & White, D. (2019). Big Data, Ethics, and Social Impact Theory – A Conceptual Framework. *Journal of Management & Engineering Integration*, *12*(1), 9–15.

Wiener, N. (1988). *The human use of human beings: Cybernetics and society*. Da Capo Press.

Wiljer, D. (2019). Understanding the patient privacy perspective on health information exchange: A systematic review. *International Journal of Medical Informatics*, *125*, 1–12. doi:10.1016/j.ijmedinf.2019.01.014 PMID:30914173

Wilson, E. J., & Vlosky, R. P. (1997). Partnering Relationship Activities: Building Theory From Case Study Research. *Journal of Business Research*, *39*(1), 59–70. doi:10.1016/S0148-2963(96)00149-X

Winick, B. J. (1992). On autonomy: Legal and psychological perspectives. *Vill. L. Rev.*, *37*, 1705. PMID:11654414

Winter, C., Hollman, N., & Manheim, D. (2023). Value alignment for advanced artificial judicial intelligence. *American Philosophical Quarterly*, *60*(2), 187–203. doi:10.5406/21521123.60.2.06

Wirtz, B. W., Weyerer, J. C., & Geyer, C. (2019). Artificial intelligence and the public sector—Applications and challenges. *International Journal of Public Administration*, *42*(7), 596–615. doi:10.1080/01900692.2018.1498103

Wirtz, J., Patterson, P. G., Kunz, W. H., Gruber, T., Lu, V. N., Paluch, S., & Martins, A. (2018). Brave new world: Service robots in the frontline. *Journal of Service Management*, *29*(5), 907–931. doi:10.1108/JOSM-04-2018-0119

Wolfram, D. A. (1995). An appraisal of INTERNIST-I. *Artificial Intelligence in Medicine*, *7*(2), 93–116. doi:10.1016/0933-3657(94)00028-Q PMID:7647840

Wolpert, D. H. (1996). The lack of a priori distinctions between learning algorithms. *Neural Computation*, *8*(7), 1341–1390. doi:10.1162/neco.1996.8.7.1341

World Economic Forum. (2020). *Model artificial intelligence governance framework and assessment guide.* World Economic Forum.

Wottawa, C. R., Genovese, B., Nowroozi, B. N., Hart, S. D., Bisley, J. W., Grundfest, W. S., & Dutson, E. P. (2016). Evaluating tactile feedback in robotic surgery for potential clinical application using an animal model. *Surgical Endoscopy*, *30*(8), 3198–3209. doi:10.100700464-015-4602-2 PMID:26514132

Woźniak, M., Siłka, J., & Wieczorek, M. (2021). Deep neural network correlation learning mechanism for CT brain tumor detection. *Neural Computing & Applications*, 1–16.

Wu, J., Zhang, Z., Feng, Z., Wang, Z., Yang, Z., Jordan, M. I., & Xu, H. (2022). *Sequential information design: Markov persuasion process and its efficient reinforcement learning*. ArXiv.

Xiao, B., & Benbasat, I. (2007). E-commerce product recommendation agents: Use, characteristics, and impact. *Management Information Systems Quarterly*, *31*(1), 137–209. doi:10.2307/25148784

Xie, C., Wang, Y., & Cheng, Y. (2022). Does Artificial Intelligence Satisfy You? A Meta-Analysis of User Gratification and User Satisfaction with AI-Powered Chatbots. *International Journal of Human-Computer Interaction*, 1–11. Advance online publication. doi:10.1080/10447318.2022.2121458

Xu, W., Dainoff, M. J., Ge, L., & Gao, Z. (2023). Transitioning to human interaction with AI systems: New challenges and opportunities for HCI professionals to enable human-centered AI. *International Journal of Human-Computer Interaction*, *39*(3), 494–518. doi:10.1080/10447318.2022.2041900

Yadegaridehkordi, E., Nilashi, M., Nasir, M., Momtazi, S., Samad, S., Supriyanto, E., & Ghabban, F. (2021). Customers segmentation in eco-friendly hotels using multi-criteria and machine learning techniques. *Technology in Society*, *65*, 101528. doi:10.1016/j.techsoc.2021.101528

Yang, Y., Jiang, L., & Wang, Y. (2023). Why do hotels go green? Understanding TripAdvisor GreenLeaders participation. *International Journal of Contemporary Hospitality Management*, *35*(5), 1670–1690. doi:10.1108/IJCHM-02-2022-0252

Yankoski, M., Weninger, T., & Scheirer, W. (2020). An AI early warning system to monitor online disinformation, stop violence, and protect elections. *Bulletin of the Atomic Scientists*, *76*(2), 85–90. doi:10.1080/00963402.2020.1728976

Yao, X., Ma, N., Zhang, J., Wang, K., Yang, E., & Faccio, M. (2022). Enhancing wisdom manufacturing as industrial metaverse for industry and society 5.0. *Journal of Intelligent Manufacturing*, 1–21. doi:10.100710845-022-02027-7

Yao, Y. (2023). Human-machine co-intelligence through symbiosis in the SMV space. *Applied Intelligence*, *53*(3), 2777–2797. doi:10.100710489-022-03574-5

Yeo, Y. H., Samaan, J. S., Ng, W. H., Ting, P. S., Trivedi, H., Vipani, A., & Kuo, A. (2023). Assessing the performance of ChatGPT in answering questions regarding cirrhosis and hepatocellular carcinoma. medRxiv, 2023-02.

Yin, D., Bond, S. D., & Zhang, H. (2014). Anxious or angry? Effects of discrete emotions on the perceived helpfulness of online reviews. *Management Information Systems Quarterly*, *38*(2), 539–560. doi:10.25300/MISQ/2014/38.2.10

Youssef, A., El-Telbany, M., & Zekry, A. (2017). The role of artificial intelligence in photo-voltaic systems design and control: A review. *Renewable & Sustainable Energy Reviews*, *78*, 72–79. doi:10.1016/j.rser.2017.04.046

Yue, W., Wang, Z., Chen, H., Payne, A., & Liu, X. (2018). Machine learning with applications in breast cancer diagnosis and prognosis. *Designs*, *2*(2), 13. doi:10.3390/designs2020013

Yu, T. H. K., & Huarng, K. H. (2010). A neural network-based fuzzy time series model to improve forecasting. *Expert Systems with Applications*, *37*(4), 3366–3372. doi:10.1016/j.eswa.2009.10.013

Zachlod, C., Samuel, O., Ochsner, A., & Werthmüller, S. (2022). Analytics of social media data–State of characteristics and application. *Journal of Business Research*, *144*, 1064–1076. doi:10.1016/j.jbusres.2022.02.016

Zagzebski, L. (2001). The uniqueness of persons. *The Journal of Religious Ethics*, *29*(3), 401–423. doi:10.1111/0384-9694.00090

Zanga, A., Ozkirimli, E., & Stella, F. (2022). A Survey on Causal Discovery: Theory and Practice. *International Journal of Approximate Reasoning, 151*. Doi:10.1016/j.ijar.2022.09.004

Zarsky, T. Z. (2016). Incompatible: The GDPR in the age of big data. *Seton Hall Law Review*, *47*, 995.

Zeithaml, V. A., Berry, L. L., & Parasuraman, A. (1996). The Behavioral Consequences of Service Quality. *Journal of Marketing*, *60*(2), 31–46. doi:10.1177/002224299606000203

Zezula, P., Amato, G., Dohnal, V., & Batko, M. (2006). *Similarity search: the metric space approach* (Vol. 32). Springer Science & Business Media. doi:10.1007/0-387-29151-2

Zhang, D., & Allagui, A. (2021). Fundamentals and performance of solar photovoltaic systems. *Design and Performance Optimization of Renewable Energy Systems*, 117–129. doi:10.1016/b978-0-12-821602-6.00009

Zhang, C., Wang, Z., Zhou, G., Chang, F., Ma, D., Jing, Y., Cheng, W., Ding, K., & Zhao, D. (2023). Towards new-generation human-centric smart manufacturing in Industry 5.0: A systematic review. *Advanced Engineering Informatics*, *57*, 102121. Advance online publication. doi:10.1016/j.aei.2023.102121

Zhang, H., Zang, Z., Zhu, H., Uddin, M. I., & Amin, M. A. (2022). Big data-assisted social media analytics for business model for business decision making system competitive analysis. *Information Processing & Management*, *59*(1), 102762. doi:10.1016/j.ipm.2021.102762

Zhang, J., Qu, Q., & Chen, X. B. (2023). A review on collective behavior modeling and simulation: Building a link between cognitive psychology and physical action. *Applied Intelligence*, 1–30. doi:10.100710489-023-04924-7

Zhang, L., Ling, J., & Lin, M. (2022). Artificial intelligence in renewable energy: A comprehensive bibliometric analysis. *Energy Reports*, *8*, 14072–14088. doi:10.1016/j.egyr.2022.10.347

Zhao, G., Li, Y., & Xu, Q. (2022). From emotion AI to cognitive AI. *International Journal of Network Dynamics and Intelligence*, 65-72.

Zhao, X., Xia, L., Tang, J., & Yin, D. (2019). " Deep reinforcement learning for search, recommendation, and online advertising: a survey" by Xiangyu Zhao, Long Xia, Jiliang Tang, and Dawei Yin with Martin Vesely as coordinator. *ACM Sigweb Newsletter, Spring*, 1–15.

Zhou, L., Paul, S., Demirkan, H., Yuan, L., Spohrer, J., Zhou, M., & Basu, J. (2021). Intelligence augmentation: Towards building human-machine symbiotic relationship. *AIS Transactions on Human-Computer Interaction, 13*(2), 243–264. doi:10.17705/1thci.00149

Zhu, D., Ye, Z., & Chang, Y. (2017). Understanding the textual content of online customer reviews in B2C websites: A cross-cultural comparison between the US and China. *Computers in Human Behavior, 76*(November), 483–493. doi:10.1016/j.chb.2017.07.045

Zhu, X., Guo, H., Mohammad, S., & Kiritchenko, S. (2014). An empirical study on the effect of negation words on sentiment. *Proceedings of the Annual Meeting of the Association for Computational Linguistics (ACL)* (pp. 304–313). Baltimore, MD. 10.3115/v1/P14-1029

Zohdi, M., Rafiee, M., Kayvanfar, V., & Salamiraad, A. (2022). Demand forecasting based machine learning algorithms on Customer Information: An applied approach. *International Journal of Information Technology : an Official Journal of Bharati Vidyapeeth's Institute of Computer Applications and Management, 14*(4), 1937–1947. doi:10.100741870-022-00875-3

Zouni, G., Markogiannaki, P., & Georgaki, I. (2021). A strategic tourism marketing framework for sports mega events: The case of Athens Classic (Authentic) Marathon. *Tourism Economics, 27*(3), 466–481. doi:10.1177/1354816619898074

Zuboff, S. (2019). *The age of surveillance capitalism: The fight for a human future at the new frontier of power*. Profile books.

Zulkifli, N. S. A., & Lee, A. W. K. (2019). Sentiment analysis in social media based on english language multilingual processing using three different analysis techniques. In Soft Computing in Data Science: 5th International Conference, [Springer Singapore.]. *Proceedings, 5*, 375–385.

Zwick, D., & Dholakia, N. (2004). Whose identity is it anyway? Consumer representation in the age of database marketing. *Journal of Macromarketing, 24*(1), 31–43. doi:10.1177/0276146704263920

Zwierenberg, E., Nap, H. H., Lukkien, D., Cornelisse, L., Finnema, E., Hagedoorn, M., & Sanderman, R. (2018). A lifestyle monitoring system to support (in) formal caregivers of people with dementia: Analysis of users need, benefits, and concerns. *Gerontechnology (Valkenswaard), 17*(4), 194–205. doi:10.4017/gt.2018.17.4.001.00

About the Contributors

Luiz Moutinho (BA, MA, PhD, MAE, FCIM) is a Visiting Prof. of Marketing at Suffolk BS, Univ. of Suffolk, UK, and at Marketing School, Portugal, and Adjunct Prof. at USP, Fiji. Until 2017 he was prof. of BioMarketing and Futures Research at the DCU BS, Ireland, and prior to that for 20 years the Foundation Chair of Marketing at the Adam Smith BS, Univ. of Glasgow, Scotland. In 2017 he received a degree of Prof. Honoris Causa from the Univ. of Tourism and Management Skopje, North Macedonia; in 2020 he was elected as the member of The Academia Europaea. He completed his PhD at the Univ. of Sheffield in 1982. He has been a Full Professor for 31 years, and Visiting Professor worldwide. He is the Founding Editor-in-Chief of the Journal of Modelling in Management (JM2) and Co-editor-in-Chief of the Innovative Marketing Journal. Main areas of research interest: marketing, management and tourism futurecast, AI, biometrics and neuroscience and the use of ANN in marketing, evolutionary algorithms, HCI, futures research. Google Scholar, Sept. 2020: 155 articles published in refereed academic journals, 34 books and 15,579 academic citations, the h-index of 57 and the i10-index of 212.

Luis Cavique is a tenured Assistant Professor at the Computer Science Section in the Department of Sciences and Technology at Universidade Aberta. He worked in the Polytechnic Education System from 1991 to 2008, namely as Adjunct Professor in the Setubal and in the Lisbon Polytechnic Institute. He received the degree in Computer Science Engineering from the New University of Lisbon (FCT-UNL), the MSc degree in Operational Research and Systems Engineering from the Technical Lisbon University and the PhD degree in Systems Engineering from the Technical Lisbon University (IST-UTL) in 2002. His research areas are in the intersection of Computer Science and Systems Engineering, particularly in Data Science. He is author of over 200 scientific works in peer reviewed journals and conferences. He is an integrated member of LASIGE in Lisbon University.

Enrique Bigné is Professor of Marketing at the University of Valencia since 2001. He previously occupied the same position at the University Jaume I (1996-2001). He received his Ph.D. in Economics and Business Administration, his Bachelor degree in Business Administration, and a Degree in Law, and post graduate diplomas in Market Research and in Operations Research. His research interests include advertising, e-WOM, tourism, and neuromarketing. His work has been published in European Journal of Marketing, Psychology & Marketing, Journal of Current Issues and Research in Advertising, International Journal of Advertising, Annals of Tourism Research, Tourism Management, Journal of Business Ethics, and Journal of Services Marketing, among others. Visiting Scholar at the University of Maryland and Berkeley. Editor of European Journal of Management & Business Economics; Revista de Análisis Turistico. Associated Editor of Journal of Global Marketing, Journal of Modelling in Management, and Pasos. Member of the Editorial Board in several journals. He has served as a Head of the Department, Dean of the School of Economics and Vice-chancellor for International Relationships and Communications of the University of Valencia Academic director of the International MBA and Master in Marketing y Comunicacion.

* * *

Svetlana Bialkova is an internationally recognised expert, global professor, author, speaker. She has a PhD, Radboud University, The Netherlands, and a JSPS fellowship, BSI RIKEN, Japan. She received several awards and secured funding, including PI role. She taught various academic courses and has worked on numerous projects in top international labs in The Netherlands, Belgium, Germany, Denmark, Switzerland, UK, Japan. She explores the joint interaction between human, media, and technology, and how this influences the multisensory experience. Translating fundamental knowledge into practice, she realised several industry related initiatives and projects with huge societal relevance. She also participated in various consortia advising policy implementation at EU level and across the globe.

Lars-Erik Casper Ferm is a current Ph.D. student at the University of Queensland Business School. His research focuses on value co-creation, digital marketing, privacy and AI. He has had his research published in leading marketing journals such as the Journal of Retailing and Consumer Services, Psychology & Marketing, the Journal of Strategic Marketing and the Journal of Global Scholars of Marketing Science.

Santiago Forgas-Coll, PhD in Marketing from University of Girona, is currently an Associate Professor in the Business Department at the University of Barcelona. His main research interest cover all aspects of digital marketing, Artificial Intelligence and Robotics, services marketing, sports management and tourism marketing, with a focus on consumer behavior. He has published articles in several international journals such as International Journal of Social Robotics, The Service Industries Journal, Journal of Business Research, Journal of Innovation and Knowledge, Service Business, Management Decision, Industrial Management & Data Systems, Online Information Review, Journal of Business and Industrial Marketing, Journal of Air Transport Management, Interaction Studies, Tourism Management, Journal of Travel Research, Journal of Vacation Marketing, Tourism Geographies, Journal of Travel and Tourism Marketing, International Journal of Tourism Research, Tourism Economics, Sport Management review or European Sport Management Quarterly among others.

Travis Greene is a PhD graduate of the Institute of Service Science at National Tsing Hua University in Taiwan. He is currently a tenure-track Assistant Professor in the Department of Digitalization at Copenhagen Business School. With an interdisciplinary background in philosophy and research interests in data science ethics and personalization, his work has appeared in journals such as Nature Machine Intelligence, Journal of the Association for Information Systems, The Journal of the Royal Statistical Society, and Big Data. He hopes to contribute new perspectives and ideas from the humanities into machine learning and statistics, and vice versa.

Duen-Huang Huang is an Associate Professor of the Center of Teacher Education Chaoyang University of Technology, Taichung City, Taiwan. He received his Ph.D. in Industrial Education from National Taiwan Normal University in 2014. His research areas include e-learning, digital transformation, and fuzzy mathematics..

Kun-Huang Huarng is the Vice President and Distinguished Professor of National Taipei University of Business and a Professor of Creative Technologies and Product Design of the same institution. Professor Huarng was awarded Life Fellow by International Society of Management Engineers. Meanwhile, he received Outstanding Service Award Literati Network Award for Excellence from Emerald and was an Advisory Board Member of Emerald. His teaching and research interests include e-commerce and big data analytics.

Ines Kuster is Professor in Marketing in the Department of Marketing - Fac. of Economics, University of Valencia, Spain. She gets her PhD in Marketing from the University of Valencia in 1999. Her research attention has focused on the areas of strategic marketing and sales. She has published articles in several refereed journals (i.e. JOUEC; Information&Management, JQR; European Journal of Innovation Management; Journal of Business and Industrial Marketing; Innovative Marketing; Qualitative Market Research: An International Journal; European Journal of Marketing; The Marketing Review; Marketing Intelligence and Planning; Journal of Global Marketing; Journal of Relationship Marketing; Annals of Tourism Research; Sex Roles; Equal Opportunites International, and other relevant Spanish journals). She is the author of diverse books and book chapters related to her investigation field. She has also presented papers at the European Marketing Conference and the Academy of Marketing Conference. She collaborates with several companies, helping them in marketing areas (recruiting salespeople, training sales managers, analysing commercial efforts, etc.).

Nuno Castro Lopes is a Ph.D. in Education from Universidade Aberta (Uab) and Ph.D. student in Science and Web Technologies from Uab and the Universidade de Trás-os-Montes e Alto Douro. Master in Educational Multimedia Communication from Uab and a Degree in Psychology from the Universidade do Minho. He is a Communication and Social Impact Manager at an NGO and a Lead Team in Data Science for Social Good Portugal member. His main area of interest is Data Science applied to Social and Human Behavior.

Manikandakumar Muthusamy was awarded the B.Tech., degree in Information Technology in 2009, and M.E., degree in Computer Science and Engineering in 2013 from Anna University, Chennai, India where he is pursuing a Ph.D. He is currently a Senior Assistant Professor in the Computer Science & Engineering Department, New Horizon College of Engineering, Bangalore, India. His research interests include artificial intelligence, data science and networking.

Karthikeyan P. is currently working as an Associate Professor in Thiagarajar College of Engineering, Madurai from 2007 onwards. He has completed the Ph.D. programme in Information and Communication Engineering under Anna University, Chennai, Tamilnadu, India in the year 2015. He published 25 papers in refereed international journals and conferences. He received the B.E. degree in Computer Science and Engineering from Madurai Kamaraj University, Madurai, Tamilnadu, India in the year

2002. He also received his M.E. degree in Computer Science and Engineering from Anna University, Chennai, Tamilnadu, India in the year 2004. His research interests include computational intelligence and educational technology.

Sara Quach is a Senior Lecturer in the Department of Marketing, Griffith University, Australia. Her research interests are in the areas of relationship marketing, services marketing, marketing technology, privacy and digital marketing. She has won several awards at top marketing conferences such as Best Paper Award at The 2016 Mystique of Luxury Brands Conference, People's Choice Award (Best Poster) at 2016 Australian & New Zealand Marketing Academy Mid-Year Doctor Colloquium, and the Best Paper Award at 2015 Australian & New Zealand Marketing Academy and Global Alliance of Marketing and Management Associations Joint Symposium. In addition, Sara has been recognised as the Rising Star in the Marketing Discipline 2020 by The Australian, Australia and New Zealand Marketing Academy's 2022 Emerging Marketing Researcher Award, and Griffith Business School PVC's Research Excellence Award (ECR) in 2022.

Laura Sáez Ortuño is currently an Associate Professor in the Business Department at the University of Barcelona, Spain. She is teaching business in Computer Science and Math faculty (University of Barcelona). Laura is a PhD Student in Marketing from University of Barcelona. Her main research interest covers all aspects of digital marketing, social media and data collection, with a focus on user behavior. From several years, she has and still leading R&D projects (CDTI) related with big data, artificial intelligence and machine learning in the scope of data collection in digital marketing.

Francisco Sarabia-Sanchez is Professor of Marketing and Market Research at Miguel Hernández University. He has been a lecturer at the University of Murcia and visiting professor at the University of Lancashire (UK) and ESAI (Portugal). He specializes in consumer analysis, research methods, and social marketing. He has published books on values and lifestyles and methodology for marketing research. He has been the director of the Department of Economic and Financial Studies, the degree coordinator, and the director of the Ph.D. in Marketing and Strategic Business Management.

Sara Shawky is a Lecturer and Research Fellow at Griffith University. Her research interests focus on digital media, social change and relationship marketing. Her work is published in leading peer-reviewed journals including Journal of Business Research, The International Journal of Market Research, Journal of Social Marketing, and Health Marketing Quarterly. Sara has been engaged in a variety of externally funded projects and has more than fifteen years of experience across well-renowned companies and academic institutions.

Galit Shmueli is a Tsing Hua distinguished professor at the Institute of Service Science, College of Technology Management, National Tsing Hua University, in Taiwan. Her research focuses on statistical and machine-learning methodology with applications in information systems and healthcare and an emphasis on human behavior. She is the author of multiple books and over 100 peer-reviewed publications. She teaches and designs business analytics courses on machine learning, forecasting, and more. She is also the inaugural editor in chief of the INFORMS Journal on Data Science and an IMS Fellow and ISI elected member.

Nikolaos Stylos is Senior Lecturer/Associate Professor in Marketing, Innovation & Digitalisation research group Lead and the Smart Networks 4 Sustainable Futures faculty research group Founder/Lead at the University of Bristol Business School, University of Bristol, UK. He has also been appointed as Honorary Professor of Hospitality Management at TUT, Taiwan, in recognition of his substantial and multi-faceted contributions to the research of Hospitality Management and Tourism Marketing. He specializes on Marketing and Innovation. His research projects merely focus on consumer decision-making as influenced by new technologies and applied in the service industries. Dr Stylos has published in some of the top marketing, tourism and innovation journals. Dr Stylos has an internationally recognised research profile and has given Keynote speeches and invited talks at international academic and industry conferences in the UK, Spain, France, UAE, Greece, Portugal, and Turkey. He has gained research funding from British research organisations (including British Academy and ESRC-UKRI), worked on EU funded projects and has participated in international research consortia.

Park Thaichon is an Associate Professor of Marketing at the University of Southern Queensland, and an Adjunct Associate Professor of Marketing at the Griffith Asia Institute, Griffith University. Park is the Australian most published author (marketing) by Scholarly Output in Australia over the period 2017 to >2022 and 2019 to >2022 (SciVal, 2022). He achieved a hg-index over 30 before his promotion to an Associate Professor (ranked at the highest possible score percentile at the Senior Lecturer level academics, Soutar et al., 2015 - ANZMAC November 2018).

Natalia Vila López is in Marketing in the Department of Marketing - Fac. of Economics, University of Valencia, Spain. She gets her PhD in Marketing from the University of Valencia in 1999. Her research attention has focused on the areas of strategic marketing and brand. She has published articles in several refereed journals (i.e. Tourism Economics, Journal of Destination Marketing, Information & Management, European Journal of Innovation Management; Journal of Business and Industrial Marketing; Innovative Marketing; Qualitative Market Research: An International Journal; European Journal of Marketing; The Marketing Review; Marketing Intelligence and Planning; Journal of Global Marketing; Journal of Relationship Marketing; Sex Roles; Equal Opportunites International, and other relevant Spanish journals). She is the author of diverse books and book chapters related to her investigation field. She has also presented papers at the European Marketing Conference and the Academy of Marketing Conference. She collaborates with several companies, helping them in marketing areas (brand positioning, competitors analysis, analysing commercial efforts, etc.).

Elizabeth Wang is a professor of Computer Science. She received her Ph.D. in computer science from North Dakota State University, Master's degree from Minnesota State University of St. Cloud, and Bachelor's degree from Beijing University of Science and Technology. Her research interests include data mining, bioinformatics, and high performance computing. As professional activities, she serves as a reviewer and/or a committee member of many international conferences and journals.

Jason Wissinger finished his Bachelor of Science in Computer Science from Waynesburg University in December of 2022. "Mind Uploading in Artificial Intelligence" is his first published contribution.

Tiffany Hui-Kuang Yu received her Ph.D. from Texas A&M University, U.S.A. She is a Professor of Public Finance, Feng Chia University, Taiwan. Currently, she also serves as the Dean of Student Affairs of Feng Chia University. Her research interests cover big data analytics, time series, health economics etc.

Index

M

N

O

P

R

S

Printed in the United States
by Baker & Taylor Publisher Services

Printed in the United States
by Baker & Taylor Publisher Services